# CliffsNotes®

## AP® Chemistry 2021 Exam

# CliffsNotes®

# AP® Chemistry 2021 Exam

*by*
*Angela Woodward Spangenberg, M.S.*

Houghton Mifflin Harcourt
Boston • New York

*About the Author*

**Angela Woodward Spangenberg** has a Master of Science in Chemistry from the University of North Carolina. She is an experienced teacher and tutor who currently works as a research scientist in Austin, Texas.

*Editorial*

**Executive Editor:** Greg Tubach

**Senior Editor:** Christina Stambaugh

**Production Editor:** Jennifer Freilach

**Copy Editor:** Lynn Northrup

**Technical Editor:** Kelly Rosier

**Proofreader:** Susan Moritz

**CliffsNotes® AP® Chemistry 2021 Exam**



Library of Congress Control Number: 2020938457
ISBN: 978-0-358-35353-9 (pbk)

Printed in the United States of America
DOO 10 9 8 7 6 5 4 3 2 1     4500804005

For information about permission to reproduce selections from this book, write to Permissions, Houghton Mifflin Harcourt Publishing Company, 3 Park Avenue, 19th Floor, New York, New York 10016.

www.hmhbooks.com

# Acknowledgments

I am grateful to the following colleagues and contributors to this book: Charles Griste, Ben Shoulders, Alan Campion, Ian Riddington, John Ketcham, Kelly Rosier, and Joy Gilmore. This work, as always, is dedicated to Kyle, Mom, Dad, Eddie, and Billy.

# Table of Contents

# Introduction

Thank you for choosing *CliffsNotes AP Chemistry 2021 Exam* to assist with your preparation for the 2021 AP Chemistry exam. This study guide is based on the *AP Chemistry Course and Exam Description* and on the most recent released versions of the AP Chemistry exam; this information is located at https://apstudents.collegeboard.org/courses/ap-chemistry. The intent of this study guide is to provide you with explanations, clarification, and plenty of opportunities for self-assessment so that you can pinpoint your weaknesses and overcome them.

Each practice question in this book was designed specifically to assess a content component of the current course in a way that closely approximates the style of questioning on the latest released versions of the AP Chemistry exam. There is ample coverage of recently introduced topics such as spectroscopy, chromatography, and Coulomb's law. On the other hand, colligative properties, organic reaction prediction, and quantum numbers, to name a few, are important and interesting topics, but since they are excluded from the current curriculum, they are also excluded from this guide. An exception has been made for the Nernst equation, which proved to be advantageous to students in recent administrations of the AP Chemistry exam and has made a reappearance on the *AP Chemistry Equations and Constants* sheet.

Of the resources available to you, no textbook, study guide, website, or YouTube tutorial is as important as your classroom teacher. Your teacher has pursued years of training both in chemistry and in teaching, and has undergone a stringent course audit in order to call the course he or she is teaching "AP Chemistry." Your teacher is your greatest ally. He or she has a vested interest in your success because his or her professional accomplishment is intertwined with student achievement. When you feel that there are shortcomings in your understanding, never hesitate to approach your teacher for clarification. You will both be glad that you did.

Whether you are using this book as a supplement throughout your school year or you are picking it up a few weeks or days before the exam, it is my hope that *CliffsNotes AP Chemistry 2021 Exam* will prove a valuable supplement to your teacher's instruction.

## Format of the AP Chemistry Exam

The AP Chemistry exam is divided into two sections. Calculators are not permitted on Section I, but you may use your calculator on Section II. Throughout the exam, you will have access to the *AP Chemistry Equations and Constants* sheet. You should familiarize yourself with this list of important equations and constants and refer to it frequently as you study for the exam.

| | Section Weight | Type of Question | Number of Questions | Recommended Time per Question | Total Time Allowed |
|---|---|---|---|---|---|
| Section I | 50% | Multiple choice | 60 | 1.5 minutes | 90 minutes |
| Section II | 50% | Long free response<br>Short free response | 3<br>4 | 22 minutes<br>9 minutes | 105 minutes |

Exams are graded on a scale of 1 to 5, with 5 being the highest score. Most colleges require a minimum score of 3 or 4 to offer a student credit for chemistry or allow the student to skip one or two semesters of introductory chemistry. Check with the chemistry departments at the universities that you will apply to in order to determine whether they award credit for the exam.

The distributions of student scores on recent administrations of the exam are shown below.

| Exam Grade | Credit Recommendation | College Grade Equivalent |
|---|---|---|
| 5 | Extremely well qualified | A |
| 4 | Well qualified | A–, B+, B |
| 3 | Qualified | B–, C+, C |
| 2 | Possibly qualified | Not applicable |
| 1 | No recommendation | Not applicable |

## Breakdown of the AP Chemistry Exam

The exam is broken down into nine units. Their weighting on the multiple-choice section is shown below, and the practice exams in this study guide have been written to reflect this weighting.

| Unit | Exam Weighting |
|---|---|
| 1: Atomic Structure and Properties | 7–9% |
| 2: Molecular and Ionic Compound Structure and Properties | 7–9% |
| 3: Intermolecular Forces and Properties | 18–22% |
| 4: Chemical Reactions | 7–9% |
| 5: Kinetics | 7–9% |
| 6: Thermodynamics | 7–9% |
| 7: Chemical Equilibrium | 7–9% |
| 8: Acids and Bases | 11–15% |
| 9: Applications of Thermodynamics | 7–9% |

## Strategies for Multiple-Choice Questions

- **Carefully read the background information when answering question set items.** There are two types of questions in the multiple-choice section: stand-alone questions and question sets. Stand-alone questions are single questions with four choices lettered A–D. Question sets begin with a prompt indicating which question numbers belong to the set. A prompt might read, "Use the following information to answer questions 2–4." Make sure that you refer to the background information while answering all of the questions in the set.
- **Do not leave any answers blank.** There is no penalty for guessing, so eliminate any choices that are obviously wrong and choose one of the remaining choices. If you cannot eliminate any choices, just make a guess. You have one chance in four of guessing correctly.
- **Do not spend too much time on one question.** Even if you are sure you can figure out the answer to a difficult question, put a mark by the question in your booklet, strike through any incorrect choices, and come back after you have worked quickly through the rest of the exam. You may be able to answer several easy questions and earn multiple points in the time it takes to answer one difficult question for a single point.
- **Don't overthink easy questions.** The questions range in difficulty from very easy to difficult. If a question seems too easy, do not assume that it is a trick. Don't second-guess yourself.
- **Review true and false items carefully.** Some questions may ask "which of the following is true" or "which of the following is false." Be aware of the word *not*. "Which of the following is not false" can be tricky if it isn't approached with care. Write a T or an F beside each answer choice to avoid confusion.
- **Check your bubbling carefully and obsessively.** Every time you turn the page, check to make sure that the question number at the top corresponds to the correct bubble that you are filling in on your answer sheet. Rather than bubbling circles completely as you work through a page, place a light hash mark across the correct answer. At the end of the page, check your question numbers and then darken the bubbles

completely. This way, if you do make a bubbling error, you can erase it quickly and completely before correcting your work.

## Multiple-Choice Math Strategies

Calculators are not allowed on the multiple-choice section of the exam, but *calculations* may still be required. This can be one of the most intimidating aspects of the AP Chemistry exam, but it does not have to be. Practice working problems without a calculator. Start early in the school year to gain comfort doing this. The math is simple. You just need to know the tricks: estimating, working with exponents, and working with logarithms.

### Estimating

Suppose a question asks for the pressure exerted by a 10-gram sample of $F_2$ in a 4-L container at 385 K. You know that this is an ideal gas law problem and that $PV = nRT$. The number of moles, $n$, is equal to the number of grams divided by the molar mass, 38.00 $g{\cdot}mol^{-1}$. The value of the gas constant, $R$, is found on the *AP Chemistry Equations and Constants* sheet. Begin by solving the equation for $P$ and plugging in the appropriate values for $V$, $n$, $R$, and $T$.

$$P = \frac{nRT}{V} = \frac{\left(\frac{10}{38}\right)(0.08206)(385)}{4}$$

Notice that in the numerator, 385 divided by 38 is approximately 10.

$$\frac{\left(\frac{10}{\cancel{38}}\right)(0.08206)\left(\cancel{385}^{10}\right)}{4} \approx \frac{(10)(0.08206)(10)}{4}$$

Next, recognize that 0.08 divided by 4 is 0.02.

$$(10)(0.02)(10) = 2$$

If you had used your calculator, you would have determined the answer to be 2.078. The estimated numerical value will only be close to one of the answer choices.

### Working with Exponents

Be comfortable with the rules for working with exponents.

- When two numbers are multiplied, their exponents are added.

$$(6 \times 10^{-11})(3 \times 10^{8}) = (6)(3) \times 10^{(-11+8)} = 18 \times 10^{-3} = 1.8 \times 10^{-2}$$

- When two numbers are divided, the exponent of the number in the denominator is subtracted from the exponent of the number in the numerator.

$$\frac{3\times10^{8}}{6\times10^{-11}} = 0.5\times10^{(8-(-11))} = 0.5\times10^{19} = 5\times10^{18}$$

- To take a square root of an exponent, divide the exponent by 2.

$$\sqrt{1\times10^{-18}} = 1\times10^{-9}$$

- In order to raise an exponent to the power of another exponent, multiply the two exponents.

$$(1 \times 10^{-6})^{3} = 1 \times 10^{(-6\times3)} = 1 \times 10^{-18}$$

### Working with Logarithms

Logarithms are another operation that appears on the exam, usually in the context of pH or pK. The log of a number is the power to which 10 must be raised in order to give that number.

$$\log(1000) = \log(10^3) = 3$$
$$\log(0.001) = \log(10^{-3}) = -3$$

Sometimes it is necessary to estimate the log of a number that is not a multiple of 10.

$$\log(2.3 \times 10^{-4})$$

The value $2.3 \times 10^{-4}$ is greater than $1 \times 10^{-4}$ and smaller than $1 \times 10^{-3}$, so the value of the log is larger than $\log(1 \times 10^{-4})$ and smaller than $\log(1 \times 10^{-3})$. The value can be estimated to lie between –3 and –4. The actual value is –3.63.

## Strategies for Free-Response Questions

The free-response section is 105 minutes long, and calculators *are* allowed.

**TIP: If you have time, you may want to consider honing your estimating skills by estimating the answers to practice free-response questions instead of using a calculator, too. After you have made your estimation, check your work with a calculator. Was your answer close?**

### Significant Figures

Free-response calculations should be reported with the correct numbers of significant figures. When counting significant figures, zeros to the left of the first nonzero value are not significant. (The underlined digits are significant.)

$0.000\underline{1055}$ Four significant figures

Zeros at the end of a number and to the left of a decimal point are not significant unless there is a decimal point.

$\underline{1{,}055{,}}000$ Four significant figures

$\underline{1{,}055{,}000}.$ Seven significant figures (more precise)

### Rounding

When two measured values are multiplied or divided, the result should be rounded to reflect the number of significant figures as the least precise measurement.

$$\frac{10{,}553(0.023)}{31.8526} = 7.6$$

When two measurements are added or subtracted, the result should be rounded to reflect the left-most decimal place that contains an uncertain digit.

$$84 + 38{,}\underline{6}00 + 9.005 = 38{,}700$$

This may be easier to visualize using a stacked equation in which all values are aligned at the decimal point.

$$\begin{array}{rl} 84 & \\ 38{,}\underline{6}00 & \\ 9.005 & \\ \hline 38{,}\underline{6}93.005 & \text{(unrounded answer)} \\ 38{,}\underline{7}00 & \text{(correct number of significant figures)} \end{array}$$

When performing a series of calculations, it is always best to carry out all the arithmetical operations before rounding to avoid propagation of error.

### Example

Divide 25.398 by 142.04. Divide 1.38 by 58.44. Add the two results and divide the answer by 0.10000. Report the final answer with the correct number of significant figures.

The first division operation results in five significant figures.

$$\frac{25.398}{142.04} = 0.17881 \quad \text{(unrounded value to be carried to the next step: 0.1788087863)}$$

The second division operation results in three significant figures because it is limited by the value 1.38.

$$\frac{1.38}{58.44} = 0.0236 \quad \text{(unrounded value to be carried to the next step: 0.023613963)}$$

The two results are added, and the result has four significant figures because it is limited by the value 0.0236.

$$\begin{array}{ll} 0.1788\underline{0}87863 & \\ 0.023\underline{6}13963 & \\ \hline 0.202\underline{4}227493 & \text{(unrounded answer)} \\ 0.202\underline{4} & \text{(correct number of significant figures)} \end{array}$$

When the result is divided by 0.10000, the final answer can be reported to four significant figures because it is limited by the value 0.2024.

$$\frac{0.202\underline{4}227493}{0.10000} = 2.024$$

There are four significant figures in the final answer. Note that no rounding occurred until the last step of the calculation.

## Additional Free-Response Strategies

Free-response questions are not traditional essay questions. You should answer each question with a sentence or two, a calculation, or a quick sketch.

- **Read each question carefully.** Circle key information and jot key words in the margins. Do not assume that you know what the question is asking without *thinking* carefully.
- **Number your responses.** Clearly label your responses with the correct question number and question part.
- **Write clearly and neatly.** You are not penalized for poor grammar, spelling, or handwriting, but if it interferes with the AP Readers' ability to interpret your response, you may lose valuable points.
- **Follow the directions precisely.** When a question asks for an answer and an explanation, you must explain your answer in order to earn credit. This also applies to questions that ask you to justify your responses by

referring to data, calculations, and drawings. Refer to the data explicitly. If a question asks for a justification and you do not refer to the data, you *may* fail to earn points, even if your answer is technically correct.

- **Include a sketch when helpful or when requested.** Some questions may be best answered with a sketch. If you feel that your answer would be clarified for the AP Reader with an illustration, provide one. Make sure to neatly label the parts of your sketch and explain how the drawing answers the question. If, on the other hand, a sketch is *requested,* you must provide one or you will not earn credit.
- **Show your work.** Be sure to use the factor-label method of dimensional analysis to check your work and show the AP Reader that you know the material. For example, a question might ask you to calculate the number of moles of hydrochloric acid required to react with 30.0 grams of calcium carbonate. Show your work, as demonstrated below.

$$2\,HCl + CaCO_3 \rightarrow CaCl_2 + H_2O + CO_2$$

$$30.0\ \text{g}\ \cancel{CaCO_3} \times \frac{1\ \cancel{\text{mol}\ CaCO_3}}{100.09\ \cancel{\text{g}\ CaCO_3}} \times \frac{2\ \text{mol HCl}}{1\ \cancel{\text{mol}\ CaCO_3}} = 0.599\ \text{mol HCl}$$

- **Do not answer more than what is asked.** You will not get extra credit, and you will waste valuable time.
- **Read and attempt to answer every part of each question.** Even if you do not know how to answer an early part, you can still earn credit for knowing how to answer the parts that come later. If you need to make up numbers for the early parts of the question in order to proceed to the parts that you do know how to work, state what you have done and carry the number through the rest of the calculations. Glean every available point.

Finally, believe in what you have learned. Sleep well the night before the exam, eat well the morning of the exam, and breathe deeply. Read carefully, think carefully, write clearly, and use this opportunity to show how much you have practiced!

# Chapter 1

# Atomic Structure and Properties

**The properties of the elements are based on atomic structure.**

The central concept in chemistry is that matter is composed of atoms. Atoms do not undergo a change in identity during ordinary chemical reactions. The characteristics of elements are a consequence of the structure of the atoms that make up the elements. This chapter will review the fundamental understandings as they are covered in the AP Chemistry curriculum. Some of the material in this chapter is a review of Pre-AP Chemistry concepts that are assumed to be prior knowledge on the AP Chemistry exam.

## Atoms: The Fundamental Building Blocks of Matter

**Matter,** anything that has mass and volume, is composed of atoms. Atoms are the basic units of **elements,** pure substances such as carbon or gold that cannot be decomposed into simpler substances by ordinary chemical or physical changes. An **atom** is the smallest unit of matter that retains its elemental identity through all physical and chemical processes.

Atoms are composed of three types of fundamental **subatomic particles:** protons, neutrons, and electrons.

| Subatomic Particle | Symbol | Charge | Mass |
|---|---|---|---|
| **Proton** | $p^+$ | positive | ~1 amu |
| **Neutron** | $n^0$ | neutral | ~1 amu |
| **Electron** | $e^-$ | negative | negligible |

The **nucleus** is the positively charged central region of an atom. The nucleus is comprised of protons, $p^+$, and neutrons, $n^0$.

A proton and a neutron each have a mass of approximately one **atomic mass unit** (amu), $1.66 \times 10^{-27}$ kg. The amu is defined as one-twelfth the mass of a neutral atom of carbon-12, which has six protons and six neutrons.

**Electrons** are negatively charged subatomic particles located in the space outside of the nucleus. The mass of an electron is negligible compared to the mass of a proton or a neutron.

The **atomic number,** $Z$, of an element is the number of protons in the nucleus of an atom of that element. Elements are tabulated systematically on the periodic table according to increasing atomic number, and the identity of an element is determined by its atomic number.

$$Z = p^+$$

The **mass number,** $A$, of an atom is equal to the total number of protons and neutrons in that atom. The number of neutrons is equal to the difference between the mass number and the atomic number. The mass number and the atomic number are always whole numbers.

$$A = p^+ + n^0$$

**Atomic Symbol**

Mass number

$$^{A}_{Z}\mathrm{X}$$

X = element symbol (see periodic table)

Atomic number

For a *neutral* atom, the number of protons is equal to the number of electrons.

$$p^{+} = e^{-}$$

An element's symbol on the periodic table is accompanied by the atomic number and the atomic weight of the element. The **atomic weight** is the weighted average of the masses of the naturally occurring isotopes of the element. **Isotopes** are atoms of the same element that have different numbers of neutrons. The atomic weight is numerically equal to the **molar mass,** which is the mass in grams of one mole of the atoms of an element.

**The Element Sodium as It Appears on the Periodic Table**

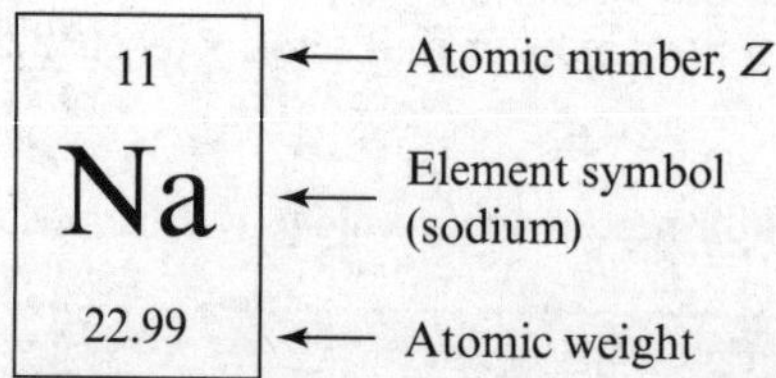


## Practice

How many protons, neutrons, and electrons are in a neutral atom of $^{26}_{12}\mathrm{Mg}$?

There are 12 protons, 14 neutrons, and 12 electrons in a neutral atom of $^{26}_{12}\mathrm{Mg}$. The atomic number, *Z*, at the lower left, is equal to the number of protons, 12. The number at the upper left is the mass number, *A*, which equals the number of protons plus the number of neutrons. The mass number is 26, so there are 14 neutrons.

$$A = 12 + n^{0} = 26$$

$$n^{0} = 14$$

In a neutral atom, the number of protons equals the number of electrons, so the number of electrons in a neutral atom of magnesium is 12.

# The Composition of Pure Substances and Mixtures

A **pure substance** is matter that cannot be broken down into simpler substances by any physical means. Elements and compounds are pure substances.

A **compound** is a pure substance composed of two or more different elements in a fixed ratio. It can be broken down into elements by chemical means. Some examples of compounds are $H_2O$, $CH_4$, and $Cr_2(SO_4)_3$.

A **molecule** is the simplest group of bonded atoms making up a pure substance that can exist independently and retain all the characteristics of the pure substance. Examples of molecules include simple ones like $H_2$, $S_8$, $H_2O$, and $CH_4$ and very complicated ones like chlorophyll ($C_{55}H_{72}N_4O_5Mg$) or proteins and nucleic acids. Notice that the **subscripts** give the number of atoms of each element in a molecule. There are eight atoms of sulfur in a molecule of $S_8$, and there are two atoms of hydrogen and one oxygen atom in a molecule of $H_2O$.

Some compounds exist as **hydrates.** They contain fixed numbers of associated water molecules. The general formula for a hydrate is

$$\text{Compound} \cdot n\ H_2O$$

For example, there are seven molecules of water associated with each unit of magnesium sulfate heptahydrate, $MgSO_4{\cdot}7\ H_2O$. A substance that has no associated water is said to be **anhydrous.**

**Mixtures** are obtained by mixing two or more pure substances. Mixtures may be either homogeneous or heterogeneous. **Homogeneous mixtures,** such as salt water or air, have the same properties throughout. **Heterogeneous mixtures,** such as oatmeal cookies or marble, do not look uniform and have different regions with different properties.

A mixture is usually separable into its component atoms or compounds by physical means. Common physical methods for separating mixtures are filtration, evaporation, distillation, and chromatography.

**Filtration** is a technique for separating heterogeneous mixtures such as sand and water. A liquid component is passed through a filter and a solid is retained in the filter. In the filtration apparatus shown below, a sidearm flask is connected to a vacuum source. The mixture is poured into the Büchner funnel that has been fitted with filter paper. The liquid, or **filtrate,** is pulled through the funnel into the flask, while the solid remains on the paper in the funnel.

**Filtration Apparatus**

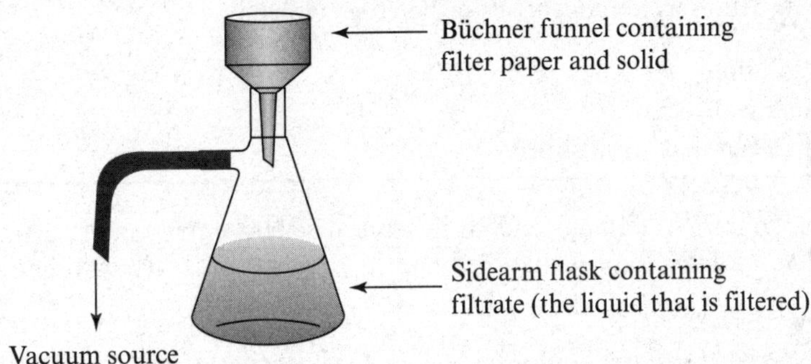


**Evaporation** is a separation technique in which a liquid is boiled away or allowed to evaporate away from a solid. Salt may be separated from a homogeneous saltwater solution by boiling away the water. Both homogeneous and heterogeneous mixtures may be separated in this way.

Distillation and chromatography are methods for separating mixtures that will be discussed in the context of intermolecular forces in Chapter 3, "Intermolecular Forces and Properties." To summarize, matter is divided into two categories, pure substances and mixtures. Pure substances are further classified as elements and compounds, while mixtures can be homogeneous and heterogeneous.

**Classification of Matter**

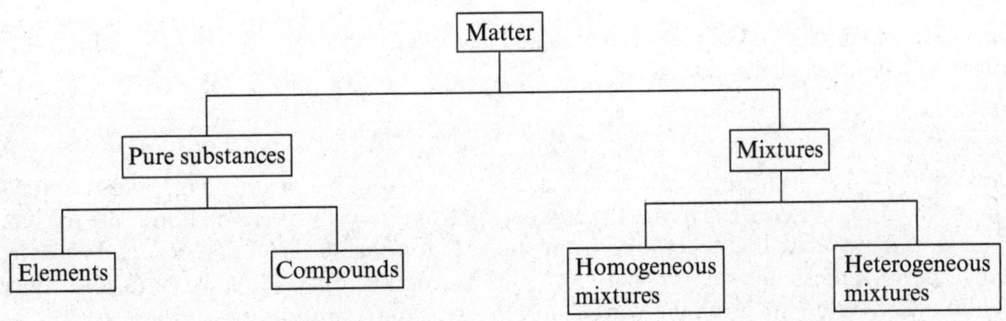

### Practice

Classify each of these substances as an element, a compound, or a mixture.

**a.** carbon dioxide
**b.** water
**c.** oxygen
**d.** tea

**(a)** Carbon dioxide, $CO_2$, is a compound in which each molecule is composed of two atoms of oxygen and one atom of carbon.

**(b)** Water, $H_2O$, is a compound in which each molecule is composed of two atoms of hydrogen and one atom of oxygen.

**(c)** Oxygen, $O_2$, is an element, since each molecule is composed of only one kind of atom.

**(d)** Tea is a mixture of water and a number of compounds infused from tea leaves.

### Practice

Classify each of these mixtures as homogeneous or heterogeneous.

**a.** unsaturated sugar solution
**b.** carbonated water in a sealed can
**c.** carbonated water in a can that has just been opened

**(a)** An unsaturated sugar solution is a homogeneous mixture. The concentration of sugar is the same throughout the water, and the word *unsaturated* implies that the quantity of sugar present is within the solubility limit at a particular temperature.

**(b)** Carbonated water in a sealed can is a homogeneous mixture because in a sealed can, the carbon dioxide gas is distributed evenly throughout the water.

**(c)** Carbonated water in a can that has just been opened is a heterogeneous mixture. When the can is opened, the pressure inside the can is reduced and carbon dioxide comes out of the solution, forming heterogeneous bubbles of gas in the water. These bubbles are distributed unevenly throughout the mixture.

## Law of Conservation of Mass

Matter cannot be created or destroyed in ordinary physical processes. This **law of conservation of mass** is the reason chemical equations must be balanced, a skill that is discussed further in Chapter 4, "Chemical Reactions." The law of conservation of mass does not apply to nuclear transformations.

Consider the reaction between sodium, a soft metal, and chlorine, a diatomic gas, to form sodium chloride. This may be represented as a chemical equation.

$$2\ Na(s) + Cl_2(g) \rightarrow 2\ NaCl(s)$$

The **reactants** are the reacting chemical species to the left of the arrow, $Na(s)$ and $Cl_2(g)$, and the **products** are the species formed, $NaCl(s)$. The **coefficients** are the numbers in front of each species that indicate the relative number of units of that species that react. The arrow shows that a chemical reaction occurs. The **state symbols** following each species indicate the state of that species in the reaction. Solids, liquids, gases, and aqueous solutions (solutions in water) are given the symbols (*s*), (*l*), (*g*), and (*aq*), respectively.

Notice that the same number of sodium atoms (2) and chlorine atoms (2) appear in both the reactants and the products. This illustrates the law of conservation of mass.

### Practice

Determine if each of the following chemical equations is correct. Justify your answers by referring to the law of conservation of mass.

**a.** $2\,C + 2\,O_2 \rightarrow 2\,CO_2$
**b.** $2\,H_2 + 2\,O_2 \rightarrow 2\,H_2O$
**c.** $4\,C + 4\,H_2 + O_2 \rightarrow C_3H_7COOH$

**(a)** This equation is correct. It obeys the law of conservation of mass, with two carbon atoms and four oxygen atoms in both the reactants and the products.

**(b)** This equation is incorrect. There are four hydrogen atoms in both reactants and products, but there are four oxygen atoms in the reactants and only two oxygen atoms in the products. The law of conservation of mass is not obeyed since there are different numbers of oxygen and hydrogen atoms on the two sides of the equation.

**(c)** This equation is correct. It obeys the law of conservation of mass, with four carbon atoms, eight hydrogen atoms, and two oxygen atoms in both the reactants and the products.

## Law of Definite Proportions

The **law of definite proportions** states that a particular compound always contains the same mass percentages of elements. Every gram of water, $H_2O$, is always composed of 0.89 gram of oxygen and 0.11 gram of hydrogen. In other words, it is 89% oxygen and 11% hydrogen by mass.

The fixed percent of each element by mass for a compound is called the **percent composition.**

$$\text{Mass percent} = \frac{\text{mass of one element}}{\text{total mass of compound}} \times 100\%$$

Experimentally determined percent composition can also be an indicator of the purity of a sample. If the sample of the compound in the practice problem below, $C_2H_4O_2$, was determined experimentally to have different mass percentages than the one shown, this would be good evidence for the presence of impurities in the sample.

### Practice

A 52.1-gram sample of the compound $C_2H_4O_2$ is composed of 20.8 grams of carbon, 3.5 grams of hydrogen, and 27.8 grams of oxygen. What is the percent composition of the sample?

The percent composition of each element is its percentage of the total mass.

$$\text{Percent carbon} = \frac{20.8\ \cancel{\text{grams}}\ \text{C}}{52.1\ \cancel{\text{grams}}\ \text{total}} \times 100\,\% = 39.9\,\%\ \text{C}$$

$$\text{Percent hydrogen} = \frac{3.5\ \cancel{\text{grams}}\ \text{H}}{52.1\ \cancel{\text{grams}}\ \text{total}} \times 100\,\% = 6.7\,\%\ \text{H}$$

$$\text{Percent oxygen} = \frac{27.8\ \cancel{\text{grams}}\ \text{O}}{52.1\ \cancel{\text{grams}}\ \text{total}} \times 100\,\% = 53.4\%\ \text{O}$$

## Law of Multiple Proportions

John Dalton was a teacher, chemist, and physicist who made observations about the ratios of the masses of elements that combine to form compounds. His contributions to chemistry include the **law of multiple proportions,** which states that when two elements form more than one compound, such as $H_2O$ and $H_2O_2$, the ratio of the masses of the second element that combine with a fixed amount of the first element is always a small, whole-number ratio.

In the example of $H_2O$ and $H_2O_2$, 1 gram of hydrogen combines with 8 grams of oxygen to form $H_2O$, and 1 gram of hydrogen combines with 16 grams of oxygen to form $H_2O_2$. The two masses of oxygen that combine with 1 gram of hydrogen can be simplified to a small, whole-number ratio.

$$\frac{\text{Mass of oxygen in } H_2O_2}{\text{Mass of oxygen in } H_2O} = \frac{16 \cancel{g}}{8 \cancel{g}} = \frac{2}{1}$$

### Practice

One gram of carbon is found to combine with 0.336 gram of hydrogen in Compound A and with 0.252 gram of hydrogen in Compound B. Show that these compounds obey the law of multiple proportions.

To demonstrate that two compounds adhere to the law of multiple proportions, show that the masses of hydrogen that combine with 1 gram of carbon can be simplified to a whole-number ratio. It is necessary to recognize that $0.3\overline{3}$ is $\frac{1}{3}$. The ratio must be multiplied by $\frac{3}{3}$ in order to obtain the small whole-number ratio.

$$\frac{\text{Mass of hydrogen in Compound A}}{\text{Mass of hydrogen in Compound B}} = \frac{0.336 \cancel{gH}}{0.252 \cancel{gH}} = \frac{1.3\overline{3}}{1} \times \frac{3}{3} = \frac{4}{3}$$

## Counting Particles: The Mole

Atoms, molecules, and other submicroscopic particles are incomprehensibly small. They are so tiny that a special method is necessary to count them. Consider a tablespoon (roughly 15 mL) of water. At ambient temperature and pressure, that tablespoon contains on the order of $5 \times 10^{23}$ molecules of $H_2O$. This enormous number cannot possibly be counted directly. The **mole,** which is the number of carbon atoms found in exactly 12 grams of carbon-12, is used to circumvent this problem. A mole, abbreviated "mol," contains a number of particles equal to Avogadro's constant, $N_A$. It works just like "a dozen."

$$N_A = 6.022 \times 10^{23} \text{ particles} = 1 \text{ mole of particles}$$

Consider the following examples:

$$1 \text{ mol of Na} = 6.022 \times 10^{23} \text{ atoms of Na}$$
$$1 \text{ mol of } H_2O = 6.022 \times 10^{23} \text{ molecules of } H_2O$$

These can be written as ratios for use in **dimensional analysis,** a technique by which units for a value are interconverted by multiplication of the value by unit factors. A **unit factor** is a ratio in which the numerator and denominator are equal, so multiplying by a unit factor is the same as multiplying by 1; the only change is in the units of the result.

From the equalities above, we can write the following unit factors and choose the one by which multiplication gives the desired units:

$$\frac{6.022 \times 10^{23} \text{ atoms of Na}}{1 \text{ mol Na}} \text{ or } \frac{1 \text{ mol Na}}{6.022 \times 10^{23} \text{ atoms of Na}} \text{ or } \frac{1 \text{ mol } H_2O}{6.022 \times 10^{23} \text{ molecules of } H_2O} \text{ or } \frac{6.022 \times 10^{23} \text{ molecules of } H_2O}{1 \text{ mol } H_2O}$$

As an example, in order to determine the number of sodium atoms in 0.150 mole of sodium, we multiply the number of moles of sodium by the unit factor that relates the number of atoms to moles.

$$\frac{6.022\times10^{23}\text{ atoms of Na}}{1\text{ mol Na}}=1$$

$$0.150\ \cancel{\text{mol Na}}\times\frac{6.022\times10^{23}\text{ atoms of Na}}{1\ \cancel{\text{mol Na}}}=9.03\times10^{22}\text{ atoms of Na}$$

### Practice

How many $CO_2$ molecules make up 1.5 moles of $CO_2$?

One mole of $CO_2$ = 6.022 × $10^{23}$ molecules of $CO_2$. In order to find the number of $CO_2$ molecules in 1.5 moles of $CO_2$, the number of moles is multiplied by the unit factor relating moles of $CO_2$ to molecules of $CO_2$.

$$\frac{6.022\times10^{23}\text{ molecules }CO_2}{1\text{ mol }CO_2}=1$$

$$1.5\ \cancel{\text{mol }CO_2}\times\frac{6.022\times10^{23}\text{ molecules }CO_2}{1\ \cancel{\text{mol }CO_2}}=9.0\times10^{23}\text{ molecules }CO_2$$

## Atomic Weight and Molar Mass

Macroscopic quantities such as mass and volume can be measured using such instruments as a balance and a graduated cylinder. Mass and volume can be related to the number of particles of a substance via the mole.

Just as a dozen bowling balls weigh more than a dozen tennis balls, a mole of iron atoms weighs more than a mole of lithium atoms.

**Atomic weight** is the average weight of a particular kind of atom in amu. It is a weighted average of the exact masses of the naturally occurring isotopes of an element. The mass of one mole of atoms of an element in grams is called the **molar mass.** It has the same numerical value as its atomic weight and is expressed in g·mol$^{-1}$. It is used to convert mass, a quantity that can be measured in a laboratory, to moles.

The mass in grams of one mole of a substance is called its molar mass, ***M***, or **molecular weight.** The molar mass of a mole of molecules is given by the sum of the molar masses of the constituent atoms of that molecule. To determine the molar mass for a compound (or **formula mass/formula weight** for an ionic formula), the number of moles of each element is multiplied by the molar mass of that element.

Suppose a student needs to determine the molar mass of glucose, $C_6H_{12}O_6$. In one molecule of glucose, there are 6 carbon atoms, 6 oxygen atoms, and 12 hydrogen atoms. That means that in one mole of glucose, there are 6 moles of carbon atoms, 6 moles of oxygen atoms, and 12 moles of hydrogen atoms.

$$M_{C_6H_{12}O_6}=\left(\frac{6\ \cancel{\text{mol C}}}{1\text{ mol }C_6H_{12}O_6}\right)\left(\frac{12.01\text{ g}}{1\ \cancel{\text{mol C}}}\right)+\left(\frac{12\ \cancel{\text{mol H}}}{1\text{ mol }C_6H_{12}O_6}\right)\left(\frac{1.01\text{ g}}{1\ \cancel{\text{mol H}}}\right)+\left(\frac{6\ \cancel{\text{mol O}}}{1\text{ mol }C_6H_{12}O_6}\right)\left(\frac{16.00\text{ g}}{1\ \cancel{\text{mol O}}}\right)$$

$$M_{C_6H_{12}O_6}=180.18\ \frac{\text{g}}{\text{mol}}$$

### Practice

Find the mass in amu of 10 atoms of sodium.

The number of amu of a small number of atoms of sodium can be determined using the relationship between moles and number of atoms and using the atomic weight of sodium found on the periodic table.

$$1 \text{ atom of Na} = 22.99 \text{ amu}$$

We can multiply the number of sodium atoms by the appropriate unit factor ratio.

$$10 \cancel{\text{ atoms Na}} \times \frac{22.99 \text{ amu}}{1 \cancel{\text{ atom Na}}} = 229.9 \text{ amu}$$

## Practice

What is the mass in grams of 0.15 mole of helium atoms?

The mass is determined by multiplying the number of moles of helium by the molar mass.

$$0.15 \cancel{\text{ mol He}} \times \frac{4.00 \text{ g He}}{1 \cancel{\text{ mol He}}} = 0.60 \text{ g He}$$

## Practice

Express the molar masses or formula masses of the following compounds in grams per mole.

**a.** $H_2O$
**b.** $(NH_4)_3PO_4$
**c.** $CaSO_4 \cdot 2\,H_2O$

**(a)** There are 2 moles of hydrogen atoms and 1 mole of oxygen atoms per mole of water.

$$\boldsymbol{M}_{H_2O} = \left(\frac{2 \cancel{\text{ mol H}}}{1 \text{ mol } H_2O}\right)\left(\frac{1.01 \text{ g H}}{1 \cancel{\text{ mol H}}}\right) + \left(\frac{1 \cancel{\text{ mol O}}}{1 \text{ mol } H_2O}\right)\left(\frac{16.00 \text{ g O}}{1 \cancel{\text{ mol O}}}\right)$$

$$\boldsymbol{M}_{H_2O} = 18.02 \frac{\text{g}}{\text{mol}}$$

**(b)** The parentheses indicate that there are three ammonium ions, $NH_4^+$, for every phosphate ion, $PO_4^{3-}$. This means that there are 3 moles of nitrogen atoms, 12 moles of hydrogen atoms, 1 mole of phosphorus atoms, and 4 moles of oxygen atoms per mole of the formula.

$$\boldsymbol{M}_{(NH_4)_3PO_4} = \left(\frac{3 \cancel{\text{ mol N}}}{1 \text{ mol } (NH_4)_3PO_4}\right)\left(\frac{14.01 \text{ g N}}{1 \cancel{\text{ mol N}}}\right) + \left(\frac{12 \cancel{\text{ mol H}}}{1 \text{ mol } (NH_4)_3PO_4}\right)\left(\frac{1.01 \text{ g H}}{1 \cancel{\text{ mol H}}}\right) + \left(\frac{1 \cancel{\text{ mol P}}}{1 \text{ mol } (NH_4)_3PO_4}\right)\left(\frac{30.97 \text{ g P}}{1 \cancel{\text{ mol P}}}\right) + \left(\frac{4 \cancel{\text{ mol O}}}{1 \text{ mol } (NH_4)_3PO_4}\right)\left(\frac{16.00 \text{ g O}}{1 \cancel{\text{ mol O}}}\right)$$

$$\boldsymbol{M}_{(NH_4)_3PO_4} = 149.12 \frac{\text{g}}{\text{mol}}$$

**(c)** This compound is a **dihydrate.** The prefix *di-* indicates that there are 2 water molecules associated with each calcium sulfate formula unit.

$$\mathcal{M}_{CaSO_4 \cdot 2H_2O} = \left(\frac{1 \cancel{\text{mol Ca}}}{1 \text{ mol CaSO}_4 \cdot 2\text{H}_2\text{O}}\right)\left(\frac{40.08 \text{ g Ca}}{1 \cancel{\text{mol Ca}}}\right) + \left(\frac{1 \cancel{\text{mol S}}}{1 \text{ mol CaSO}_4 \cdot 2\text{H}_2\text{O}}\right)\left(\frac{32.06 \text{ g S}}{1 \cancel{\text{mol S}}}\right)$$

$$+ \left(\frac{6 \cancel{\text{mol O}}}{1 \text{ mol CaSO}_4 \cdot 2\text{H}_2\text{O}}\right)\left(\frac{16.00 \text{ g O}}{1 \cancel{\text{mol O}}}\right) + \left(\frac{4 \cancel{\text{mol H}}}{1 \text{ mol CaSO}_4 \cdot 2\text{H}_2\text{O}}\right)\left(\frac{1.01 \text{ g H}}{1 \cancel{\text{mol H}}}\right)$$

$$\mathcal{M}_{CaSO_4 \cdot 2H_2O} = 172.18 \frac{\text{g}}{\text{mol}}$$

## Density

Volume is another quantity that can be measured in a laboratory. The volume of a substance is related to the number of grams of that substance by its **density,** the mass of the substance per unit volume.

$$\text{Density} = \frac{\text{mass}}{\text{volume}}$$

The density of $PCl_3$ is 1.57 g·mL$^{-1}$ at 25 °C. There are 1.57 grams of $PCl_3$ in every milliliter of $PCl_3$. The density can be written as two different ratios that can be used in dimensional analysis.

$$\frac{1.57 \text{ g PCl}_3}{1 \text{ mL PCl}_3} = 1 \quad \text{or} \quad \frac{1 \text{ mL PCl}_3}{1.57 \text{ g PCl}_3} = 1$$

If a pipette is used to deliver 5.0 mL of $PCl_3$ to a reaction vessel, the number of grams that has been added is the product of the volume and the density. The mass can then be converted to moles using the molar mass.

$$5.0 \cancel{\text{mL PCl}_3} \times \frac{1.57 \cancel{\text{g PCl}_3}}{1 \cancel{\text{mL PCl}_3}} \times \frac{1 \text{ mol PCl}_3}{137.32 \cancel{\text{g PCl}_3}} = 0.057 \text{ mol PCl}_3$$

The densities of fluids dictate which immiscible (unable to mix) fluids will float on top of others. Vegetable oil is less dense than water, so it forms a layer above water in a container. It does not mix with water for reasons that will be reviewed in Chapter 3, "Intermolecular Forces and Properties."

**TIP: One milliliter is equal to one cubic centimeter. That is, 1 mL = 1 cm$^3$.**

### Practice

A chemist needs to use 0.030 mole of $Br_2$ in a reaction. How many milliliters are required? The density of $Br_2$ is 3.01 g·mL$^{-1}$.

The number of moles is first converted to grams. The density is then used to determine the volume.

$$0.030 \cancel{\text{mol Br}_2} \times \frac{159.80 \cancel{\text{g Br}_2}}{1 \cancel{\text{mol Br}_2}} \times \frac{1 \text{ mL Br}_2}{3.01 \cancel{\text{g Br}_2}} = 1.6 \text{ mL Br}_2$$

## Empirical and Molecular Formulas

The **molecular formula** of a compound is an expression of the number of atoms of each element in a molecule of that compound. Glucose has a molecular formula of $C_6H_{12}O_6$.

The **empirical formula** of a compound is the lowest whole-number ratio of atoms in that compound. When each subscript in the molecular formula of glucose is divided by 6, the molecular formula simplifies to the empirical formula, $CH_2O$.

## Practice

The molecular formula of crotonic acid is $C_4H_6O_2$. What is the empirical formula?

Division of each of the subscripts in the formula by 2 gives the simplest possible whole-number ratio of the atoms in the formula, $C_2H_3O$.

## Empirical and Molecular Formula Determination

Empirical formulas can be determined from the mass percent, or percent composition of a compound. Suppose a compound is found to be 54.5% carbon, 9.2% hydrogen, and 36.3% oxygen by mass. We can use this information to determine the empirical formula.

The first step is to determine the number of moles of each element in 100.0 grams of the compound. In 100.0 grams of the compound there are 54.5 grams of carbon, 9.2 grams of hydrogen, and 36.3 grams of oxygen. These masses are multiplied by atomic weight unit factors obtained from the periodic table.

$$54.5\ \cancel{\text{g C}}\left(\frac{1\ \text{mol C}}{12.01\ \cancel{\text{g C}}}\right) = 4.54\ \text{mol C}$$

$$9.2\ \cancel{\text{g H}}\left(\frac{1\ \text{mol H}}{1.01\ \cancel{\text{g H}}}\right) = 9.1\ \text{mol H}$$

$$36.3\ \cancel{\text{g O}}\left(\frac{1\ \text{mol O}}{16.00\ \cancel{\text{g O}}}\right) = 2.27\ \text{mol O}$$

The next step is to find the simplest whole-number ratio by dividing each number of moles by the smallest number. In this case, divide by 2.27 moles.

$$\frac{4.54\ \text{mol C}}{2.27\ \text{mol}} \approx 2$$

$$\frac{9.1\ \text{mol H}}{2.27\ \text{mol}} \approx 4$$

$$\frac{2.27\ \text{mol O}}{2.27\ \text{mol}} = 1$$

These integer values give an empirical formula of $C_2H_4O$.

Now that the empirical formula has been determined, we can find the molecular formula using the experimentally determined molar mass of the compound, 264.24 g·mol$^{-1}$, and the empirical formula molar mass. We can determine the molar mass of the empirical formula as discussed above.

$$\boldsymbol{M}_{C_2H_4O} = \left(\frac{2\ \cancel{\text{mol C}}}{1\ \text{mol C}_2\text{H}_4\text{O}}\right)\left(\frac{12.01\ \text{g C}}{1\ \cancel{\text{mol C}}}\right) + \left(\frac{4\ \cancel{\text{mol H}}}{1\ \text{mol C}_2\text{H}_4\text{O}}\right)\left(\frac{1.01\ \text{g H}}{1\ \cancel{\text{mol H}}}\right) + \left(\frac{1\ \cancel{\text{mol O}}}{1\ \text{mol C}_2\text{H}_4\text{O}}\right)\left(\frac{16.00\ \text{g O}}{1\ \cancel{\text{mol O}}}\right)$$

$$\boldsymbol{M}_{C_2H_4O} = 44.06\frac{\text{g}}{\text{mol}}$$

Next, the molar mass of the compound is divided by the empirical formula molar mass.

$$\frac{264.24\ \text{g} \cdot \cancel{\text{mol}^{-1}}}{44.06\ \text{g} \cdot \cancel{\text{mol}^{-1}}} \approx 6$$

The subscripts in the empirical formula are multiplied by the result to give the molecular formula, $C_{12}H_{24}O_6$.

Percent composition may be used to draw conclusions about the percent purity of a sample. It is very important to understand that any two compounds with the same empirical formula must have identical percent composition.

## Practice

A research group discovers and purifies a small molecule from a marine plant species in the Caribbean Sea. It has an empirical formula of $C_{10}H_{11}O_2$. They name it *Caribbeanium*. Another research group attempts to reproduce these results. They isolate a 0.892-mg sample that has the elemental composition shown in the table below. What conclusions can be drawn from these results regarding the identity and purity of the sample isolated by the second group?

| Element | Mass (mg) |
|---|---|
| C | 0.573 |
| H | 0.064 |
| O | 0.255 |
| **Total** | **0.892** |

If the sample isolated by the second group is a pure sample of *Caribbeanium,* the percent composition should match that of *Caribbeanium.*

The percent composition of *Caribbeanium* is determined first. When using an empirical formula to determine percent composition, start by finding the molar mass of the empirical formula.

$$M_{C_{10}H_{11}O_2} = \left(\frac{10\ \cancel{\text{mol C}}}{1\ \text{mol}\ C_{10}H_{11}O_2}\right)\left(\frac{12.01\ \text{g C}}{1\ \cancel{\text{mol C}}}\right) + \left(\frac{11\ \cancel{\text{mol H}}}{1\ \text{mol}\ C_{10}H_{11}O_2}\right)\left(\frac{1.01\ \text{g H}}{1\ \cancel{\text{mol H}}}\right) + \left(\frac{2\ \cancel{\text{mol O}}}{1\ \text{mol}\ C_{10}H_{11}O_2}\right)\left(\frac{16.00\ \text{g O}}{1\ \cancel{\text{mol O}}}\right)$$

$$= 163.21\frac{\text{g}}{\text{mol}}$$

The mass of each element present in 163.21 grams (1 mole) of *Caribbeanium* is

$$\left(\frac{10\ \cancel{\text{mol C}}}{1\ \text{mol}\ C_{10}H_{11}O_2}\right)\left(\frac{12.01\ \text{g C}}{1\ \cancel{\text{mol C}}}\right) = 120.10\ \text{g C}$$

$$\left(\frac{11\ \cancel{\text{mol H}}}{1\ \text{mol}\ C_{10}H_{11}O_2}\right)\left(\frac{1.01\ \text{g H}}{1\ \cancel{\text{mol H}}}\right) = 11.11\ \text{g H}$$

$$\left(\frac{2\ \cancel{\text{mol O}}}{1\ \text{mol}\ C_{10}H_{11}O_2}\right)\left(\frac{16.00\ \text{g O}}{1\ \cancel{\text{mol O}}}\right) = 32.00\ \text{g O}$$

The percent composition of *Caribbeanium* is

$$\text{Percent C} = \frac{120.10 \not{g} \text{ C}}{163.21 \not{g} \text{ C}_{10}\text{H}_{11}\text{O}_2} \times 100\% = 73.6\% \text{ C}$$

$$\text{Percent H} = \frac{11.11 \not{g} \text{ H}}{163.21 \not{g} \text{ C}_{10}\text{H}_{11}\text{O}_2} \times 100\% = 6.8\% \text{ H}$$

$$\text{Percent O} = \frac{32.00 \not{g} \text{ O}}{163.21 \not{g} \text{ C}_{10}\text{H}_{11}\text{O}_2} \times 100\% = 19.6\% \text{ O}$$

The percent composition of the unknown sample is determined and compared.

$$\text{Percent C} = \frac{0.573 \not{mg} \text{ C}}{0.892 \not{mg} \text{ unknown}} \times 100\,\% = 64.2\% \text{ C}$$

$$\text{Percent H} = \frac{0.064 \not{mg} \text{ H}}{0.892 \not{mg} \text{ unknown}} \times 100\,\% = 7.2\% \text{ H}$$

$$\text{Percent O} = \frac{0.255 \not{mg} \text{ O}}{0.892 \not{mg} \text{ unknown}} \times 100\,\% = 28.6\% \text{ O}$$

The percent composition of the sample isolated by the second group does not match that of *Caribbeanium.* Two conclusions may be drawn about the identity and purity of the unknown sample:

1. The unknown sample *may* contain *Caribbeanium,* but if it does, it must also contain impurities that cause the percent composition to differ.
2. The unknown sample *may not* contain *Caribbeanium*. No conclusion may be drawn about the purity of the sample if this is the case.

## Combustion Analysis

In a combustion reaction, a compound reacts rapidly with oxygen while generating heat and light.

**Combustion analysis** is a method for determining the percent composition and empirical formula of a compound. When the data from a combustion analysis are used in conjunction with molar mass information determined from an experimental technique called mass spectrometry, the molecular formula of a compound can be determined.

In a combustion reaction of a **hydrocarbon,** a compound that contains carbon and hydrogen atoms bound together, carbon dioxide, $CO_2$, and water vapor, $H_2O$, are generated according to the equation below.

$$C_xH_y + z\,O_2 \rightarrow x\,CO_2 + \frac{y}{2}\,H_2O$$

Although balancing chemical reactions is a topic that will be addressed in Chapter 4, "Chemical Reactions," there are two very important features of this reaction to note in the context of combustion analysis. First, the number of moles of carbon dioxide produced, $x$, is equal to the number of moles of carbon that were initially present in the sample. Second, the number of moles of $H_2O$ produced is equal to half the number of moles of hydrogen atoms, $y$, that were present in the original sample. When oxygen atoms are present in the unknown compound, the difference between the mass of the original sample and the mass that can be attributed to carbon and hydrogen is equal to the mass of the oxygen in the sample. All the atoms in the reactants are also present in the products, and *the law of conservation of mass is obeyed.*

## Practice

An unknown compound contains only carbon, hydrogen, and oxygen. Combustion of a 30.0-gram sample of the compound in excess oxygen yields 80.1 grams of $CO_2$ and 12.6 grams of $H_2O$. What is the empirical formula of the compound?

The number of moles of carbon in the original sample can be determined from the grams of $CO_2$ obtained. The number of moles of hydrogen can be determined from the mass of water.

$$\text{moles of carbon} = 80.1\ \cancel{\text{g CO}_2} \times \frac{1\ \cancel{\text{mol CO}_2}}{44.01\ \cancel{\text{g CO}_2}} \times \frac{1\ \text{mol C}}{1\ \cancel{\text{mol CO}_2}} = 1.82\ \text{mol C}$$

$$\text{moles of hydrogen} = 12.6\ \cancel{\text{g H}_2\text{O}} \times \frac{1\ \cancel{\text{mol H}_2\text{O}}}{18.02\ \cancel{\text{g H}_2\text{O}}} \times \frac{2\ \text{mol H}}{1\ \cancel{\text{mol H}_2\text{O}}} = 1.40\ \text{mol H}$$

The masses of carbon and hydrogen can be used to find the mass of oxygen in the sample since the total mass is known.

$$\text{mass of carbon} = 1.82\ \cancel{\text{mol C}} \times \frac{12.01\ \text{g C}}{1\ \cancel{\text{mol C}}} = 21.9\ \text{g C}$$

$$\text{mass of hydrogen} = 1.40\ \cancel{\text{mol H}} \times \frac{1.01\ \text{g H}}{1\ \cancel{\text{mol H}}} = 1.41\ \text{g H}$$

$$\text{Total mass of sample} = \text{mass of carbon} + \text{mass of hydrogen} + \text{mass of oxygen}$$
$$30.0\ \text{g} = 21.9\ \text{g} + 1.41\ \text{g} + \text{mass of oxygen}$$
$$\text{Mass of oxygen} = 6.7\ \text{g}$$

The mass of oxygen is used to find the number of moles of oxygen in the sample.

$$\text{moles of oxygen} = 6.7\ \cancel{\text{g O}} \times \frac{1\ \cancel{\text{mol O}}}{16.00\ \cancel{\text{g O}}} = 0.42\ \text{mol O}$$

The final step in determining the empirical formula is division of the number of moles of each element by the smallest number of moles.

$$\text{Carbon:}\quad \frac{1.82}{0.42} = 4.33$$
$$\text{Hydrogen:}\quad \frac{1.40}{0.42} = 3.33$$
$$\text{Oxygen:}\quad \frac{0.42}{0.42} = 1$$

The number 0.33 should be recognized as $\frac{1}{3}$, so the simplest ratio can be determined by multiplying each number by 3.

$$\text{Carbon:}\quad 4.33 \times 3 = 13$$
$$\text{Hydrogen:}\quad 3.33 \times 3 = 10$$
$$\text{Oxygen:}\quad 1 \times 3 = 3$$

The empirical formula, therefore, is $C_{13}H_{10}O_3$.

# The Evolution of Scientific Models

Scientific models are based on experimental evidence and are refined over time as new data becomes available.

John Dalton's work supported the concept that compounds were composed of atoms, and his discoveries prompted him to present a comprehensive atomic theory in the early nineteenth century. The postulates of atomic theory have been refined since Dalton's original proposal, but the fundamental accuracy of most of his ideas has been verified experimentally.

**Dalton's Atomic Theory**

- All matter is composed of atoms.
- Atoms cannot be created or destroyed.*
- The atoms of a given element are identical in mass, but different from other elements.**
- Atoms combine in simple, whole-number ratios to form compounds.
- Atoms are reorganized, not fundamentally changed, in chemical reactions.

Chemists' understanding of the structure of the atom has changed dramatically since Dalton published his atomic theory in 1808. At that time, Dalton imagined the atom to be an indivisible particle, not unlike a small marble. Subsequent discoveries and observations led to the adaptation of scientists' model of the internal structure of the atom. The history of the atomic model is used here to illustrate how a scientific model evolves to reflect new experimental results.

# The Quantum Nature of the Atom

The model of the atom has evolved from one in which electrons orbited in elliptical paths around the nucleus at the center of an atom like planets orbiting a star. The structure of the atom is currently understood to be a more complicated picture of electrons behaving as waves.

**Electronic orbitals** can be visualized as three-dimensional regions surrounding the nucleus where the probability of finding an electron is high. They are not elliptical. Instead, orbitals are understood to be shaped like diffuse spheres, petals, and donuts. This section will briefly review the evolution of scientists' understanding of atomic structure.

## Electromagnetic Radiation

**Electromagnetic radiation,** which played a crucial role in shaping current ideas of atomic structure, is radiant energy that is composed of oscillating electric and magnetic fields. Visible light, radio waves, X-rays, gamma rays, ultraviolet light, and infrared radiation are all electromagnetic radiation. A **photon** is a unit of electromagnetic radiation that has a **quantum,** or fixed amount, of energy. It behaves as both a wave and a particle.

Electromagnetic radiation can be described in terms of its wavelength, amplitude, frequency, speed, and energy.

The **wavelength,** $\lambda$, is the distance between two consecutive crests or troughs. The **amplitude** is the maximum displacement of the wave from the equilibrium position.

* It was later discovered that mass and energy could be interconverted in nuclear transformations.

** Dalton proposed that atoms of the same kind of element were identical. It was shown later, with mass spectrometry and the discovery of isotopes, that this was incorrect.

**Wavelength and Amplitude**

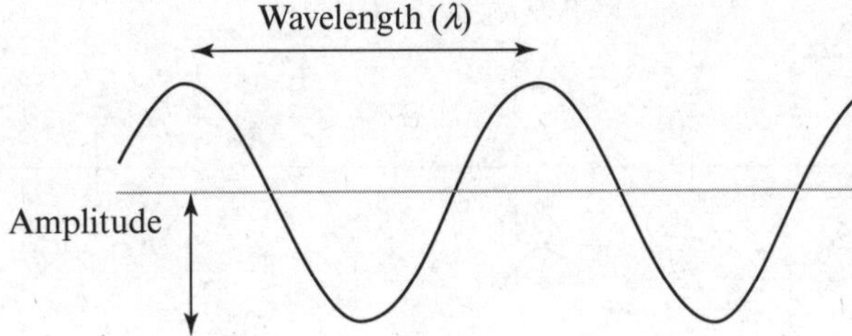


Electromagnetic radiation can have a continuous spectrum of wavelengths, but the common units for wavelengths of visible light are the nanometer (1 nm = $10^{-9}$ m) and the Ångstrom (1 Å = $10^{-10}$ m).

The **frequency** of a wave, ν, is the number of waves that pass a fixed point in space per second. The units of frequency are hertz (abbreviated Hz). Hertz are cycles per second, also represented as $s^{-1}$.

All electromagnetic radiation travels at the **speed of light,** *c*, and this speed is related to the wavelength and frequency. This equation, in addition to the value for the speed of light, is included on the *AP Chemistry Equations and Constants* sheet.

$$c = \lambda\nu$$

$$c = 2.998 \times 10^{8}\,\text{m}\cdot\text{s}^{-1}$$

It follows that wavelength and frequency are inversely proportional. The higher the frequency, the shorter the wavelength.

The **energy,** *E*, of a photon is related to its frequency by the Planck relation and Planck's constant, $h = 6.626 \times 10^{-34}$ J·s. Both the energy equation and the value of Planck's constant are provided on the exam.

$$E = h\nu$$

Substitution for ν gives the relationship among the energy, the wavelength, and the speed of light.

$$E = \frac{hc}{\lambda}$$

The energy of a photon is directly proportional to its frequency and inversely proportional to its wavelength.

The electromagnetic spectrum should be remembered in terms of the energy ranking of the different types of electromagnetic radiation. Memorization of specific wavelengths or frequencies is not necessary, but students should know the order of decreasing energy (increasing wavelength) of electromagnetic radiation.

Gamma rays < X-rays < Ultraviolet < Visible < Infrared < Microwaves < Radio
(highest energy, lowest $\lambda$) (lowest energy, highest $\lambda$)

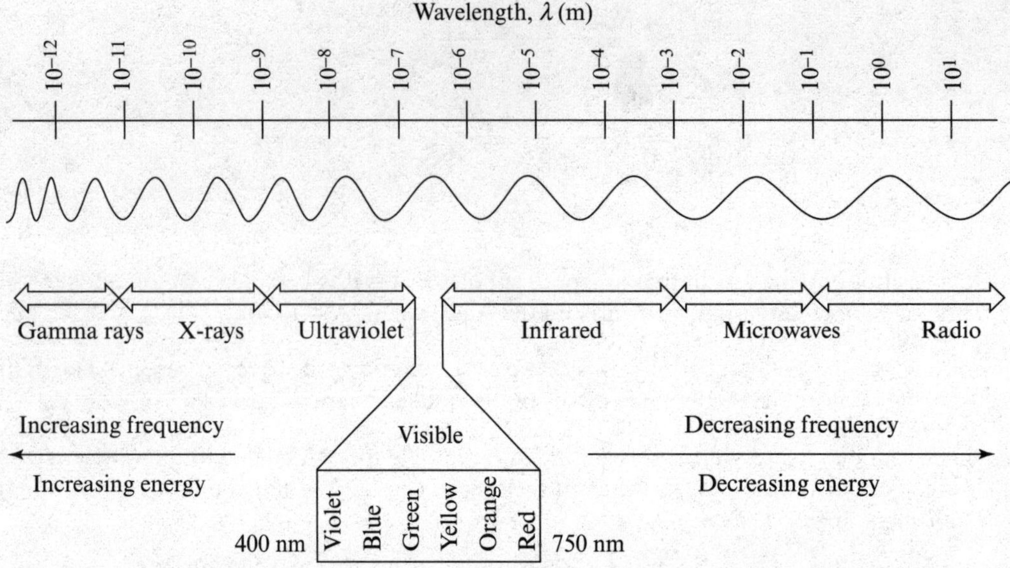


In the sections and chapters that follow, the chemistry applications of several of these classes of electromagnetic radiation, including microwave, infrared, UV-visible, and X-ray radiation, will be reviewed.

## Practice

Find the frequency of an infrared photon that has a wavelength of 750 nm.

The equation for the speed of light ($c = \lambda\nu$) can be rearranged to solve for frequency.

$$\nu = \frac{c}{\lambda} = \frac{2.998 \times 10^{8}\ \cancel{m} \cdot s^{-1}}{750\ \cancel{nm}} \times \frac{10^{9}\ \cancel{nm}}{1\ \cancel{m}} = 4.0 \times 10^{14}\ s^{-1}$$

## Practice

What is the energy of a photon that has a wavelength of 620 nm?

The energy can be found using the Planck relationship.

$$E = \frac{hc}{\lambda} = \frac{(6.626 \times 10^{-34}\ J \cdot \cancel{s})(2.998 \times 10^{8}\ \cancel{m \cdot s^{-1}})}{620\ \cancel{nm}} \times \frac{10^{9}\ \cancel{nm}}{1\ \cancel{m}} = 3.2 \times 10^{\ 19}\ J$$

## Emission and Absorption Spectra: The Bohr Model of the Atom

The study of the emission of ultraviolet and visible light by excited hydrogen atoms was the foundation for the quantum mechanical model of the atom. When white light, which is comprised of the continuous spectrum of visible wavelengths, is passed through a prism, it is separated into a rainbow of wavelengths.

**Separation of White Light into the Visible Spectrum**

When a high voltage is applied across a sample of hydrogen gas in a discharge tube, the hydrogen atoms are excited. This means that they are in a high-energy state. These excited hydrogen atoms emit light as they relax to lower-energy states. When the light emitted by excited hydrogen is passed through a prism, only discrete wavelengths are observed. These wavelengths comprise the emission spectrum of hydrogen.

**The Hydrogen Emission Spectrum**

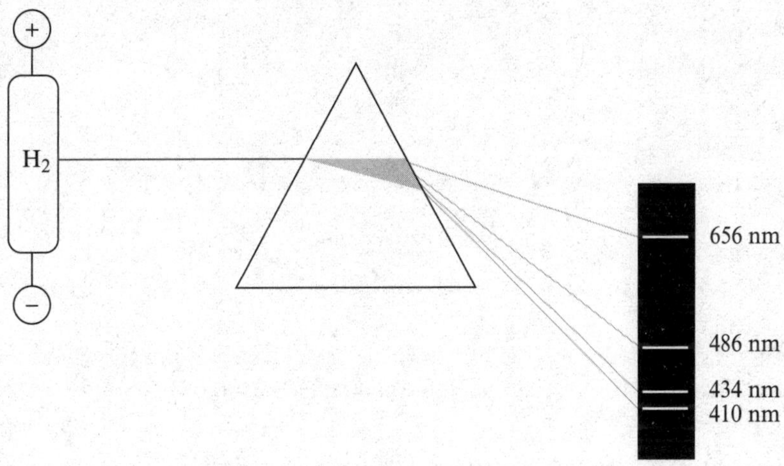


Furthermore, when white light is passed through a sample of hydrogen gas, some of the wavelengths are absorbed. The missing wavelengths in the light that emerges from the sample coincide exactly with the wavelengths that are produced in the emission spectrum. The spectrum in which the absorbed frequencies are absent is called the **absorption spectrum** of hydrogen.

**The Hydrogen Absorption Spectrum**

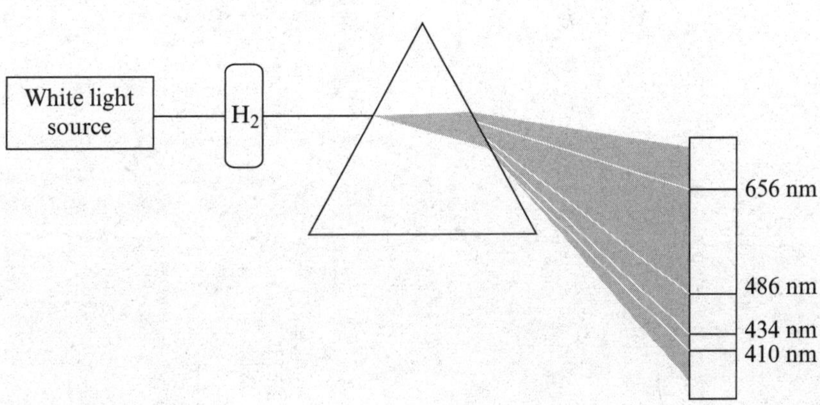

The **Bohr model** of the atom, presented in 1913 by Danish physicist Niels Bohr, is a quantum model that provides an early explanation for the absorption and emission spectra of hydrogen. According to the Bohr model, an electron orbits the nucleus in a fixed, circular path. Orbits are only located at discrete distances from the nucleus, and each orbit has an associated energy. In other words, the energy is **quantized.** The electrons orbiting the nucleus either occupy the lowest-energy **ground state,** $n = 1$, which is the configuration closest to the nucleus, or they occupy an excited, higher-energy state, $n = 2, 3, 4, \ldots \infty$. The positive integer, $n$, is called the **principal quantum number,** and it describes the main energy level that an electron occupies.

When the electron in the ground state of a hydrogen atom is excited by an applied voltage, it is promoted to a higher energy level. The electron relaxes back to the ground state, and the transition gives off a photon that is equal in energy to the energy difference between the excited state and a lower-energy state. Since only certain energies are allowed, this accounts for the characteristic wavelengths that are observed in the emission spectrum of hydrogen.

**The Bohr Model of a Hydrogen Atom: Atomic Emission**

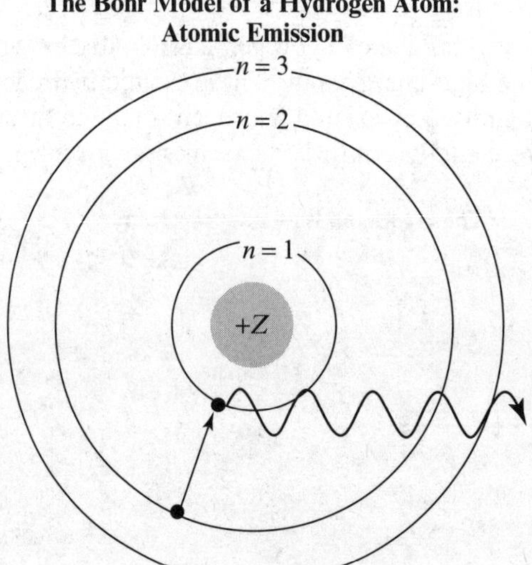


The absorption spectrum is generated when the hydrogen electron absorbs a photon with an energy exactly equal to the difference in energy between two levels. The light that passes through the sample is missing the wavelengths that correspond to photons that are absorbed.

**The Bohr Model of a Hydrogen Atom: Atomic Absorption**

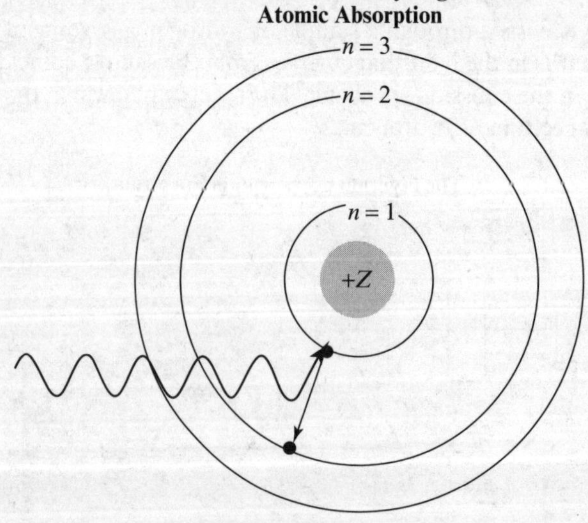

Higher-energy levels in a hydrogen atom become increasingly close in energy. In other words, the energy difference between the $n = 2$ level and the $n = 3$ level is less than the energy difference between the $n = 1$ and $n = 2$ energy levels.

Transitions from higher-energy levels to the $n = 1$ energy level emit wavelengths in the ultraviolet region of the electromagnetic spectrum. This set of spectral lines is referred to as the **Lyman series.** Transitions from higher-energy levels to the $n = 2$ level emit lower-energy wavelengths. These are referred to as the **Balmer series,** and four of these lines are in the visible region of the spectrum.

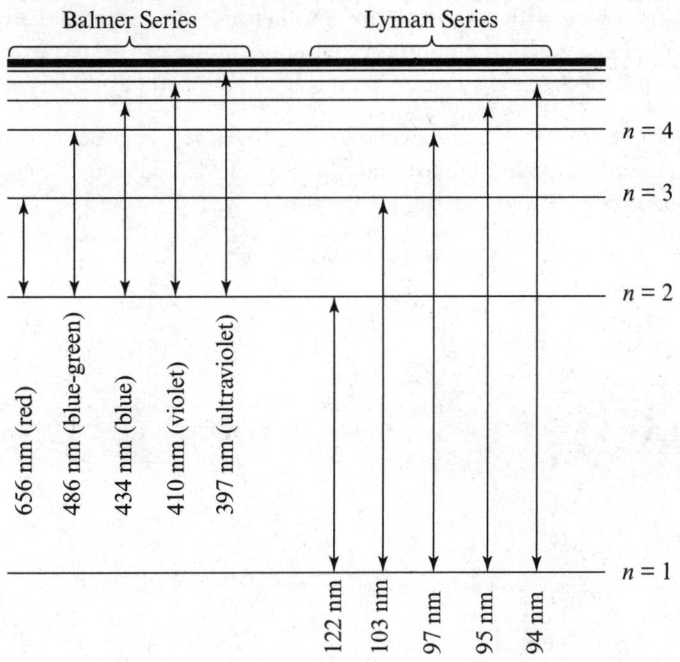


The Bohr model works well for the electronic energy transitions in hydrogen, but it does not work when applied to atoms with more than one electron. The problem is that *electrons do not orbit the nucleus in fixed circular or elliptical orbits.*

The circular orbit picture is based on classical mechanics, which describe the motion of large objects such as projectiles, roller coasters, and planets. An acceleration in classical mechanics is a change in the speed or direction of an object in motion. An orbiting electron would be in a continuous state of acceleration due to its circular path, and an accelerating electron would be expected to constantly emit electromagnetic radiation. In this way, it would lose energy and spiral into the nucleus. Because this does not happen, it is clear that the Bohr model is not fully correct, although it is still a useful visualization of the quantization of electronic energy in single-electron atoms.

## The Quantum Mechanical Atom

A complete revolution in atomic theory occurred in the 1920s with the development of **quantum mechanics,** the study of the energy and motion of subatomic particles. Quantum mechanical descriptions are applied to very small particles that can only exist in certain energy states and can change their energy by absorbing or emitting electromagnetic radiation.

The energies and positions of electrons in atoms are described by wave functions. **Wave functions** are solutions to the **Schrödinger equation,** a mathematical model of electronic motion that was based on the treatment of electrons as waves. Wave functions describe **atomic orbitals.** Each atomic orbital is composed of a maximum of two electrons.

For each wave function, there is a certain probability of finding an electron at a certain distance from the nucleus. The squares of the wave functions are related to the probability of finding an electron in each point in space surrounding the nucleus.

The lowest-energy atomic orbitals in each main energy level, or **shell,** beginning with $n = 1$, are the $s$ orbitals. ***S* orbitals** are spherically symmetrical orbitals surrounding the nucleus. The radius at which the electron has the highest probability of existing increases with the energy of the electron.

The second principal energy level, $n = 2$, contains one $s$ orbital and three $p$ orbitals, $p_x$, $p_y$, and $p_z$. The ***p* orbitals** are oriented along the $x$, $y$, and $z$ axes with respect to one another and are shaped like two equal-sized balloons tied together at their ends. Since each orbital is composed of a maximum of two electrons, the entire $n = 2$ shell can contain a maximum of eight electrons (two electrons in the $s$ orbital and a total of six in the $p$ orbitals).

Each shell beginning with $n = 3$ has a set of five $d$ orbitals. The ***d* orbitals** are either shaped like four equal-sized balloons tied together at their ends (a three-dimensional four-leaf clover) or, as in the case of the $d_z^2$ orbital, like two equal-sized balloons surrounded by a donut-shaped region.

**The Shapes of Atomic Orbitals**

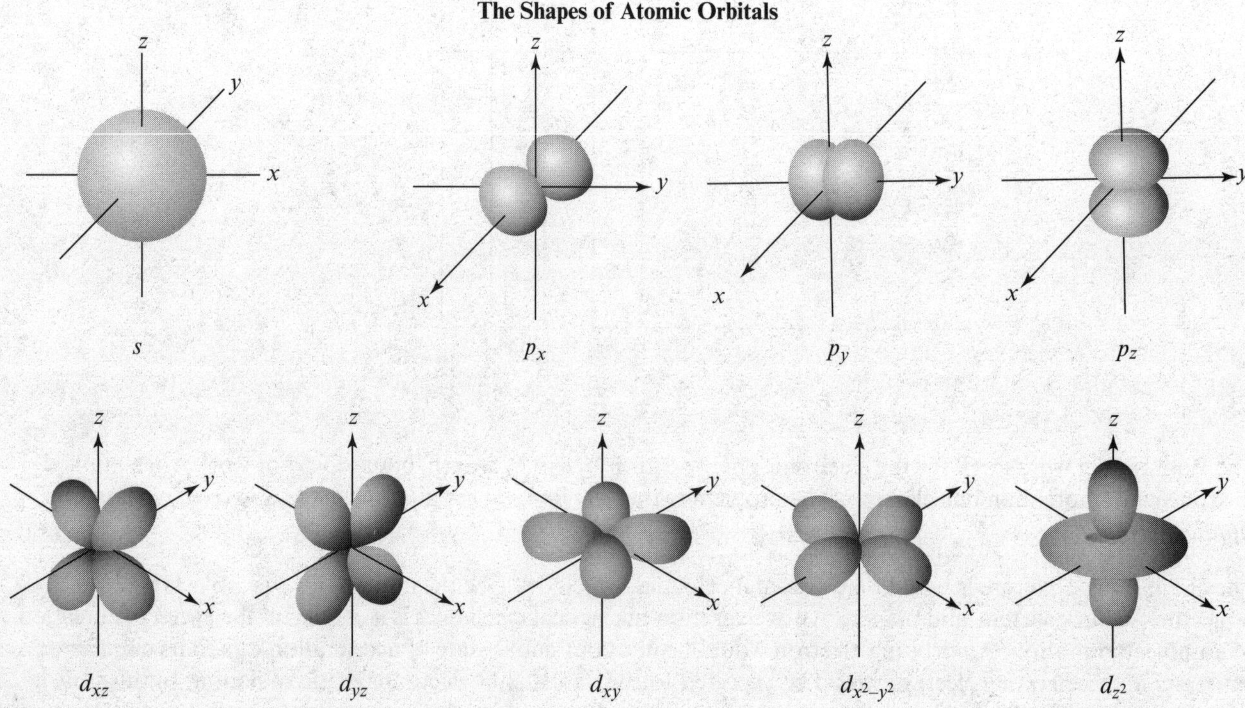


The $f$ orbitals have complicated shapes and names. Students will *not* need to know the shapes or names of the $f$ orbitals for the AP Chemistry exam. They will only need to know that there are seven $f$ orbitals in each shell, starting with $n = 4$.

Each type of orbital designation, $s$, $p$, $d$, or $f$, present within a shell represents a **subshell** or **sublevel.** The number of subshells within a particular shell is equal to the shell number, $n$. The number of orbitals in the shell is equal to $n^2$, and since two electrons may occupy each orbital, the total possible number of electrons in a shell is equal to $2n^2$.

This pattern is summarized below.

| Shell, $n$ | Orbitals Present | Number of Subshells (n) | Number of Orbitals in the Shell ($n^2$) | Total Possible Number of Electrons in the Shell ($2n^2$) |
|---|---|---|---|---|
| 1 | 1$s$ | 1 ($s$) | 1 | 2 |
| 2 | 2$s$<br>$2p_x$, $2p_y$, $2p_z$ | 2 ($s$ and $p$) | 4 | 8 |
| 3 | 3$s$<br>$3p_x$, $3p_y$, $3p_z$<br>$3d_{xy}, 3d_{yz}, 3d_{xz}, 3d_{x^2-y^2}, 3d_{z^2}$ | 3 ($s$, $p$, and $d$) | 9 | 18 |
| 4 | 4$s$<br>$4p_x$, $4p_y$, $4p_z$<br>$4d_{xy}, 4d_{yz}, 4d_{xz}, 4d_{x^2-y^2}, 4d_{z^2}$<br>4$f$ orbitals (There are seven 4$f$ orbitals, for a maximum total of 14 electrons in the 4$f$ subshell.) | 4 ($s$, $p$, $d$, and $f$) | 16 | 32 |

# Electron Configuration and Periodicity

The **electron configuration** of an atom provides a shorthand description of the energies and positions of the atom's electrons. Electron configurations can often be written by following the periods of the periodic table from left to right and from top to bottom. As can be seen from the diagram of the periodic table below, the order of filling is as follows:

$$1s^2 2s^2 2p^6 3s^2 3p^6 4s^2 3d^{10} 4p^6 5s^2 4d^{10} 5p^6 6s^2 4f^{14} 5d^{10} 6p^6 7s^2 5f^{14} 6d^{10}$$

## The Aufbau Principle

According to the **aufbau principle,** each electron in a ground state atom occupies the lowest-energy available orbital. The order of orbital filling for many atoms can easily be remembered by dividing the periodic table into the $s$ block, the $p$ block, the $d$ block, and the $f$ block.

**The $s$, $p$, $d$, and $f$ Blocks of the Periodic Table**

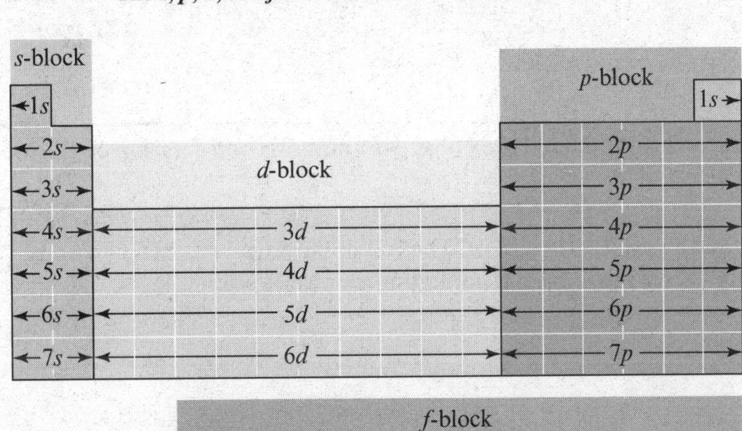

A hydrogen atom's single electron is in the lowest-energy ($n$ = 1) $s$ orbital in the ground state. The electron configuration for the ground state of hydrogen is $1s^1$, where the superscript denotes the number of electrons occupying the orbital.

A helium atom has two electrons in the $n$ = 1 shell. The electron configuration of helium is $1s^2$.

A neutral atom of lithium has three electrons. In the ground state, two of the electrons are in the $n$ = 1 shell and one is in the $n$ = 2 shell. The electron configuration of lithium is $1s^22s^1$.

Suppose we would like to write the electron configuration for the ground state of arsenic. Arsenic is located in the fourth row of the periodic table within the $p$ block, as shown below. Counting the positions of hydrogen and helium gives $1s^2$. Next are lithium and beryllium: $2s^2$. Continued counting in order of atomic number from boron through zinc results in $2p^63s^23p^64s^23d^{10}$. Arsenic is the third element in the $4p$ sublevel, so the last three electrons are denoted $4p^3$. The full electron configuration for arsenic is $1s^22s^22p^63s^23p^64s^23d^{10}4p^3$. (In order to emphasize the increase in orbital energy as a function of the main energy level, some teachers prefer the following grouping: $1s^22s^22p^63s^23p^63d^{10}4s^24p^3$. Both configurations will earn credit on the AP Chemistry exam.)

**The Electron Configuration of Arsenic Based on Periodic Table Position**

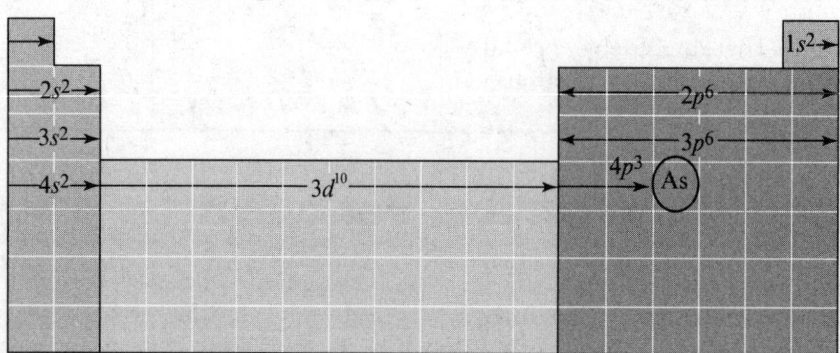


There are exceptions to this filling order. Students are not required to write electron configurations on the AP Chemistry exam for elements that do not follow this pattern, but they should recognize when an electron configuration does not follow the conventional order and understand why these deviations occur. To illustrate, the electron configuration of copper is expected to be $1s^22s^22p^63s^23p^6\mathbf{4s^23d^9}$. Instead, the actual configuration of copper is $1s^22s^22p^63s^23p^6\mathbf{4s^13d^{10}}$. An element's order of orbital filling is the electronic arrangement that minimizes electron-electron repulsions and achieves the lowest-energy configuration overall.

**Noble gas notation** is a shorthand method of writing electron configurations. For noble gas notation, the symbol of the noble gas encountered just prior to the element in question on the periodic table is written in brackets, and then the electron configuration is continued. The noble gas configuration for the arsenic example is $[Ar]4s^23d^{10}4p^3$.

## Practice

---

Write the full electron configuration and the noble gas notation configuration of the ground state of cobalt, Co.

---

Cobalt is found in the fourth period of the table. The full notation is $1s^22s^22p^63s^23p^64s^23d^7$. Counting back to the previous noble gas, argon, gives the noble gas configuration $[Ar]4s^23d^7$.

## Hund's Rule and the Pauli Exclusion Principle

Electron configurations can be represented using **orbital notation** in which the electrons are depicted by arrows. Two electrons in an orbital must spin in opposite directions. Arrows pointing up and down denote spin-up and spin-down electrons, respectively. The orbital notation for oxygen is shown below.

**Oxygen Orbital Notation**

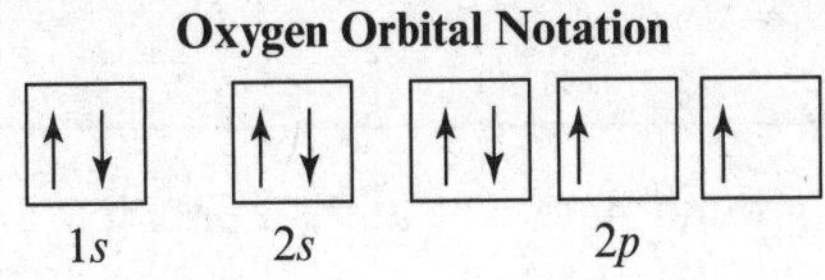


This example illustrates both Hund's rule and the Pauli exclusion principle. **Hund's rule** states that in the ground state, electrons occupy each orbital in a subshell singly before pairing up. In addition, the first electrons in the separate orbitals of a subshell have parallel spins. In other words, each orbital of a subshell is occupied by a spin-up electron before any orbital in the subshell gets a spin-down electron.

According to the **Pauli exclusion principle,** a single orbital cannot hold two spin-up or two spin-down electrons. A pair of electrons may only exist together in an orbital if they have opposite spins.

Exceptions to the aufbau principle and Hund's rule occur when an electron is in an **excited state,** or higher-energy state. Below, carbon is shown in two excited states. The first is a violation of the aufbau principle and the second is a violation of Hund's rule.

**Excited States of Carbon**

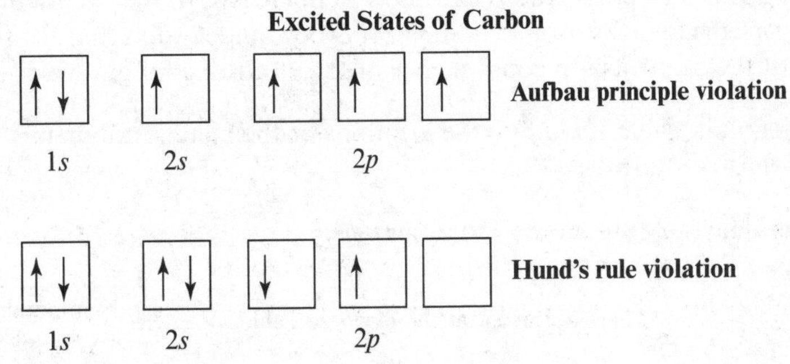


Violations of the Pauli exclusion principle are forbidden.

### Practice

Write the electron configuration for sulfur using orbital notation.

The electron configuration for sulfur is $1s^2 2s^2 2p^6 3s^2 3p^4$.

**Sulfur Orbital Notation**

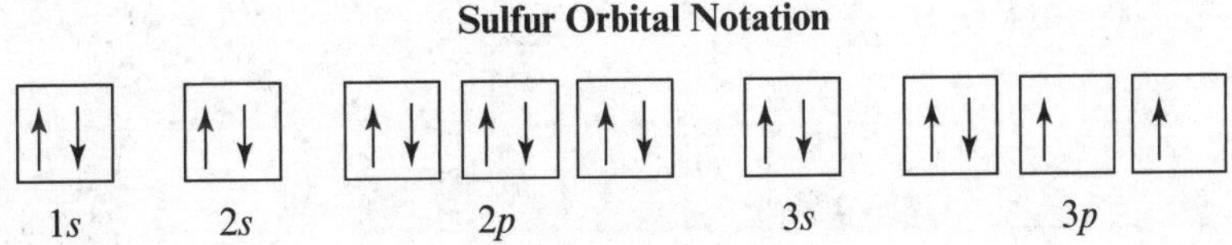

## Practice

Which of the following is a ground state, which is an excited state, and which is forbidden?

**a.** $1s^22s^22p^5$
**b.** $1s^22s^22p^23s^3$
**c.** $1s^22s^12p^6$

**(a)** The configuration $1s^22s^22p^5$ is in the ground state. The aufbau principle and the Pauli exclusion principle are presumably obeyed.

**(b)** The configuration $1s^22s^22p^23s^3$ is forbidden. The 3*s* orbital holds more than two electrons, so the Pauli exclusion principle is violated.

**(c)** The configuration $1s^22s^12p^6$ is an excited state since one of the electrons in the 2*s* sublevel has been promoted to the 2*p* sublevel. The Pauli exclusion principle is presumably obeyed.

# The Periodic Table

The periodic table was first arranged by Dmitri Mendeleev in 1869. He systematically organized the elements according to atomic weight and realized that there was a regular, periodic repetition of the chemical and physical properties of the elements. Mendeleev even left spaces in the periodic table for elements that had yet to be discovered. This regular repetition of properties is called the **periodic law.** In the twentieth century, Henry Moseley reorganized the periodic table by atomic number. It is now understood that the regular repetition of properties, or **periodicity,** of the elements is a consequence of the electron configurations of the elements.

The general layout of the periodic table, including the locations of the metals, nonmetals, and metalloids, should be very familiar to AP Chemistry students.

Horizontal rows on the periodic table are referred to as **periods.**

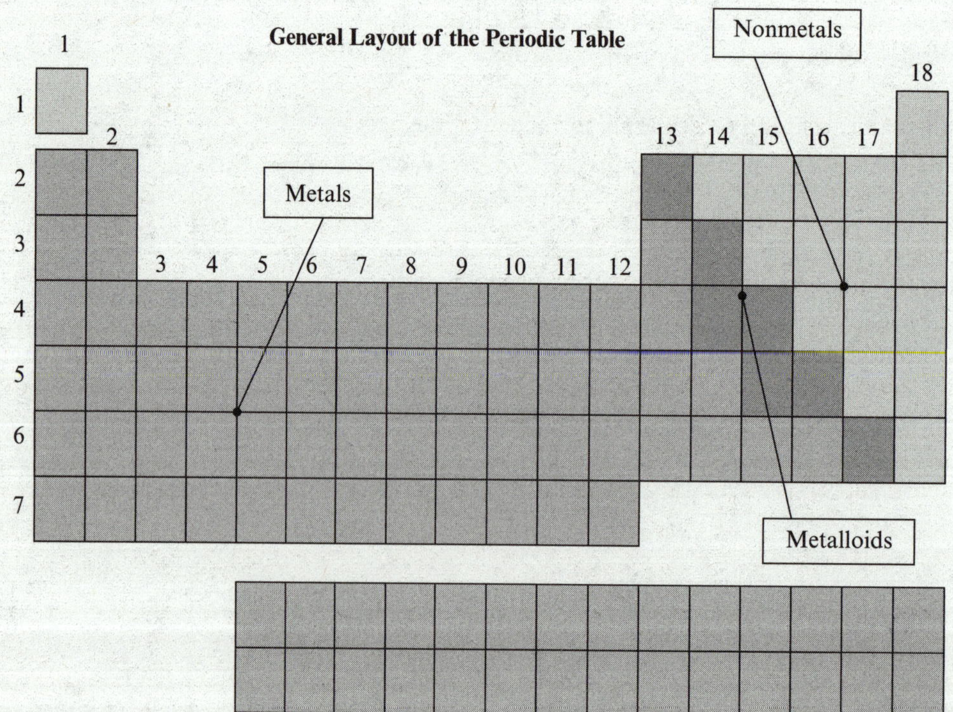

The properties of any element on the periodic table are a direct consequence of that element's electron configuration. Electron configurations of the elements in **groups** (vertical columns) of the periodic table are similar, so the elements in a group may display similar chemical properties.

The **main groups** are groups 1, 2, and 13 through 18. The older designations for these groups are IA, IIA, and IIIA through VIIIA.

**The Locations of Main Groups, Transition Metals, and Inner Transition Metals**

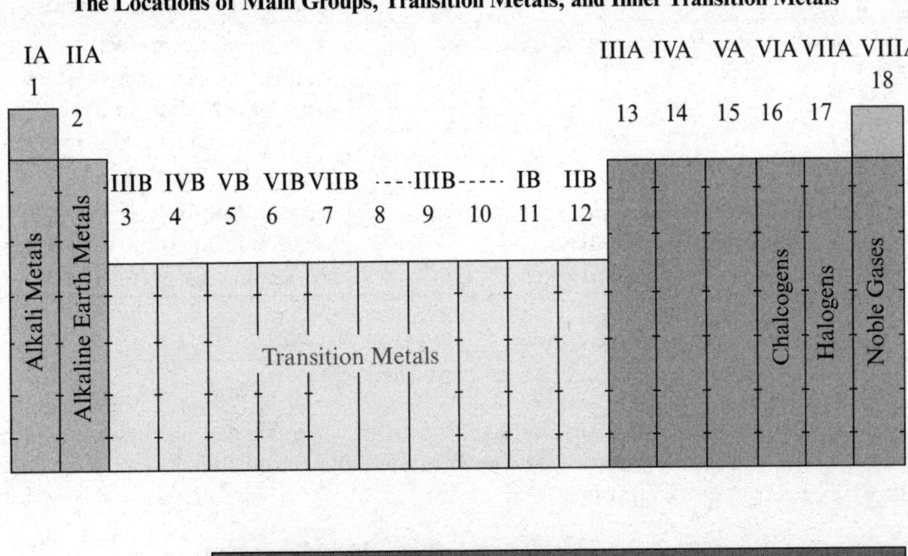


Elements within the same group have the same number of valence electrons. The **valence electrons** are the electrons in the outermost shell of an atom. These valence electrons are responsible for the reactivity of the elements. For example, a neutral atom of phosphorus has an electron configuration of $1s^22s^22p^63s^23p^3$. Phosphorus has five electrons in the highest energy level ($n = 3$), and these are its valence electrons.

When an atom loses electrons, it becomes a positively charged species called a **cation.**

When an atom (or group of atoms) gains electrons, it becomes a negatively charged species called an **anion.**

The groups and their properties are summarized below.

| Group | Valence Electron Configuration | Elements | Characteristics |
|---|---|---|---|
| 1 (IA) Alkali metals | $ns^1$ for $n \geq 2$ | Li, Na, K, Rb, Cs, Fr | Highly reactive elements that tend to lose a single electron to form $M^+$ cations with noble gas electron configurations. This tendency increases down the group. |
| 2 (IIA) Alkaline earth metals | $ns^2$ for $n \geq 2$ | Be, Mg, Ca, Sr, Ba, Ra | Less reactive than the alkali metals. Tend to lose two electrons to form $M^{2+}$ cations with noble gas electron configurations. This tendency increases down the group. |

*continued*

| Group | Valence Electron Configuration | Elements | Characteristics |
|---|---|---|---|
| 16 (VIA) Chalcogens | $ns^2np^4$ | O, S, Se, Te, Po | Tend to gain two electrons to form $X^{2-}$ anions with noble gas electron configurations. This tendency decreases down the group. Form a large variety of molecular compounds with nonmetals. |
| 17 (VIIA) Halogens | $ns^2np^5$ | F, Cl, Br, I, At | Tend to gain one electron to form $X^-$ anions. This tendency decreases down the group. Form diatomic molecules, $X_2$, but are not found in elemental form in nature due to their high reactivity. |
| 18 (VIIIA) Noble gases | $ns^2np^6$ for $n \geq 2$ ($1s^2$ for helium) | He, Ne, Ar, Kr, Xe, Rn | Do not usually form compounds except under special conditions. They exist in their monatomic form in nature. |

Certain elements always occur as diatomic molecules, $X_2$. These are $H_2$, $N_2$, $O_2$, $F_2$, $Cl_2$, $Br_2$, and $I_2$; the locations of these elements on the periodic table can be used to memorize them.

The most stable electron configuration for many main group atoms is a **noble gas configuration,** either $1s^2$ or $ns^2np^6$. When the main group elements form ions, as shown in the table below, they lose or gain electrons as necessary to achieve this electron configuration.

**Electron Configurations of Common Main Group Ions**

| 1 IA | 2 IIA | 15 VA | 16 VIA | 17 VIIA | 18 VIIIA |
|---|---|---|---|---|---|
| $H^+$ | | | | | He $1s^2$ |
| $Li^+$ $1s^2$ | $Be^{2+}$ $1s^2$ | $N^{3-}$ $2s^22p^6$ | $O^{2-}$ $2s^22p^6$ | $F^-$ $2s^22p^6$ | Ne $2s^22p^6$ |
| $Na^+$ $2s^22p^6$ | $Mg^{2+}$ $2s^22p^6$ | $P^{3-}$ $3s^23p^6$ | $S^{2-}$ $3s^23p^6$ | $Cl^-$ $3s^23p^6$ | Ar $3s^23p^6$ |
| $K^+$ $3s^23p^6$ | $Ca^{2+}$ $3s^23p^6$ | | $Se^{2-}$ $4s^24p^6$ | $Br^-$ $4s^24p^6$ | Kr $4s^24p^6$ |
| $Rb^+$ $4s^24p^6$ | $Sr^{2+}$ $4s^24p^6$ | | | $I^-$ $5s^25p^6$ | Xe $5s^25p^6$ |
| $Cs^+$ $5s^25p^6$ | $Ba^{2+}$ $5s^25p^6$ | | | | Rn $6s^26p^6$ |
| $Fr^+$ $6s^26p^6$ | $Ra^{2+}$ $6s^26p^6$ | | | | |

The **transition metals** make up the *d* block of the periodic table. Transition metals generally lose electrons to form cations, but ionization patterns for the transition metals are not as regular and predictable as they are for the main group elements. From left to right across a period, electrons are being added to an inner shell rather than to an outer shell, so the transition metals do not demonstrate the striking group characteristics that the main group elements display. In addition, many transition metals can have more than one oxidation state. For instance, when copper ionizes, it can form either $Cu^+$ or $Cu^{2+}$.

Cobalt, for example, can form either $Co^{2+}$ cations or Co3$^{+}$ cations. The *s* electrons, which are highest in energy, are lost first.

Electron configuration of $Co^{0}$: $[Ar]4s^2 3d^7$
Electron configuration of $Co^{2+}$: $[Ar]3d^7$
Electron configuration of $Co^{3+}$: $[Ar]3d^6$

### Practice

How many valence electrons are in a neutral atom of silicon?

The electron configuration of silicon is $1s^2 2s^2 2p^6 3s^2 3p^2$. The total number of electrons in the $n = 3$ shell is four.

### Practice

Anions with a –2 charge are most likely to be formed from which main group elements on the periodic table?

When elements with the electron configuration $ns^2 np^4$ gain two electrons, they achieve a noble gas configuration ($ns^2 np^6$), so the main group elements are most likely to be the chalcogens (group 16 or VIA).

## Periodic Trends

Trends in atomic properties should be understood in terms of coulombic arguments, as discussed below. Shielding and effective nuclear charge are key ideas that students should be comfortable using in their discussions of the periodic trends. A trend itself should never be offered as an explanation for observed properties.

According to **Coulomb's law,** the energy required to remove an electron from an atom is proportional to the magnitude of the product of the charge of the electron and the effective nuclear charge, and inversely proportional to the distance from the nucleus to the electron.

$$E \propto \frac{q_1 q_2}{r}$$

In this equation, $q_1$ and $q_2$ are the charge of the electron and the effective nuclear charge, and $r$ is the distance between the charges. The higher the effective nuclear charge, the more energy is required to remove the oppositely charged electron.

Electrons that are more distant from the nucleus require less energy to remove.

**Shielding** from nuclear attraction accounts for the ease with which a valence electron is removed from a neutral atom when compared with that of an inner shell electron, or core electron. **Core electrons** are all nonvalence electrons in an atom.

Since like charges repel, electrons experience repulsion from one another. The larger the number of core electrons between a valence electron and the nucleus, the easier the valence electron is to remove because it is shielded from the positive charge of the nucleus by the core electrons.

**Effective nuclear charge, $Z_{eff}$,** is the apparent nuclear charge experienced by an electron after inner shell electron repulsions have been accounted for. It is the difference between the atomic number, $Z$, and the number of inner shell electrons or core electrons.

$$Z_{eff} = Z - (\text{number of core electrons})$$

As we have seen, shells are further divided into subshells. For atoms with multiple electrons, the subshells within a shell have slightly different energies. This can be understood in terms of the **penetration effect.** A comparison of the radial distributions of orbitals within the second shell shows that the 2*s* electrons penetrate the cloud of 1*s* core electrons more than the 2*p* electrons. Due to their greater penetration and greater degree of exposure to the positive charge of the nucleus, the 2*s* electrons experience more attraction to the nucleus and are slightly lower in energy than the 2*p* electrons.

**Orbital Penetration**

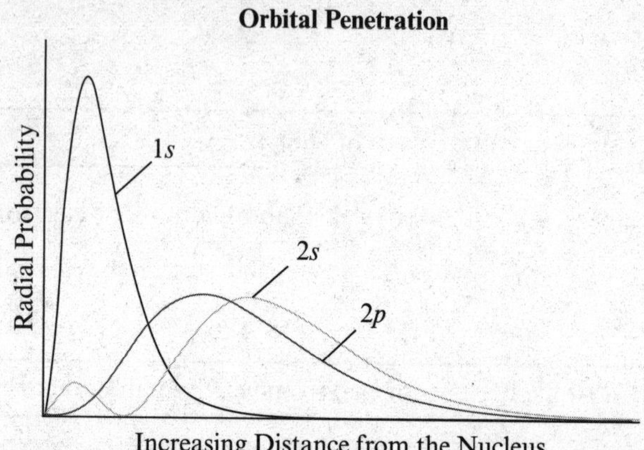


This pattern holds true for the *d* and *f* subshells, too. The *p* subshell electrons experience a greater coulombic attraction to the nucleus and are lower in energy than the *d* subshell electrons in the same shell due to a higher degree of penetration. Likewise, the *d* subshell electrons are lower in energy than the *f* subshell electrons in the same shell.

### Practice

Which electron in phosphorus is easier to remove from the atom, a 2*p* electron or a 3*p* electron?

Two effects make a 3*p* electron easier to remove than a 2*p* electron. The 3*p* electron is more distant from the nucleus than the 2*p* electron, so the 3*p* electron experiences less coulombic attraction to the nucleus and requires less energy to remove. The second effect is that of shielding. The 2*p* electron is shielded from the nucleus by electrons in the $n = 1$ shell, but the 3*p* electron is shielded by electrons in both the $n = 1$ and the $n = 2$ shells. For this reason, also, the 3*p* electron experiences less coulombic attraction to the nucleus and is easier to remove.

### Practice

What is the effective nuclear charge experienced by a valence electron in a phosphorus atom?

The atomic number of phosphorus, $Z$, is 15. The core electrons are in the $n = 1$ and $n = 2$ shells. This is a total of 10 core electrons. The effective nuclear charge experienced by the $n = 3$ electrons is

$$Z_{eff} = 15 - 10 = 5$$

## Ionization Energy

The energy required to remove the highest-energy outermost electron from an atom in the gas phase to form a +1 cation, as represented in the equation below, is called the **first ionization energy,** $IE_1$.

$$X(g) \rightarrow X^{+}(g) + e^{-} \qquad IE_1$$

*Across a period* on the periodic table, first ionization energy tends to increase. Increased effective nuclear charge going from left to right causes a greater coulombic attraction between the electrons and the nucleus.

*Down a group,* ionization energy tends to decrease. The highest-energy electrons occupy increasingly higher shells. With a greater number of shells shielding the valence electrons from the nucleus, more electron-electron repulsion and less nuclear attraction is experienced by the outermost electrons. Additionally, the distance of the outermost electrons from the nucleus increases down a group, so the energy required to remove these electrons decreases.

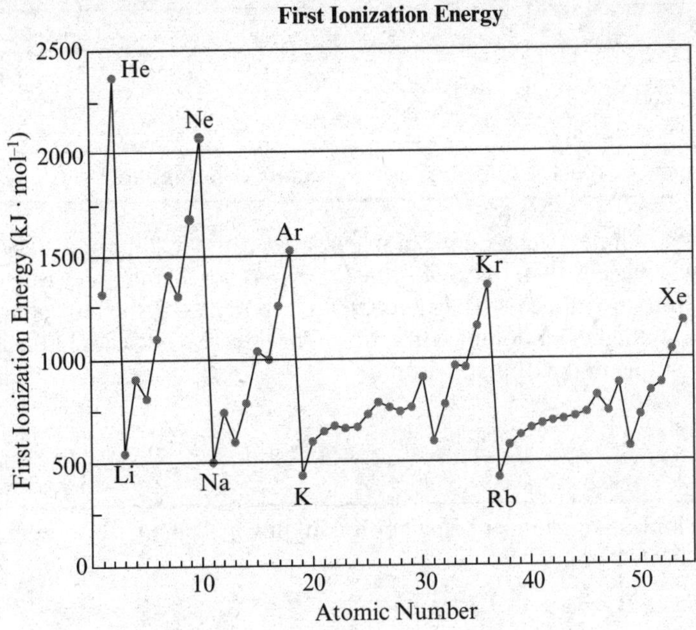


The energies required to remove subsequent electrons to form +2 and +3 cations are called **second** and **third ionization energies.** After the atom loses one electron, subsequent ionizations become increasingly difficult because of decreased electronic repulsion and increased attraction of the remaining electrons to the positively charged nucleus. Ionization energies increase substantially as more and more electrons are removed.

The equations for second and third ionization energies, $IE_2$ and $IE_3$, are given by

$$X^{+}(g) \rightarrow X^{2+}(g) + e^{-} \qquad IE_2$$
$$X^{2+}(g) \rightarrow X^{3+}(g) + e^{-} \qquad IE_3$$

Notice the increases in subsequent ionization energies in the ionization data for the first five elements, shown in the table below. Also, if we compare the effect of shells on ionization energy, we see that removal of an electron from a core configuration requires substantially more energy than electrons that occupied a higher-energy shell.

Consider helium, for instance. The second ionization energy is a little over two times the first. It requires more energy to remove a second electron from a positively charged $He^{+}$ ion.

For lithium and other elements with electrons in higher shells, the additional effect can be seen. The second ionization energy for lithium is *fourteen times* the first. Not only does it require more energy to remove a second electron from an already positively charged lithium ion, the energy increase is substantial because the second electron is being removed from a noble gas core electron configuration.

**Ionization Energies ($kJ \cdot mol^{-1}$)**

| | $IE_1$ | $IE_2$ | $IE_3$ | $IE_4$ | $IE_5$ |
|---|---|---|---|---|---|
| H | 1312 | | | | |
| He | 2372 | 5250 | | | |
| Li | 520 | 7297 | 11,810 | | |
| Be | 899 | 1757 | 14,845 | 21,000 | |
| B | 800 | 2426 | 3659 | 25,020 | 32,820 |

## Practice

Is the first ionization energy expected to be higher for sodium or potassium?

Sodium would be expected to have a higher ionization energy because its outermost electrons are less shielded from the positively charged nucleus than those of potassium, making them harder to remove. In addition, sodium's outermost electrons are in the $n = 2$ shell (compared to the $n = 3$ shell for potassium), so they are closer to the nucleus and subject to higher coulombic attraction. The effective nuclear charge is the same for sodium and potassium, so $Z_{eff}$ is inconsequential in this situation.

## Practice

Consider the first four ionization energies below for an unknown element. To what group of the periodic table does the element likely belong?

**Ionization Energies ($kJ \cdot mol^{-1}$)**

| $IE_1$ | $IE_2$ | $IE_3$ | $IE_4$ |
|---|---|---|---|
| 737 | 1450 | 7731 | 10,545 |

The second ionization energy is about two times the first ionization energy for this element. The third ionization energy is more than five times the second ionization energy. The fourth ionization energy is less than twice the third ionization energy. The relatively large increase between $IE_3$ and $IE_2$ indicates that after the second ionization, a noble gas core electron configuration has been achieved. This indicates that the element is probably a member of group 2. (The data shown is for magnesium.)

# Electron Affinity

**Electron affinity** is the energy released upon the addition of an electron to an atom in the gas phase according to the following equation.

$$X(g) + e^- \rightarrow X^-(g)$$

The more negative the electron affinity, the more energy is released when the electron combines with the atom and the more easily the electron is added to the atom. The trend is not as pronounced as that for ionization energy, but the explanation may be provided using similar arguments.

*Across a period* on the periodic table, electron affinity generally tends to become more negative. Increased effective nuclear charge from left to right across the period increases the coulombic attraction between the electrons and the nucleus, so electrons are added more easily.

*Down a group*, electron affinity becomes less negative. Down a group, electrons are being added to increasingly higher shells. With a greater number of shells between the outermost electrons and the nucleus, shielding is increased. More electron-electron repulsion and less nuclear attraction is experienced by the outermost electrons. The distance of the outermost electrons from the nucleus is again increasing, causing a decrease in coulombic attraction.

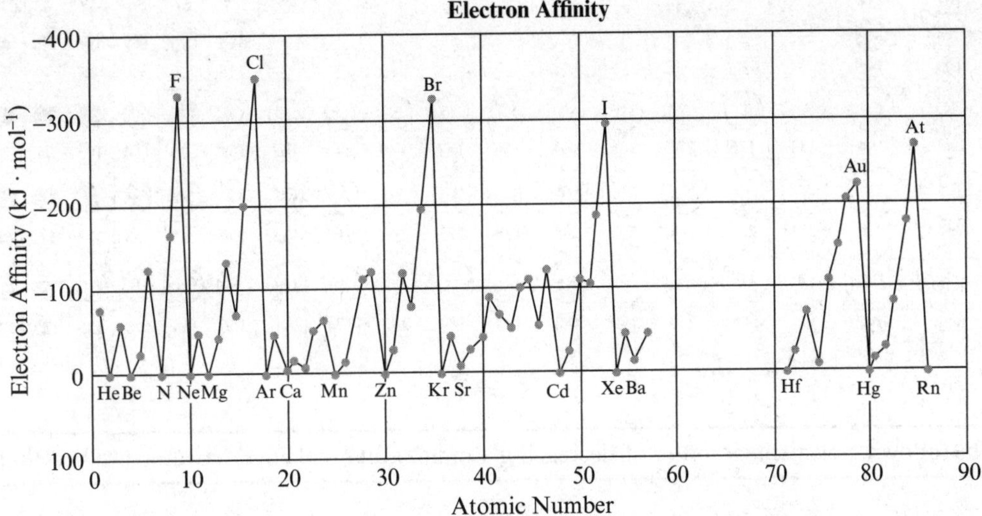


## Practice

Is more energy released with the addition of an electron to a fluorine atom or to an oxygen atom?

Adding an electron to a fluorine atom releases more energy than adding an electron to an oxygen atom. Fluorine has a more negative electron affinity because it has a higher effective nuclear charge than oxygen. An electron is being added to the same shell, $n = 2$, for both fluorine and oxygen, so shielding is irrelevant.

## Atomic Radii

The **atomic radius,** or covalent radius, is defined as one-half the distance between the nuclei of two atoms of the same element bound by a single covalent bond or a metallic bond.

**Atomic radius**

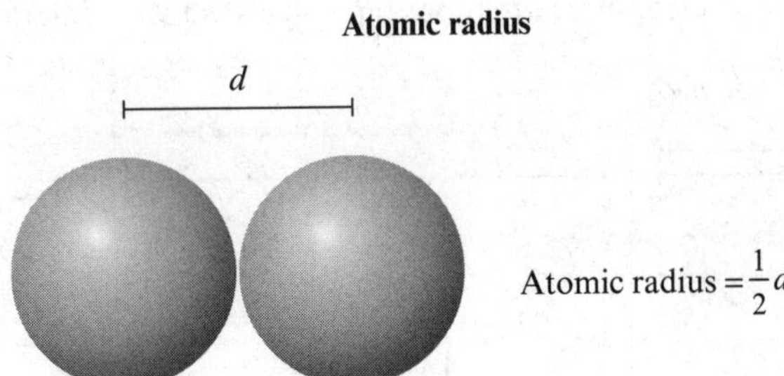


*Across a period* on the periodic table, atomic radii tend to decrease due to increasing effective nuclear charge. The electrons experience greater coulombic attraction to the nucleus, so they are held more closely.

*Down a group,* atomic radii increase due to increased shielding. The larger the number of shells between the outermost electrons and the nucleus, the greater the electronic repulsions and the greater the distance between the valence electrons and the nucleus. The electrons are held less tightly due to decreased coulombic attraction.

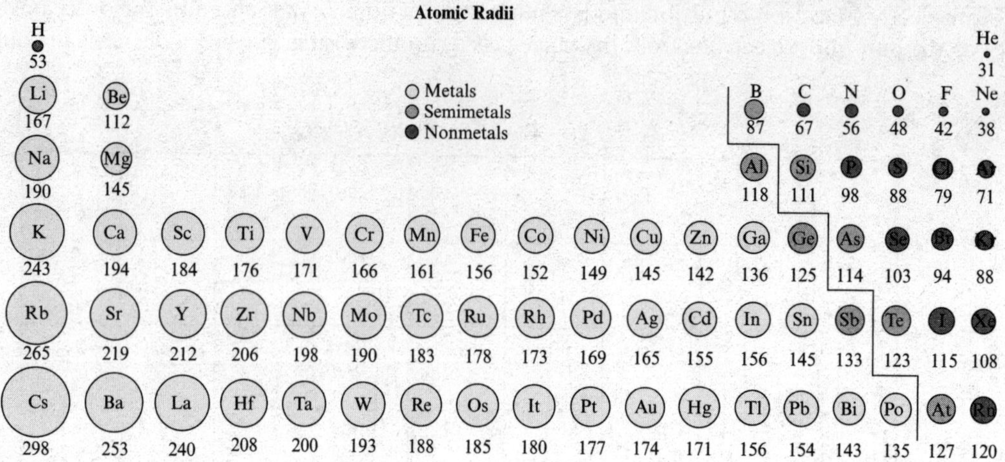


## Practice

---

Arrange the following elements in order of decreasing atomic radius: calcium, barium, and beryllium.

---

$$\text{radius}_{Ba} > \text{radius}_{Ca} > \text{radius}_{Be}$$

The radius of beryllium (Be) is the smallest, calcium (Ca) has an intermediate radius, and the radius of barium (Ba) is the largest. Since barium is closest to the bottom of the periodic table, there are more shells shielding its outer electrons from the nucleus. The electronic repulsions and the distance of the outermost electrons from the nucleus is the greatest, so the coulombic attraction is the smallest. The electrons of beryllium are the least shielded, so they are most attracted by and therefore closest to the nucleus.

## Ionic Radii

Within a period or a group, ionic radii follow the same general trend as atomic radii. They decrease from left to right, and they increase from top to bottom. The shielding and coulombic arguments are similar.

Cations, which are formed when neutral atoms lose electrons, are smaller than their parent atoms. When electrons are lost, there is less electronic repulsion and the atom contracts from its neutral size. Anions are larger than their parent atoms. When electrons are gained, there are more repulsive electronic interactions and the ion expands from its neutral size.

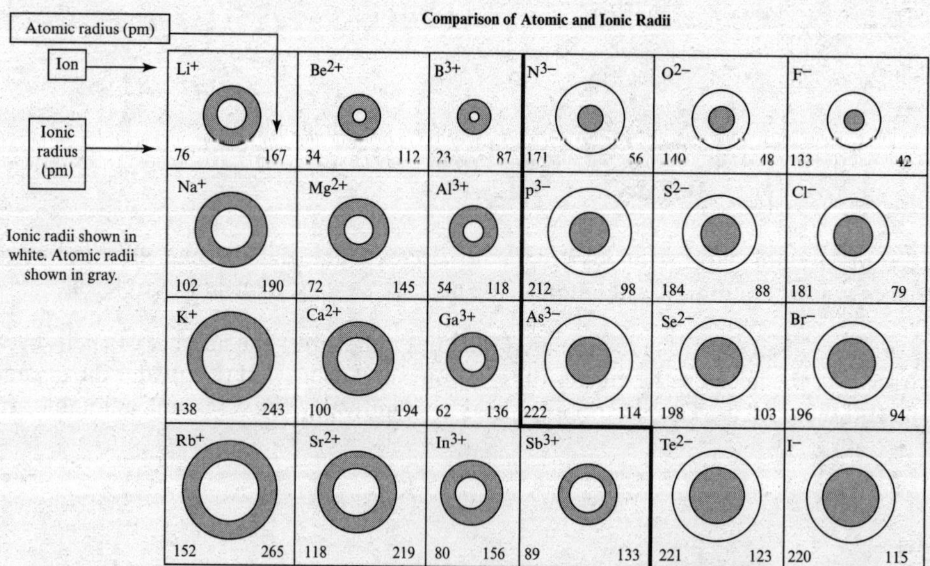

**Isoelectronic species** are atoms or ions that have equal numbers of electrons. For example, $O^{2-}$ and Ne are isoelectronic. When the radii of isoelectronic species are compared, as the nuclear charge increases, the size decreases.

## Practice

Rank the following ions from largest to smallest: $K^+$, $Cl^-$, $Ca^{2+}$, $S^{2-}$.

The correct ranking of the ions from largest to smallest is $S^{2-}$, $Cl^-$, $K^+$, $Ca^{2+}$. All the ions are isoelectronic and have a $[Ne]3s^23p^6$ electron configuration, while the nuclear charge increases with atomic number from sulfur to calcium. Increased nuclear charge causes the electrons to be held more closely to the nucleus, decreasing the ionic radius.

# Electronegativity

Electronegativity is a measure of how strongly an atom draws electrons toward itself in a covalent bond. The higher the electronegativity of an element, the more it attracts shared electrons.

The **Pauling electronegativity scale** is one that is very commonly used. Pauling electronegativities range from less than 1 for the group 1 metals, which are not very electronegative, to 4.0 for fluorine, the most electronegative element. Electronegativities of noble gases are not discussed since the group 18 elements rarely participate in covalent bonding. The consequences of electronegativity differences between bonded atoms will be discussed further in Chapter 2, "Molecular and Ionic Compound Structure and Properties."

**Pauling Electronegativities**

| | | | | | | | | | | | | | | | | |
|---|---|---|---|---|---|---|---|---|---|---|---|---|---|---|---|---|
| 1 **H** 2.2 | | | | | | | | | | | | | | | | |
| 3 **Li** 1.0 | 4 **Be** 1.5 | | | | | | | | | | | 5 **B** 2.0 | 6 **C** 2.5 | 7 **N** 3.0 | 8 **O** 3.4 | 9 **F** 4.0 |
| 11 **Na** 0.9 | 12 **Mg** 1.2 | | | | | | | | | | | 13 **Al** 1.5 | 14 **Si** 1.9 | 15 **P** 2.2 | 16 **S** 2.6 | 17 **Cl** 3.1 |
| 19 **K** 0.9 | 20 **Ca** 1.0 | 21 **Sc** 1.3 | 22 **Ti** 1.5 | 23 **V** 1.6 | 24 **Cr** 1.6 | 25 **Mn** 1.6 | 26 **Fe** 1.8 | 27 **Co** 1.9 | 28 **Ni** 1.9 | 29 **Cu** 1.9 | 30 **Zn** 1.7 | 31 **Ga** 1.8 | 32 **Ge** 2.0 | 33 **As** 2.2 | 34 **Se** 2.6 | 35 **Br** 2.9 |
| 37 **Rb** 0.8 | 38 **Sr** 1.0 | 39 **Y** 1.2 | 40 **Zr** 1.3 | 41 **Nb** 1.6 | 42 **Mo** 2.1 | 43 **Tc** 1.9 | 44 **Ru** 2.2 | 45 **Rh** 2.3 | 46 **Pd** 2.2 | 47 **Ag** 1.9 | 48 **Cd** 1.7 | 49 **In** 1.8 | 50 **Sn** 1.8 | 51 **Sb** 2.0 | 52 **Te** 2.1 | 53 **I** 2.6 |
| 55 **Cs** 0.8 | 56 **Ba** 0.9 | 57 **La** 1.1 | 72 **Hf** 1.3 | 73 **Ta** 1.5 | 74 **W** 2.3 | 75 **Re** 1.9 | 76 **Os** 2.2 | 77 **Ir** 2.2 | 78 **Pt** 2.3 | 79 **Au** 2.5 | 80 **Hg** 2.0 | 81 **Tl** 1.6 | 82 **Pb** 1.9 | 83 **Bi** 2.0 | 84 **Po** 2.0 | 85 **At** 2.2 |
| 87 **Fr** 0.7 | 88 **Ra** 0.9 | | | | | | | | | | | | | | | |

*Across a period* on the periodic table, electronegativity increases. Increasing effective nuclear charge going from left to right increases attractions between the electrons and the nucleus.

*Down a group*, electronegativity decreases. Electrons are more shielded from the nucleus, so they experience a weaker nuclear attraction. The distance of the outermost electrons from the nucleus increases down a group, causing the coulombic attraction of the outermost electrons to the nucleus to decrease.

## Practice

Which is more electronegative, sulfur or phosphorus?

Sulfur is more electronegative than phosphorus. The effective nuclear charge of sulfur is greater than that of phosphorus, which causes sulfur to attract electrons more strongly. Both sulfur and phosphorus are in the same period, so shielding arguments do not apply.

# Periodic Trend Anomalies

Some elements do not follow the periodic trends for ionization energy or electron affinity. The explanations for anomalous variations in these periodic properties are sometimes best explained using orbital notation.

For example, the first ionization energies in the second period generally increase from left to right, but two elements in the second period, beryllium and nitrogen, do not follow the trend.

Beryllium has a higher ionization energy than boron. The decrease in ionization energy between beryllium and boron is due to the ease of removal of an electron from a 2*p* orbital in boron compared with a 2*s* orbital in beryllium. The 2*s* orbital is slightly lower in energy than a 2*p* orbital. Electrons in an *s* orbital penetrate closer to the nucleus than in a *p* orbital. This leads to a higher coulombic attraction between the 2*s* electrons and the nucleus.

Nitrogen has a higher ionization energy than oxygen. When two electrons occupy the same orbital, there is a certain amount of electron-electron repulsion between them. When an electron is removed from a 2*p* orbital of an oxygen atom, the repulsion it experiences from the other electron in the 2*p* orbital is relieved. Nitrogen's 2*p* orbitals, however, are all singly occupied and the electrons do not experience the additional relief of repulsive interactions when an electron is removed.

**Ionization of Nitrogen and Oxygen**

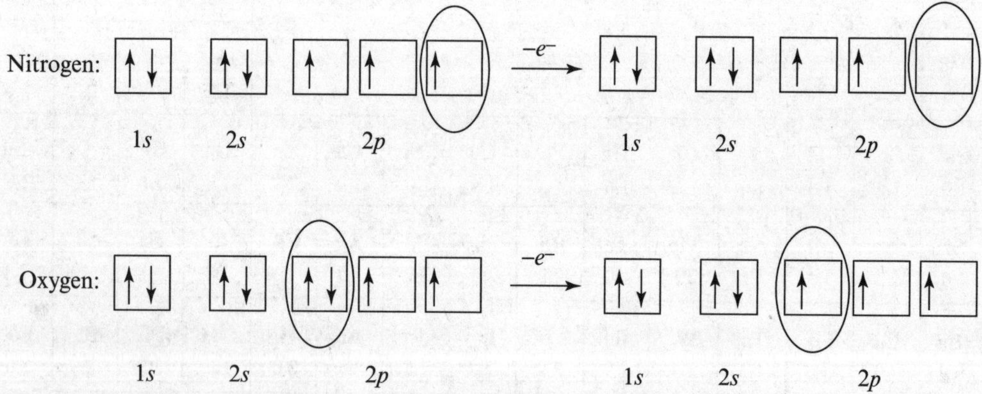


## Practice

The electron affinity of nitrogen is less negative than that of carbon. Provide an explanation for this exception to the periodic trend.

It is more difficult to add an electron to nitrogen than it is to add one to carbon; this exception to the trend has a similar explanation to the one for the anomalous ionization energies of nitrogen and oxygen.

When two electrons occupy the same orbital, there is some repulsion between them. When an electron is added to carbon, the result is three singly occupied 2*p* orbitals. When an electron is added to nitrogen, however, that electron is forced to pair with an existing electron in one of the already singly occupied 3*p* orbitals. The increased electronic repulsion results in a less negative electron affinity for nitrogen compared to carbon.

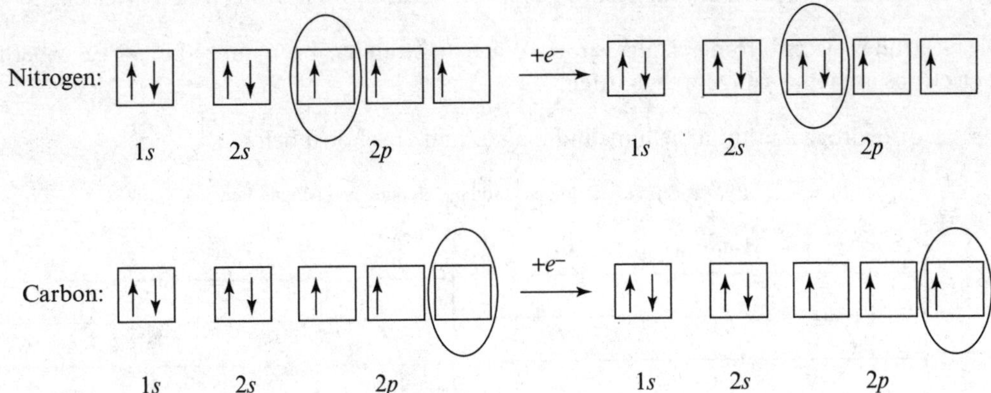


## Photoelectron Spectroscopy

Experimental evidence for the shell and subshell energies and populations of electrons in atoms comes from a technique called **photoelectron spectroscopy,** or PES.

Photoelectron spectroscopy has its basis in the photoelectric effect. The **photoelectric effect** is the ability of photons of a certain minimum energy to cause electrons to be ejected from a metal. Photons with insufficient energy are incapable of ejecting electrons.

**The Photoelectric Effect**

Insufficient energy to eject an electron from the metal

Sufficient energy to eject an electron from the metal

$e^-$

In a photoelectron spectroscopy experiment, atoms are bombarded with beams of X-ray photons with a known energy. The energy of the photons is sufficiently high that when a photon collides with any electron in a neutral atom, it causes the electron to be ejected.

The **kinetic energy,** *KE*, of the ejected electrons (their energy of motion) is measured in a detector. The difference between the energy of the photon used, $h\nu$, and the kinetic energy of the electron is equal to the ionization energy, *IE*, the amount of energy that was required to eject the electron.

$$IE = h\nu - KE$$

A useful feature of PES is the fact that any electron can be ejected by this technique. The photons are not limited to ejecting the most loosely held outer electrons. The ionization energies, also called the **binding energies,** of electrons from all of the subshells of an atom are detected. The relative number of electrons with a given energy is recorded.

Only a single electron is ejected from a neutral atom. When an atom has been ionized in a PES experiment, no subsequent electrons are removed from that atom.

The PES spectra of hydrogen, lithium, helium, and magnesium are shown below.

**PES Spectra of Hydrogen, Lithium, Helium, and Magnesium**

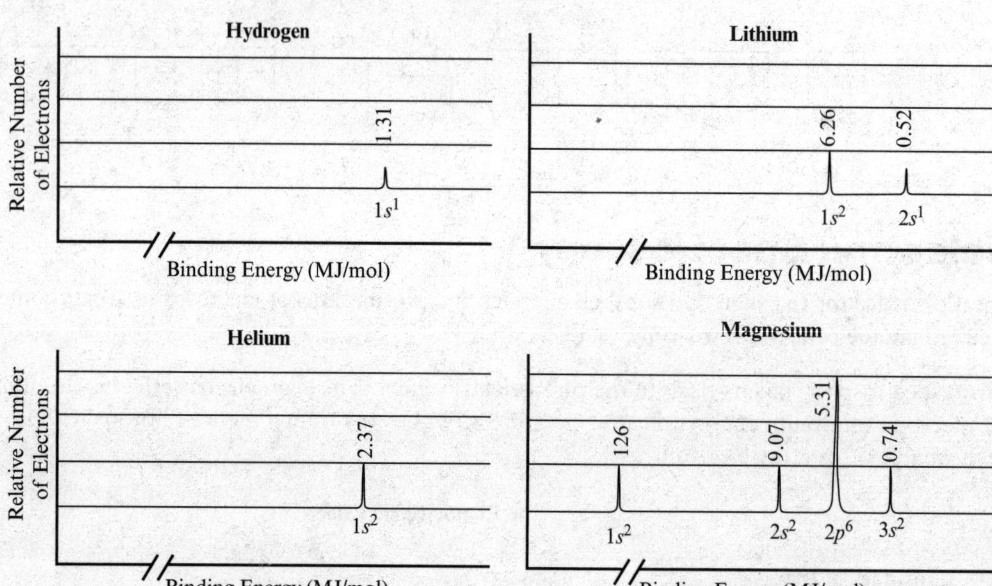


The PES signal for the $1s$ electron of hydrogen is half as intense as the signal for the $1s$ electrons of helium. This is because there are twice as many $1s$ electrons in a helium atom as there are in a hydrogen atom.

The binding energy of the $1s$ electrons in the helium atom is also greater than that of hydrogen. Helium has a higher nuclear charge than hydrogen, so its electrons are more strongly attracted to the nucleus.

When answering questions about spectra, students should draw on their knowledge of periodic trends and electron configurations. PES spectra should be viewed by students as experimental evidence for shells, subshells, and electron configurations of atoms.

## Practice

**a.** The PES data below are for which element?

**b.** The $2s$ electrons for sodium have a binding energy of 6.84 MJ/mol. Explain why that value is different from the energy of the $2s$ electrons of this element.

| Binding Energy (MJ/mol) | Relative Number of Electrons |
|---|---|
| 0.42 | 1 |
| 2.38 | 6 |
| 3.93 | 2 |
| 29.1 | 6 |
| 37.1 | 2 |
| 347 | 2 |

**(a)** The data are for potassium. The electron configuration of potassium is $1s^2 2s^2 2p^6 3s^2 3p^6 4s^1$. One peak is expected for each subshell, and the height of each peak corresponds to the number of electrons in that subshell.

**(b)** The energy of the 2*s* electrons for potassium corresponds to the second peak in the spectrum, 37.1 MJ/mol. The larger value means that the 2*s* electrons are held more tightly to potassium than they are to sodium, which has a 2*s* electron binding energy of 6.84 MJ/mol. This is because the nuclear charge experienced by the 2*s* electrons is greater for potassium than it is for sodium. This results in an increased coulombic attraction between the 2*s* electrons of potassium and the nucleus compared with that of sodium. Notice that although we are comparing atoms in the same group, we are comparing similar orbitals, so in this case, shielding arguments are irrelevant.

# Atomic Nuclei

As mentioned previously, the nucleus of an atom is the small, positively charged region at the center that is comprised of protons and neutrons. Atomic nuclei have been studied using a technique called mass spectrometry.

## Isotopes

Atoms of the same element that have different numbers of neutrons are called **isotopes.** The **atomic weight** of an element is the weighted average of the masses of the isotopes. The atomic weight of a hypothetical element with naturally occurring isotopes A, B, and C is calculated as follows:

$$\text{Atomic weight} = (\text{Mass A} \times \text{Fraction A}) + (\text{Mass B} \times \text{Fraction B}) + (\text{Mass C} \times \text{Fraction C})$$

Naturally occurring carbon atoms exist as approximately 99% carbon-12 (which has 6 protons and 6 neutrons in its nucleus) and 1% carbon-13 (which has 6 protons and 7 neutrons in its nucleus). The atomic weight of carbon can be calculated from this data.

$$\text{Atomic weight} = [(12\text{ amu}) \times 0.99] + [(13\text{ amu}) \times 0.01] = 12.01\text{ amu}$$

Notice that this agrees with the atomic weight of carbon on the periodic table.

### Practice

Naturally occurring chlorine consists of two isotopes, $^{35}Cl$ and $^{37}Cl$. Use the periodic table (page 349) to determine the fractions of atoms in a chlorine sample that are $^{35}Cl$ and $^{37}Cl$.

The sum of all the fractions must add up to one. Since there are two isotopes, the fraction of one of the isotopes, say $^{35}Cl$, can be assigned $x$, and the fraction of the other isotope, $^{37}Cl$, can be $1 - x$. The atomic weight of chlorine according to the periodic table is 35.45 amu. The fractions can then be determined.

$$\begin{aligned}\text{Atomic weight} &= 35.45\text{ amu}\\ 35.45\text{ amu} &= [(35\text{ amu})(x)] + [(37\text{ amu})(1 - x)]\\ 35.45 &= 35x + 37 - 37x\\ 2x &= 1.55\\ x &= 0.78\end{aligned}$$

Therefore, the fraction of naturally occurring chlorine that is $^{35}Cl$ is 0.78, or 78%. The fraction of chlorine that is $^{37}Cl$ is $1 - x = 1 - 0.78 = 0.22$, or 22%.

## Mass Spectrometry

**Mass spectrometry** is an experimental technique that allows chemists to determine atomic masses, molecular masses, and isotope patterns in atoms and molecules.

In a mass spectrometer, vaporized atoms or molecules are first ionized by one of a variety of means, usually to form positive ions. They are then accelerated through a magnetic field. This causes them to separate based on their mass-to-charge ratio, m/z. Less massive species are deflected more than more massive species. The ions are detected, and their degree of deflection is reported in a **mass spectrum.**

**Mass Spectrometer**

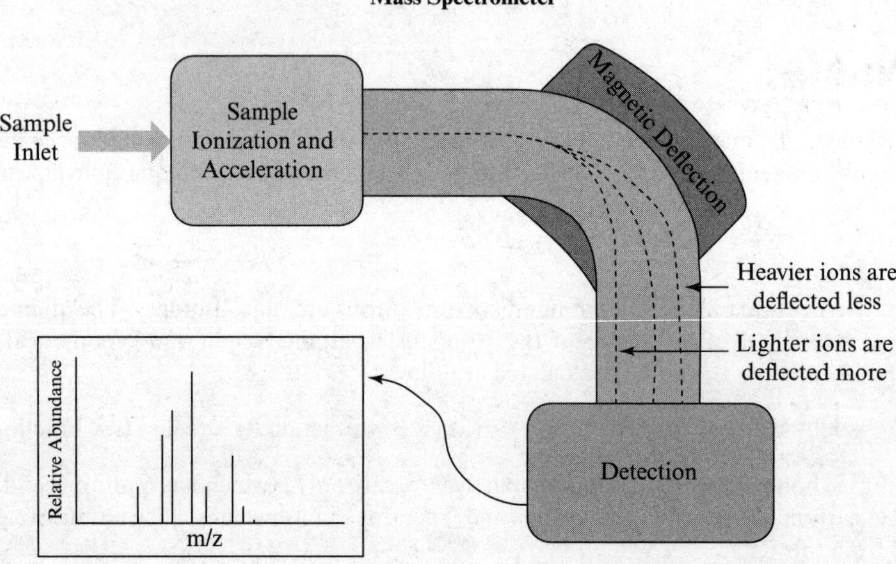


Mass spectrometry provides proof that all atoms of an element are not necessarily identical, contrary to what John Dalton believed. Instead, many elements are composed of isotopes. The isotopic compositions of different elements may be determined by inspection of mass spectra.

The simulated mass spectrum of copper is shown below. There are two naturally occurring, stable isotopes of copper; these account for the two peaks in the spectrum. The relative heights of the peaks reflect the relative number of atoms of each isotope. This data can be used to calculate the average atomic weight.

**The Mass Spectrum of Copper**

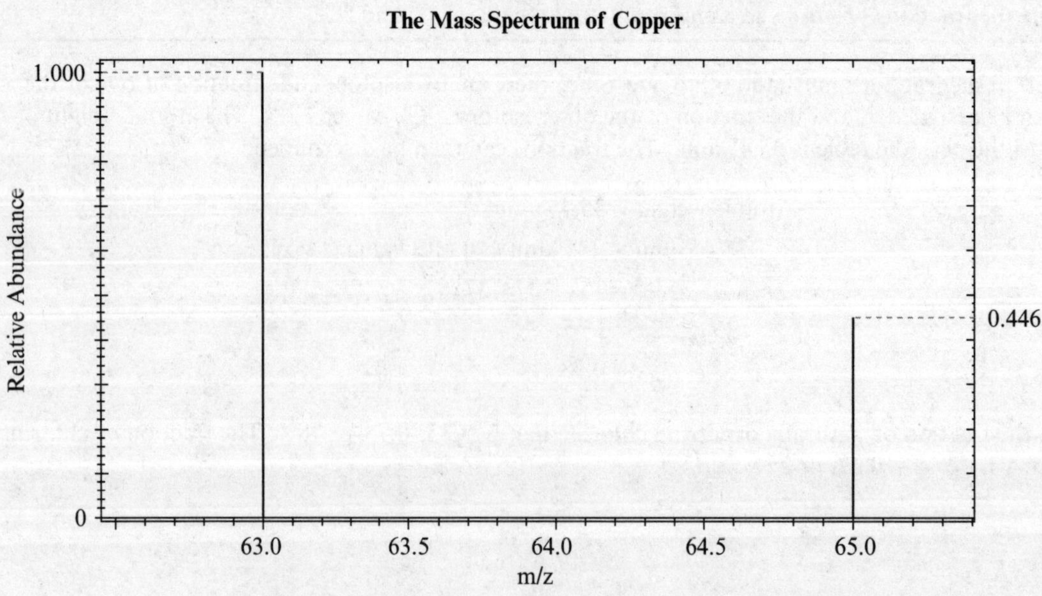

The fractions of each isotope are given by the relative abundances of each, divided by the sum of the heights of the peaks.

$$\text{fraction}_{^{63}\text{Cu}} = \frac{1.000}{1.000 + 0.446} = 0.6916$$

$$\text{fraction}_{^{65}\text{Cu}} = \frac{0.446}{1.000 + 0.446} = 0.3084$$

These fractions correspond to the natural abundance percentages of the isotopes of copper, 69.16% and 30.84%.

The average atomic mass of an element can be found by determining the weighted average of the mass-to-charge ratios expressed in units of amu. Each fraction is multiplied by the corresponding mass-to-charge ratio. The sum of the results gives the average atomic weight of the element.

Average atomic mass of copper = (0.6916 × 63.0 amu) + (0.3084 × 65.0 amu) = 63.6 amu

## Practice

Based on the following mass spectrum, determine the percent abundance of each isotope and the atomic weight of zinc.

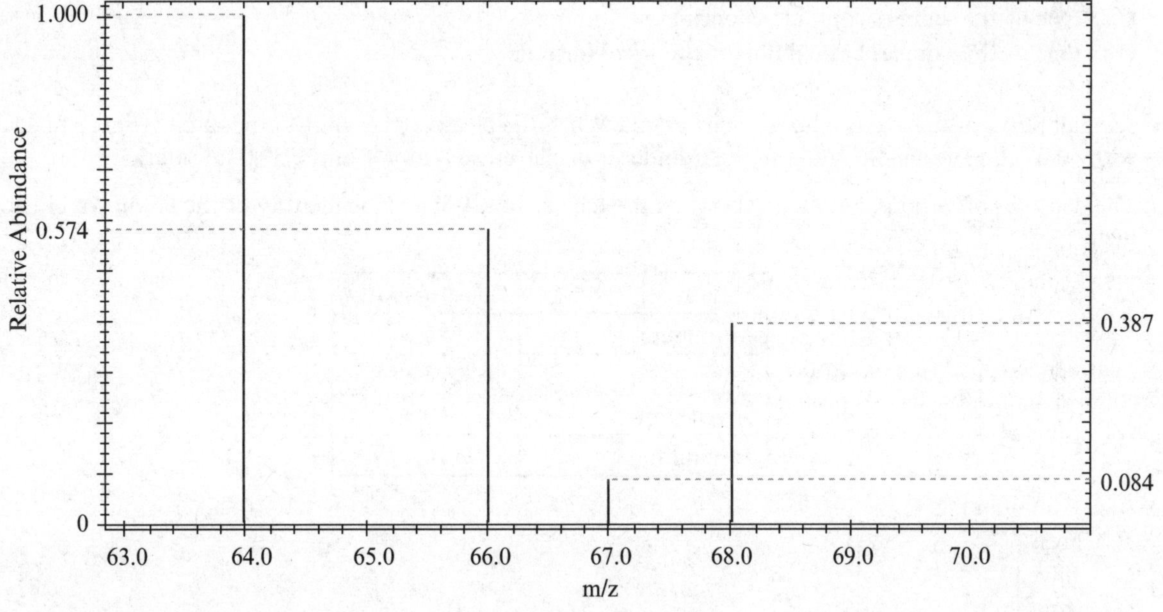


The isotope fractions are determined from the ratios of the heights of the peaks to the sum of the heights.

$$\text{fraction}_{^{64}\text{Zn}} = \frac{1.000}{2.045} = 0.489$$

$$\text{fraction}_{^{66}\text{Zn}} = \frac{0.574}{2.045} = 0.281$$

$$\text{fraction}_{^{67}\text{Zn}} = \frac{0.084}{2.045} = 0.041$$

$$\text{fraction}_{^{68}\text{Zn}} = \frac{0.387}{2.045} = 0.189$$

These fractions can be converted to percent abundances when multiplied by 100. The percentages of the isotopes of zinc are 48.9% $^{64}Zn$, 28.1% $^{66}Zn$, 4.1% $^{67}Zn$, and 18.9% $^{68}Zn$.

The atomic weight is the weighted average of the masses of the isotopes.

$$\text{Atomic weight} = \left(\frac{1.00}{2.045}\times 64.0\text{ amu}\right)+\left(\frac{0.574}{2.045}\times 66.0\text{ amu}\right)+\left(\frac{0.084}{2.045}\times 67.0\text{ amu}\right)+\left(\frac{0.387}{2.045}\times 68.0\text{ amu}\right)$$

$$= 31.296+18.525+2.752+12.868 \text{ (Additional significant figures shown for illustration.)}$$

$$= 65.4\text{ amu}$$

# Review Questions

## Multiple Choice

**1.** Two samples of potassium chloride are analyzed for chlorine content and found to contain different mass percentages of chlorine. Which of these is the best explanation?

**A.** The two samples came from different sources.
**B.** One of the samples has a different percentage of potassium isotopes.
**C.** One of the samples contains impurities.
**D.** One of the samples has a different empirical formula.

**2.** A small block of metal has a mass of 5.4 grams. When the block is placed in a graduated cylinder filled with 50.0 mL of water, the water in the cylinder is displaced so that it is at the 52.0 mL mark.

The densities of various metals are shown in the table below. What is the identity of the unknown block of metal?

| Metal | Density (g/mL) |
|---|---|
| Aluminum | 2.7 |
| Iron | 7.8 |
| Silver | 10.5 |
| Tantalum | 16.4 |

**A.** Aluminum
**B.** Iron
**C.** Silver
**D.** Tantalum

**3.** A pure sample of aluminum oxide, $Al_2O_3$, has a mass of 51 grams. What mass of oxygen is present in the sample?

**A.** 8 grams
**B.** 20 grams
**C.** 24 grams
**D.** 48 grams

**4.** The mass spectrum shown is that of which element?

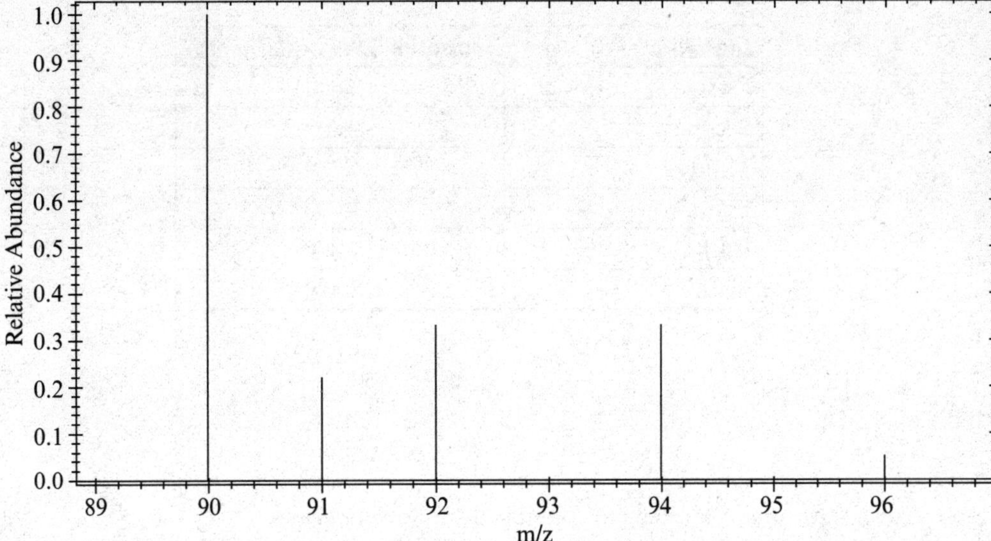


**A.** Y
**B.** Zr
**C.** Nb
**D.** Mo

**5.** The figure below represents part of the emission spectrum of a one-electron ion in the gas phase. All the lines result from electronic transitions from excited states to the ground state.

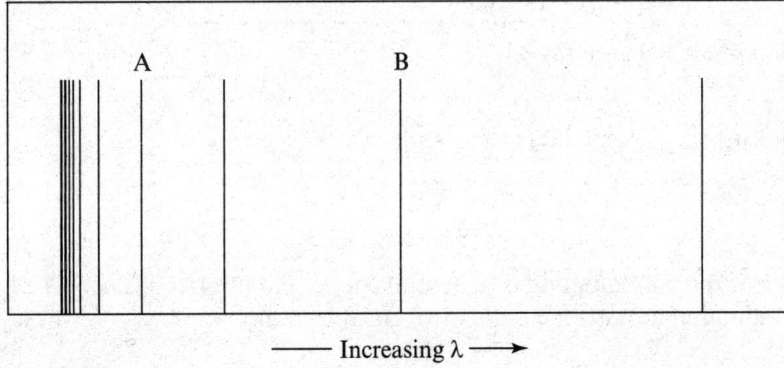


Which of the following is true?

**A.** Line A results from an electronic transition from $n = 4$ and line B from $n = 2$.
**B.** Line A results from an electronic transition from $n = 5$ and line B from $n = 3$.
**C.** Line A results from an electronic transition from $n = 8$ and line B from $n = 10$.
**D.** The large number of overlapping lines in the low wavelength region makes it impossible to tell from this figure.

**6.** The following table shows the first six ionization energies for an element. What is the identity of the element?

| Ionization | Ionization Energy ($kJ \cdot mol^{-1}$) |
|---|---|
| First | 600 |
| Second | 1800 |
| Third | 2700 |
| Fourth | 11,600 |
| Fifth | 14,800 |
| Sixth | 18,400 |

**A.** Al
**B.** Na
**C.** Mg
**D.** C

**7.** The photoelectron spectrum of an element contains the following peaks:

| MJ/mol | Relative Intensity |
|---|---|
| 433 | 2 |
| 48.5 | 2 |
| 39.2 | 6 |
| 5.44 | 2 |
| 3.24 | 6 |
| 0.77 | 1 |
| 0.63 | 2 |

This element is best described as a(an)

**A.** transition metal
**B.** element that exists in elemental form as a molecule
**C.** alkali metal
**D.** nonmetal

**8.** The ground state electron configuration of a neutral copper atom is $[Ar]4s^1 3d^{10}$. Which of the following is a ground state configuration for an ion commonly formed by copper?

**A.** $[Ar]4s^2 3d^9$
**B.** $[Ar]4s^2 3d^8$
**C.** $[Ar]4s^1 3d^9$
**D.** $[Ar]3d^{10}$

**9.** Waves A and B represent the wavelengths of photons that may be emitted when an electron in an atom relaxes from an excited state to the ground state. Which of the following best describes the waves?

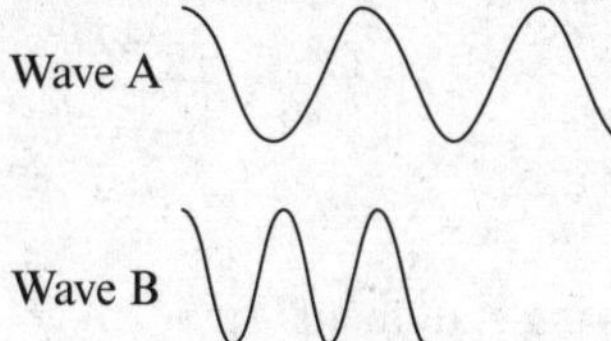


**A.** Wave A was generated by a transition from a higher energy level in the element than wave B.
**B.** Wave B was generated by a transition from a higher energy level in the element than wave A.
**C.** Wave A has greater amplitude and a longer wavelength than wave B.
**D.** Wave B has a greater amplitude and a higher frequency than wave A.

**10.** The orbital configuration below depicts

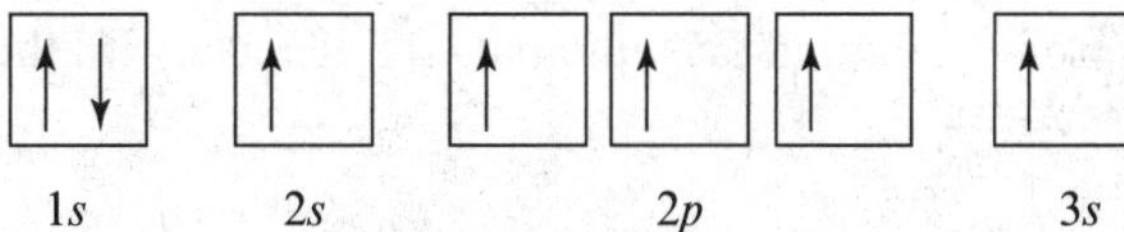


**A.** oxygen in the ground state
**B.** sodium in the ground state
**C.** nitrogen in an excited state
**D.** fluorine in an excited state

**11.** Nitrogen forms many compounds of the form $NR_3$, where R is hydrogen or a hydrocarbon group like $CH_3$. Which of the following elements is likely to form similar compounds?

**A.** Si
**B.** As
**C.** Al
**D.** S

**12.** The photoelectron spectrum of a certain pure element, X, is shown below:

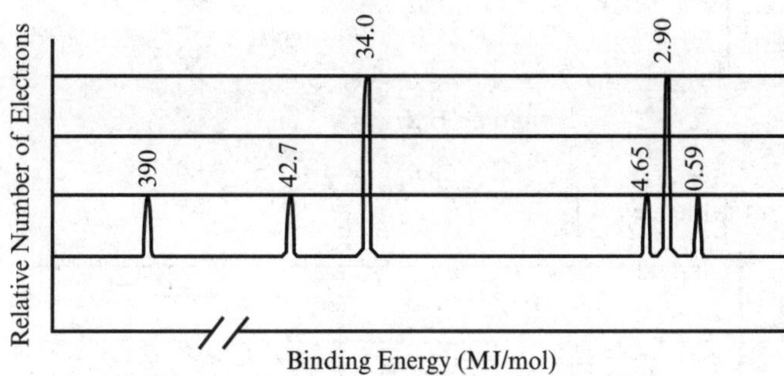


When two elements combine to form an ionic compound, the sum of their charges is zero. For example, $Mg^{2+}$ and $Cl^-$ form the neutral compound $MgCl_2$. Element X will most likely form which of the following compounds?

**A.** MgX, $KX_2$
**B.** $MgX_2$, KX
**C.** XO, $XCl_2$
**D.** $X_2O$, XCl

**13.** Which of the following arrangements shows the listed atoms in order of increasing first ionization energy?

**A.** Na < Ne < Ar
**B.** Ne < Ar < Na
**C.** Ar < Ne < Na
**D.** Na < Ar < Ne

**14.** Which phrase best completes the following statement?

The energy required to eject a 2*s* electron from a neutral atom of nitrogen in a PES experiment is __________ the fourth ionization energy, $IE_4$, of nitrogen.

**A.** greater than
**B.** less than
**C.** equal to
**D.** four times

**15.** Which of the following electron configurations represents the element with the greatest electronegativity?

**A.** $1s^22s^22p^5$
**B.** $1s^22s^22p^63s^23p^3$
**C.** $1s^22s^22p^63s^23p^4$
**D.** $1s^22s^22p^4$

# Long Free Response

**1.** The following PES spectra are for two elements that combine with each other to form binary compounds.

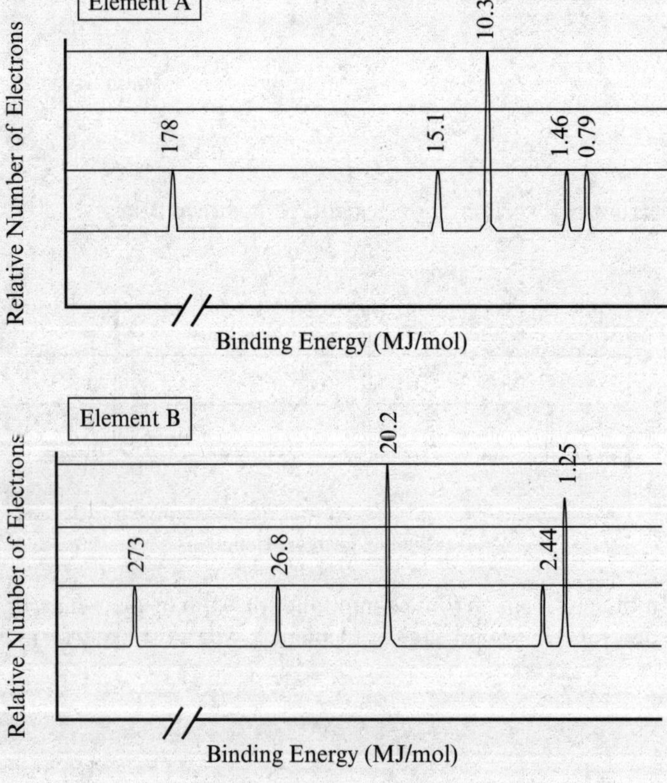


**a.** Write the electron configurations of elements A and B. Identify the two elements.
**b.** Why do the 2*s* electrons have different energies for the two elements?

c. How many moles of element A are present in 1.00 gram of element A?

d. Compound 1 contains 1.00 gram of element A and 5.05 grams of element B. Compound 2 contains 1.00 gram of element A and 3.79 grams of element B. Determine the number of moles of element B that react with 1.00 gram of element A in each of the two compounds.

e. **i.** Write the empirical formulas for the two compounds.
**ii.** Use ratios to show that the two compounds adhere to the law of multiple proportions.

f. The mass spectrum of element A contains a peak at 28.00 amu with an intensity of 1.000, a second peak at 29.00 amu with an intensity of 0.05074, and a third peak at 30.00 amu. What is the relative intensity of the third peak?

## Short Free Response

**2.** Pictured below to scale are an atom of calcium, an atom of sulfur, and a formula unit of calcium sulfide, CaS.

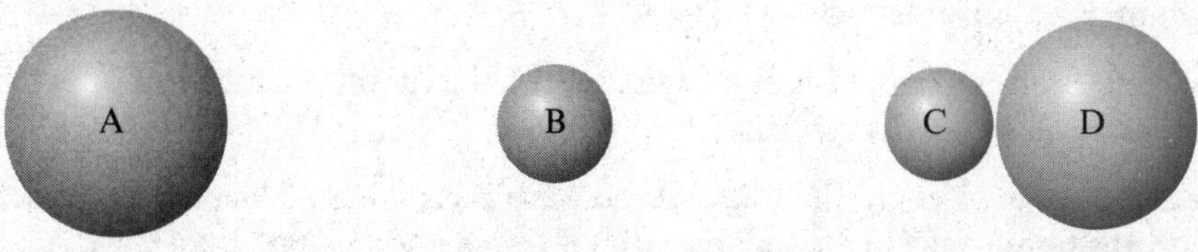


a. Clearly label A, B, C, and D as the correct atom or ion.

b. Discuss your assignment in terms of the atomic and ionic radii of the species.

c. Write the full electron configuration for a neutral atom of calcium and a neutral atom of sulfur. How many valence electrons are present in a neutral atom of calcium and a neutral atom of sulfur?

d. How many grams of sulfur are expected to be present in 180.3 grams of calcium sulfide?

# Answers and Explanations

## Multiple Choice

**1. C.** Potassium ionizes by losing a single electron to achieve a stable, noble gas electron configuration. Chlorine gains a single electron. Binary compounds of potassium and chlorine with empirical formulas other than KCl are not formed. Naturally occurring potassium should have a relatively fixed isotopic abundance, regardless of the source of the KCl. The two samples may have come from different sources, but this in itself is not the reason the compositions are different. Two samples of pure KCl from two different sources are expected to have identical composition. The most likely reason for the difference is the presence of impurities in one of the samples, choice C.

**2. A.** The volume of the metal can be determined from the water displacement.

$$52.0 \text{ mL} - 50.0 \text{ mL} = 2.0 \text{ mL}$$

The density is found by dividing the mass of the metal by the volume.

$$\text{Density} = \frac{m}{V} = \frac{5.4 \text{ g}}{2.0 \text{ mL}} = 2.7 \text{ g/mL}$$

The metal is aluminum, choice A.

**3. C.** The formula for aluminum oxide is $Al_2O_3$. The molar mass is about 102 g·mol$^{-1}$. The number of grams of oxygen present in 51 grams of a pure sample can be determined by dimensional analysis.

$$51 \cancel{\text{g } Al_2O_3} \times \frac{1 \cancel{\text{mol } Al_2O_3}}{102 \cancel{\text{g } Al_2O_3}} \times \frac{3 \cancel{\text{mol O}}}{1 \cancel{\text{mol } Al_2O_3}} \times \frac{16.00 \text{ g O}}{1 \cancel{\text{mol O}}} = 24 \text{ g O, choice C}$$

**4. B.** The weighted average of the masses can be estimated to obtain the atomic weight of the element. It is greater than 90 amu, so yttrium (88.91 amu) cannot be the correct answer. The average mass is significantly less than about 93 amu, so the answer is not niobium or molybdenum (92.91 and 95.94 amu, respectively). The correct answer must be zirconium (91.22 amu), choice B.

**5. B.** As wavelength increases, energy decreases, so the transition to the far right of the plot results from the lowest energy transition from $n = 2$ (the first excited state) to $n = 1$ (the ground state). Line B results from the next transition, from $n = 3$ to $n = 1$, and line A results from the transition from $n = 5$ to $n = 1$, making choice B correct.

**6. A.** The ionization energies for the first three electrons are much smaller than the ionization energy for the fourth. Removing a fourth electron disrupts the noble gas configuration of the ion, so the element must be in group 13 of the periodic table. The element is Al, choice A.

**7. A.** There are seven distinct energy sublevels in the atom. The element is in the *d* block of the periodic table and has the electron configuration $1s^22s^22p^63s^23p^64s^23d^1$. Notice that the 4*s* electrons are higher in energy than the 3*d* electron. The principal quantum number determines the relative energy of the sublevel. This element, scandium, is a transition metal, choice A.

**8. D.** The ground state electron configuration of $Cu^+$, one of the cations formed by copper, is that in which the 4*s* electron has been lost, $[Ar]3d^{10}$, choice D.

**9. B.** Wave B has a shorter wavelength than wave A, so wave B is higher in energy, choice B. It is emitted during a larger transition from a higher-energy orbital to the ground state than wave A. (Both waves have the same amplitude.)

**10. C.** The electron configuration shows seven electrons. The element is a neutral nitrogen atom. Instead of a $1s^22s^22p^3$ electron configuration, one of the electrons has been promoted to the 3*s* orbital. This indicates an excited state, choice C.

**11. B.** Arsenic, As (choice B), is in the same group on the periodic table as nitrogen. It displays similar chemical behavior due to its similar $np^3$ electron configuration.

**12. C.** Electrons with higher binding energy are located in energy levels closer to the nucleus. The electrons with a binding energy of 290 MJ/mol correspond to the 1*s* electrons. The relative height of that peak corresponds to two electrons. The resulting electron configuration, $1s^22s^22p^63s^23p^64s^2$, indicates that the element, X, is calcium. When calcium ionizes, it loses its two valence electrons to form a cation. Calcium cation, $Ca^{2+}$, forms calcium oxide with $O^{2-}$, so CaO corresponds to XO, choice C. Calcium forms calcium chloride with $Cl^-$, so $CaCl_2$ corresponds to $XCl_2$.

**13. D.** Neon's (Ne) outermost electron is in a lower shell than the outermost electrons of argon (Ar) and sodium (Na). Neon will be less shielded and experience a relatively high coulombic attraction to the nucleus. It has the highest first ionization energy. Sodium has the smallest effective nuclear charge and has the lowest ionization energy of the three elements. The correct order of increasing first ionization energy is shown in choice D, Na < Ar < Ne.

**14. B.** Less energy, choice B, is required to remove a 2*s* electron from a neutral atom than to remove a $2s^1$ electron from a $N^{3+}$ ion. The coulombic attraction of the remaining electrons with the +3 charge of the ion makes removal of a fourth electron more difficult than removal of a single 2*s* electron from a neutral atom of nitrogen.

**15. A.** The electron configuration in choice A corresponds to a fluorine atom. Fluorine is the most electronegative element because it has a very high effective nuclear charge, a high coulombic attraction between its outermost electrons and the nucleus, and low shielding by inner shell electrons relative to atoms of the other elements.

## Long Free Response

**1.** **a.** Element A has the electron configuration $1s^22s^22p^63s^23p^2$. Element A is silicon. Element B has the electron configuration $1s^22s^22p^63s^23p^5$. Element B is chlorine.

**b.** The 2*s* electron for silicon has a lower ionization energy than that of chlorine because chlorine has a higher effective nuclear charge. The larger effective nuclear charge results in a greater coulombic attraction between the outermost electrons and the nucleus.

**c.** The molar mass is used in the determination of the number of moles of silicon in 1.00 gram.

$$1.00\ \cancel{\text{g Si}} \times \frac{1\ \text{mol Si}}{28.09\ \cancel{\text{g Si}}} = 0.0356\ \text{mol Si}$$

**d.** The number of moles of chlorine that combine with 1.00 gram of silicon in each compound is determined as follows:

$$5.05\ \cancel{\text{g Cl}} \times \frac{1\ \text{mol Cl}}{35.45\ \cancel{\text{g Cl}}} = 0.142\ \text{mol Cl in Compound 1}$$

$$3.79\ \cancel{\text{g Cl}} \times \frac{1\ \text{mol Cl}}{35.45\ \cancel{\text{g Cl}}} = 0.107\ \text{mol Cl in Compound 2}$$

**e.** **i.** The smallest whole-number ratio for the first compound is determined by dividing each number of moles by the smallest number of moles, 0.0356 mol Si.

$$\text{Silicon: } \frac{0.0356}{0.0356} = 1$$

$$\text{Chlorine: } \frac{0.142}{0.0356} = 4$$

The empirical formula for the first compound is $SiCl_4$.

The smallest whole-number ratio for the second compound is determined the same way.

$$\text{Silicon: } \frac{0.0356}{0.0356} = 1$$

$$\text{Chlorine: } \frac{0.107}{0.0356} = 3$$

The empirical formula for the second compound is $SiCl_3$.

**ii.** The law of multiple proportions states that when two elements form two different binary compounds, the ratio of the masses of the second element that combine with a fixed amount of the first element is always a small whole-number ratio. The ratio of the masses of the chlorine that combine with 1.00 gram of silicon must be determined and converted to a whole-number ratio.

$$\frac{5.05\ \text{g}}{3.79\ \text{g}} = \frac{1.33}{1} \times \frac{3}{3} = \frac{4}{3}$$

Chlorine combines with silicon in a 4:3 mass ratio for the two compounds.

**f.** Let $x$ be the unknown intensity. The atomic weight of silicon, 28.09 amu, can be found on the periodic table. This atomic weight is the weighted average of the masses in the spectrum. The sum of the intensities for this calculation is 1.000 + 0.05074 + $x$.

$$28.09 = \left(28.00 \times \frac{1.000}{1.000 + 0.05074 + x}\right) + \left(29.00 \times \frac{0.05074}{1.000 + 0.05074 + x}\right) + \left(30.00 \times \frac{x}{1.000 + 0.05074 + x}\right)$$

$$28.09 = \frac{28.00 + 1.4714 + 30.00x}{1.05074 + x}$$

$$x = 0.02298$$

# Short Free Response

**2. a.** Species A represents a neutral calcium atom, species B represents a neutral sulfur atom, species C represents a $Ca^{2+}$ cation, and species D represents an $S^{2-}$ anion.

**b.** The neutral calcium atom is larger than the neutral sulfur atom because its outermost electrons are in a higher principal energy level and experience more shielding from the nucleus. In addition, the effective nuclear charge is less for calcium than it is for sulfur. For these reasons, the outermost electrons are held more closely to the nucleus in sulfur than they are in calcium. The two ions, $Ca^{2+}$ and $S^{2-}$, are isoelectronic. Ionic radii of isoelectronic species decrease with increasing atomic number and nuclear charge, so $Ca^{2+}$ is smaller than $S^{2-}$.

**c.** The electron configuration for a neutral atom of calcium is $1s^2 2s^2 2p^6 3s^2 3p^6 4s^2$. Calcium has two valence electrons.

The electron configuration for a neutral atom of sulfur is $1s^2 2s^2 2p^6 3s^2 3p^4$. Sulfur has six valence electrons.

**d.** The number of grams of sulfur present in 180.3 grams of CaS is determined using the molecular formula and the molar mass.

$$M_{CaS} = (1 \times 40.08 \text{ g·mol}^{-1}) + (1 \times 32.06 \text{ g·mol}^{-1}) = 72.14 \text{ g·mol}^{-1}$$

$$180.3 \cancel{\text{g CaS}} \times \frac{1 \cancel{\text{mol CaS}}}{72.14 \cancel{\text{g CaS}}} \times \frac{1 \cancel{\text{mol S}}}{1 \cancel{\text{mol CaS}}} \times \frac{32.06 \text{ g S}}{1 \cancel{\text{mol S}}} = 80.13 \text{ g S}$$

# Chapter 2

# Molecular and Ionic Compound Structure and Properties

**The physical and chemical properties of compounds and molecules are determined by the nature of the bonds between constituent atoms.**

Atoms and ions interact with one another to form molecules and compounds with properties that can be attributed to these interactions. These attractive interactions between atoms and ions are referred to as **bonds.** The impact of bonding on the physical and chemical properties of solids, liquids, and gases is discussed in this chapter and in Chapter 3, "Intermolecular Forces and Properties."

## Nomenclature

Naming compounds is not a skill that will be tested directly on the AP Chemistry exam, but it is a prerequisite for success. This section provides a review of basic chemical nomenclature.

## Inorganic Nomenclature

**Compounds** are groups of atoms composed of more than one type of element; they can be broadly classified as ionic or covalent. The ionic or covalent nature of a compound, in turn, determines how the compound is named.

### Ionic Nomenclature

**Ionic compounds** are those that are formed between positively charged ions, or **cations,** and negatively charged ions, or **anions.** Ionic compounds are often formed between metal cations and nonmetal anions. They may also be formed from **polyatomic ions,** which are ions that are composed of more than one atom. The number of cations and anions that comprise a formula unit of an ionic compound is such that the sum of the charges for the formula unit is zero.

For example, consider the ionic compound formed from potassium and sulfur atoms. Since potassium is in group 1 of the periodic table, it forms $K^+$ cations. Sulfur is in group 16, so it forms $S^{2-}$ anions. The sum of the charges for an ionic compound must equal zero, so there must be two potassium cations and one sulfur anion in each formula unit. The formula is $K_2S$.

The names of ionic compounds begin with the name of the cation and end with the name of the anion. The ending of the name of a monatomic anion is changed to *-ide*. The name of $K_2S$ is potassium sulfide.

In order to understand the nomenclature of ionic compounds, it is helpful to memorize some important polyatomic ions and their charges.

| Polyatomic Ion Name | Formula |
|---|---|
| Acetate | $C_2H_3O_2^-$ |
| Ammonium | $NH_4^+$ |
| Carbonate | $CO_3^{2-}$ |
| Hydrogen carbonate (also known as bicarbonate) | $HCO_3^-$ |
| Perchlorate | $ClO_4^-$ |
| Chlorate | $ClO_3^-$ |
| Chlorite | $ClO_2^-$ |

*continued*

| Polyatomic Ion Name | Formula |
|---|---|
| Hypochlorite | $ClO^-$ |
| Chromate | $CrO_4^{2-}$ |
| Dichromate | $Cr_2O_7^{2-}$ |
| Dihydrogen phosphate | $H_2PO_4^-$ |
| Hydrogen phosphate | $HPO_4^{2-}$ |
| Phosphate | $PO_4^{3-}$ |
| Hydrogen sulfate | $HSO_4^-$ |
| Sulfate | $SO_4^{2-}$ |
| Sulfite | $SO_3^{2-}$ |
| Hydroxide | $OH^-$ |
| Nitrate | $NO_3^-$ |
| Nitrite | $NO_2^-$ |
| Oxalate | $C_2O_4^{2-}$ |
| Permanganate | $MnO_4^-$ |
| Peroxide | $O_2^{2-}$ |

Many metals form multiple cations with different charges. Transition metals and some others exhibit this behavior. For instance, iron may lose two electrons to result in $Fe^{2+}$ or it may lose three electrons to form $Fe^{3+}$.

In naming a compound that contains a metal that may form more than one cation, a Roman numeral is used to show the charge of the ion. For example, the name of $Fe(OH)_2$ is iron(II) hydroxide and the name of $Fe(OH)_3$ is iron(III) hydroxide.

| Metal | Common Ion(s) |
|---|---|
| Antimony | $Sb^{3+}$, $Sb^{5+}$ |
| Bismuth | $Bi^{3+}$, $Bi^{5+}$ |
| Cadmium | $Cd^{2+}$ |
| Cobalt | $Co^{2+}$, $Co^{3+}$ |
| Copper | $Cu^+$, $Cu^{2+}$ |
| Iron | $Fe^{2+}$, $Fe^{3+}$ |
| Lead | $Pb^{2+}$, $Pb^{4+}$ |
| Mercury* | $Hg_2^{2+}$, $Hg^{2+}$ |
| Nickel | $Ni^{2+}$ |
| Silver | $Ag^+$ |
| Tin | $Sn^{2+}$, $Sn^{4+}$ |
| Zinc | $Zn^{2+}$ |

**Note that mercury(I) is $Hg_2^{2+}$ and mercury(II) is $Hg^{2+}$.*

Ionic hydrates are named as usual for ionic compounds, and the number of water molecules is specified with a Greek prefix. For instance, there are five water molecules in copper(II) sulfate pentahydrate, $CuSO_4 \cdot 5\ H_2O$.

| Number | Greek Prefix |
|---|---|
| 1 | *mono-* |
| 2 | *di-* |
| 3 | *tri-* |
| 4 | *tetra-* |
| 5 | *penta-* |
| 6 | *hexa-* |
| 7 | *hepta-* |
| 8 | *octa-* |
| 9 | *nona-* |
| 10 | *deca-* |

## Practice

Name and write the formulas for the ionic compounds formed between the following:

**a.** lithium and nitrogen
**b.** strontium and bromine

**(a)** Lithium nitride is the name of the ionic compound formed between lithium and nitrogen. Lithium is in group 1, so it forms $Li^+$; nitrogen is in group 15, so it forms $N^{3-}$. The formula for lithium nitride that leads to an overall charge of zero is $Li_3N$.

**(b)** Strontium bromide is the name of the ionic compound formed between strontium and bromine. Strontium is in group 2, so it forms $Sr^{2+}$; bromine is in group 17, so it forms $Br^-$. The formula for strontium bromide that gives an overall charge of zero is $SrBr_2$.

## Practice

Write formulas for the following polyatomic ionic compounds:

**a.** aluminum sulfate
**b.** sodium peroxide
**c.** iron(II) nitrate
**d.** mercury(I) acetate

The formulas are as follows:

**(a)** $Al_2(SO_4)_3$,
**(b)** $Na_2O_2$
**(c)** $Fe(NO_3)_2$
**(d)** $Hg_2(C_2H_3O_2)_2$

## Practice

Write the formula for cobalt(II) chloride hexahydrate.

The *hexa-* prefix indicates six water molecules are associated with the cobalt(II) chloride. The formula is $CoCl_2{\cdot}6\ H_2O$.

## Binary Covalent Nomenclature

**Binary covalent compounds** are neutral compounds composed of two nonmetals. The same Greek prefixes used in hydrate nomenclature denote the number of atoms of each element in a binary covalent molcculc. Usually, the less electronegative element is named first, followed by the more electronegative element. The end of the name of the second element is changed to *-ide*.

When there is only one atom of the first element, however, the prefix *mono-* is not used. For example, $CO_2$ is called "carbon dioxide," not "monocarbon dioxide."

When a prefix ending in an "a" or an "o" precedes the name of an element that begins with a vowel, the "a" or "o" of the prefix is dropped. For example, CO is called "carbon monoxide," not "carbon monooxide."

### Practice

---

Name the following compounds:

**a.** $N_2O_5$
**b.** $P_4O_{10}$
**c.** $SiCl_4$
**d.** $CS_2$

---

The names of the compounds are

**(a)** dinitrogen pentoxide
**(b)** tetraphosphorus decoxide
**(c)** silicon tetrachloride
**(d)** carbon disulfide

## Organic Nomenclature

**Organic compounds** are broadly defined as compounds containing carbon and (usually) hydrogen. The term *organic* encompasses a vast number of molecules that may or may not be associated with life processes. The definition is generally understood to exclude alloys like steel, which is composed of carbon atoms interspersed among iron atoms, and simple carbon-oxygen molecules like carbon dioxide and carbonate ions.

**Hydrocarbons** are a class of organic compounds that consist of only hydrogen and carbon.

**Alkanes,** the simplest hydrocarbons, consist of singly bonded carbon atoms. These are also called **saturated hydrocarbons.** The molecular formula for a saturated hydrocarbon is $C_nH_{2n+2}$.

The names of alkanes end in *-ane*. The names of the first ten unbranched alkanes are shown in the table below. It is not necessary to name organic compounds on the AP Chemistry exam; however, familiarity with organic compounds is advantageous.

**The Simplest Unbranched Saturated Hydrocarbons**

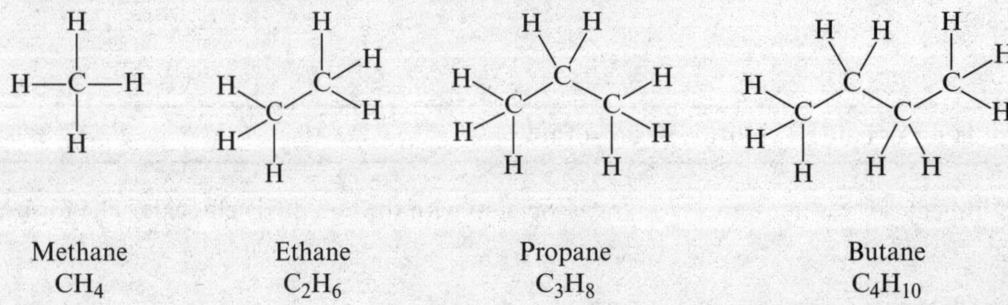

Methane $CH_4$ | Ethane $C_2H_6$ | Propane $C_3H_8$ | Butane $C_4H_{10}$

| Name of Alkane | Molecular Formula | Name of Alkane | Molecular Formula |
|---|---|---|---|
| Methane | $CH_4$ | Hexane | $C_6H_{14}$ |
| Ethane | $C_2H_6$ | Heptane | $C_7H_{16}$ |
| Propane | $C_3H_8$ | Octane | $C_8H_{18}$ |
| Butane | $C_4H_{10}$ | Nonane | $C_9H_{20}$ |
| Pentane | $C_5H_{12}$ | Decane | $C_{10}H_{22}$ |

Organic compounds may also contain **functional groups,** which are groups of atoms in a molecule that give the molecule characteristic physical and chemical properties. There is no need to memorize the systematic nomenclature for these compounds, but familiarity with common organic functional groups facilitates compound property prediction. Many of the characteristics mentioned here will be reviewed in greater detail later in this chapter, so it is a good idea to refer to this table later.

| Functional Group | Structure | Characteristics | Example |
|---|---|---|---|
| Alcohol | 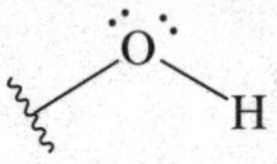 | Alcohols are relatively polar since they are capable of donating and accepting hydrogen bonds. | Methanol 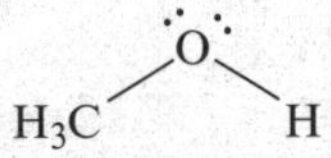 |
| Amine | 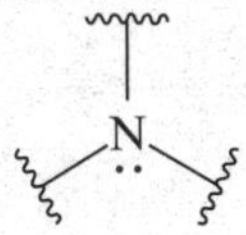 | Amines are important in acid-base chemistry because they behave as weak bases. | Ethylamine 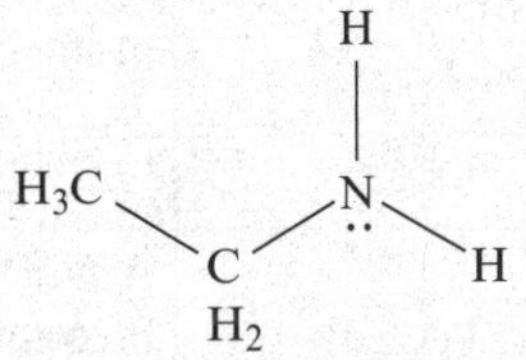 |
| Alkyl halide | 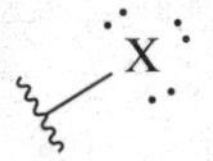 Where X = F, Cl, Br, or I | Alkyl halides are important reactants in organic chemistry. | 2-Chlorobutane 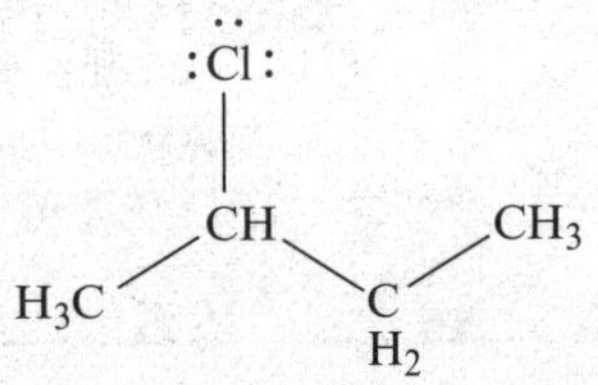 |
| Ether | 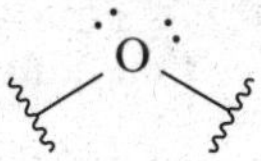 | Ethers tend to be volatile compounds with dipole moments. | Diethyl ether 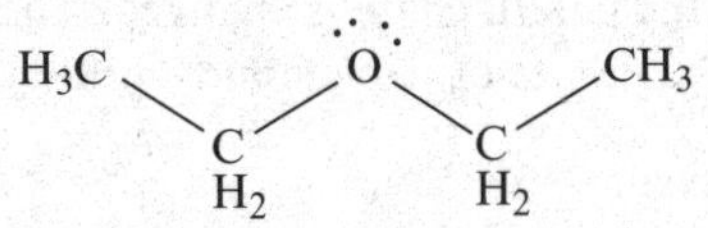 |
| Aldehyde | 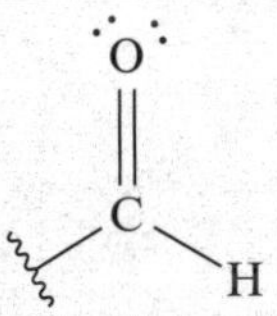 | Aldehydes tend to be volatile compounds with dipole moments. | Pentanal 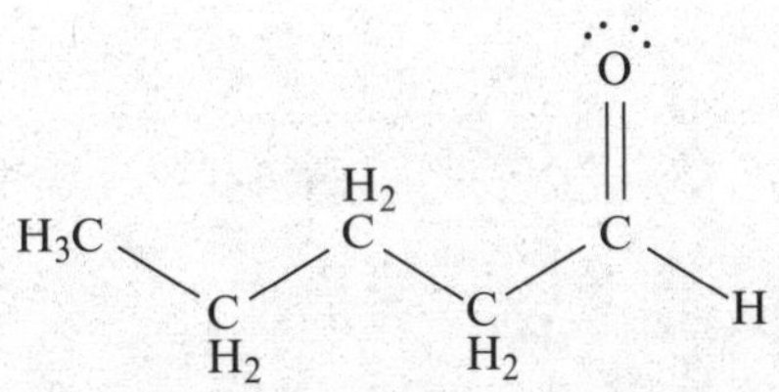 |

*continued*

| Functional Group | Structure | Characteristics | Example |
|---|---|---|---|
| Ketone | 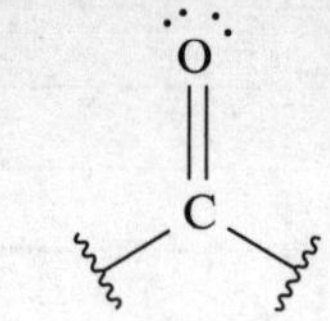 | Ketones tend to be volatile compounds with dipole moments. | 3-Hexanone 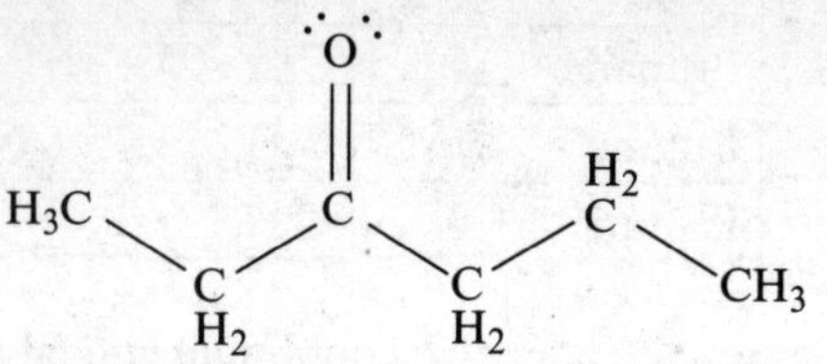 |
| Carboxylic acid | 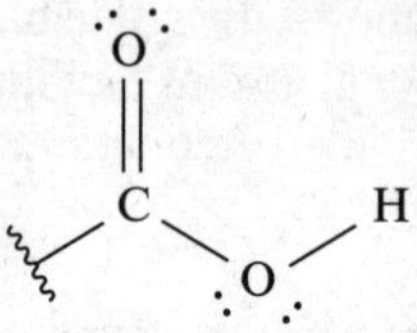 | Carboxylic acids are important in acid-base chemistry because they are weak acids. They are capable of accepting and donating hydrogen bonds. | Octanoic acid 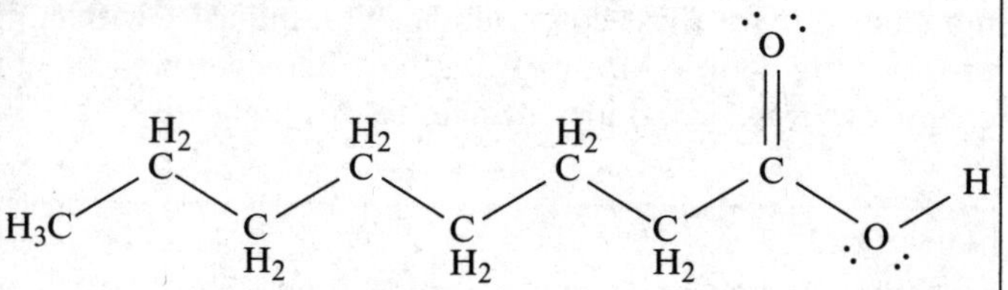 |
| Ester | 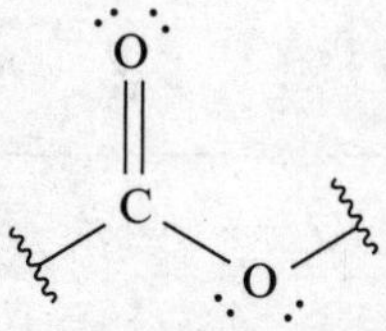 | Esters tend to be volatile compounds with dipole moments. | Methyl butanoate 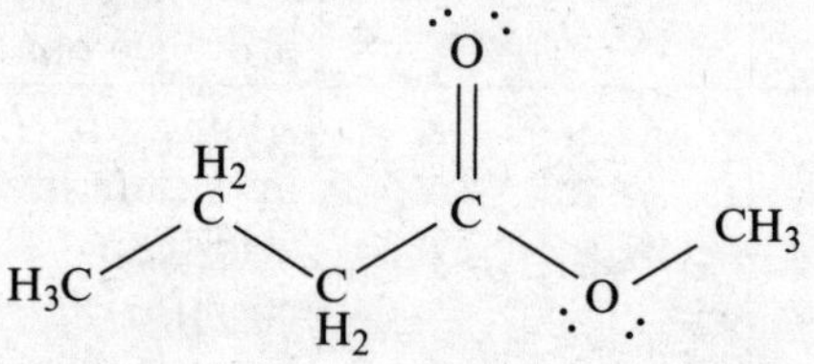 |
| Amide | 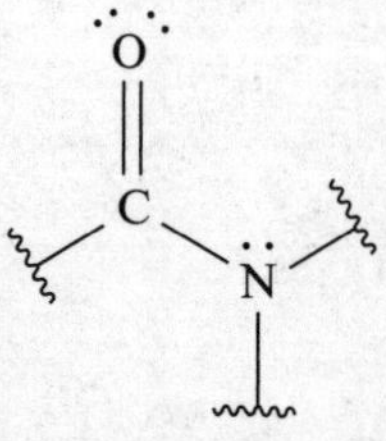 | Amide groups link amino acids into long polypeptide chains that comprise proteins. | Methyl propanamide 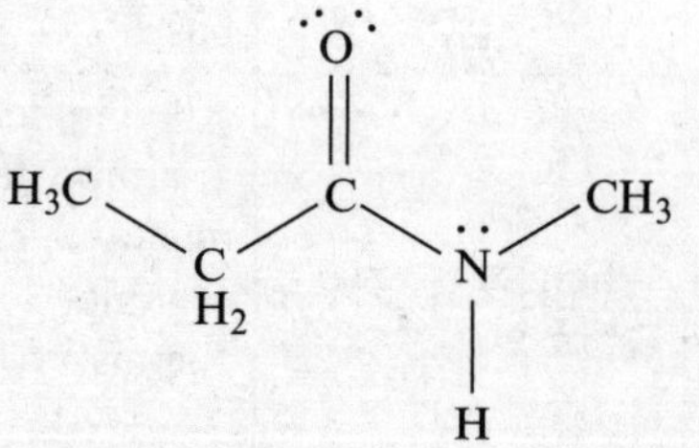 |

Hydrocarbons may also contain carbons that are double-bonded or triple-bonded to one another. In these cases, the ending of the parent chain is changed. The names of alkenes, which are molecules that contain double bonds, end in *-ene*. The names of triple bond–containing alkynes are changed to *-yne*.

**Double and Triple Bonds**

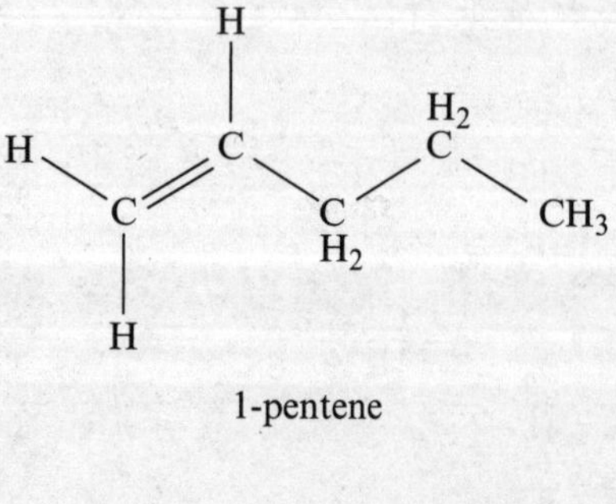

1-pentene

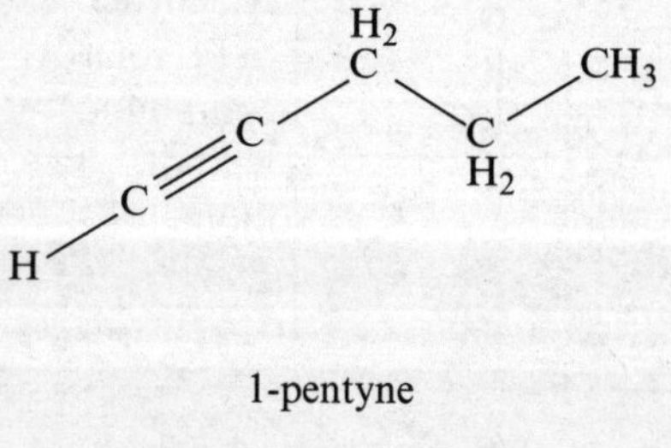

1-pentyne

## Practice

Refer to the earlier table (pages 59–60), if necessary, to match the structures below with one of the following names: 1-butene, triethylamine, 1-heptyne, propanoic acid, 2-hexanol.

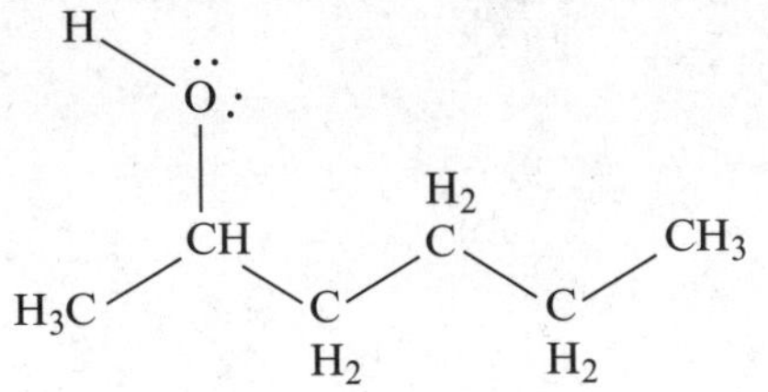

A

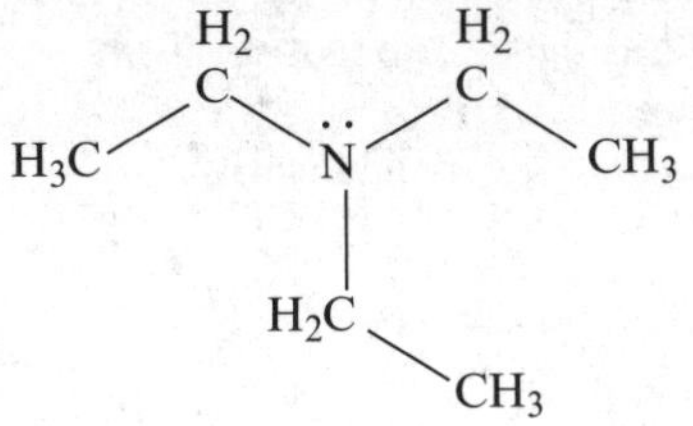

B

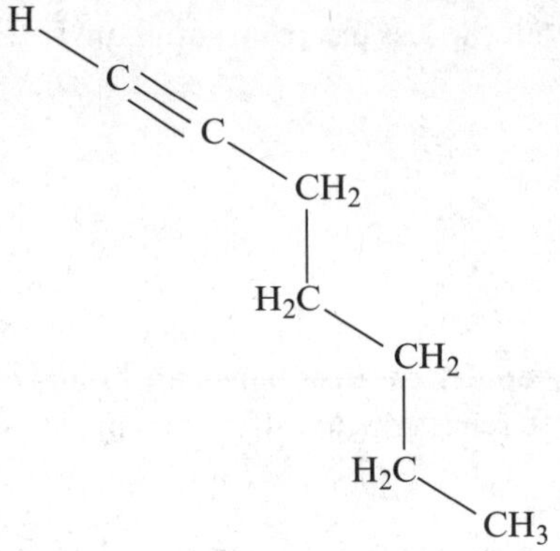

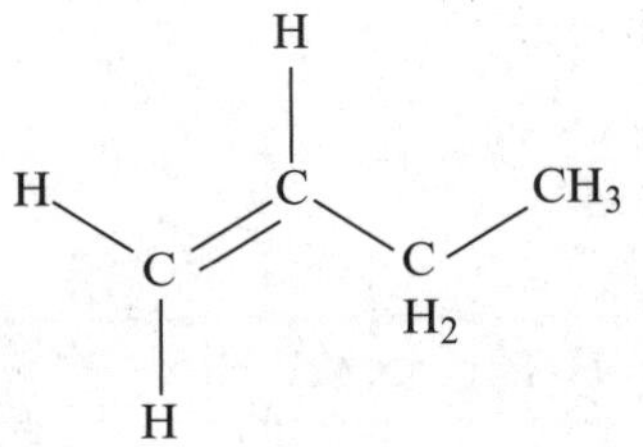

C

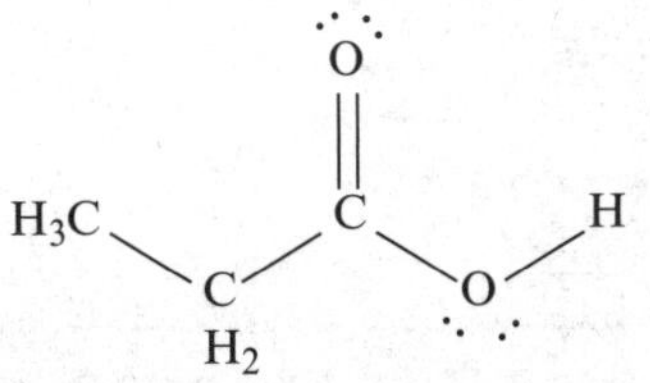

D

E

Structure A is 2-hexanol, a 6-carbon alcohol. It can be recognized by the –OH functional group.

Structure B is triethylamine, a nitrogen-containing molecule (amine) with three 2-carbon chains.

Structure C is 1-butene, a 4-carbon alkene that contains double-bonded carbon atoms.

Structure D is propanoic acid, a 3-carbon carboxylic acid that can be recognized by its $-CO_2H$ functional group.

Structure E is 1-heptyne, a 7-carbon alkyne that contains triple-bonded carbon atoms.

# Chemical Bonds

**Chemical bonds** are forces that hold atoms together in compounds. A bond results from the sharing or transfer of electrons between two atoms. The distinction between the two major kinds of bonds, covalent and ionic, is not concrete. Rather, bonding character can be classified on a continuum between covalent and ionic based on the electronegativity differences between the two atoms involved in the bonding.

# Covalent Bonding

**Covalent bonding** occurs when two nonmetallic atoms share at least one pair of electrons. In general, for a bond to be classified as covalent, the distribution of electron density between the two atoms participating in the bond is not unequal to the extent that the atoms would be considered ionized. Usually the two atoms have an electronegativity difference of about 1.7 or less. (Refer to Chapter 1, "Atomic Structure and Properties," for a review of electronegativity.)

The Pauling electronegativity of a hydrogen atom is about 2.2. The electronegativity of chlorine is about 3.1. The difference in electronegativity, ΔEN, between chlorine and hydrogen is 0.9. Since this difference is less than 1.7, the bond between hydrogen and chlorine in a molecule of hydrochloric acid, HCl, is considered to be covalent.

$$\Delta EN = 3.1 - 2.2 = 0.9$$

## Polar and Nonpolar Covalent Bonds

**Polar covalent bonds** are covalent bonds in which the electron pair is shared unequally by the two atoms due to differences in electronegativity greater than about 0.4 and less than 1.7. The hydrogen-chlorine bond discussed previously is a polar covalent bond. The greater effective nuclear charge of chlorine draws the electron density of the bond toward the chlorine nucleus. The polarity of the bond is shown by an arrow with a crossed tail pointing from the less electronegative atom to the more electronegative atom.

**Bond Polarity**

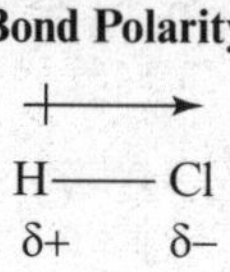


**Nonpolar covalent bonds** are bonds in which electron pair is shared equally by the two atoms due to very small differences or no differences in electronegativity between the bonded atoms. The covalent bonds in $H_2$, $O_2$, and $Cl_2$ are nonpolar because there is no difference in electronegativity between two atoms of the same element.

Carbon-hydrogen bonds are classified as nonpolar covalent.

### Practice

---

Rank the following single covalent bonds in order of increasing polarity: H–F, H–P, H–O, and H–S.

---

Phosphorus is the least electronegative of the four heavy atoms, F, P, O, and S. It experiences more shielding of the nucleus by core electrons than fluorine and oxygen, and it has the lowest effective nuclear charge, so the H–P bond can be assumed to be the least polar. By contrast, the fluorine atom is the most electronegative because it has the highest effective nuclear charge. Its valence electrons are less shielded than those of sulfur and phosphorus. The H–F bond is the most polar. Oxygen is more electronegative than sulfur because its valence electrons experience less shielding, so the correct order of increasing bond polarity is H–P < H–S < H–O < H–F.

## Bond Length

The **bond length,** or the average distance between the nuclei of two bonded atoms, is the distance at which the potential energy of the two atoms is at a minimum. This is often illustrated in potential energy diagrams like the one shown below.

The lower the potential energy, the more stable the arrangement and the stronger the bond. When the atoms are very far apart, as shown on the right of the diagram, there is no interaction between them and the potential energy is zero. At closer distances, the nucleus of each atom begins to attract the electron cloud of the other atom. As the atoms approach one another (moving to the left of the diagram), their potential energy decreases. At the equilibrium distance, the attractive and repulsive forces are balanced and the energy is at a minimum. For two hydrogen atoms, this distance is 0.074 nm. From the diagram, we can see that the energy required to separate the two atoms completely is 458 kJ/mol.

At distances closer than the bond length, on the other hand, the repulsive forces between electron clouds and nuclei result in a sharp increase in the potential energy.

**Bond Length: Potential Energy versus Internuclear Distance**

0.074 nm

Potential Energy (kJ/mol): 0, –458

d

Internuclear Distance (nm)

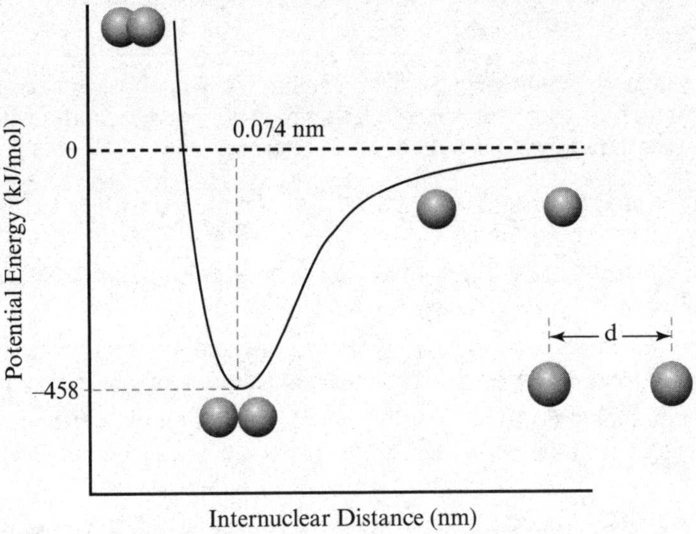

The energy required to separate the two atoms completely is 458 kJ/mol.

## Lewis Dot Structures

The number of valence electrons is an important factor in the bonding behavior of the elements because they are the electrons that are available to be transferred or shared between atoms.

**Lewis dot structures** for elements consist of the symbol of the element surrounded by dots representing the valence electrons.

Recall from Chapter 1 that the number of valence electrons for the main group elements is equal to the A group number from the older periodic table numbering system, as shown in the figure below. The only exception is helium, which has two valence electrons. Notice that the positions of the first four electrons are spaced as far apart as possible before the electrons are paired.

**Lewis Dot Diagrams of the Main Group Elements**

| I | II | III | IV | V | VI | VII | VIII |
|---|---|---|---|---|---|---|---|
| H· | | | | | | | He: |
| Li· | ·Be· | ·B· | ·C· | ·N· | :O· | :F· | :Ne: |
| Na· | ·Mg· | ·Al· | ·Si· | ·P· | :S· | :Cl· | :Ar: |
| K· | ·Ca· | ·Ga· | ·Ge· | ·As· | :Se· | :Br· | :Kr: |
| Rb· | ·Sr· | ·In· | ·Sn· | ·Sb· | :Te· | :I· | :Xe: |
| Cs· | ·Ba· | ·Tl· | ·Pb· | ·Bi· | :Po· | :At· | :Rn: |

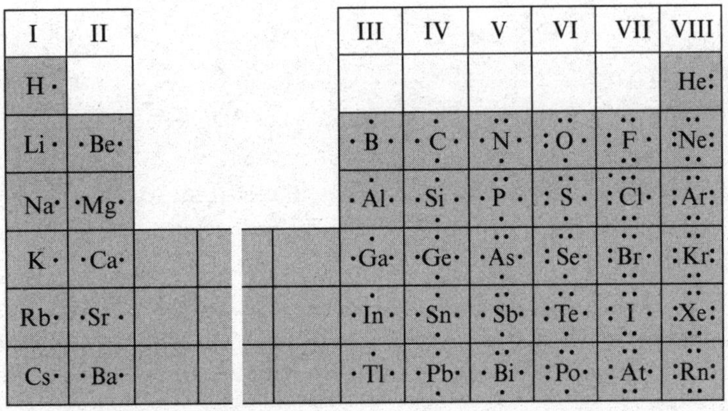

The Lewis dot structures for the *d* and *f* block elements are not evaluated on the AP Chemistry exam.

### Drawing Lewis Structures of Molecules and Polyatomic Ions

The Lewis structures of covalent molecules and polyatomic ions consist of the symbols for the atoms surrounded by dots and lines. The lines represent the bonds, or shared pairs of electrons, and the dots represent the unshared electrons.

A large number of molecules and polyatomic ions follow the **octet rule.** This is the tendency for main group elements to achieve noble gas electron configurations. Hydrogen follows the **duet rule.** It can have a maximum of two electrons.

When the octet rule is obeyed, each atom except for hydrogen has eight valence electrons in the most stable configuration. Drawing correct Lewis structures is a skill that requires practice, and it will be assessed on the AP Chemistry exam. There are six steps in writing a correct Lewis structure:

1. Arrange the atom labels in a reasonable, symmetrical skeleton, with the least electronegative elements in the central positions. Hydrogen is never central.
2. Calculate the total number of valence electrons, $N$, needed in the molecule. Each atom except for hydrogen needs eight valence electrons. Hydrogen needs two valence electrons.
3. Calculate the number of available electrons, $A$, by adding the number of valence electrons each atom contributes. Add an electron for each negative charge, and subtract an electron for each positive charge.
4. Calculate the number of shared electrons, $S$, using $S = N - A$. The number of bonds, indicated by lines, is half the number of shared electrons since there are two shared electrons per bond. Add the bonds to the structure.
5. Place the rest of the available electrons around the atoms in order to achieve an octet for each atom except hydrogen. These electrons are called **lone pairs.**
6. Assign **formal charges** to each atom in the molecule. The formal charges on the atoms in a molecule describe the distribution of charge in a molecule.
   - The sum of the formal charges in a *neutral molecule* is zero.
   - The sum of the formal charges in a *polyatomic ion* equals the overall charge of the ion.
   - The formal charge of each atom in the structure is determined by subtracting the number of bonds and unshared electrons from the main group number for that atom:

   Formal charge = number of valence electrons – number of bonds – number of unshared electrons

The six-step method can be used to draw the Lewis structure of a carbonate ion:

1. Carbonate ion, $CO_3^{2-}$, has one carbon and three oxygens. Carbon is the least electronegative, so it is drawn as the center of a symmetrical skeleton.

```
O   O
  C

  O
```

2. There are four atoms in the ion, and each atom needs eight electrons, so

$$N = 4 \times 8 = 32$$

3. The number of available electrons is the sum of the valence electrons contributed by each atom plus the two extra electrons indicated by the –2 charge. Carbon is in group IVA, so it contributes four valence electrons. Oxygen is in group VIA, so each of the three oxygens contributes six valence electrons.

$$A = (1 \text{ carbon} \times 4) + (3 \text{ oxygens} \times 6) + 2 = 24$$

4. The number of shared electrons is

$$S = N - A = 32 - 24 = 8$$

The number of bonds is four, since each bond is composed of two electrons ($8 \div 2 = 4$). For a carbonate ion, this corresponds to two single bonds and one double bond. It does not matter which bond is drawn as a double bond.

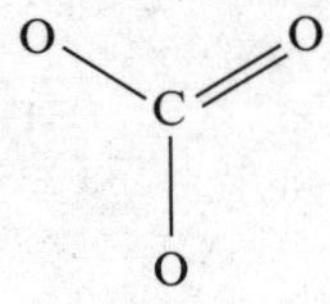

5. The difference between the number of available electrons, 24, and the number of shared electrons, 8, is the number of electrons that occur as lone pairs. There are 16 nonbonded electrons, or 8 lone pairs.

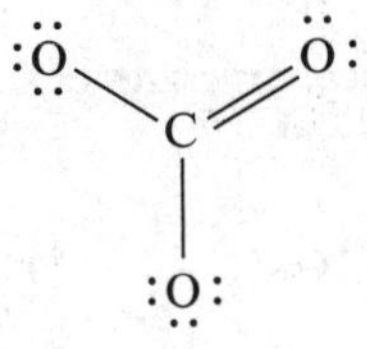

It is equally correct to draw the bonds as pairs of dots rather than lines. Both methods will earn credit on the AP Chemistry exam.

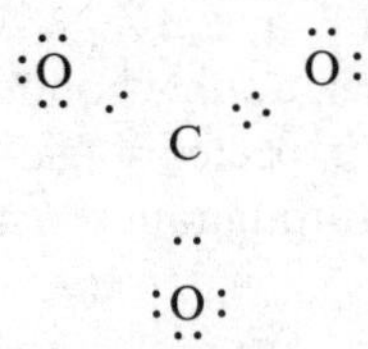

6. The sum of the formal charges in a carbonate ion is –2. To assign a formal charge (FC) to each atom, we take the number of valence electrons for each atom and subtract the number of bonds and the number of unshared electrons. We find that each single bonded oxygen atom has a formal charge of –1 and that the carbon and double bonded oxygen atoms have formal charges of zero.

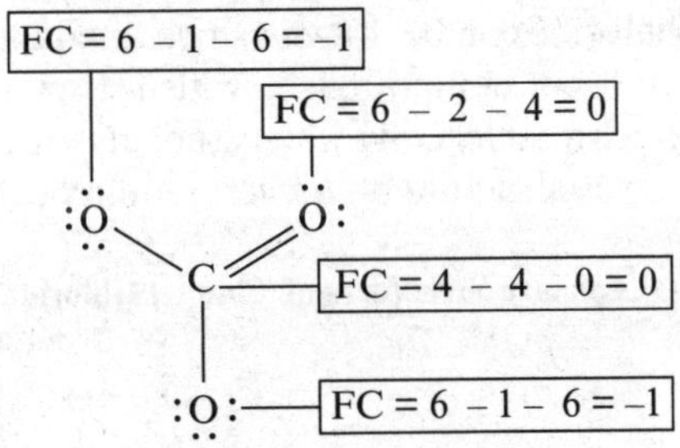

FC = 6 – 1 – 6 = –1

FC = 6 – 2 – 4 = 0

FC = 4 – 4 – 0 = 0

FC = 6 – 1 – 6 = –1

## Practice

Draw Lewis structures for the following species. Be sure to include formal charges.

**a.** ONF
**b.** NOF
**c.** HCN
**d.** HNC

Notice that in ONF and HNC, nitrogen is the central atom. In NOF, oxygen is central, and in HCN, carbon is central.

The correct Lewis structures are shown below. The position of the central atom has an effect on the formal charges of the atoms, as can be seen when formal charges are assigned.

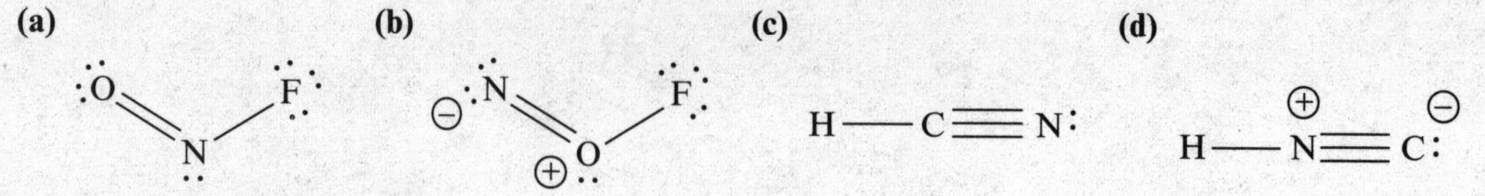

(a) (b) (c) (d)

## Resonance Structures

Some molecules, such as the carbonate ion shown above, can be drawn as several different valid structures. These are called resonance structures.

**Resonance Structures of Carbonate**

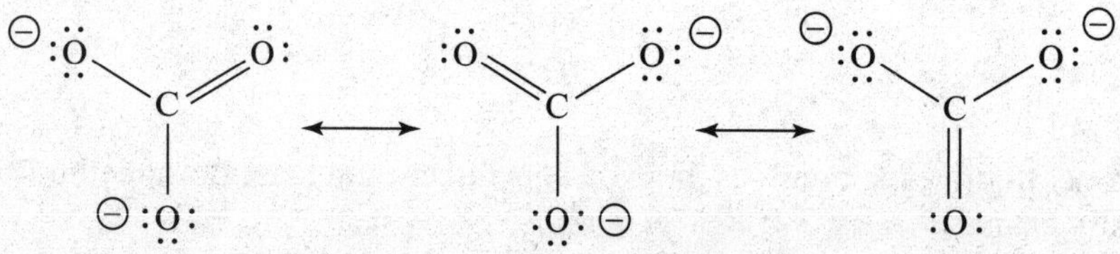

In general, as will be discussed later in this chapter, double bonds are shorter than single bonds. The bond lengths for a carbonate ion are intermediate between the expected single- and double-bond lengths. In addition, the four pairs of shared electrons are distributed equally throughout the molecule. This is called **delocalization,** and the structure is said to exhibit **resonance.**

In order to fully describe the bonding in such **hybrid structures,** all resonance structures must be drawn with the resonance relationship indicated by double-headed arrows. It is important to recognize that the double bonds are not exchanging positions. In the carbonate ion, all three bonds are of equal length. The bonds are shorter than carbon-oxygen single bonds, but they are longer than carbon-oxygen double bonds.

In some cases, not all resonance structures are equal, and some resonance forms contribute more to the overall hybrid structure than others. Thionyl chloride can be drawn as two unequal resonance forms. ***Note:*** When students are asked to draw Lewis structures of compounds in which strict adherence to the octet rule by all atoms is possible, the structure on the right is acceptable. Only when students are asked to compare the relative contributions of resonance structures is the structure on the left required. It violates the octet rule.

**Resonance Structures of Thionyl Chloride**

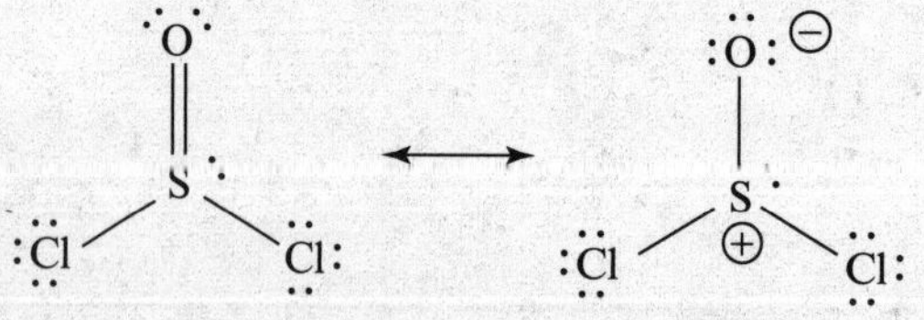

The structure that makes the larger contribution to the hybrid is the one in which:

- each second period element has a complete octet
- the formal charge on each atom is as close to zero as possible (referred to as "least separation of charge")
- negative charges occur on more electronegative elements

In the case of thionyl chloride, the structure on the left makes the greater contribution to the hybrid structure because it has the least separation of charge.

Many common organic compounds contain a six-membered ring. Benzene, $C_6H_6$, can be represented as two resonance structures that make equal contributions to the hybrid. In the bottom representation, it is understood that each vertex of the hexagon is a carbon atom bound to one hydrogen and two carbons.

**Resonance Structures of Benzene**

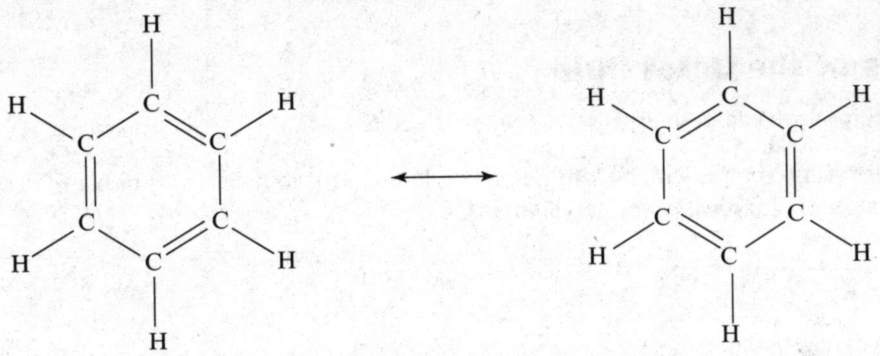

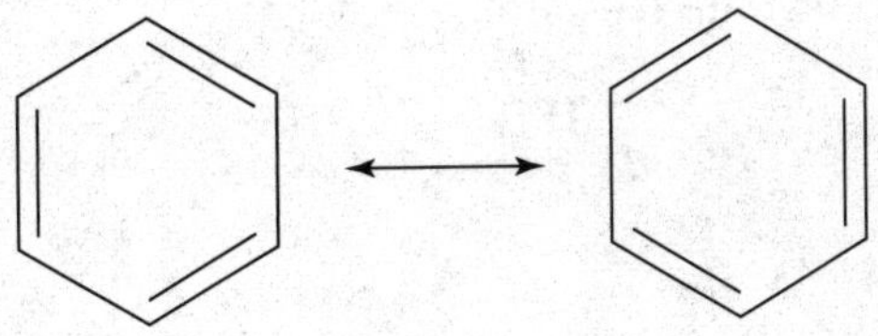

The structure of benzene is an intermediate between the two resonance forms shown above.

## Practice

For each of the pairs of resonance structures below, select the structure in each box that makes the greater contribution to the hybrid.

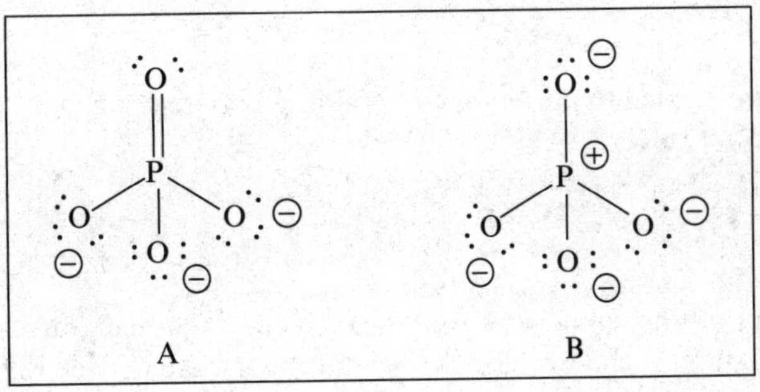

A B

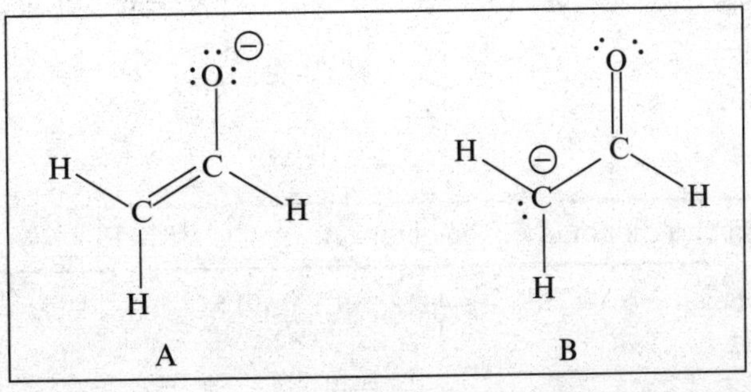

A B

In the first pair, structure A makes the greater contribution to the $PO_4^{-3}$ resonance hybrid. There is less separation of charge in this structure than there is in structure B, and two of the atoms, P and O, have formal charges of zero.

In the second pair, structure A makes the greater contribution to the $H_3C_2O^-$ resonance hybrid. The negative formal charge occurs on oxygen, the more electronegative element.

## Limitations of the Octet Rule

In some cases, atoms form compounds without achieving octets.

- Beryllium forms two bonds with no lone pairs. Beryllium needs four electrons rather than eight, so the number of needed electrons, *N*, for beryllium is four.

  :F̤̈—Be—F̤̈:

- Group IIIA elements, including boron and aluminum, tend to form three covalent bonds. These elements need six electrons rather than eight, so the number of needed electrons, *N*, is six. The compounds formed have no lone pairs on the central group IIIA atom.

  H, H, B, H

- Some *p* block elements in period 3 and higher form compounds that have more electrons available than the number needed to satisfy the octet rule. If the number of shared electrons, *S*, is less than the number needed to bond all the atoms, *S* is changed to the number of electrons needed to bond all the atoms. After all other octets have been satisfied, any remaining available electrons are placed around the central atom. In the example below, $XeF_2$, there are three atoms, so $N = 8 \times 3 = 24$. The number of available electrons, *A*, is 7 for each fluorine atom and 8 for the xenon atom ($A = [(7 \times 2) + (8 \times 1)] = 22$). The number of shared electrons, *S*, is determined by subtracting *A* from *N*.

  $$S = N - A = 24 - 22 = 2$$

  Because four electrons are needed to participate in bonding between xenon and the fluorine atoms, this compound violates the octet rule.

  :F̤̈—Xe—F̤̈:

Lewis structures are limited in that they only work well for molecules that have an even number of valence electrons. Other models of bonding, such as molecular orbital theory, are needed to describe the bonding in molecules with odd numbers of valence electrons, but the details of molecular orbital theory are excluded from the AP Chemistry curriculum.

### Practice

Which of the following molecules do not obey the octet rule: $AlCl_3$, $PCl_3$, $PCl_5$, $SF_2$, $SO_4^{2-}$?

The compounds that do not obey the octet rule are $AlCl_3$, $PCl_5$, and $SO_4^{2-}$.

Aluminum is in group IIIA. It only needs six valence electrons.

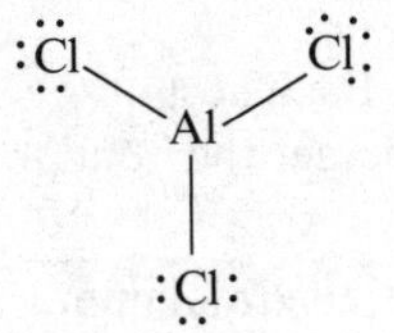

Phosphorus pentachloride needs to share ten electrons in order to bond all the chlorine atoms to the central phosphorus. The calculated number of shared electrons (8) is less than the minimum number needed to bond all the chlorine atoms to the phosphorus atom. In this case, five bonds are drawn, representing the minimum number of shared electrons. The available electrons ($A$) are then added to provide each chlorine atom with an octet.

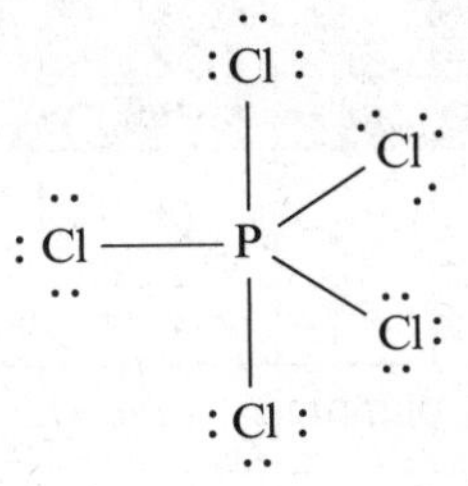

$N = 6 \times 8 = 48$

$A = (5 \times 7) + (1 \times 5) = 40$

$S = 48 - 40 = 8$

Some species, such as the sulfate ion, can be drawn with strict adherence to the octet rule by all atoms (and this is permissible on the AP Chemistry exam if a question merely asks for a Lewis structure), but the resulting structure has a great deal of separation of charge. All four oxygen atoms in this structure have negative charges, and the sulfur atom has a +2 charge. The more important contributors to the hybrid structure are those that have negative charges on only two of the oxygen atoms.

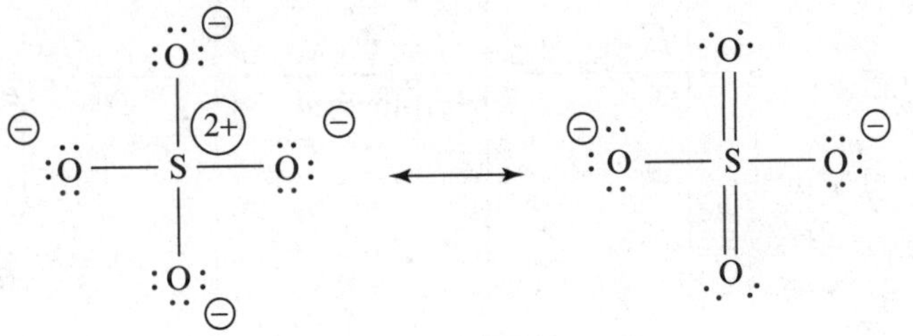

The other compounds, $PCl_3$ and $SF_2$, obey the octet rule.

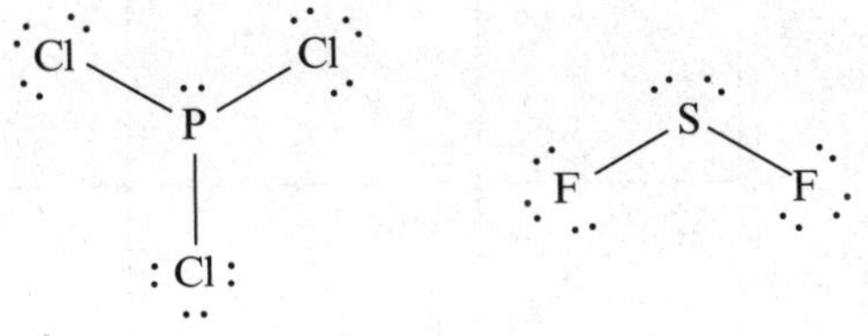

## The VSEPR Model

Once the Lewis structure of a molecule has been established, the molecular shape and polarity can be determined using the **valence shell electron pair repulsion (VSEPR) model.** The VSEPR model is a system for prediction of molecular geometry, and it requires memorization.

- Bonds and electron pairs are treated as regions of electron density around the nucleus of an atom.
- These regions of electron density are oriented as far apart as possible in order to minimize repulsions.
- Lone pairs of electrons on a central atom take up more space than bonding pairs of electrons.

**Steps in Predicting the Molecular Geometry of a Molecule**

1. Draw the Lewis structure.
2. Count the regions of high electron density. Each atom, X, bound to the central atom, A, is counted as one region of electron density, regardless of whether the bond to that atom is single, double, or triple. Each lone pair, E, counts as one region.
3. Determine the **electronic geometry** based on the total number of regions of electron density, X + E.
4. Determine the **molecular geometry** by considering only the bonding pairs of electrons, X.

The following electronic geometries and molecular geometries must be memorized. (Hybridization about the central atom will be discussed below in the section entitled "Orbital Hybridization.")

| Electronic Geometry | Hybridization About the Central Atom (*sp*, $sp^2$, and $sp^3$ are tested) | Formula | Molecular Geometry | Structure | Examples |
|---|---|---|---|---|---|
| Linear | *sp* | $AX_2$ | Linear | 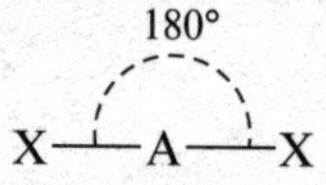 180° | $BeCl_2$, $CO_2$, HCN |
| Trigonal planar | $sp^2$ | $AX_3$ | Trigonal planar | 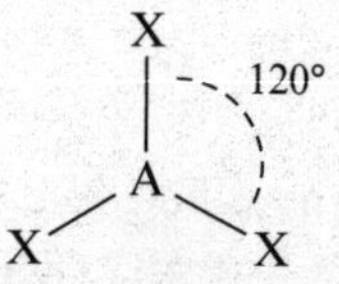 120° | $BF_3$, $SO_3$, $CO_3^{2-}$ |
| | | $AX_2E$ | Bent | 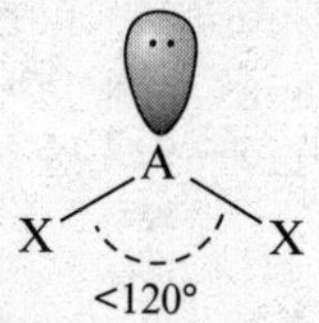 <120° | $SO_2$, $O_3$, $NO_2^-$ |
| Tetrahedral | $sp^3$ | $AX_4$ | Tetrahedral | 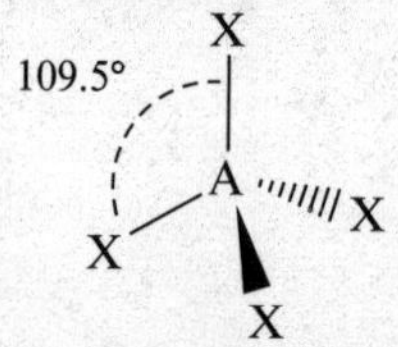 109.5° | $CH_4$, $NH_4^+$, $SO_4^{2-}$ |
| | | $AX_3E$ | Trigonal pyramidal | 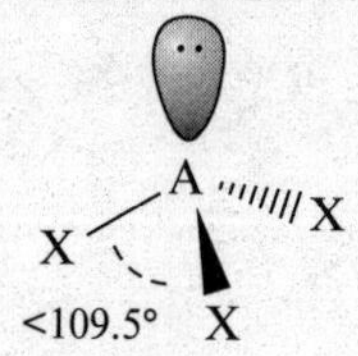 <109.5° | $NH_3$, $PCl_3$, $SO_3^{2-}$ |
| | | $AX_2E_2$ | Bent | 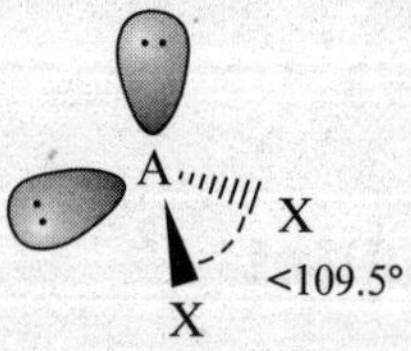 <109.5° | $H_2O$, $OF_2$, $NH_2^-$ |
| Trigonal bipyramidal | $sp^3d$ (Hybridization involving *d* orbitals is not tested on the AP Chemistry exam. It is shown for information only.) | $AX_5$ | Trigonal bipyramidal | 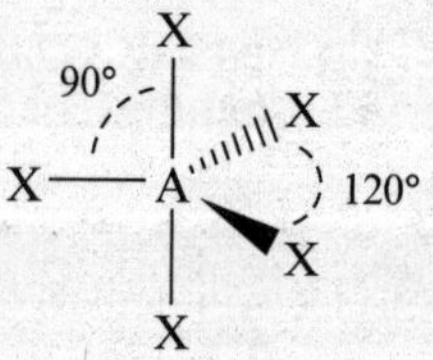 90°, 120° | $PCl_5$, $PF_5$ |

| Electronic Geometry | Hybridization About the Central Atom ($sp$, $sp^2$, and $sp^3$ are tested) | Formula | Molecular Geometry | Structure | Examples |
|---|---|---|---|---|---|
| | | $AX_4E$ | Seesaw-shaped (diphenoidal) | | $SF_4$, $XeF_4$, $IF_4^+$ |
| | | $AX_3E_2$ | T-shaped | | $BrF_3$, $ICl_3$ |
| | | $AX_2E_3$ | Linear | | $I_3^-$, $XeF_2$ |
| Octahedral | $sp^3d^2$ (Hybridization involving *d* orbitals is not tested on the AP Chemistry exam. It is shown for information only.) | $AX_6$ | Octahedral | 90° | $SF_6$, $PCl_6^-$ |
| | | $AX_5E$ | Square pyramidal | | $IF_5$, $XeOF_4$ |
| | | $AX_4E_2$ | Square planar | | $XeF_4$, $ICl_4^-$ |

The molecular geometry of more complicated organic structures can be elucidated by considering each central atom in turn. For example, the molecular geometry of acetic acid, $CH_3CO_2H$, can be determined by considering the two carbon atoms individually. The geometry with respect to the $-CH_3$ carbon is tetrahedral ($AX_4$), while the geometry of the $-CO_2H$ carbon is trigonal planar ($AX_3$).

**A Three-Dimensional Model of an Acetic Acid Molecule**

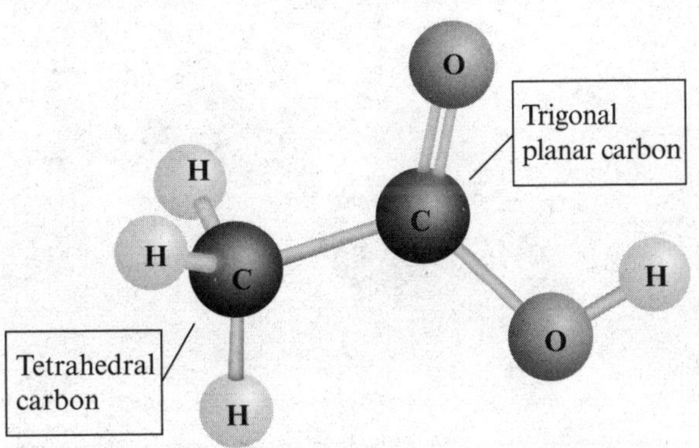

## Practice

Predict the electronic and molecular geometries around each of the central atoms in $NO_2^-$, $N_2O$, $XeO_3$, and $XeO_4^{2-}$. Notice that each species can be correctly drawn as more than one resonance structure.

Draw each Lewis structure and count the numbers of atoms and lone pairs surrounding the central atoms.

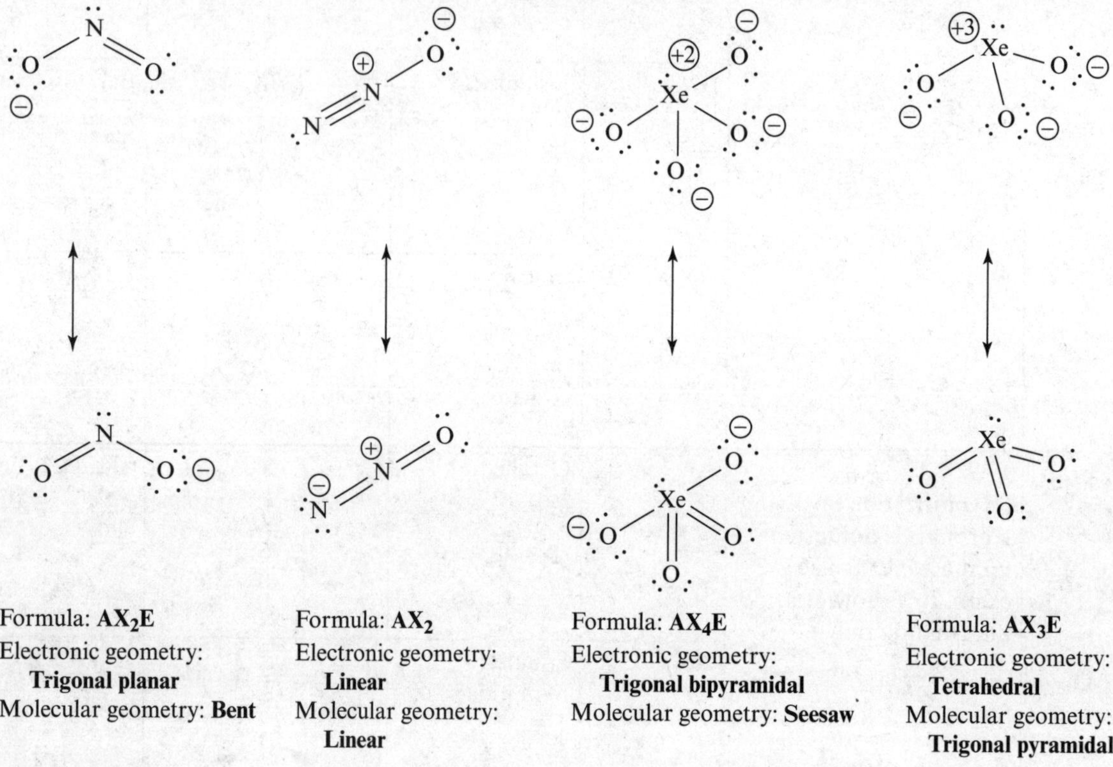


# Dipole Moment

Once the molecular geometry of a molecule or polyatomic ion has been established, the polarity of the molecule can be predicted. The **dipole moment** is a measure of the inequality of charge distribution, or polarity, of a molecule. The dipole moment is indicated by a vector that points in the direction of the more negative region of the molecule.

In the $CO_2$ molecule shown below, the oxygen atoms are more electronegative than the carbon atom, so the vectors point to the oxygen atoms.

The overall dipole moment of a molecule is the vector sum of the individual bond dipoles. In symmetric molecules, such as $CO_2$, the bond polarity vectors cancel, the dipole moment of the molecule is zero, and the molecule is nonpolar.

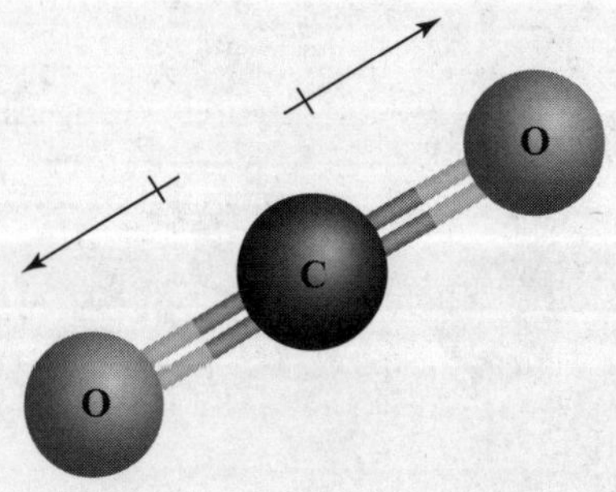

When the arrangement of bonds and lone pairs around the central atom is asymmetric, as is the case for ammonia, the overall dipole moment is nonzero and the molecule is polar.

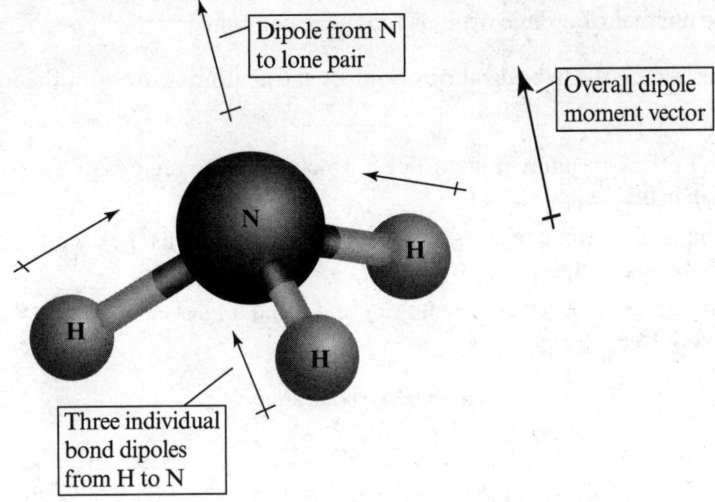


## Practice

The structures below represent *ortho*-dichlorobenzene and *para*-dichlorobenzene. Which of these molecules below is polar? The Pauling electronegativities of chlorine and carbon are 3.1 and 2.5, respectively.

:Cl: :Cl:

*ortho*-dichlorobenzene *para*-dichlorobenzene

The C–C and C–H bonds are nonpolar, so the only bonds that need to be considered are the C–Cl bonds. Chlorine is substantially more electronegative than carbon. The bond dipole vectors point from the carbon atoms toward the chlorine atoms. The vectors cancel in *para*-dichlorobenzene, so it is nonpolar. Since vectors do not cancel in *ortho*-dichlorobenzene, it is a polar molecule.

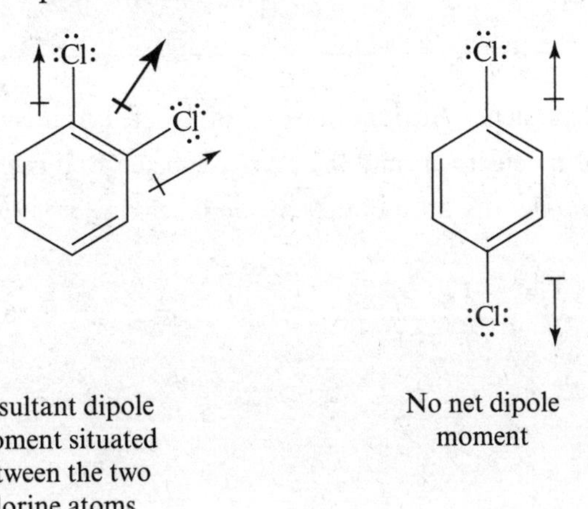


Resultant dipole moment situated between the two chlorine atoms

No net dipole moment

## Orbital Hybridization

When bonding occurs, atomic orbitals are thought to combine to form molecular **hybrid orbitals.** The number of hybrid orbitals equals the number of atomic orbitals that were combined.

A simple method for determining the hybridization about a central atom is to count the number of regions of electron density.

- Central atoms, such as the beryllium atom in $BeF_2$, which have two regions of electron density and bond angles of 180°, are ***sp* hybridized.**
- Central atoms surrounded by three regions of electron density, such as the carbon atom in carbonate, are ***sp*$^2$ hybridized** and have bond angles close to 120°.
- Central atoms with tetrahedral electronic geometry and bond angles close to 109.5°, such as the nitrogen in ammonia, are ***sp*$^3$ hybridized.**

**Orbital Hybridization**

:F—Be—F:

*sp*

*sp*$^2$

*sp*$^3$

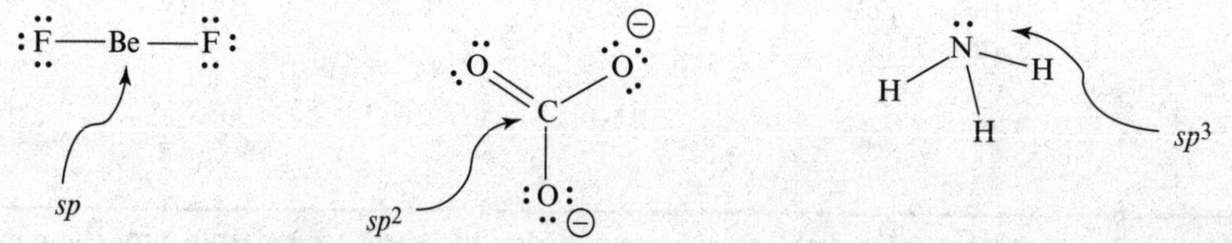

## Practice

State the hybridization (*sp*, *sp*$^2$, or *sp*$^3$) for each of the indicated carbon atoms.

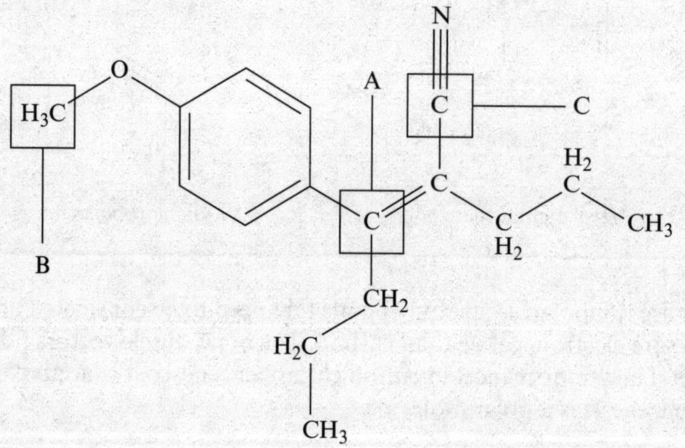

There are three regions of electron density around the carbon labeled **A;** it is *sp*$^2$ hybridized.

There are four regions of electron density around the carbon labeled **B;** it is *sp*$^3$ hybridized.

There are two regions of electron density around the carbon labeled **C;** it is *sp* hybridized.

## Sigma and Pi Bonds

**Sigma ($\sigma$) bonds** are cylindrically symmetric bonds formed from head-on overlap of two atomic orbitals. A $\sigma_{s\text{-}s}$ bond is formed between two *s* orbitals, as is the case in $H_2$. In HF, a $\sigma_{s\text{-}p}$ bond is formed between the *s* orbital of hydrogen and a *p* orbital of fluorine. A $\sigma_{p\text{-}p}$ orbital is formed between two *p* orbitals, as in $F_2$. It is acceptable to refer to any of these simply as a "$\sigma$ bond."

**Sigma Bonding**

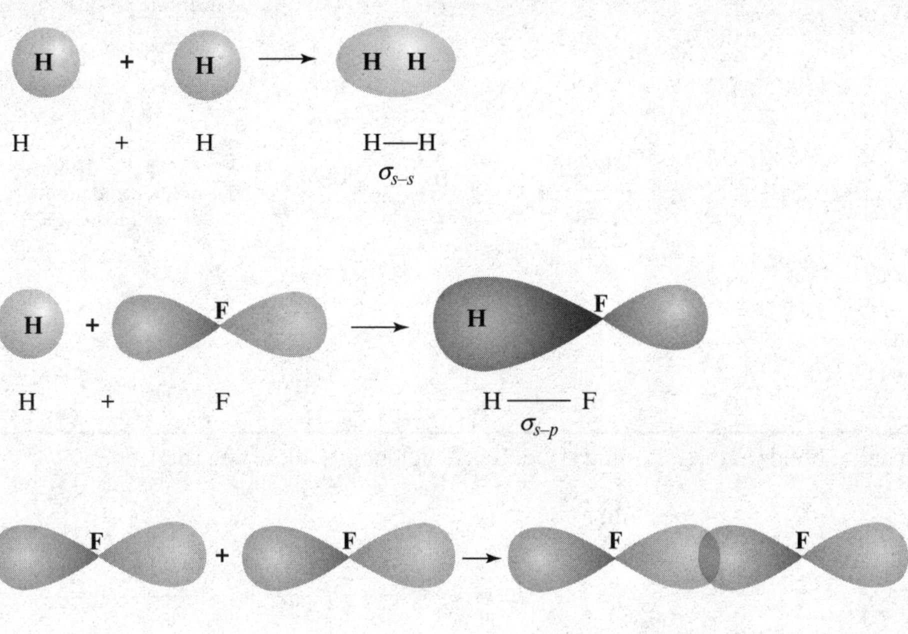


**Pi ($\pi$) bonds** result from side-by-side overlap of *p* orbitals. Each $\pi$ bond contains a pair of electrons and is perpendicular to a line drawn between the nuclei of the two bonded atoms (the internuclear axis). Pi bonds can only form between atoms that are already participating in $\sigma$ bonding. They are weaker than $\sigma$ bonds because there is not as much overlap between the $\pi$-bonded *p* orbitals.

A $\pi$ bond occupies a region both above and below the internuclear axis. This representation should not be confused with two separate $\pi$ bonds.

**Pi Bonding**

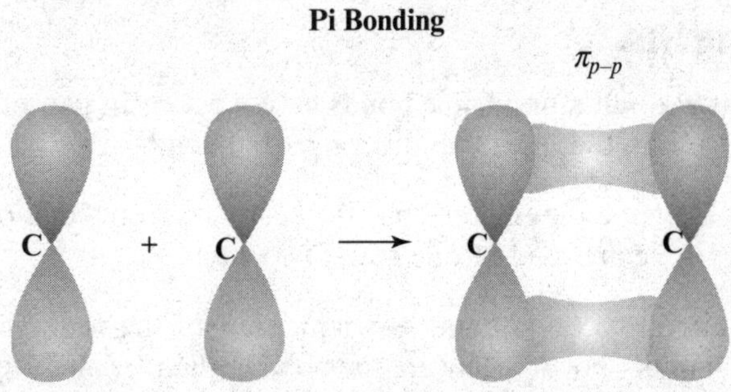


In atoms that participate in double and triple bonding, the first bond formed between the atoms is a $\sigma$ bond, and the second and third bonds are $\pi$ bonds.

Atoms joined by only $\sigma$ bonds can rotate around those bonds, but atoms joined by $\pi$ bonds cannot rotate without breaking the bonds. For this reason, a molecule like 1-chloropropane, which is connected by only $\sigma$ bonds, can be drawn several different ways without changing the structure. A molecule like 1-chloropropene, however, which has a $\sigma$ and a $\pi$ bond between carbons 1 and 2, exists in two discrete forms called **geometric isomers.**

**Geometric Isomerism**

1-chloropropane

C–C bond rotation

The C–C sigma bond allows free rotation. No geometric isomers are possible.

1-chloropropane

No C=C bond rotation

The C–C pi bond hinders free rotation and results in two geometric isomers.

## Practice

Identify the $\sigma$ and $\pi$ bonds in the following structure. Which bonds allow free rotation?

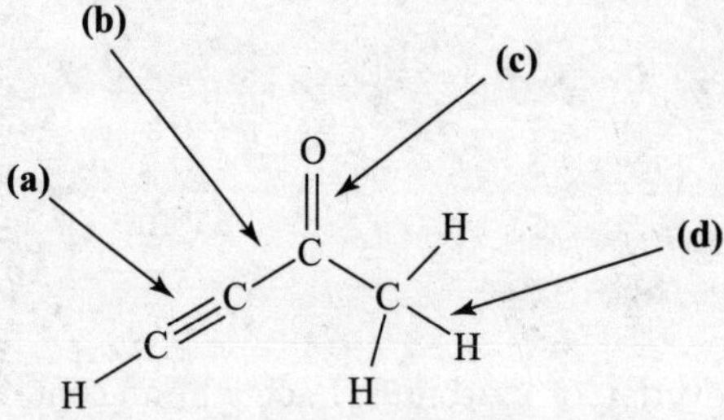


Bond **(a)** is a triple bond composed of one $\sigma$ and two $\pi$ bonds. Bonds **(b)** and **(d)** are single bonds, each composed of one $\sigma$ bond. Bond **(c)** is a double bond composed of one $\sigma$ and one $\pi$ bond.

Only **(b)** and **(d)**, the single bonds, allow free rotation.

## Multiple Bond Strengths

In general, when comparing single, double, and triple bonds between a given pair of atoms, the larger the number of electrons shared between the atoms, the shorter and stronger the bond.

In general, single bonds are the longest and weakest since they consist of only one $\sigma$ bond. Triple bonds are the strongest and shortest since they are composed of one $\sigma$ and two $\pi$ bonds.

**Bond enthalpy** is the energy required to break a bond. It is a measure of the strength of a bond. The higher the bond enthalpy, the stronger the bond. Shown below are the bond enthalpies and average bond lengths of carbon-carbon single, double, and triple bonds for comparison.

| Bond | Bond Enthalpy (kJ/mol) | Approximate Bond Length (pm) |
|---|---|---|
| C–C | 348 | 154 |
| C=C | 614 | 147 |
| C≡C | 839 | 137 |

## Ionic Bonding

**Ionic bonds** are electrostatic attractions between positive and negative ions. Ionic bonds are formed when electrons are transferred between two or more atoms. The resulting charged species are attracted to one another by **coulombic attractions,** the forces that attract positively charged particles to negatively charged particles.

The electronegativity difference between atoms participating in an ionic bond is usually greater than 1.7.

The energy released when a cation and an anion in the gas phase come together to form an ionic solid is called the **lattice energy.** It is the measure of the strength of an ionic bond. It can be represented using the form of Coulomb's law that was introduced in Chapter 1, "Atomic Structure and Properties":

$$E \propto \frac{q_1 q_2}{r}$$

The lattice energy, $E$, is proportional to the product of the ionic charges, $q_1$ and $q_2$. It is inversely proportional to the distance, $r$, between the centers of the ions. When the ions have charges that are opposite in sign, the lattice energy is negative, indicating an attractive force and a release of energy when the ionic solid forms. The more negative the lattice energy, the stronger the ionic bond.

Ionic solids form in such a way as to maximize attractive forces between oppositely charged ions and minimize repulsive forces. This leads to orderly crystal structures in which each cation is surrounded by anions and each anion is surrounded by cations. The regular positions occupied by the ions are referred to as **lattice points.**

**The Structure of an Ionic Crystal**

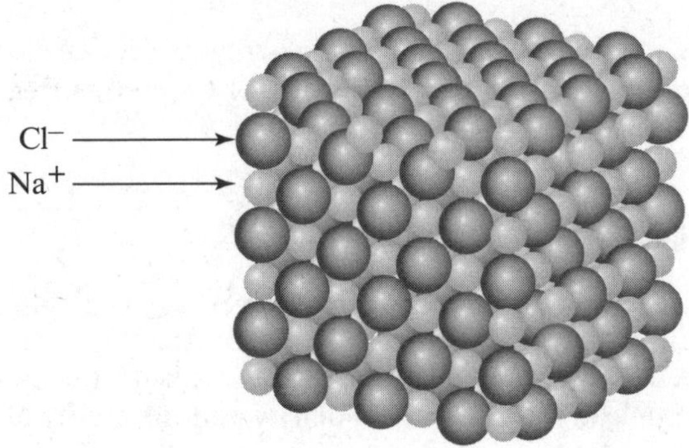


### Practice

Predict the order of increasingly negative lattice energy for $CaCl_2$, $Fe_2O_3$, KCl, and FeO.

There are two factors to consider when answering this question. The first is the charge of each ion, and the second is the radius. Since the radii of the cations (ranging from 140 pm for $K^+$ to 60 pm for $Fe^{3+}$) and the radii of the anions (140 pm for $O^{2-}$ and 180 pm for $Cl^-$) have similar orders of magnitude, the product of the charges, $q_1q_2$, will be the more important determinant in this example.

| Compound | Cation | Anion | $q_1q_2$ | Lattice Energy ($kJ \cdot mol^{-1}$) |
|---|---|---|---|---|
| $CaCl_2$ | $Ca^{2+}$ | $Cl^-$ | $(2 \times -1) = \mathbf{-2}$ | –2268 |
| $Fe_2O_3$ | $Fe^{3+}$ | $O^{2-}$ | $(3 \times -2) = \mathbf{-6}$ | –14,309 |
| KCl | $K^+$ | $Cl^-$ | $(1 \times -1) = \mathbf{-1}$ | –701 |
| FeO | $Fe^{2+}$ | $O^{2-}$ | $(2 \times -2) = \mathbf{-4}$ | –3795 |

Since the value of $q_1q_2$ becomes more negative progressing from KCl to $Fe_2O_3$, the order of increasing lattice energy magnitude is KCl < $CaCl_2$ < FeO < $Fe_2O_3$. Actual reported values for the lattice energies are shown for comparison.

## Metallic Bonding

Metallic bonding is characterized by a regular three-dimensional array of metal cations surrounded by freely moving valence electrons. This is sometimes referred to as the "electron sea" model of bonding. The outermost electrons are delocalized, which accounts for many of the observable physical properties of solid metals, such as conductivity and malleability. The properties will be discussed in greater detail in Chapter 3, "Intermolecular Forces and Properties."

**The Electron Sea Model of Metallic Bonding**

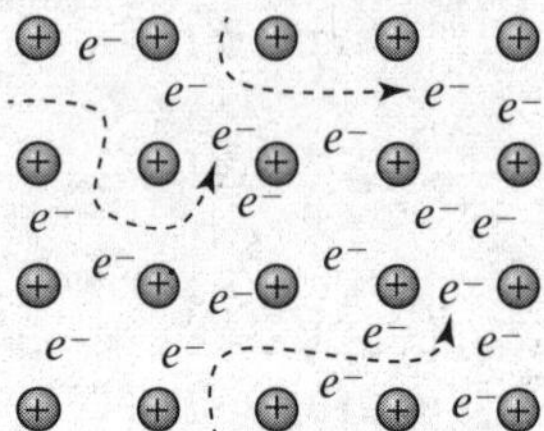

# Spectroscopy and the Electromagnetic Spectrum

**Spectroscopy** is the measurement of the interactions of light with matter. The interaction of molecules with photons depends on the atoms and bonds present in the molecules and the energy of the photons. In an AP Chemistry course, a basic understanding of the interactions of microwaves, infrared waves, ultraviolet/visible waves, and X-rays is expected.

In Chapter 1, we saw that X-ray radiation is high enough in energy to eject an electron completely from an atom. UV/visible, infrared, and microwaves are lower in energy than X-rays, so they generally are not energetic enough to ionize atoms or remove electrons from molecules.

## Microwave Radiation

Microwave radiation is absorbed by a metallic substance and the energy heats the substance.

For polar covalent molecules, microwave rotational spectroscopy measures the interaction of microwave radiation with the dipole moment of a molecule in the gas phase. Polarity and the ability of molecules to interact through dipole-dipole intermolecular forces will be reviewed in Chapter 3, "Intermolecular Forces and Properties." Interaction with the electric field of the radiation causes rotational transitions in a polar molecule.

**Interaction of Microwaves with Water**

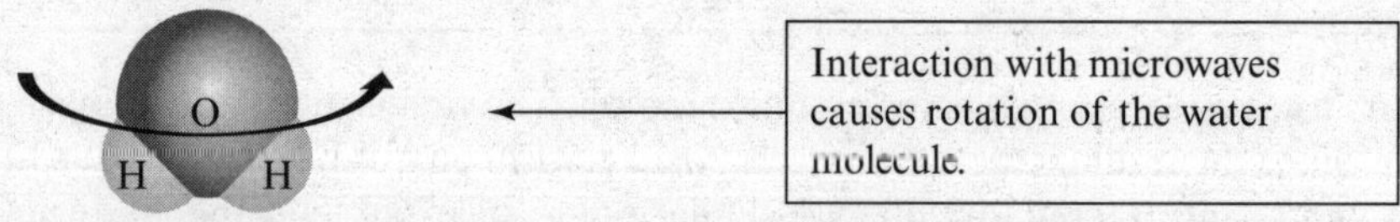


This can be visualized and remembered by thinking of a microwave oven. The water in foods is polar, so it heats up in a microwave by transitioning to higher rotational states. A microwave is a relatively inefficient way of heating a pure substance like vegetable oil because pure oil is nonpolar. On the other hand, you probably know that you should never put aluminum foil in the microwave. This is because metals can absorb and erratically reflect the microwave energy, heating up dangerously and potentially damaging the microwave.

## Infrared Radiation

Covalent bonds in molecules vibrate. Radiation in the infrared portion of the electromagnetic spectrum interacts with molecules to induce bond vibrational changes.

Two covalently bonded atoms can be visualized as suspended marbles connected by a spring. The spring stretches and recoils with a certain frequency. For a covalent bond, this frequency lies in the infrared region of the electromagnetic spectrum. In general, bonds between large atoms vibrate with lower frequencies than bonds between small atoms.

**Marbles Connected by a Spring**

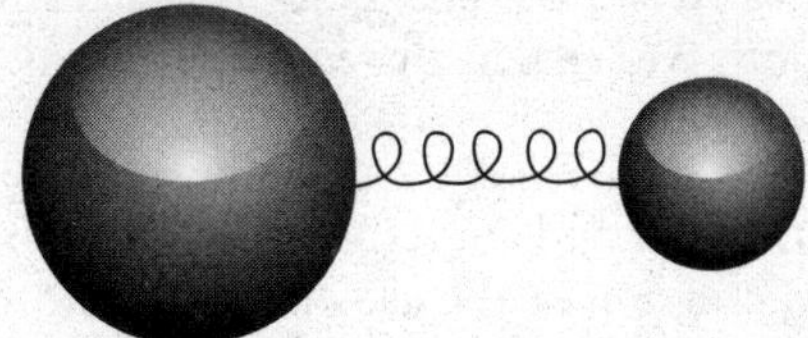

When photons matching the vibrational frequencies of bonds are absorbed, the absorbed energy causes molecules to transition from vibrational ground states to vibrational excited states. The vibrational states are quantized, which means that the vibrations of bonds can have only certain allowed frequencies. This is the basis for **infrared spectroscopy,** a technique that provides chemists with one method for determining the types of functional groups present in a molecule.

As discussed previously, a **functional group** is a group of atoms that determines the chemical properties of a compound. Some examples of the types of bonds that are easily identified by infrared spectroscopy are C–H, C–C, C=C, C–N, C–O, C=O, and O–H bonds. The different types of bonds have different characteristic vibrational frequencies in the infrared region.

An **infrared spectrometer** is a device that measures the frequencies of infrared radiation that are absorbed by a sample. An **infrared spectrum,** which is characteristic of the molecules in the sample, is produced.

The infrared region encompasses photon frequencies of $4.3 \times 10^{14} - 3.0 \times 10^{11}\ s^{-1}$. These frequencies are very large, so they are generally expressed as **wavenumbers,** $cm^{-1}$. The wavenumber is just the frequency divided by the speed of light in $cm \cdot s^{-1}$. Students will *not* be expected to calculate wavenumber, but they should know that wavenumber is proportional to frequency.

The percent transmittance, which shows the portion of infrared radiation that is not absorbed by a sample, is plotted on the vertical axis of an infrared spectrum.

The infrared spectrum below is the spectrum of methanol, $CH_3OH$. From a table of characteristic IR frequencies, some of the peaks can be identified. Students are only required to know that different kinds of bonds absorb different characteristic IR frequencies and that these absorptions can help in the identification of the functional groups present in a compound.

**The Infrared Spectrum of Methanol**

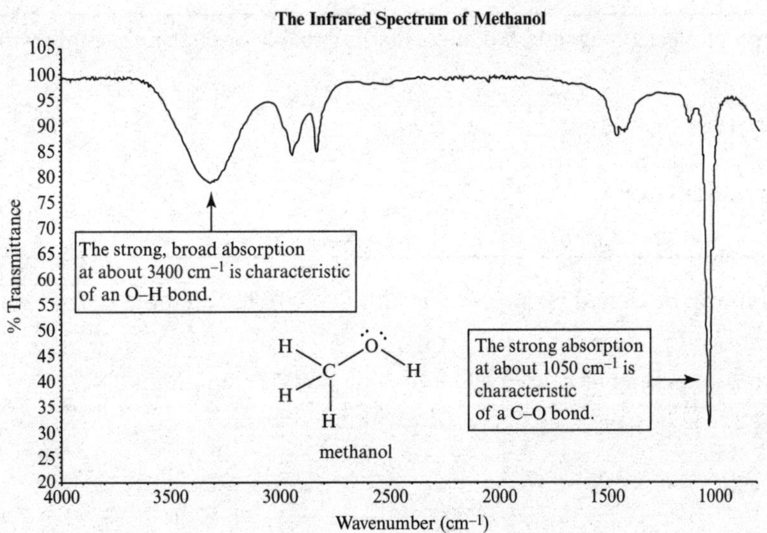

## UV/Visible Radiation

Ultraviolet or visible radiation can be absorbed by a molecule when the molecule has an electronic energy gap that matches the energy of the ultraviolet or visible photon.

As shown in the figure below, the photon absorbed causes a ground state electron to be promoted to a higher energy level, or excited state. When the electron loses energy to relax back to the ground state, a photon is emitted that possesses the energy of the transition. We see that there is a large peak at 245 nm, the wavelength of the photons that are absorbed.

For a review of the Bohr model of absorbance and emittance of photons by atoms, refer to Chapter 1, "Atomic Structure and Properties." Additional discussion of the use of spectrophotometry for the measurement of solution concentration will follow in Chapter 3, "Intermolecular Forces and Properties."

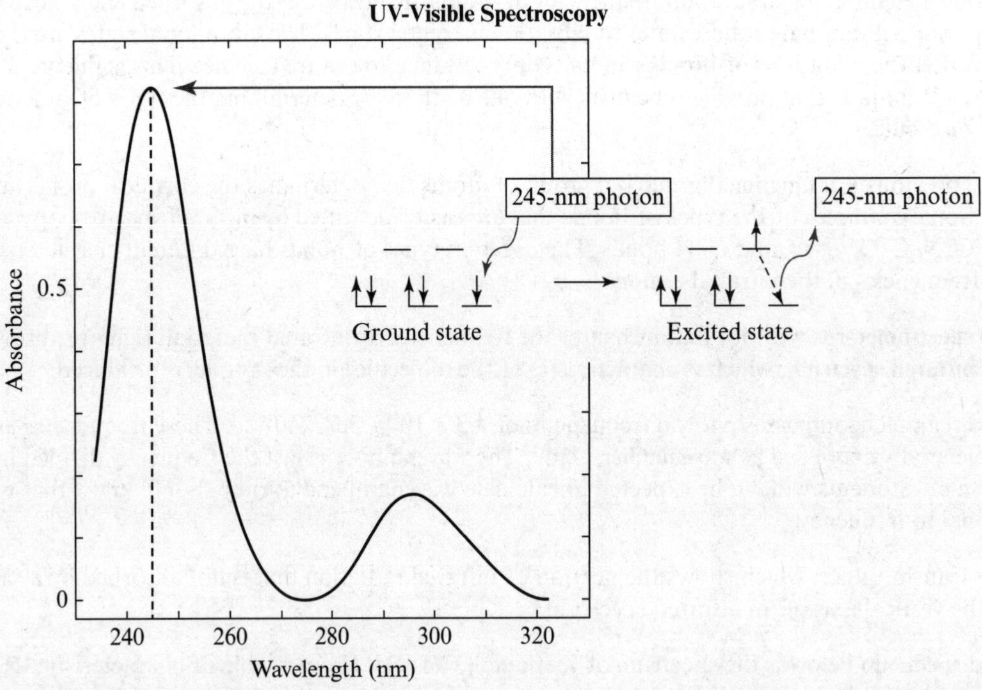


## Practice

State the classification of electromagnetic radiation that is capable of causing the following processes in atoms or molecules:

**a.** Electronic transitions
**b.** Ionization
**c.** Bond vibrational transitions
**d.** Rotation

**(a)** Electronic transitions can be caused by interaction with UV/visible radiation.

**(b)** Ionization can be caused by interaction with X-rays.

**(c)** Bond vibrational transitions can be caused by interaction with IR radiation.

**(d)** Rotational transitions can be caused by interaction with microwaves.

# Review Questions

## Multiple Choice

*Refer to the following table of electronegativities in order to answer questions 1 and 2.*

| Element | Pauling Electronegativity |
|---|---|
| C | 2.5 |
| Si | 1.9 |
| S | 2.6 |
| O | 3.4 |

**1.** In which type of bond are the electrons shared most equally?

**A.** C–Si
**B.** C–S
**C.** C–O
**D.** Si–O

**2.** Select the choice that correctly ranks the bonds from least polar to most polar.

**A.** C–S < C–O < C–Si < Si–O
**B.** Si–O < C–Si < C–O < C–S
**C.** Si–O < C–O < C–Si < C–S
**D.** C–S < C–Si < C–O < Si–O

**3.** Which choice best reflects the internuclear distance between two covalently bonded carbon atoms?

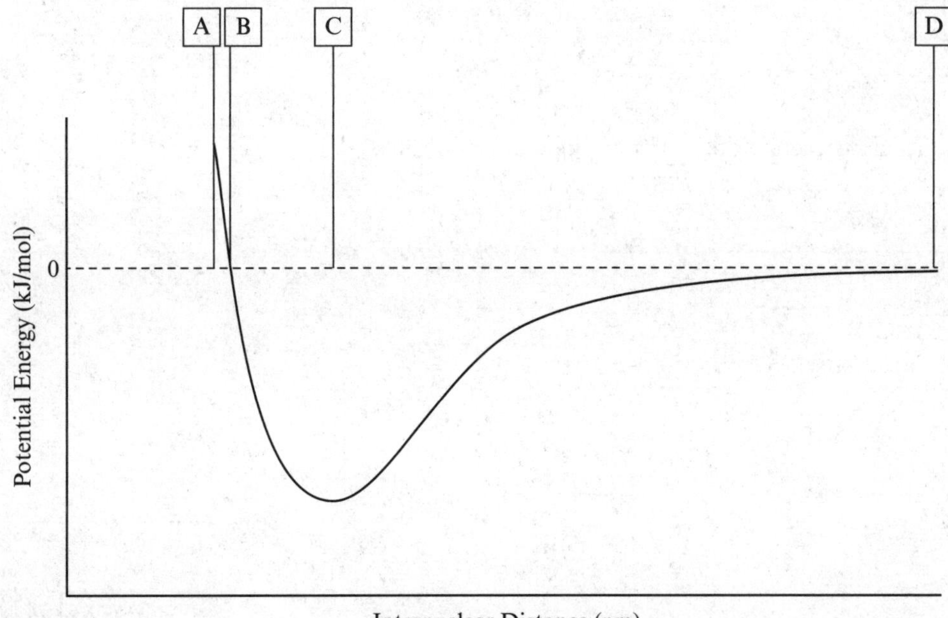


**A.** Length A
**B.** Length B
**C.** Length C
**D.** Length D

**4.** Bonds in molecules are promoted to higher vibrational states due to the absorption of photons in which region of the electromagnetic spectrum?

**A.** X-ray
**B.** Ultraviolet
**C.** Visible
**D.** Infrared

**5.** Which of the following is the most important contributing structure to the thiocyanate resonance hybrid? The Pauling electronegativities of carbon, nitrogen, and sulfur are about 2.5, 3.0, and 2.6, respectively.

**A.** $S{=}C{=}\overset{\ominus}{N}$

**B.** $\overset{\oplus}{S}{\equiv}C{-}\overset{2-}{N}$

**C.** $\overset{\ominus}{S}{-}C{\equiv}N$

**D.** $S{=}\overset{\oplus}{C}{-}\overset{2-}{N}$

**6.** Which of the following molecules is nonpolar?

**A.** $SO_3$
**B.** $XeO_3$
**C.** $ClF_3$
**D.** $IF_5$

**7.** For which of the following molecules does the central atom obey the octet rule?

**A.** $BH_3$
**B.** $NH_3$
**C.** $BrF_3$
**D.** $AlF_3$

**8.** Which molecular structure does *not* have bond angles of 90°?

**A.** $CCl_4$
**B.** $XeF_4$
**C.** $PCl_5$
**D.** $SF_6$

*Refer to the structure below to answer questions 9 and 10.*

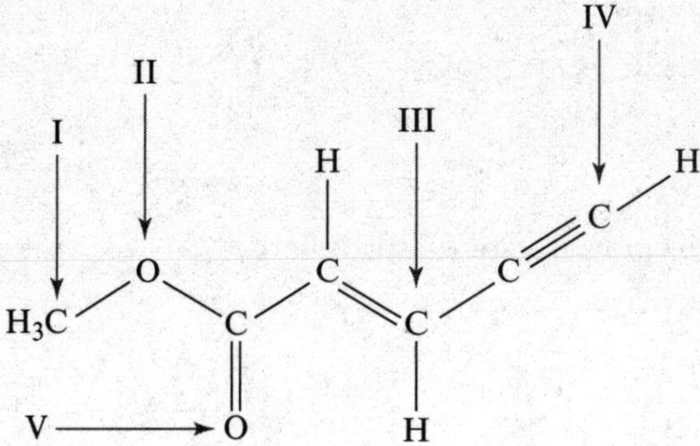


**9.** Which atom(s) in the structure are $sp^2$ hybridized?

**A.** I only
**B.** I and II
**C.** III and V
**D.** IV and V

**10.** Which atom(s) in the structure are involved in $\sigma$ bonding only?

**A.** I only
**B.** I and II
**C.** III and V
**D.** IV and V

**11.** What is the formal charge on the central nitrogen atom in the structure shown below?

$H_3C$–$\ddot{N}$–$CH_3$

**A.** +1
**B.** 0
**C.** –1
**D.** –2

**12.** Which of the following bond types is expected to be the shortest in length?

**A.** N≡N
**B.** N=N
**C.** N–N
**D.** C–N

**13.** Consider the Lewis structures of $H_3CCN$ and $H_3CNH_2$. Which statement best describes the bonding in these two molecules?

**A.** The C–N bond in $H_3CNH_2$ allows free rotation. It is shorter and stronger than the C–N bond in $H_3CCN$.
**B.** The C–N bond in $H_3CNH_2$ hinders free rotation. It is longer and weaker than the C–N bond in $H_3CCN$.
**C.** The C–N bond in $H_3CNH_2$ allows free rotation. It is longer and weaker than the C–N bond in $H_3CCN$.
**D.** The C–N bond in $H_3CNH_2$ hinders free rotation. It is shorter and stronger than the C–N bond in $H_3CCN$.

**14.** Which of the following choices best describes metallic bonding?

A. Ionic
B. Covalent
C. Nonpolar covalent
D. Delocalized

**15.** Which of the following has the most negative lattice energy?

A. NaCl
B. $Na_2O$
C. $MgCl_2$
D. MgO

## Long Free Response

**1. a.** Two Lewis structures can be drawn for the $HNO_2$ molecule. Assign formal charges to atoms O–1, O–2, N–1, and N–2.

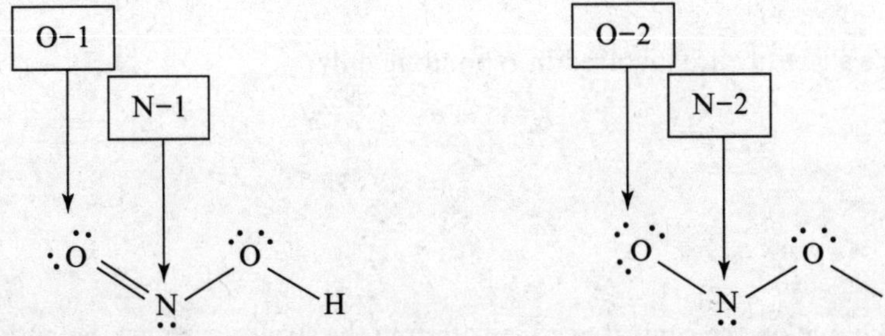


**b.** Which one of the two structures, the one on the left or the one on the right, best represents the $HNO_2$ molecule? Explain your choice by referencing both structures.

**c.** What is the orbital hybridization for the nitrogen atom in $HNO_2$?

**d.** What is the molecular geometry at the nitrogen atom in $HNO_2$?

**e.** Which bond indicated, A or B, is shorter in length? Justify your response.

**f.** Provide an estimation of the indicated bond angles, ∠ONO and ∠NOH, in the figure below.

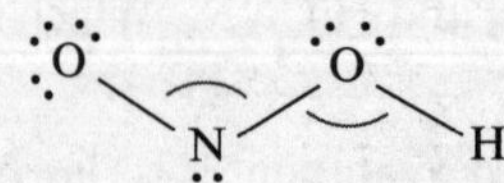

## Short Free Response

**2.** Hypochlorous acid is composed of one atom of chlorine, one atom of oxygen, and one atom of hydrogen.

**a.** In the box on the left, draw a complete Lewis structure for the compound in which chlorine is central. In the box on the right, draw a complete Lewis structure for the compound in which oxygen is central. Each of your chlorine and oxygen atoms should be drawn with an octet of electrons.

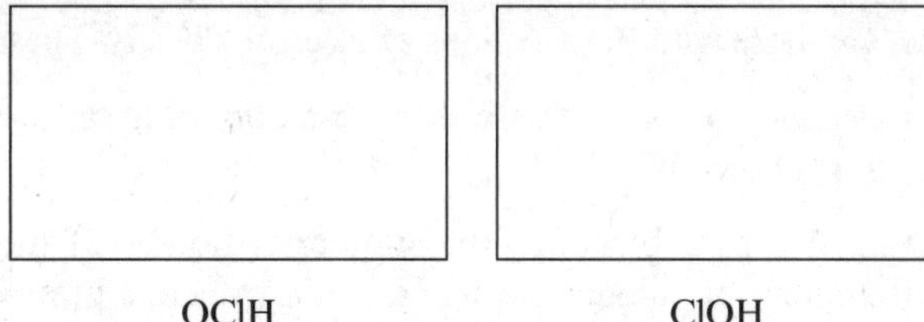

OClH ClOH

**b.** Which of these two compounds is the correct structure? Use the two Lewis structures to justify your response.

**c.** What is the electronic geometry of each Lewis structure drawn in part **a**?

**d.** What is the molecular geometry of each Lewis structure drawn in part **a**?

# Answers and Explanations

## Multiple Choice

**1.** **B.** From the table of electronegativities, we can see that the two elements that are closest in electronegativity are C (2.5) and S (2.6), choice B. These two elements form a bond in which the electrons are shared most equally.

**2.** **D.** The electronegativity differences are as follows:

C–S: $2.6 - 2.5 = 0.1$

C–Si: $2.5 - 1.9 = 0.6$

C–O: $3.4 - 2.5 = 0.9$

Si–O: $3.4 - 1.9 = 1.5$

The least polar bond is the one with the smallest difference in electronegativity between the atoms, and the most polar bond is the one with the largest electronegativity difference between the atoms. Therefore, the ranking from least polar to most polar is shown in choice D: C–S < C–Si < C–O < Si–O.

**3.** **C.** The bond length corresponds to the point on the graph where the potential energy is the lowest, choice C.

**4.** **D.** Infrared spectroscopy, choice D, is used to determine the types of functional groups present in a molecule by measuring the absorption of electromagnetic radiation due to the vibrations of covalent bonds.

**5.** **A.** Choices B and D can be eliminated based on the large separation of charge within their structures. They make very little contribution to the hybrid structure. Choices A and C contribute to the hybrid, but choice A is predicted to make a greater contribution since the negative charge is located on the more electronegative element.

**6.** **A.** $SO_3$ has a trigonal planar electronic geometry and a trigonal planar molecular geometry. The molecule has no net dipole moment, so choice A is correct. The rest of the molecules listed have net dipole moments since their molecular geometries are asymmetric. $XeO_3$ has a tetrahedral electronic geometry and a trigonal pyramidal molecular geometry. $ClF_3$ has a trigonal bipyramidal electronic geometry and a T-shaped molecular geometry. $IF_5$ has an octahedral electronic geometry and a square pyramidal molecular geometry.

**7.** **B.** $NH_3$ has a tetrahedral electronic geometry; the nitrogen atom has 6 bonded electrons and 2 nonbonded electrons, so it obeys the octet rule, choice B. Boron in $BH_3$ (choice A) and aluminum in $AlF_3$ (choice D) each have 6 bonded electrons only. Bromine in $BrF_3$ (choice C) has 6 bonded electrons and 4 nonbonded electrons.

**8.** **A.** $CCl_4$, choice A, has tetrahedral electronic and molecular geometries; its bond angles are about 109.5°. $XeF_4$ (choice B) has square planar electronic geometry with bond angles of 90°. $PCl_5$ (choice C) has trigonal bipyramidal molecular geometry with bond angles of 120° and 90°. $SF_6$ (choice D) has octahedral molecular geometry with bond angles of 90°.

**9.** **C.** The $sp^2$ hybridized atoms are those with three regions of electron density. These are atoms III and V, choice C.

**10.** **B.** The atoms that only have single bonds are those with sigma bonding only. Atoms I and II are only participating in single bonding, choice B.

**11.** **C.** The formal charge is calculated by subtracting the number of bonds (2) and the number of unshared electrons (4) from the total number of valence electrons. For nitrogen, a group VA element, there are five valence electrons, so the formal charge is –1, choice C.

$$FC = 5 - 2 - 4 = -1$$

**12.** **A.** Multiple bonds are stronger and shorter than single bonds. The triple bond, choice A, is the strongest and the shortest.

**13.** **C.** The C–N bond in $H_3CNH_2$ allows free rotation because it is a single bond. The C–N bond in $H_3CCN$ is a triple bond. Single bonds tend to be longer and weaker than triple bonds. Therefore, choice C is correct.

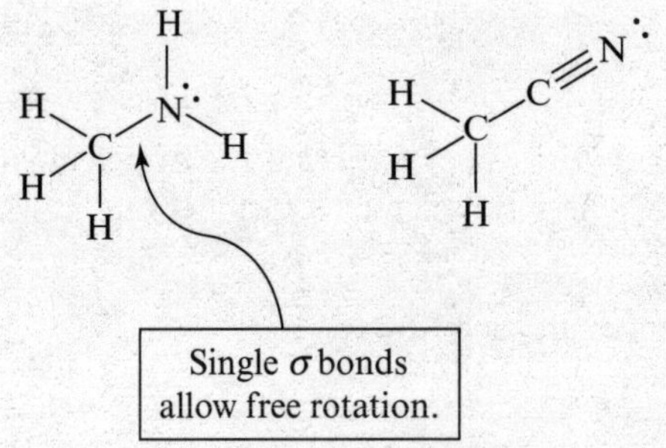

Single σ bonds allow free rotation.

**14.** **D.** Metallic bonding is characterized by a regular three-dimensional array of metal cations surrounded by freely moving, or *delocalized* (choice D), valence electrons. This is sometimes referred to as the "electron sea" model of bonding.

**15.** **D.** In order to achieve octets, Na becomes $Na^+$, Mg becomes $Mg^{2+}$, Cl becomes $Cl^-$, and O becomes $O^{2-}$. The coulombic attraction between ions is proportional to the magnitude of the product of the charges of the atoms. The product of the charges of $Mg^{2+}$ and $O^{2-}$, choice D, is the greatest of the choices given (2 × 2 = 4).

## Long Free Response

**1.** **a.** Formal charge is calculated by subtracting the number of bonds and the number of unshared electrons from the number of valence electrons (group A number) for the atom.

O–1 FC: 6 – 2 – 4 = 0

N–1 FC: 5 – 3 – 2 = 0

O–2 FC: 6 – 1 – 6 = –1

N–2 FC: 5 – 2 – 2 = +1

**b.** The structure on the left makes the greater contribution to the resonance hybrid since it has the least separation of charge.

**c.** The nitrogen atom has three regions of electron density, so it is $sp^2$ hybridized.

**d.** The molecular geometry at the nitrogen atom is bent.

**e.** The A bond is shorter in length since it is a hybrid between a single and a double bond. The more double-bond character an atom has, the shorter it is.

**f.** The bond angles are slightly compressed due to the lone pairs on the central atoms. The angle ∠ONO is slightly less than 120°. The angle ∠NOH is slightly less than 109.5°.

## Short Free Response

**2.** **a.** The correct Lewis structures and their formal charges are shown below.

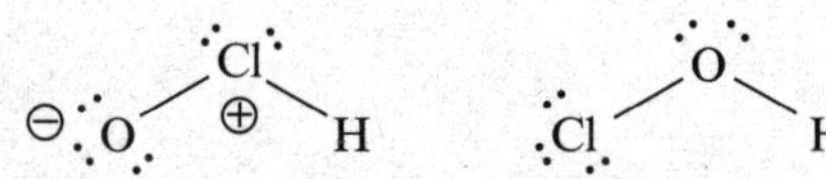


**b.** The compound on the right is the correct structure since it has no formal charges. The structure on the left has a greater separation of charge.

**c.** The central atom for each structure has $AX_2E_2$ tetrahedral electronic geometry.

**d.** The molecular geometry for each structure is bent.

# Chapter 3

# Intermolecular Forces and Properties

**Intermolecular interactions between molecules influence the physical properties of matter.**

In Chapter 2, interactions between atoms *within* molecules were discussed. In this chapter, the interactions *between* molecules and the impact of these interactions on the physical properties and the states of matter are discussed.

## Intermolecular Forces

**Intermolecular forces,** also known as **IMFs,** are weak attractions between covalent molecules or discrete atoms. *They should not be confused with bonds.* Bonds are intramolecular forces within molecules. When bonds are broken or formed, the identity of the molecule is changed. When intermolecular forces between molecules are broken or formed, the molecules themselves remain unchanged.

When molecular solids such as ice melt, for instance, some of the intermolecular forces (called hydrogen bonds) between the $H_2O$ molecules are broken, but the covalent bonds between hydrogen and oxygen are not. Additional intermolecular forces are overcome in order to separate the $H_2O$ molecules completely into the gas phase when water evaporates or boils; but again, no covalent bonds are broken. The amount of energy required to break a covalent O–H bond is much greater than the amount of energy required to overcome the intermolecular forces between molecules.

Many physical properties of compounds are a direct consequence of the types of intermolecular forces present between molecules. In general, because stronger intermolecular forces require more energy to overcome than weaker intermolecular forces, they lead to higher boiling points, higher melting points, higher viscosities, higher surface tensions, and lower vapor pressures.

There are four general classes of intermolecular forces that will be discussed:

- London dispersion forces
- dipole-dipole attractions
- dipole-induced dipole attractions
- hydrogen bonds

### London Dispersion Forces

**London dispersion forces** are the weakest of the intermolecular attractions. They are present between any molecules that are in proximity to one another. They are a result of **instantaneous dipoles** caused by the momentary and reversible accumulation of electron density in one region of a molecule at the expense of another region. This fleeting dipole moment causes the electron density in surrounding molecules to be attracted to the temporary partial positive region. This results in additional instantaneous dipoles in the neighboring molecules. The end result is a net attraction between the molecules.

London dispersion forces are the only intermolecular forces acting between nonpolar molecules and single atoms such as noble gases.

**Polarizability** refers to how easily electrons can be displaced within a molecule. The more polarizable the electron density is in a molecule, the stronger the London dispersion forces experienced by the molecule. A molecule with greater surface area and with many electrons is more polarizable than a molecule with less surface area and fewer electrons. This is illustrated by the phases of the halogens at room temperature and atmospheric pressure. Moving down the group, each halogen has more electrons and is more polarizable. The London dispersion forces become increasingly strong, and the halogens have more of a tendency to stick together. $F_2$ and $Cl_2$ are gases at room temperature, $Br_2$ is a liquid, and $I_2$ is a solid. The $I_2$ molecule has the most electrons and is the most polarizable, so its London dispersion forces are the strongest.

Shown below are two organic molecules called neopentane and pentane. Although both molecules are composed of 5 carbons and 12 hydrogens, pentane has a higher boiling point than neopentane because of its larger surface area and greater polarizability.

**Comparison of the London Dispersion Forces in Pentane and Neopentane**

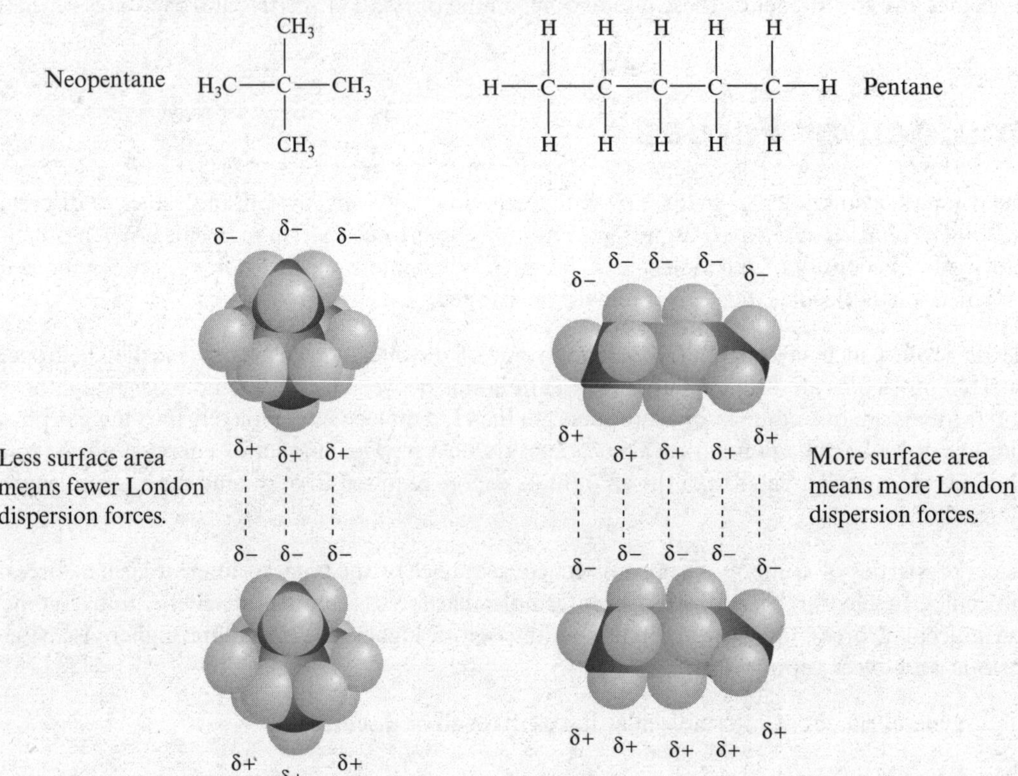


## Dipole-Dipole Attractions

**Dipole-dipole attractions** are intermolecular forces between polar molecules. The partial positive portion of one molecule is attracted to the partial negative end of a neighboring molecule. The more polar the molecules, the stronger the dipole-dipole attraction.

**Dipole-Dipole Attractions between HCl Molecules**

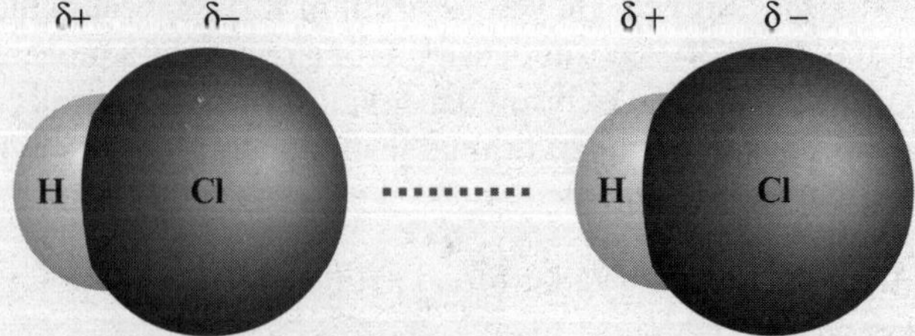


## Dipole-Induced Dipole Attractions

**Dipole-induced dipole attractions** are interactions between polar and nonpolar molecules. Proximity to a polar molecule causes an instantaneous dipole in the polarizable electrons of a nonpolar molecule, resulting in an attraction. Nonpolar oxygen molecules are capable of dissolving in polar water molecules because of dipole-induced dipole interactions.

## Hydrogen Bonds

**Hydrogen bonding,** a subset of dipole-dipole attractions, occurs in molecules that have a hydrogen atom covalently bound to N, O, or F. These elements are so electronegative and remove so much of the electron density from the hydrogen atom that the hydrogen begins to behave like a proton—a bare positively charged particle. It experiences a strong attraction to the electron pairs of the N, O, or F atoms of adjacent molecules, and the resulting attraction is the strongest of the intermolecular forces. Hydrogen bonds are still very weak, however, when compared to ionic or covalent bonds.

**Despite their name, hydrogen bonds are *not* bonds. They are intermolecular forces.**

**Hydrogen Bonding between $H_2O$ Molecules**

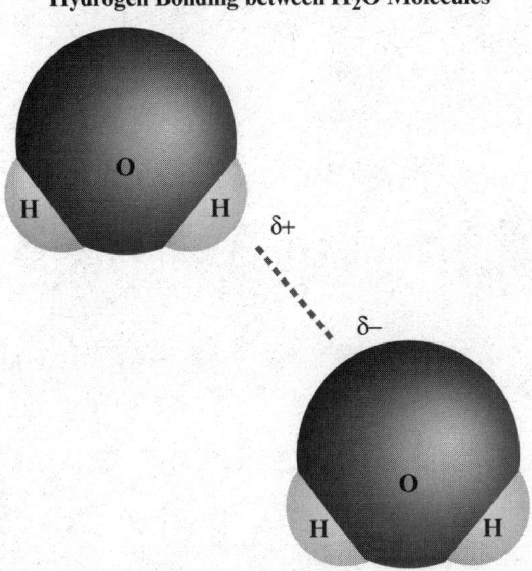


Hydrogen bonding results in boiling point trend anomalies for the second-period hydrides HF, $H_2O$, and $H_3N$. In general, boiling points for the hydrides increase down a group on the periodic table due to increased London dispersion forces. For groups 15, 16, and 17, though, the boiling point of the period 2 hydride is much larger than expected. Because hydrogen bonding is a much stronger intermolecular force than dipole-dipole interactions or London dispersion forces, the energy required to convert $NH_3$, HF, or $H_2O$ from the liquid to the gas phase is higher than would be predicted based on the trend.

**Boiling Point Trends for the Groups 14–17 Hydrides**

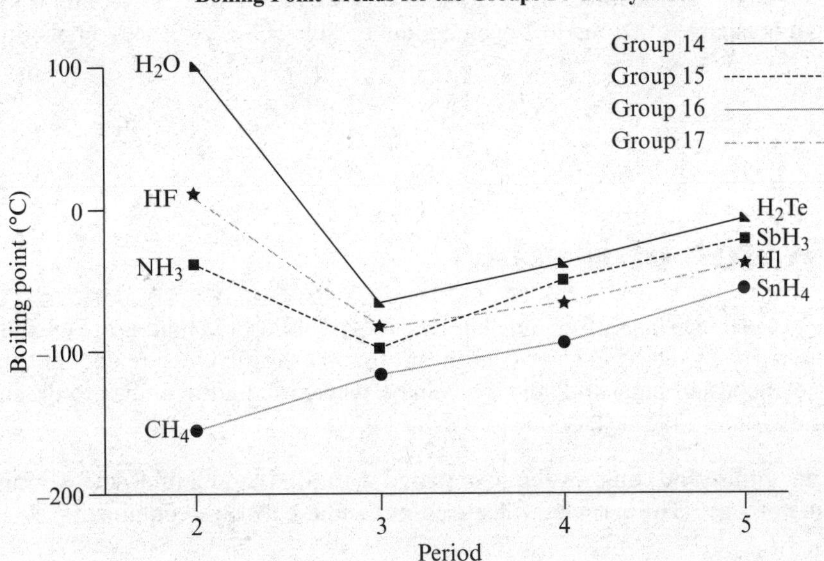

Hydrogen bonding is very important in biology, and there is emphasis on this fact in the AP Chemistry curriculum. Noteworthy examples include hydrogen bonding between the base pairs of the DNA molecule, as shown below. Adenine and thymine form two hydrogen bonds, and cytosine and guanine form three hydrogen bonds.

**The Role of Hydrogen Bonding in the Structure of DNA**

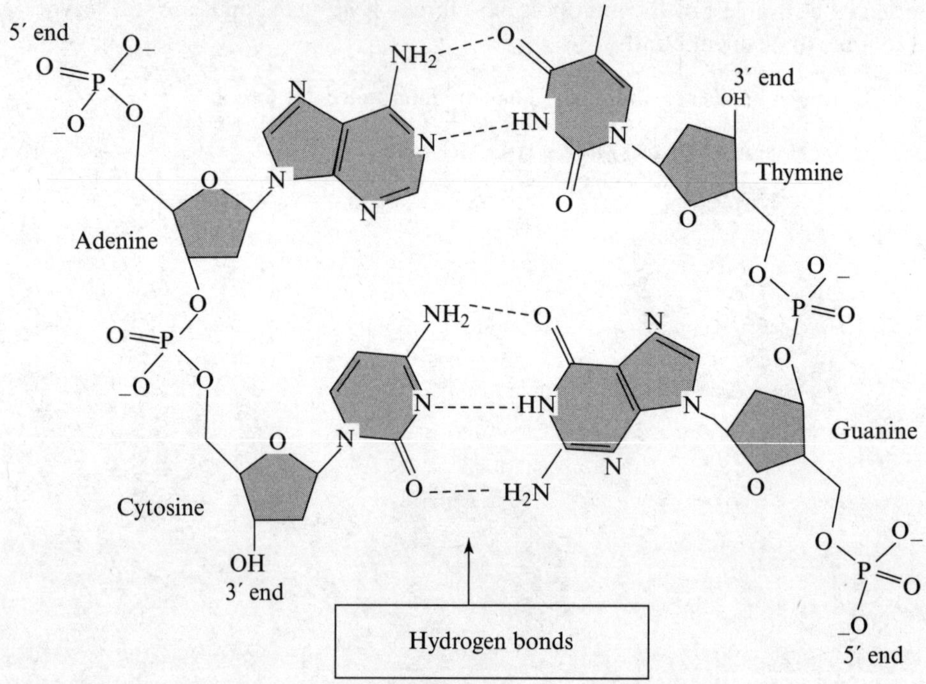

## Practice

---

Rank the listed substances in order of increasing boiling point. Justify your ranking.

$$H_2S, H_2O, C_2H_2$$

---

The correct order of increasing boiling point is $C_2H_2 < H_2S < H_2O$. Acetylene is a nonpolar molecule, so it experiences weak London dispersion forces only. $H_2S$ and $H_2O$ are both polar. $H_2O$ contains hydrogen atoms bound to oxygen, so it is capable of strong hydrogen bonding, while $H_2S$ experiences dipole-dipole attractions only.

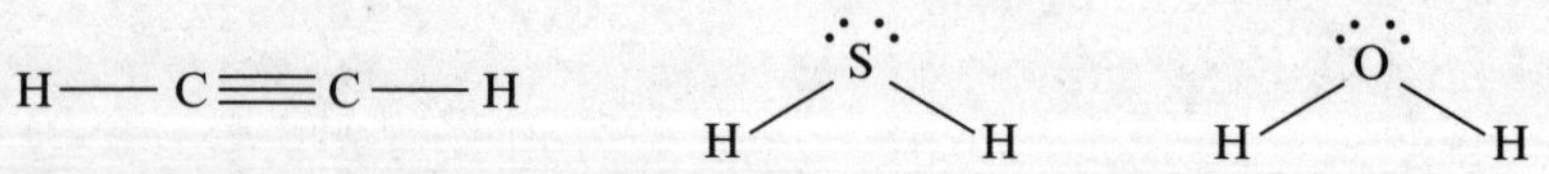

# Physical States of Matter

Solids and liquids are **condensed states** of matter. The densities of solids and liquids are much higher than those of gases, and they are relatively incompressible (this means that increasing pressure does not substantially decrease their volume). Solids are rigid and hold their shape, whereas liquids conform to the shape of the bottom of their container and assume a flat surface due to gravity.

Gases, by contrast, are diffuse and compressible. Compared with solids and liquids, there is much more space between the molecules of a gas. The particles will spread out and evenly fill a container.

**Particulate Comparison of the Solid, Liquid, and Gas States**

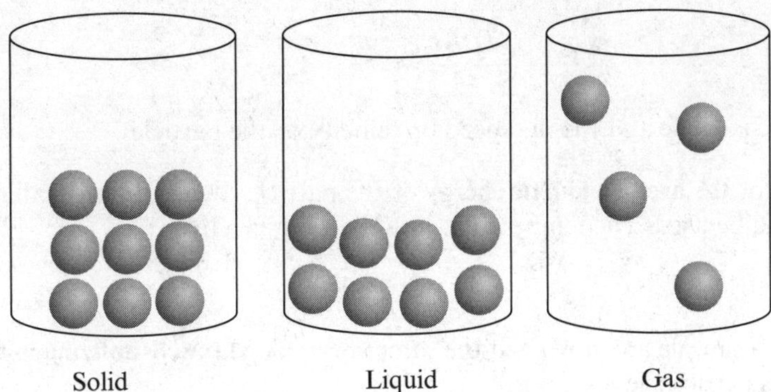


# Gases

## Kinetic Molecular Theory

**Ideal gases** are those that behave in a way that is well-described by **kinetic molecular theory.** At sufficiently high temperatures and low pressures, all gases behave ideally.

**Kinetic molecular theory postulates:**

- The atoms or molecules that make up a gas are in continuous, random motion.
- All collisions between gas particles or between gases and the walls of their container are perfectly elastic (no energy is lost due to the collisions).
- The pressure exerted by a gas is due to collisions of the particles with the walls of the container.
- The volume of individual gas particles is negligible compared to the volume of the empty space between the particles.
- The temperature of a gas is a measure of the average kinetic energy of the particles.

Samples of gases can be described by four variables: pressure, temperature, volume, and number of moles.

**Pressure** is a measure of the force exerted by a gas per unit area. At sea level, the height of mercury in a mercury barometer that is supported by the atmosphere, or the **barometric pressure,** is 760 mm Hg. This is equal to 760 torr, 1 atmosphere (atm), or 101 kiloPascals (kPa). The necessary conversion factors are provided on the *AP Chemistry Equations and Constants* sheet.

**Mercury Barometer**

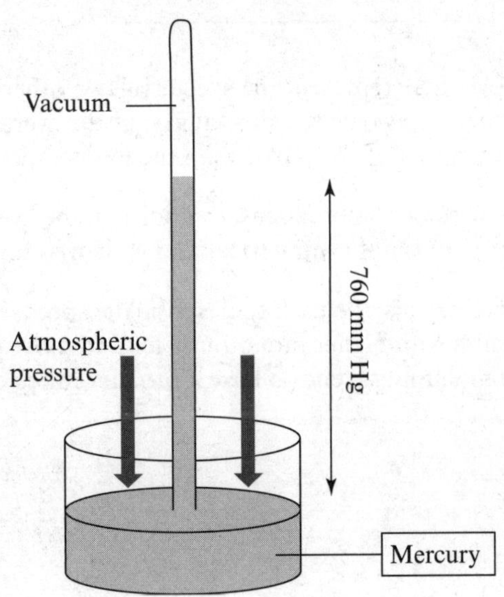

**Kinetic energy** is energy of motion. The kinetic energy, *KE*, for a single particle is given by

$$KE = \frac{1}{2}mv^2$$

where $m$ is the mass of the particle and $v$ is the speed or velocity of the particle.

**Temperature** is a measure of the **average kinetic energy** of the particles in a sample. In all gas law calculations, the temperature in **Kelvin** must be used. The temperature in Kelvin is given by

$$K = °C + 273$$

Not all of the particles in a sample are moving at the same speed. A **Maxwell-Boltzmann distribution** plot shows the range of speeds of gas particles in a sample.

**The Maxwell-Boltzmann Distributions of Two Samples of Gas at Different Temperatures**

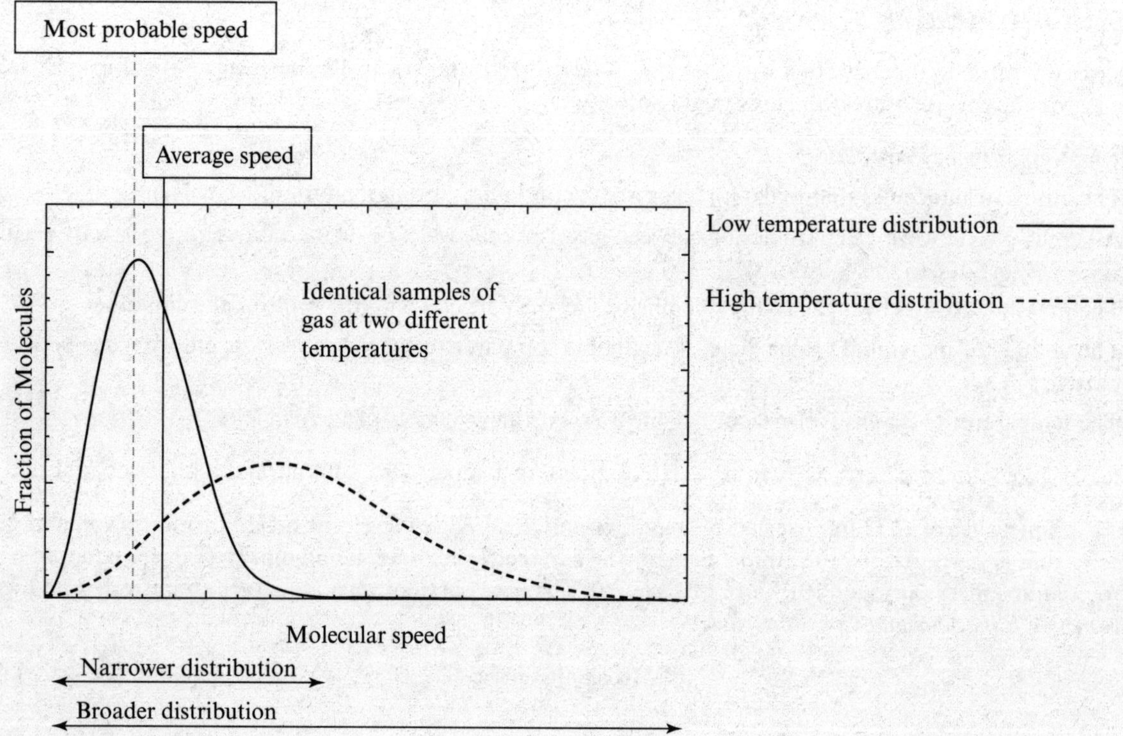


The two distributions, or curves, shown here represent the speeds of two otherwise identical gas samples at two different temperatures. The areas under the two curves are equal, but the average speed of the particles in the high-temperature sample is greater than that of the particles in the low-temperature sample.

As temperature increases, the range of velocities broadens. In addition, the average velocity is slightly higher than the most probable velocity since the curve is not symmetrical and is skewed toward higher speeds.

The Maxwell-Boltzmann distribution may also be used to describe the speeds of different types of molecules at one temperature. In this case, molecules with higher molar mass tend to move at speeds closer to their average velocity, while molecules with lower molar mass tend to have a broader range of velocities.

**The Maxwell-Boltzmann Distributions of Two Samples of Gas with Different Molar Masses**

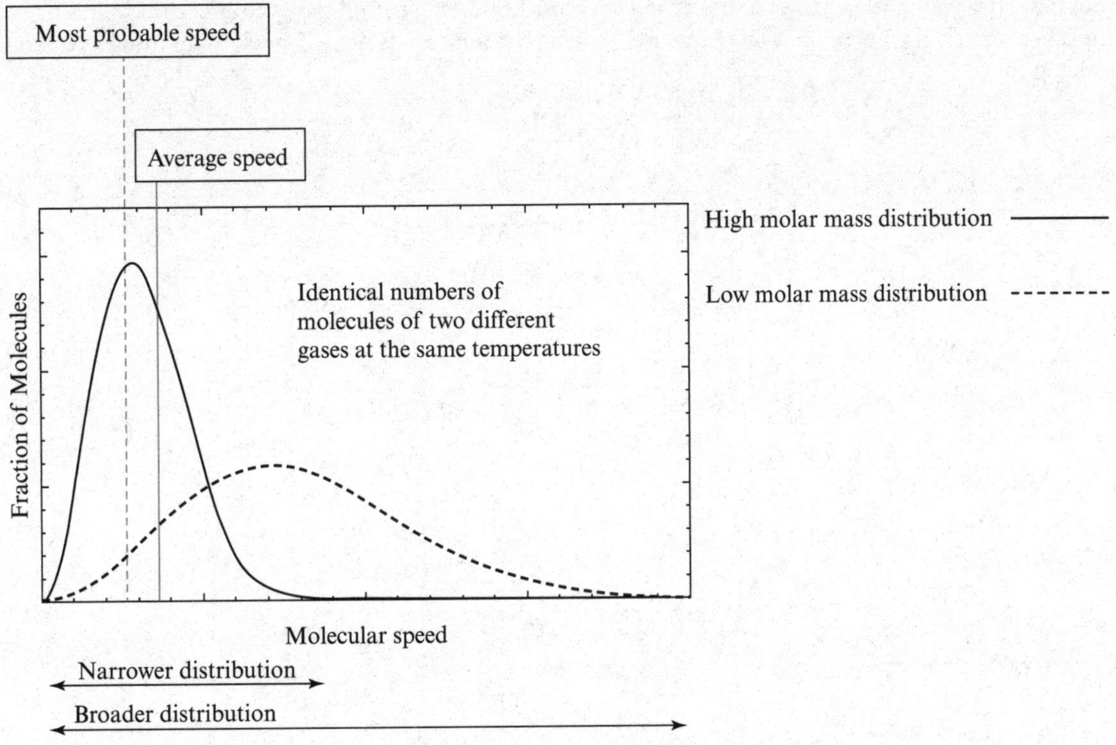


In general, the average velocity of gas molecules in a sample increases with increasing temperature and decreases with increasing molar mass, ***M***.

$$\text{Average velocity} \propto \sqrt{\frac{T}{M}}$$

**Diffusion** is the dispersal of gas particles throughout a container. **Effusion** is the escape of gas particles through tiny pores in a barrier from regions of higher pressure to regions of lower pressure. Average velocity of a gas is related to the rates of diffusion and effusion of the gas because the faster the particles are moving, the more quickly they fill a container or pass through an opening in a barrier. All gases diffuse and effuse more quickly at higher temperatures than at lower temperatures.

**Diffusion and Effusion**

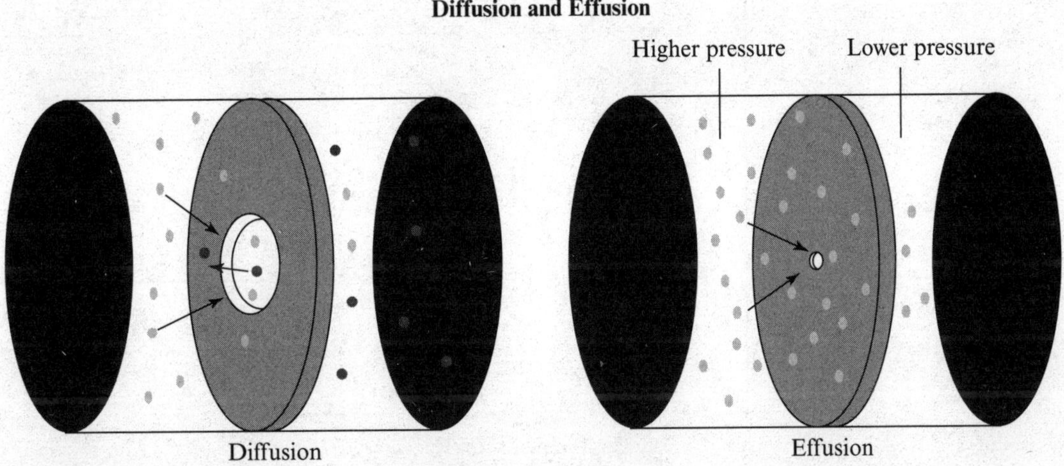


Smaller molecular weight gases effuse and diffuse faster than higher molecular weight gases.

## Boyle's Law

The product of the volume and the pressure of a gas is constant. At constant temperature, the volume, $V$, of a gas sample is inversely proportional to its pressure, $P$. This relationship is named **Boyle's law** after its discoverer, Robert Boyle.

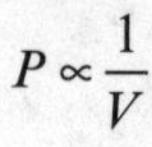

$$P \propto \frac{1}{V}$$

**Boyle's Law**

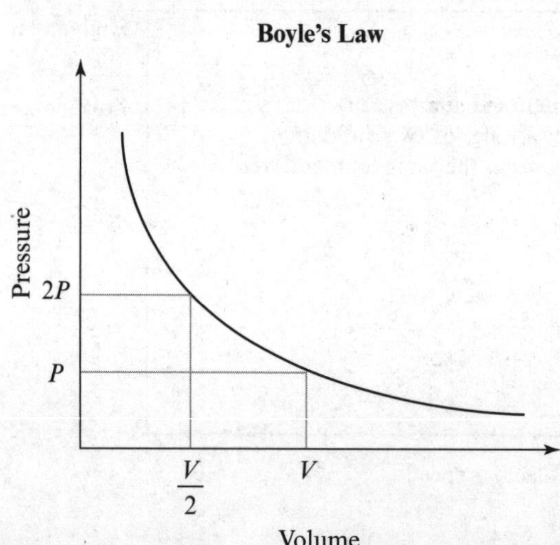


## Charles' Law

**Charles' law** states that the volume of a sample gas is directly proportional to the Kelvin temperature of the gas at constant pressure.

$$V \propto T$$

The graph of volume versus temperature can be extrapolated to a temperature at which volume is theoretically zero. This temperature, 0 Kelvin or **absolute zero,** is –273 °C.

**Charles' Law**

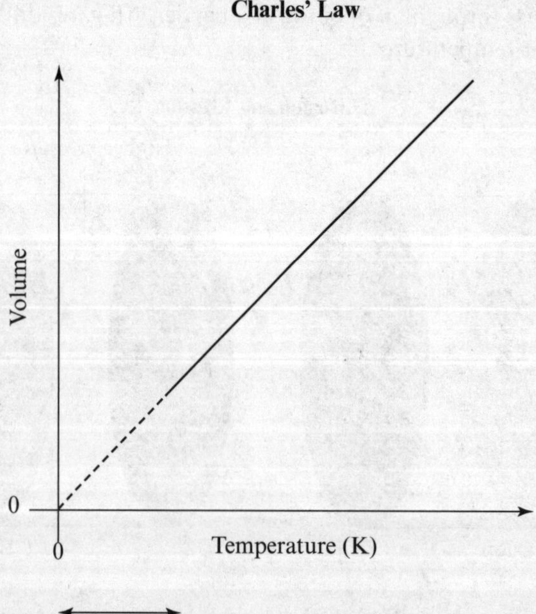

## Avogadro's Law

Equal volumes of gases at the same temperature and pressure contain identical numbers of moles of gas molecules. This relationship is known as **Avogadro's law:** The number of moles of gas, *n*, is directly proportional to the volume.

$$n \propto V$$

## Ideal Gas Law

The experimental observations of Boyle, Charles, and Avogadro are summarized by the **ideal gas law,** the equation from which any ideal gas conditions can be determined.

$$PV = nRT$$

The **universal gas constant, *R*,** is 0.08206 L·atm·mol$^{-1}$·K$^{-1}$. The ideal gas law and the value of *R* are given on the *AP Chemistry Equations and Constants* sheet.

For example, the pressure exerted by 0.50 mole of an ideal gas occupying 3.0 L at 200. K can be found by rearranging the ideal gas equation.

$$P = \frac{nRT}{V}$$

$$P = \frac{0.50\ \cancel{\text{mol}} \times 0.08206\ \cancel{\text{L}} \cdot \text{atm} \cdot \cancel{\text{mol}^{-1}} \cdot \cancel{\text{K}^{-1}} \times 200\ \cancel{\text{K}}}{3.0\ \cancel{\text{L}}} = 2.7\ \text{atm}$$

Furthermore, since *R* is constant, rearranging the ideal gas equation to solve for *R* gives a form that allows for many useful calculations.

$$R = \frac{PV}{nT}$$

The initial conditions of a gas, $P_i$, $V_i$, $n_i$, and $T_i$, are related to the final conditions, $P_f$, $V_f$, $n_f$, $T_f$, by

$$R = \frac{P_iV_i}{n_iT_i} = \frac{P_fV_f}{n_fT_f}$$

Any of the four variables that remain constant can be cancelled out of the equation. For instance, a problem may provide an initial volume and pressure. The question may then ask for a final pressure at a new volume, assuming the number of moles and temperature are held constant. The number of moles and the temperature cancel out of the equation since $n_i = n_f$ and $T_i = T_f$.

$$\frac{P_iV_i}{\cancel{n_i}\cancel{T_i}} = \frac{P_fV_f}{\cancel{n_f}\cancel{T_f}}$$

$$P_iV_i = P_fV_f$$

$$P_f = \frac{P_iV_i}{V_f}$$

In this way, any of the ideal gas laws can be quickly derived from the ideal gas equation given on the *AP Chemistry Equations and Constants* sheet.

### Practice

Find the volume in liters occupied by 0.22 mole of an ideal gas at a temperature of 42 °C and 3800 torr.

The pressure must be converted to atmospheres.

$$3800 \cancel{\text{torr}} \times \frac{1 \text{ atm}}{760 \cancel{\text{torr}}} = 5.0 \text{ atm}$$

The temperature must be expressed in Kelvin.

$$42\ °\text{C} + 273 = 315 \text{ K}$$

The ideal gas equation is rearranged to solve for volume.

$$V = \frac{nRT}{P} = \frac{(0.22 \cancel{\text{mol}})(0.08206 \text{ L} \cdot \cancel{\text{atm}} \cdot \cancel{\text{mol}^{-1}} \cdot \cancel{\text{K}^{-1}})(315 \cancel{\text{K}})}{5.0 \cancel{\text{atm}}} = 1.1 \text{ L}$$

### Practice

A rigid 5.0-L container holds 2.0 moles of gas at 800.0 torr. What is the new pressure if an additional 1.5 moles of gas are added at constant temperature?

Since both the volume and the temperature are held constant, the needed equation relates pressure to number of moles. The final number of moles, $n_f$, is equal to the initial number of moles, $n_i$, plus 1.5 moles, which is the number that has been added.

$$\frac{P_i \cancel{V_i}}{n_i \cancel{T_i}} = \frac{P_f \cancel{V_f}}{n_f \cancel{T_f}}$$

$$P_i n_f = P_f n_i$$

$$P_f = \frac{P_i n_f}{n_i}$$

$$P_f = \frac{800.0 \text{ torr} \times (2.0 \cancel{\text{mol}} + 1.5 \cancel{\text{mol}})}{2.0 \cancel{\text{mol}}} = 1400 \text{ torr}$$

## Standard Temperature and Pressure

By convention, **standard temperature and pressure,** STP, for the purposes of gas law calculations, are 0 °C (273 K) and 1 atm.

### Practice

Find the volume occupied by 1.0 mole of an ideal gas at STP.

Standard temperature is 273 K, and standard pressure is 1 atm.

Rearranging the ideal gas equation gives the expression for volume.

$$V = \frac{nRT}{P} = \frac{1.0 \cancel{\text{mol}} \times 0.08206 \text{ L} \cdot \cancel{\text{atm}} \cdot \cancel{\text{mol}^{-1}} \cdot \cancel{\text{K}^{-1}} \times 273 \cancel{\text{K}}}{1.0 \cancel{\text{atm}}} = 22.4 \text{ L}$$

This is a very useful relationship, because **at STP, the volume of one mole of *any* ideal gas is 22.4 L.**

## Gas Density

The density, $D$, of a gas is equal to the mass of gas that will occupy a given volume at a certain temperature and pressure.

$$D = \frac{m}{V}$$

A relationship can be derived between the temperature, pressure, density, and molar mass using the ideal gas law. Since molar mass, $\mathbf{M}$, is equal to grams per mole, the relationship between the number of moles, $n$, and the mass, $m$, is given by the following relationship.

$$n = \frac{m}{\mathbf{M}}$$

This relationship can be substituted into the ideal gas law.

$$PV = \left(\frac{m}{\mathbf{M}}\right)RT$$

Rearrangement gives an equation for the molar mass, and $D$ can be substituted for $\frac{m}{V}$.

$$\mathbf{M} = \frac{mRT}{VP}$$

$$\mathbf{M} = \frac{DRT}{P}$$

A common laboratory investigation in AP Chemistry involves determination of the molar mass of a volatile liquid by elucidation of its density at a known temperature and pressure. A container with a known volume is weighed together with a small piece of aluminum foil punctured with a pinhole. A small amount of the liquid is placed in the bottom of the container and the container is covered with the foil. The flask is gently heated to a recorded temperature until the entire sample of liquid is vaporized. It is assumed that the pressure reached by the vapor is equal to the barometric pressure in the room and that the entire volume of the container is filled with vapor. The flask is allowed to cool to room temperature so that the vapor that filled the flask condenses. The mass of the flask and foil is subtracted from the mass of the flask, foil, and liquid to give the mass of the vapor.

### Practice

A student obtains a 10-mL sample of an unknown liquid. She weighs an Erlenmeyer flask and a square of aluminum foil on a balance and obtains a combined mass of 55.21 grams. The student places the liquid into the flask, covers the flask with the aluminum foil, and then places the flask into a water bath at 90. °C. Once the liquid has completely vaporized, the student removes the flask and allows it to reach room temperature. Droplets of the volatile liquid form on the walls of the flask. The flask, foil, and liquid are weighed and found to have a mass of 55.31 grams. The student rinses out the flask and fills it to the brim with water. She carefully pours the water into a graduated cylinder and finds that the water has a volume of 57.4 mL. The atmospheric pressure in the room is 1.00 atm. What is the molar mass of the liquid?

The mass of the vapor is the difference between the empty flask with foil and the flask with foil and the condensate.

$$m = 55.31\text{ g} - 55.21\text{ g} = 0.10\text{ g}$$

The temperature is expressed in Kelvin.

$$T = 90.\ °\text{C} + 273 = 363\text{ K}$$

The density of the vapor is the mass per unit volume. The mass of the re-condensed liquid is the same as the mass of the vapor that occupied the flask at the high temperature.

$$D = \frac{0.10\ \text{g}}{57.4\ \text{mL}} = 0.0017\ \text{g}\cdot\text{mL}^{-1}$$

(0.00174216)

Finally, the molar mass is determined using the equation derived from the ideal gas law. It's best to plug the entire output from the calculator into the next calculation (0.00174216).

$$M = \frac{DRT}{P} = \frac{(0.00174126\ \text{g}\cdot\cancel{\text{mL}^{-1}})(0.08206\ \cancel{\text{L}}\cdot\cancel{\text{atm}}\cdot\text{mol}^{-1}\cdot\cancel{\text{K}^{-1}})(363\ \text{K})}{1.00\ \cancel{\text{atm}}} \times \frac{1000\ \cancel{\text{mL}}}{1\ \cancel{\text{L}}} = 52\ \text{g}\cdot\text{mol}^{-1}$$

## Dalton's Law of Partial Pressures

**Dalton's law** states that the total pressure of gas in a mixture of gases is equal to the sum of the pressures that the individual gases would exert by themselves. For gases A, B, and C in a mixture, the total pressure, $P_{total}$, is equal to the sum of the partial pressures.

$$P_{total} = P_A + P_B + P_C$$

The **partial pressure** of a gas is equal to the total pressure multiplied by the **mole fraction,** *X*, of that gas in the mixture. The following equation is included on the *AP Chemistry Equations and Constants* sheet.

$$P_A = P_{total} \times X_A,\ \text{where } X_A = \frac{\text{moles A}}{\text{total moles}}$$

$$X_A = \frac{n_A}{n_A + n_B + n_C}$$

### Practice

A 50-mL tube inverted over a dish of water contains $CO_2$ and water vapor at STP. The partial pressure of $H_2O$ is 24 torr. What is the mole fraction of $CO_2$ in the tube?

The partial pressure of $CO_2$ is given by the difference between the total pressure (atmospheric pressure) and the partial pressure of water.

$$P_{CO_2} = P_{total} - P_{H_2O}$$

$$P_{CO_2} = 760\ \text{torr} - 24\ \text{torr} = 736\ \text{torr}$$

Since the partial pressure of $CO_2$ is the mole fraction of $CO_2$ times the total pressure, the mole fraction can be calculated.

$$X_{CO_2} = \frac{P_{CO_2}}{P_{total}} = \frac{736\ \cancel{\text{torr}}}{760\ \cancel{\text{torr}}} = 0.968$$

Questions like the one in the previous example apply to laboratory situations in which the gaseous product(s) of a reaction (usually a decomposition) are collected in a tube over water. The gas that displaces water in the inverted tube contains water vapor with a partial pressure equal to the vapor pressure of water at that temperature. The rest of the gases collected in the inverted tube are products of the reaction.

**Collection of the Gaseous Products of a Decomposition Reaction**

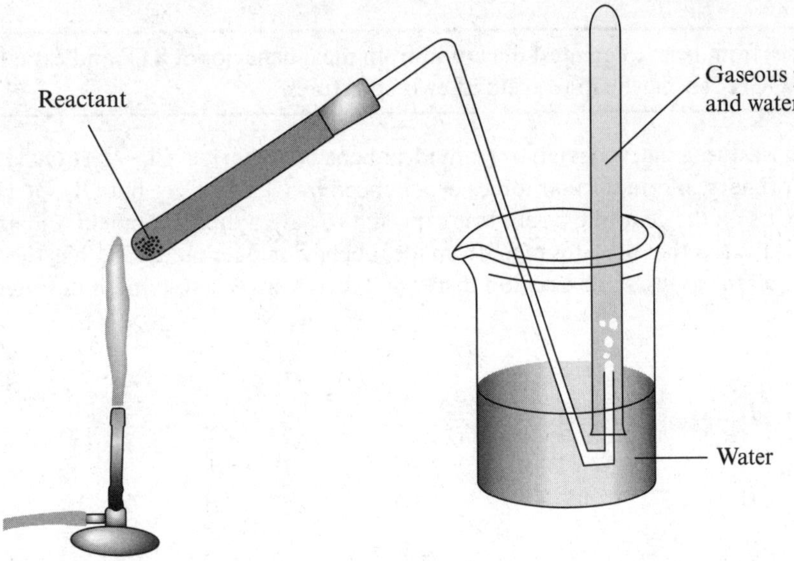

## Real Gases

The particles of an *ideal* gas are assumed to have no volume and no interaction with the particles around them other than perfectly elastic collisions. They experience no intermolecular forces.

*In reality, no gases are ideal.* **Real gas** particles occupy volume and experience intermolecular attractions (and repulsions at sufficiently close distances). Gas particles are attracted to other gas particles because of intermolecular forces. The stronger the intermolecular forces between the molecules of the gas, the more the behavior of the gas deviates from that predicted by the ideal gas law. In addition, the larger the molecule, the less accurately the behavior of the gas is predicted by the ideal gas law.

Some gases behave more ideally than others, however. Helium (He), for instance, is a smaller molecule and occupies less volume than $CH_4$. It is less polarizable than $CH_4$, so its intermolecular forces are weaker and its behavior is better described by the ideal gas law.

As the average kinetic energy (or temperature) of a sample of gas particles increases, the particles move faster. The faster the gas molecules move, the more energetic their collisions are. Faster-moving gas particles are better able to overcome intermolecular attractions, and so at higher temperatures gas behavior is better predicted by the ideal gas law.

At very low temperatures, some of the collisions between the gas molecules are not energetic enough to overcome the strength of the intermolecular forces. Collisions are more inelastic (in other words, the molecules stick together). As temperatures approach those at which a gas is expected to condense to the liquid state, the gas increasingly deviates from ideal behavior.

At higher pressures, molecules are closer together, and intermolecular attractions have an increased influence on the behavior of the gas than at lower pressures. As pressures increase, gases deviate more from ideal behavior.

It is important to remember that gases deviate from ideal behavior as their temperature is decreased and their pressure is increased. In other words, the ideal gas law becomes less useful for predicting gas behavior as the gas approaches the liquid state.

## Practice

Rank the following gases from least to greatest deviation from ideal behavior at STP and explain your ranking: $Cl_2$, $CH_4$, $CH_3OCH_3$. (It may help to sketch Lewis structures.)

The correct ranking from least to greatest deviation from ideal behavior is $CH_4 < Cl_2 < CH_3OCH_3$. London dispersion forces are the primary intermolecular forces experienced by $Cl_2$ and $CH_4$, but $CH_3OCH_3$ is a polar molecule with two lone pairs on the central oxygen. It is expected to have a dipole moment and experience dipole-dipole interactions, so it is the gas that deviates most from ideal behavior. Methane, $CH_4$, has the fewest electrons and is least polarizable of all the gases, so its London dispersion forces are weakest and it deviates least from ideal behavior.

# Solids

Solids are substances in which the atoms or molecules are packed tightly and have fixed locations. They can be classified as either **amorphous** or **crystalline.** Amorphous solids are solids like glass that are relatively disordered on the particulate level. **Crystalline solids** are regular three-dimensional arrangements of atoms, ions, or molecules. Sodium chloride, for instance, is a crystalline solid. Each particle is said to occupy a **lattice point** in the three-dimensional array. For sodium chloride, each lattice point is occupied by a sodium or a chloride ion.

**Crystalline and Amorphous Solids**

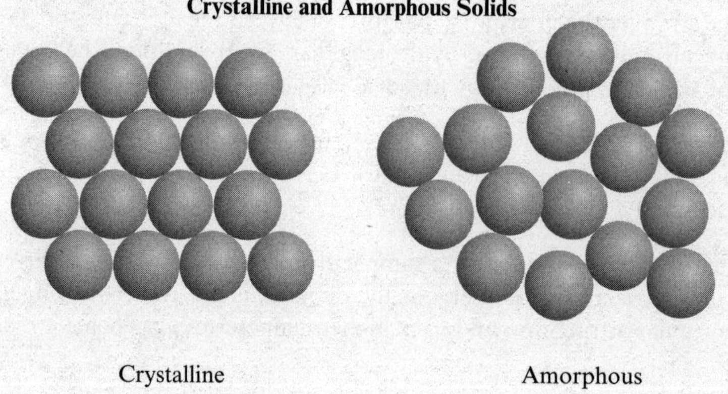


Solids can be further subcategorized as four general types: ionic, network covalent, metallic, and molecular. Each type of solid has properties that can be explained based on the type of bonding and intermolecular forces that dominate.

## Ionic Solids

**Ionic solids** are three-dimensional arrays of alternating cations and anions held together by coulombic attractions. These electrostatic attractions are relatively strong and give ionic solids high melting points and boiling points.

Ionic solids are hard and brittle. When an ionic solid is struck hard enough for the ions to slip past one another, repulsive attractions between like charges cause the crystal to break.

**Breaking an Ionic Crystal**

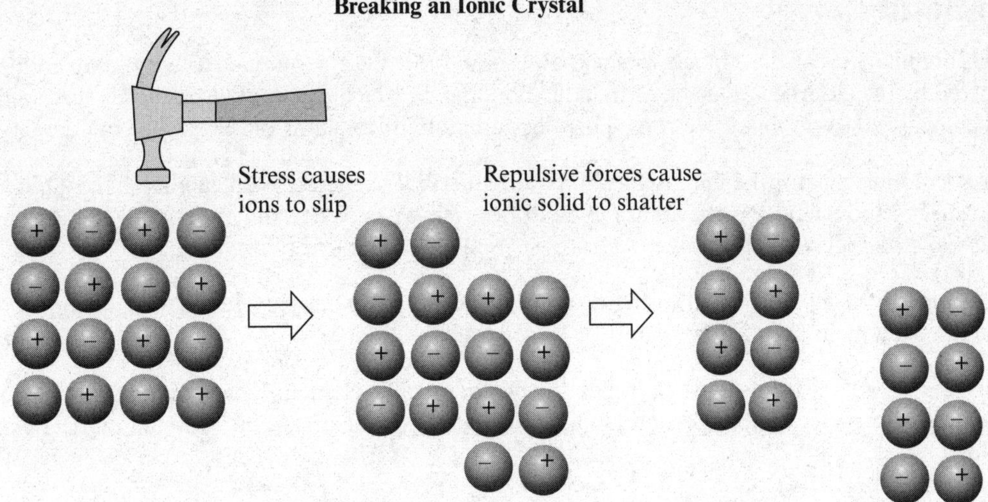


Electrical conductivity is a measure of how well a material transmits an electric current. Ions or electrons must be able to move in order for a substance to conduct electricity. The ions in ionic solids are fixed in place, so ionic solids are nonconductive. (By contrast, molten, or melted, ionic compounds as well as dissociated ions in aqueous solutions are mobile and conductive.)

## Network Covalent Solids

**Network covalent solids** are arrays of atoms held together by covalent bonds. An example is quartz, a crystalline form of $SiO_2$.

Network covalent solids have extremely high melting points since disruption of the atoms requires breaking covalent bonds. They also tend to be extremely hard and insoluble in water.

**Allotropes** are different bonding forms of the same element in the same phase. Diamond and graphite are both allotropes of carbon. Diamond is composed of $sp^3$ hybridized, $\sigma$-bonded carbon atoms. It is not electrically conductive since the electrons are localized between covalently bound atoms. Graphite is a different allotrope of carbon, consisting of fused six-carbon rings with delocalized single and double bonds. It is a conductive network solid because it is composed of layers of two-dimensional sheets of rings in which the $\pi$-bonded electrons are free to move. The attractive forces between the graphite sheets are relatively weak London dispersion forces.

**The Structures of Diamond and Graphite**

**Diamond**

Each carbon atom is $sp^3$ hybridized and bound to four other carbon atoms.

**Graphite**

Intermolecular forces between layers.

Each carbon atom is $sp^2$ hybridized and bound to three other carbon atoms.

Elemental silicon is another important network covalent solid. Silicon is a **semiconductor.** Since its electrons can escape their covalent bond arrangement at room temperature, its electrical conductivity is greater than that of an electrical insulator like diamond, but less than that of a conductor like pure copper.

## Metallic Solids

Metallic solids are made up of the *s* block, *d* block, and some *p* block elements such as aluminum and lead. They are characterized by the **electron sea model** of metallic bonding, in which a crystal of cations is surrounded by loosely held, delocalized (or mobile) electrons. This electron mobility explains the electrical conductivity of metals.

The electron sea model explains the high malleability and ductility of metals. **Malleability** is the ability to be pounded into sheets, and **ductility** is the ability to be drawn into wires. When the material is pounded or stretched, the loose electron cloud allows the metallic nuclei to easily slide past one another.

The melting points of metals vary widely, ranging from very low to quite high. Mercury, for example, melts at –39 °C, while tungsten melts at 3422 °C.

**Alloys** are mixtures of metal atoms and other metallic or nonmetallic elements. They are prepared by mixing molten elements that produce a homogeneous solid solution when cooled. The electron sea model of bonding applies to alloys.

In general, the disruption of the regular crystal lattice of pure metal atoms by atoms of another element makes it more difficult for the layers of nuclei to slip past one another. In other words, most alloys are harder, less malleable, and less ductile than pure metals.

The two major classes of alloys are substitutional and interstitial.

**Substitutional alloys** result when a metal is mixed with an element of similar atomic radius. Usually the atomic radii of the atoms in the mixture are within 15% of one another.

**Brass,** a mixture of copper and zinc, is an example of a substitutional alloy. Copper and zinc atoms are both transition metals, and they have similar radii. The densities of substitutional alloys are intermediate between those of their constituent elements. Different compositions of brass have densities between 8.4 and 8.7 $g{\cdot}mL^{-1}$. Brass is less dense than pure copper ($d = 8.96\ g{\cdot}mL^{-1}$) and more dense than pure zinc ($d = 7.13\ g{\cdot}mL^{-1}$).

**Interstitial alloys** result when much smaller atoms, usually nonmetals such as carbon, boron, and nitrogen, are mixed with a metal with a larger atomic radius.

**Steel,** a mixture of iron and carbon, is an interstitial alloy.

**Substitutional and Interstitial Alloys**

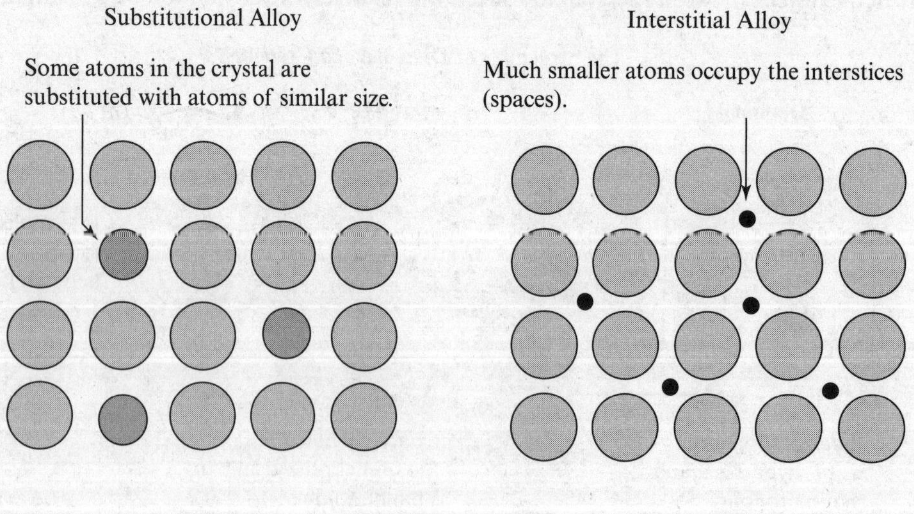

### Practice

Sterling silver is composed of about 93% silver and 7% copper. Is this a substitutional or interstitial alloy?

Since silver and copper atoms are both metals with fairly similar atomic radii, sterling silver is a substitutional alloy.

## Molecular Solids

Molecular solids are composed of discrete covalent molecules packed closely together. The individual molecules are held together by intermolecular forces, so molecular solids are softer and lower-melting than the other three types of solids. They may be amorphous, like wax, or crystalline, like ice or solid sucrose (table sugar). Other examples include solid $CO_2$ (dry ice), $I_2$ (iodine crystals), and $C_9H_8O_4$ (aspirin).

Although most solids tend to be denser than their corresponding liquid phase, solid water (ice) is a rare and notable exception. Each water molecule experiences hydrogen bonding with two additional water molecules. Due to the empty spaces in the crystal structure of ice, water is a rare example of a solid substance that is less dense than its liquid phase.

**The Crystal Structure of Ice**

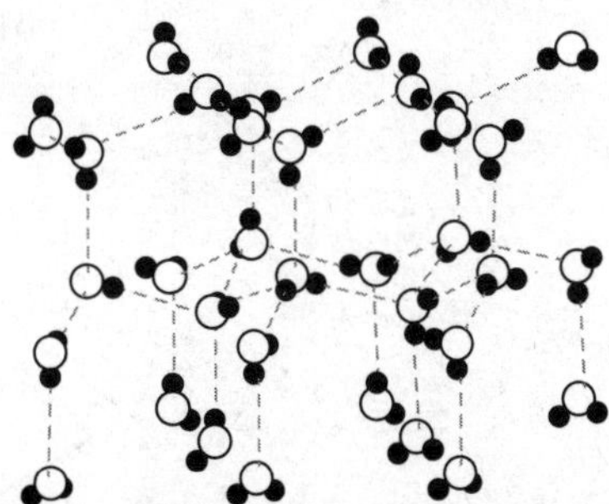

**Polymers** are sequences of small molecules called **monomers** that are covalently bound in long chains. Polymers are a class of very large molecules that do not fit neatly into the network covalent category since discrete polymer molecules usually do not span the entire solid. They are held together by weak intermolecular forces, and many have relatively low melting points. Polymers are better described as molecular solids.

**Polysaccharides** are **biopolymers,** polymers that come from biological sources. They are composed of long chains of simple sugar monomers such as glucose. The glucose biopolymer shown here, amylose, is a component of starch. Amylose is a solid powder isolated from vegetable matter. It is soluble in water and can be used to detect triiodide ions in laboratory solutions.

**Amylose, a Polymer of Glucose**

Glucose monomers

Amylose, a polysaccharide

Some polymers are **synthetic,** which means they are produced in a lab or commercial setting. Common useful examples are **polypropylene,** made from monomers of propene, and **acrylic,** made from monomers of methyl methacrylate.

**Synthetic Polymers**

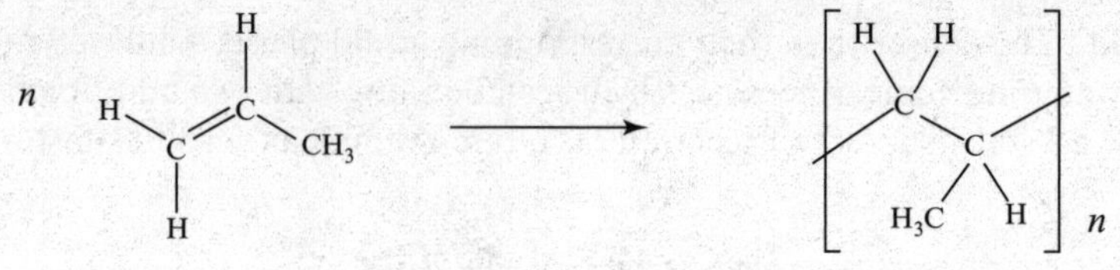

Propene monomer

Polypropylene polymer

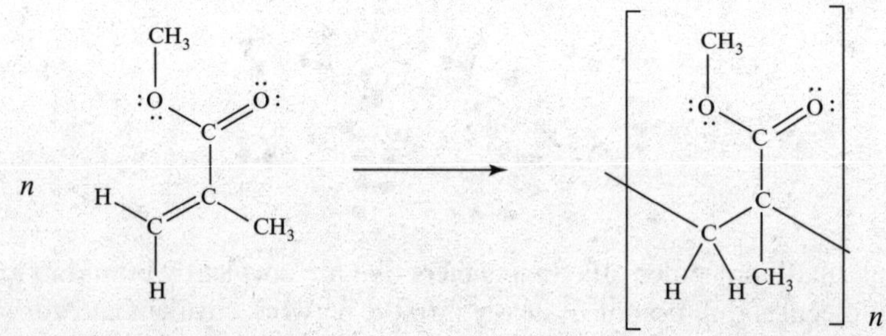

Methyl methacrylate monomer

Acrylic polymer

These are **thermoplastic** materials, which are tough and rigid solids at low temperatures due to a large number of intermolecular forces. These materials become flexible and pliable at high temperatures due to disruption of their intermolecular forces.

Polymers can be synthesized with a variety of properties to serve many different purposes. Students should be familiar with the conventional polymer structure notation (*n* repeating units, with repeating units enclosed in brackets or parentheses). The specific names of these molecules and mechanisms for their formation are beyond the scope of the AP Chemistry curriculum.

## Liquids

In the **liquid** state, molecules have interactions with their nearest neighbors, but these interactions are dynamic and the molecules are continually moving past one another. In contrast to solids, the locations of individual molecules in liquids are not fixed.

The shape of a liquid substance is also variable. Particles tend to arrange themselves in such a way as to minimize the surface area of the liquid. Liquids form flat surfaces in open containers and spherical droplets when suspended.

The volume of a liquid is relatively constant; it is only minimally compressible.

### Surface Tension

**Surface tension** refers to the tendency of a liquid to resist an increase in surface area. A molecule within the liquid interacts with surrounding molecules on all sides. These intermolecular forces are called **cohesive forces,** attractions between like molecules.

Molecules on the surface, by contrast, only experience cohesive forces with the molecules below them in the liquid. This unbalanced attraction of the surface molecules causes them to be constantly drawn down into the liquid, resulting in flattening of a liquid surface and formation of spherical droplets. The stronger the intermolecular forces between the liquid molecules, the greater the surface tension.

**Surface Tension**

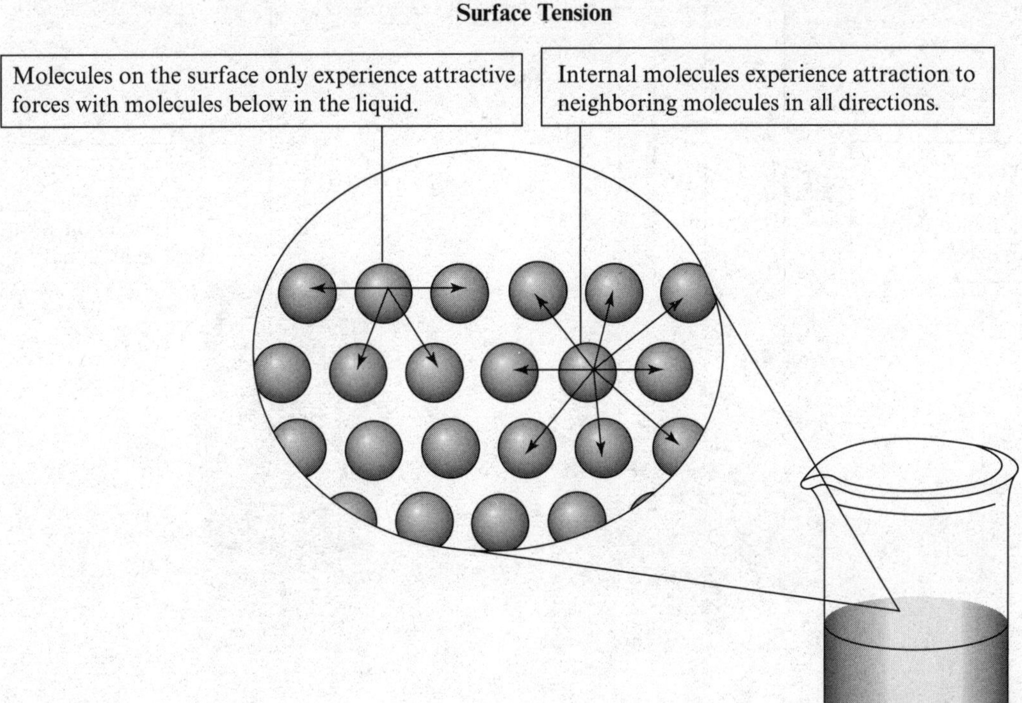


**Surfactants** are molecules that accumulate at the surface of a liquid and disrupt the surface tension. They are usually molecules that are **amphiphilic,** meaning they contain both a water-soluble, or **hydrophilic,** component and a water-insoluble, or **hydrophobic,** component. Soaps, detergents, and anti-fogging agents all behave as surfactants.

## Viscosity

**Viscosity** is the resistance of a liquid to flow. The stronger the intermolecular forces between the liquid molecules, the more viscous they are. Honey, which is a mixture of polar sugar molecules and water, is more viscous than pure water. In addition, long molecules such as polymers are more likely to become entangled with one another, resulting in high viscosity.

The viscosity of liquids is reduced at higher temperatures. When the temperature is raised, the average kinetic energy of the molecules is increased. This leads to greater motion and a greater ability to overcome intermolecular forces.

## Capillary Action

**Capillary action** refers to the behavior of liquids in narrow tubes.

When a narrow glass tube is placed in water, capillary action causes the water to rise in the tube. This is due to **adhesive forces,** which are intermolecular forces between the molecules of the water and the surface of the glass. Since adhesive forces between water and glass are stronger than cohesive forces between the molecules, the meniscus is concave.

Mercury, by contrast, has a convex meniscus since the cohesive forces between the mercury atoms are stronger than the adhesive forces between the mercury and the glass.

**Capillary Action**

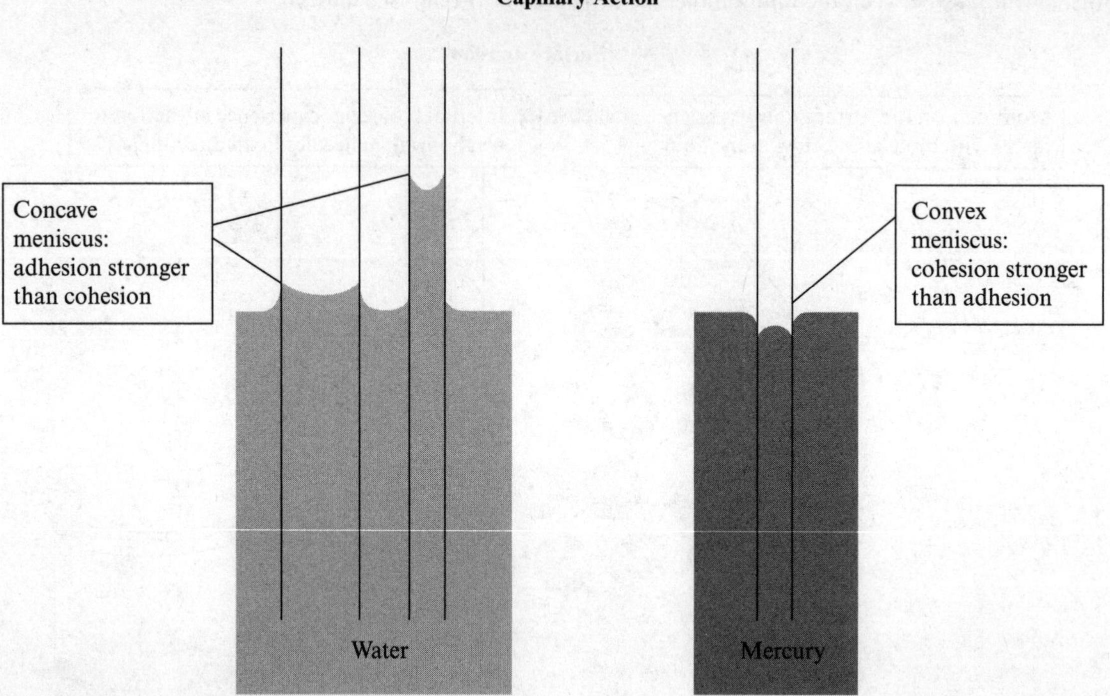


## Vapor Pressure

**Vaporization,** or **evaporation,** is a phase change from liquid to gas that occurs at the surface of a liquid. In a closed container, liquid begins to evaporate: Molecules at the surface are released into the gas phase. The rate of evaporation is greater than the rate of condensation until the temperature-dependent **equilibrium vapor pressure** is reached. Also referred to as simply the **vapor pressure,** this is the pressure of a gas in equilibrium above its liquid phase at a given temperature. At the equilibrium vapor pressure, the rate of evaporation and the rate of condensation are equal.

**Vapor Pressure**

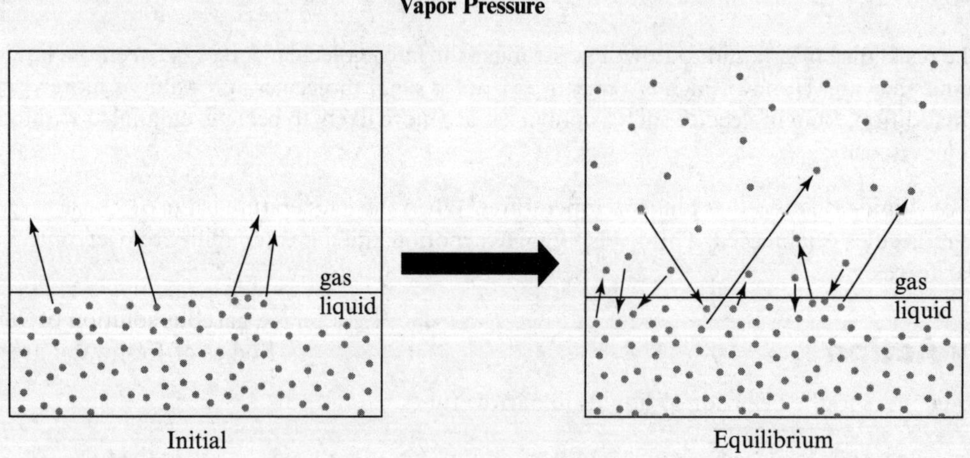


Vapor pressure depends on the strength of the intermolecular forces between the liquid molecules. The stronger the intermolecular forces, the more difficult it is for a liquid molecule to escape the attraction of its neighbors and transition to the vapor phase. For this reason, a molecule capable of participating in two hydrogen bonds will have a lower vapor pressure at room temperature than ethanol, $CH_3CH_2OH$, a molecule capable of a single hydrogen bond, or diethyl ether, $CH_3CH_2OCH_2CH_3$, a polar molecule that is incapable of hydrogen bonding. Diethyl ether, which experiences the weakest intermolecular forces of the three substances, has the highest vapor pressure.

**The Vapor Pressures of Three Substances as Functions of Temperature**

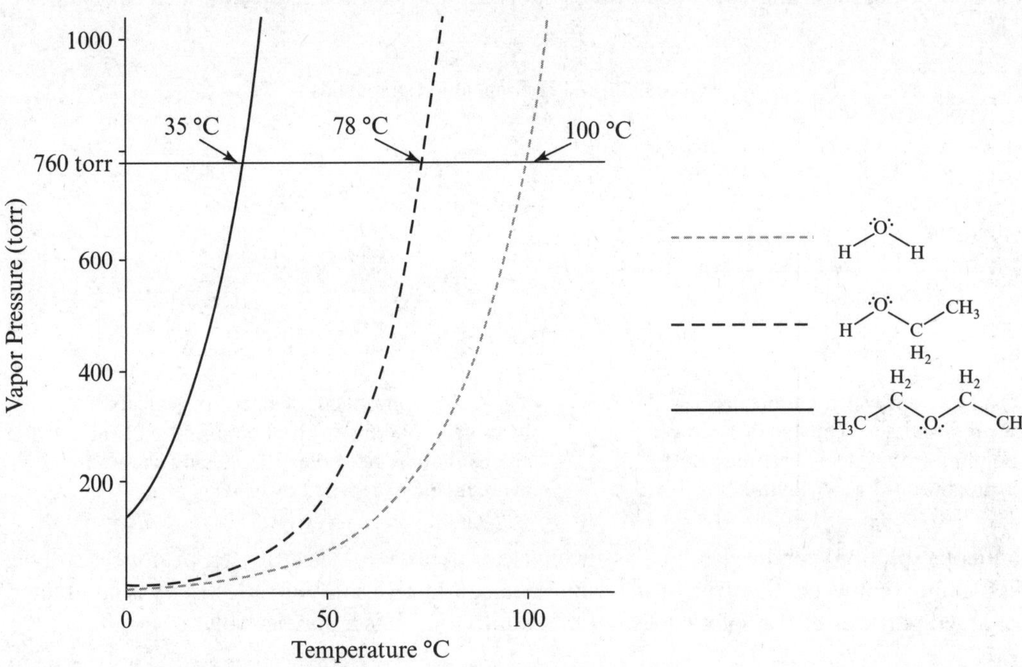


As temperature increases, the vapor pressure of a liquid increases. This is because at higher temperatures, a higher fraction of the molecules has sufficient energy to overcome the intermolecular forces at the surface of the liquid in order to enter the vapor phase.

The **boiling point** is the temperature at which the vapor pressure of the liquid equals the external pressure, which is usually the atmospheric pressure. Notice from the curves above that diethyl ether, the least polar of the three molecules shown, has the highest vapor pressure at all temperatures. Its boiling point at 760 torr or 1 atm is the lowest at 35 °C. It requires the least amount of energy to overcome the intermolecular forces. The boiling point of water at 760 torr is 100 °C. Its intermolecular forces are the strongest of the three substances shown, so more energy is required to overcome these attractions in order for water to enter the gas phase.

## Solutions

A **solution** is a homogeneous mixture of a major component and one or more minor components. The major component is called the **solvent** and the minor components are called **solutes.**

Solutions may be solids, liquids, or gases. An alloy such as bronze, which is a homogeneous mixture composed of 88% copper and 12% tin, is an example of a solid solution. Vinegar is an example of a liquid solution. Vinegar is a solution of 5% acetic acid dissolved in water. The atmosphere at sea level is a gaseous solution of 78% nitrogen, 21% oxygen, and smaller amounts of carbon dioxide, neon, and other gases. Bodies of freshwater are solutions of dissolved oxygen and other gases.

Many AP Chemistry questions will deal with **aqueous solutions,** those in which water is the solvent.

### Solubility and Intermolecular Forces

When an element or compound is capable of being dissolved in a particular solvent under a particular set of conditions, it is said to be **soluble** in that solvent. Solubility depends on the strength of attractive interactions between solute and solvent particles. The degree of solubility of a solute varies with temperature in a particular solvent.

In general, molecules with similar intermolecular forces are **miscible** (capable of being mixed). This is the principle behind the phrase "like dissolves like." Ionic and polar solutes tend to be soluble in polar solvents. Oxalic acid, shown below, is soluble in water because it is capable of forming hydrogen bonds with water.

Nonpolar solutes tend to be soluble in nonpolar solvents. Fats, oils, and other nonpolar compounds, such as the major component of beeswax, are insoluble in water but soluble in nonpolar solvents such as hexane, $C_6H_{12}$, and benzene, $C_6H_6$.

**Hydrophilic and Hydrophobic Compounds**

$H_3C(H_2C)_{29}$–C(=O)–O–$(CH_2)_{14}CH_3$

Oxalic acid is water-soluble because the molecules are capable of hydrogen bonding with water. This molecule is **hydrophilic** or "water-loving."

The major component of beeswax is incapable of hydrogen bonding. It also contains long hydrocarbon chains that are nonpolar. These long chains are **hydrophobic** or repelled by water.

Solutes in aqueous solutions can be classified as strong electrolytes, weak electrolytes, or nonelectrolytes. An **electrolyte** is a solute that when dissolved in water makes a conductive solution. Electricity is conducted by the motion of charged particles. If the solute does not break into ions, it is a nonelectrolyte.

**Strong electrolytes** are strong acids, strong bases, and soluble salts. When these compounds are added to water, they dissociate completely into ions. An additional intermolecular force, the **ion-dipole interaction,** is responsible for the solvation of the ions by water molecules. As the name suggests, ion-dipole interactions are coulombic attractions between cations and the electron-rich regions of polar molecules, and between anions and the electron-poor regions of polar molecules.

Sodium chloride is a soluble ionic compound. When a crystal of NaCl is added to water, the $Na^+$ and $Cl^-$ ions dissociate. They are completely solvated by water. The ability to produce a simplified particulate sketch similar to the one shown below is a valuable skill for AP Chemistry students.

**The Solvation of Sodium Chloride by Water**

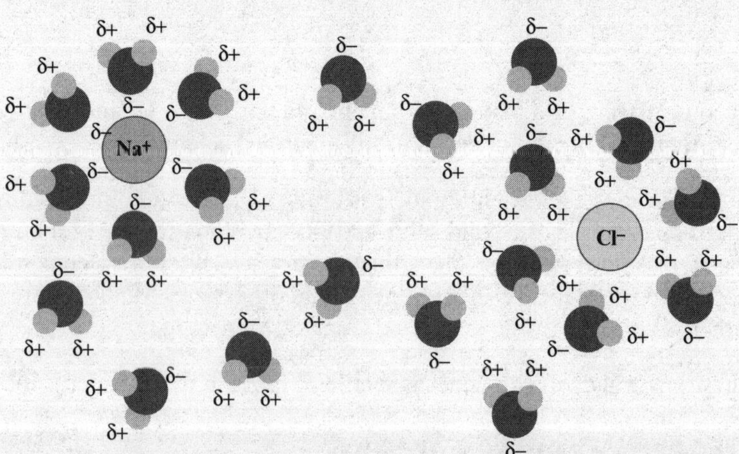

Molecular compounds that do not dissociate into ions in aqueous solutions are **nonelectrolytes.** Covalent molecules like sucrose (white cane sugar), which break up into discrete neutral molecules in water but do not dissociate into ions, are nonelectrolytes.

**Weak electrolytes** are compounds that undergo incomplete dissociation into ions in aqueous solutions. In other words, a few of the molecules ionize, but the rest do not. Weak electrolytes include weak acids and bases and slightly soluble salts.

Strong electrolytes are highly conductive in aqueous solutions. They allow a current to flow through the solution. A complete circuit with a power source that passes through a strong electrolyte solution allows a light bulb to shine brightly. Weak electrolytes allow only some of the current to flow, so the light bulb shines dimly. Nonelectrolytes allow so little current to flow that no light can be detected from the bulb.

**Electrolytes and Nonelectrolytes**

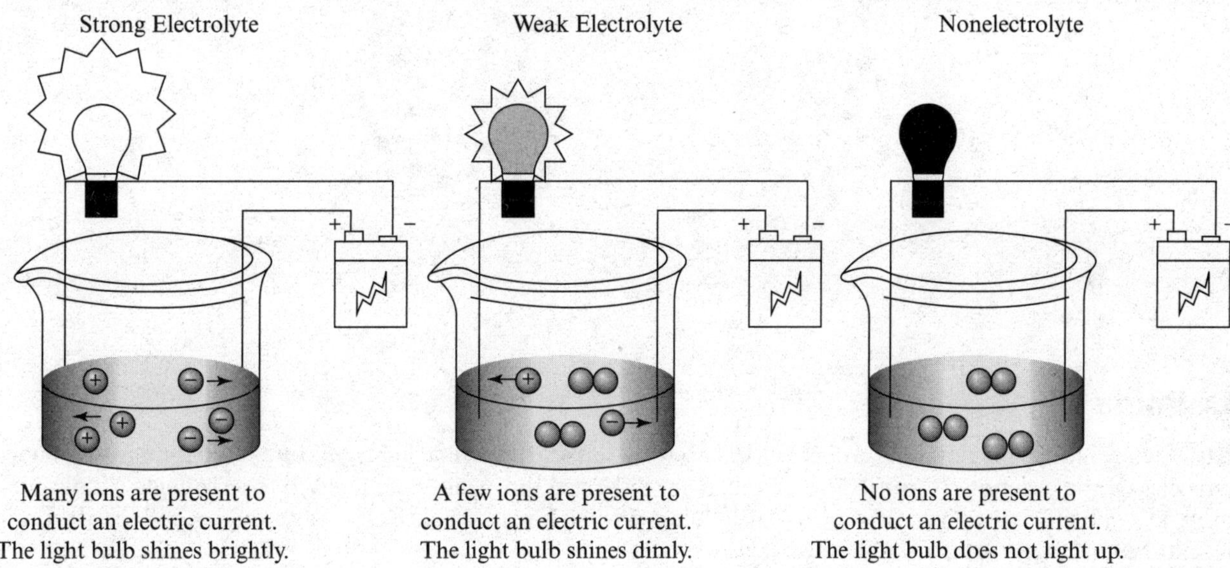


## Practice

A student would like to classify two unknown solids, A and B, as ionic or covalent. The student adds distilled water to a sample of each solid and discovers that A is soluble in water, but B is not. The student uses a conductivity meter to determine the conductivity of the aqueous solutions. Only solution A is conductive. Finally, the student finds that a sample of A is insoluble in hexane, $C_6H_{14}$, but that a sample of B is soluble. What types of bonding are most likely present in compounds A and B?

Since compound A forms a conductive solution when it dissolves in water and since it is insoluble in the nonpolar hydrocarbon solvent, it may either be a water-soluble ionic compound or a strong acid or base. Compound B's insolubility in water alone does not eliminate the possibility that it could be an ionic solid, since not all ionic compounds are water-soluble. Compound B is soluble in hexane, which is a nonpolar solvent. Since a nonpolar solvent is incapable of dissolving ionic compounds, compound B must be covalent rather than ionic.

## Biochemical Implications of Solubility

The water solubility of many biologically important molecules can have profound effects on biological structures and processes. For instance, proteins, which are polymers of amino acids, may have some portions that are hydrophobic and other portions that are hydrophilic. Since cytosol, the fluid portion of a cell, is an aqueous solution, proteins will arrange so that the hydrophilic portions are in contact with the solution and the hydrophobic portions associate with one another.

**Protein Folding in Aqueous Solution**

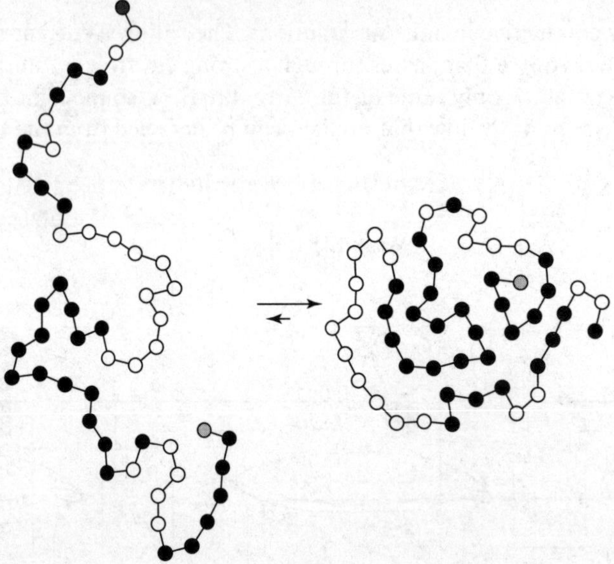


## Colloids

Although solutions are not necessarily colorless, they do not scatter visible light. Mixtures that contain suspended particles on the order of $10^{-9}$ to $10^{-6}$ m across that are dispersed throughout are referred to as **colloids.** Smoky air, milk, fog, gelatin, and whipped cream are all colloids. When a beam of light is passed through a colloid, the beam is scattered in a way that is visually detectable.

**The Scattering of a Beam of Light by a Colloid**

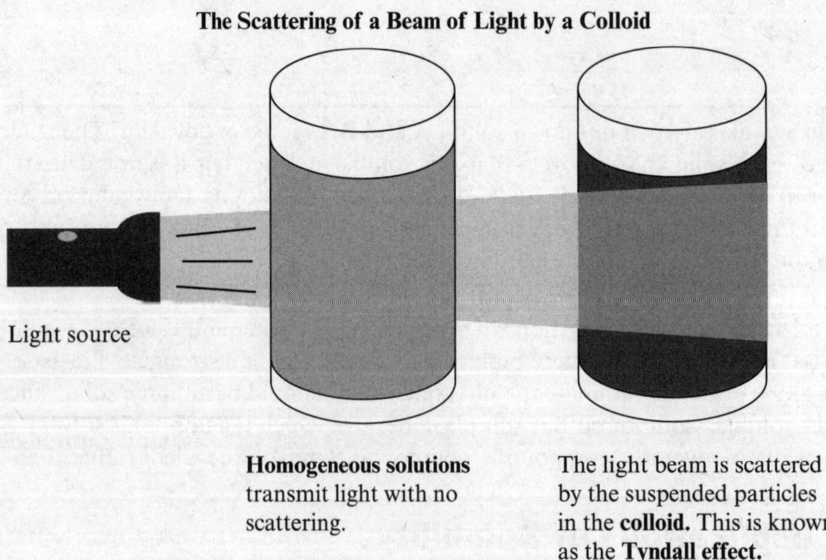


**Homogeneous solutions** transmit light with no scattering.

The light beam is scattered by the suspended particles in the **colloid.** This is known as the **Tyndall effect.**

## Effect of Temperature on the Solubility of Solids

As the temperature of a solvent increases, the rate of dissolution of solid solutes also increases. For many solids, an increase in temperature also increases the **solubility,** or the amount of solute that is capable of being dissolved in a certain volume of solvent. This is because the increased average kinetic energy leads to an increased number and strength of collisions between solvent and solute particles. These collisions disrupt intermolecular and electrostatic attractions between solute particles.

A **saturated solution** contains exactly the amount of solute that can be dissolved in a given amount of solvent at that temperature.

An **unsaturated solution** contains less than the maximum amount of solute that can be dissolved in a given amount of solvent at that temperature.

A **supersaturated solution** contains more than the maximum amount of solute that can be dissolved in a given amount of solvent at that temperature. Supersaturated solutions tend to form solid crystals of the excess solute until a saturated solution is achieved at a particular temperature.

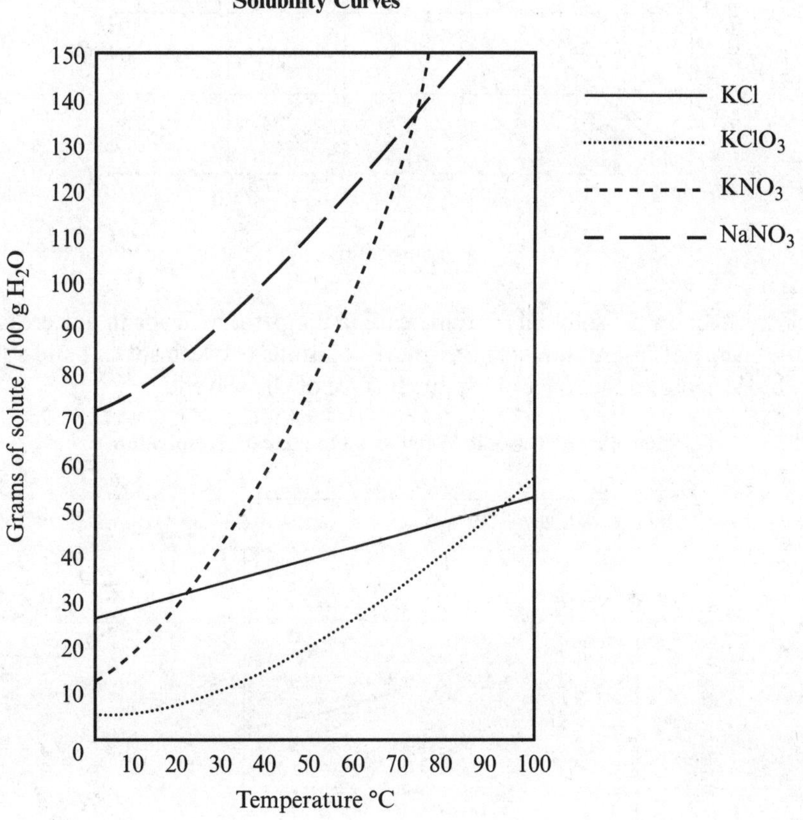


### Practice

Refer to the figure above to determine the following. A 200-gram sample of water is saturated with $KNO_3$ at 60 °C. Approximately how much $KNO_3$ crystallizes if the solution is cooled to 30 °C?

From the solubility curve for $KNO_3$, we see that 100 grams of water contains approximately 100 grams of $KNO_3$ in the saturated state at 60 °C, so 200 grams of water must contain 200 grams of $KNO_3$. At 30 °C, 100 grams of water contains approximately 40 grams of $KNO_3$, so 200 grams of water contains 80 grams of $KNO_3$. When the 200-gram water solution is cooled, the $KNO_3$ crystallizes until 80 grams are left dissolved in the water. The amount that crystallizes is

$$200\text{ g} - 80\text{ g} = 120\text{ g } KNO_3$$

## Effects of Pressure and Temperature on the Solubility of Gases

The number of moles of a gas that will dissolve in a liter of liquid solution is directly proportional to the partial pressure of the gas in the space above the solution. This is known as **Henry's law.** As gas pressure increases, more gas particles collide with the surface of the solvent and are incorporated into the solution.

**The Solubility of Gases in Water as a Function of Pressure**

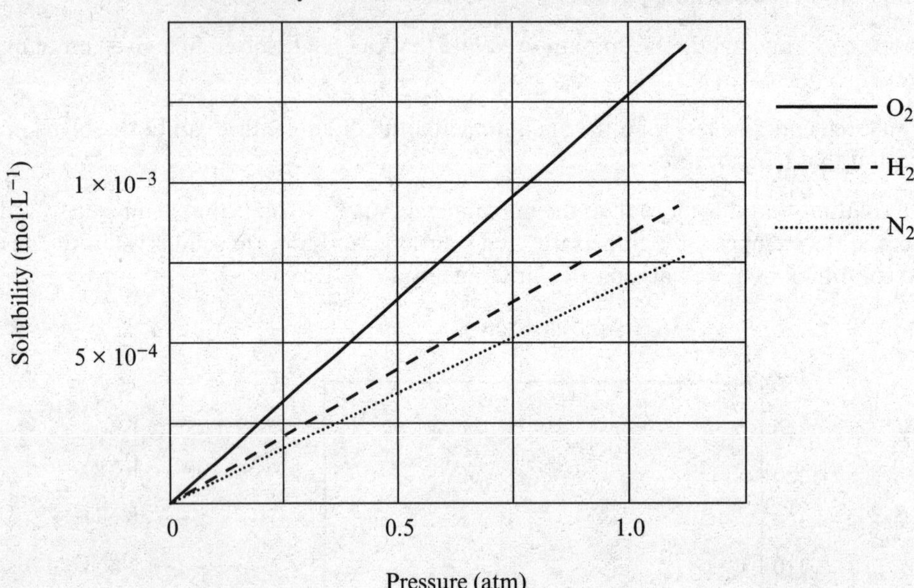


Temperature also has an effect on gas solubility. An increase in temperature leads to a decrease in the solubility of gases. As the kinetic energy of a solution increases, more collisions between solvent and solute particles cause the gaseous solute to be ejected into the space above the surface of the solvent.

**Solubility of Gases in Water as a Function of Temperature**

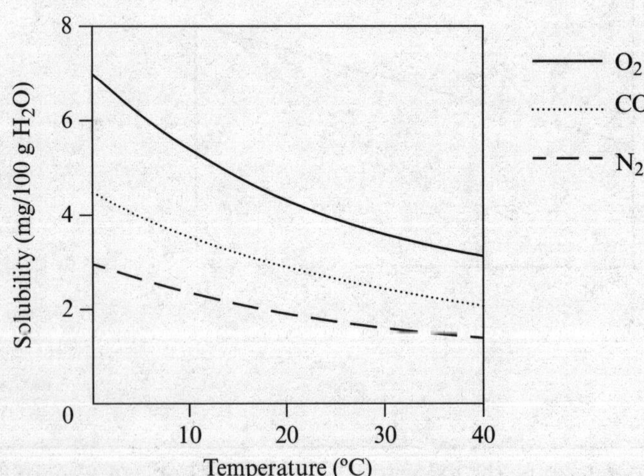


## Solution Concentration

There are several ways of expressing the concentration of a solution, but in AP Chemistry, the only ones that are needed are mole fraction, $X$, and molarity, $M$.

Mole fraction was discussed earlier in this chapter in the context of Dalton's law of partial pressures.

**Molarity** is the number of moles of solute dissolved in enough solvent to make exactly 1 liter of solution.

Any time a species is shown in brackets, molarity is implied. A 2.0 *M* solution of $OH^-$ is indicated by $[OH^-] = 2.0\ M$.

**Volumetric flasks,** flasks that are calibrated to hold a precise volume at a particular temperature, are used to prepare solutions of known molarity. In order to make a known concentration of a solution, the solid solute is weighed on a balance and placed into a volumetric flask. Solvent is added to the mark on the neck of the flask.

**Volumetric Flask**

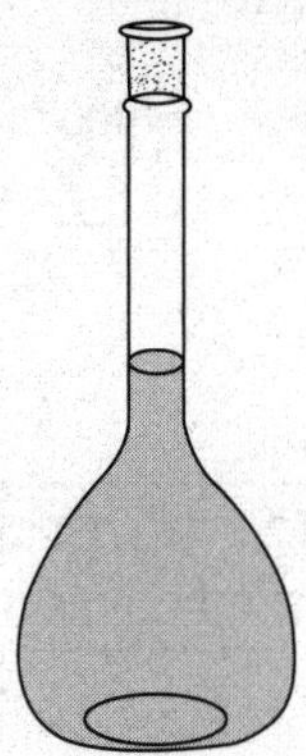

Precise volumes of liquid solutes or concentrated solutions are delivered using **pipettes.**

**Pipettes**

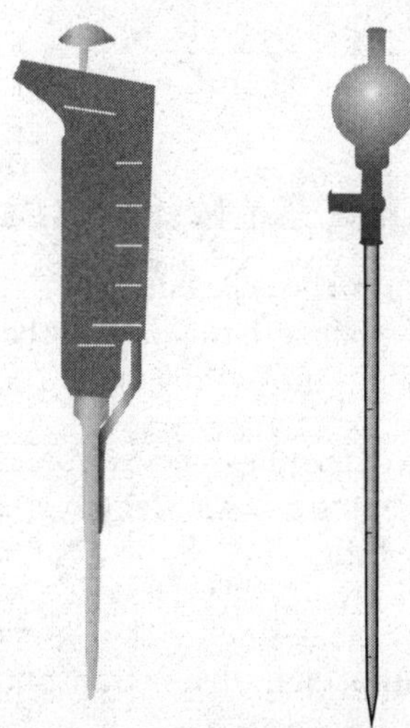

Sometimes a solution is made by dilution of a more concentrated solution. Dilution is a decrease in concentration of a solute by addition of a solvent. The dilution equation, which can be used to calculate the final concentration of a solution after dilution, is a consequence of the fact that the number of moles of solute does not change.

$$M_1 \times V_1 = M_2 \times V_2$$

## Practice

A student weighs 101.7 grams of magnesium chloride hexahydrate in a weigh boat on a balance and then pours the salt into a 100.0-mL volumetric flask. The student adds about 50 mL of distilled water, swirls the mixture until the solid dissolves completely, and then fills the flask with water to the 100.0-mL mark. What is the molarity of the resulting solution?

The molar mass of $MgCl_2 \cdot 6H_2O$ is 203.32 g·mol$^{-1}$. The number of moles is found by multiplying the mass of the salt by the molar mass.

$$101.7 \ \cancel{\text{g } MgCl_2 \cdot 6H_2O} \times \frac{1 \text{ mol } \cancel{MgCl_2 \cdot 6H_2O}}{203.32 \ \cancel{\text{g } MgCl_2 \cdot 6H_2O}} \times \frac{1 \text{ mol } MgCl_2}{1 \ \cancel{\text{mol } MgCl_2 \cdot 6H_2O}} = 0.5002 \text{ mol } MgCl_2$$

The molarity is the number of moles divided by the volume.

$$\frac{0.5002 \text{ mol } MgCl_2}{100.0 \ \cancel{\text{mL}}} \times \frac{1000 \ \cancel{\text{mL}}}{1 \text{ L}} = 5.002 \ M$$

## Practice

---

A 5.00-mL volumetric pipette is used to deliver 12.0 *M* HCl to a 25.0-mL volumetric flask. Distilled water is added to the 25.0-mL mark. What is the resulting concentration of the dilute solution?

---

The initial molarity and volume of HCl, $M_1$ and $V_1$, are 12.0 *M* and 5.00 mL, respectively. The final volume of dilute HCl solution is 25.0 mL. The final molarity can be determined by rearranging the dilution equation.

$$M_2 = \frac{M_1 \times V_1}{V_2}$$

$$M_2 = \frac{(12.0 \ M) \times (5.00 \ \cancel{\text{mL}})}{25.0 \ \cancel{\text{mL}}} = 2.40 \ M$$

# Measuring Concentration Using Spectrophotometry

Some solutions have color because the molecules they contain absorb certain wavelengths of visible light. (The color of a solution is complementary to the wavelength that is absorbed.) Information about the concentration of a colored solute can be elucidated from the amount of visible light a solution absorbs.

A **spectrophotometer** is an instrument used to determine the absorbance of a solution. It contains a source of light that can be separated into discrete wavelengths. A wavelength is selected that corresponds to an electronic energy level transition in a molecule under investigation.

The solution to be measured is placed in a **cuvette,** a small vial of known width, *b*. The cuvette with the sample is inserted into the spectrometer and the amount of light that is transmitted through the sample is detected and used to determine the amount of light that is absorbed.

**Components of a Spectrophotometer**

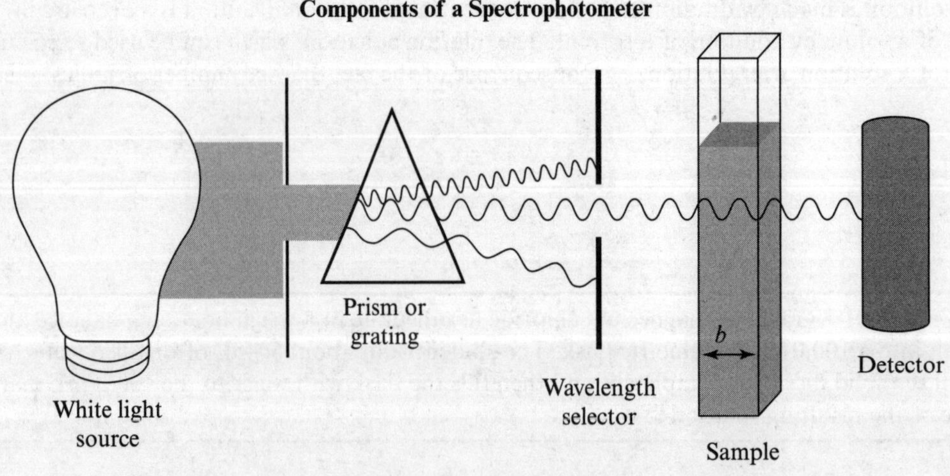

When the absorbed wavelength is passed through a solution, the amount of light that is absorbed by the sample is directly proportional to the concentration of the sample. This relationship, which is known as **Beer's law,** is provided on the *AP Chemistry Equations and Constants* sheet.

$$A = \varepsilon bc$$

In this equation, $A$ is the **absorbance,** $\varepsilon$ is the **molar absorptivity,** $b$ is the **path length** or the width of the cuvette, and $c$ is the concentration.

To determine the concentration of an unknown sample, a calibration curve using absorptions of a series of solutions of known concentration is needed. The absorptions are plotted versus their concentration. The slope of the resulting line gives the molar absorptivity for the solute. Once the absorptivity is known, the concentration of the unknown can be determined.

## Practice

The molar absorptivity, $\varepsilon$, of $[Fe(C_{12}H_8N_2)_3]^{2+}$ at its maximum absorption wavelength is $1.11 \times 10^4$ L·mol$^{-1}$·cm$^{-1}$. What is the molarity of a sample that has an absorbance, $A$, of $5.55 \times 10^2$ in a 1.0-cm cuvette?

The concentration can be determined by rearranging the Beer's law equation.

$$c = \frac{A}{\varepsilon b}$$

$$c = \frac{5.55 \times 10^2}{(1.11 \times 10^4 \text{ L} \cdot \text{mol}^{-1} \cdot \cancel{\text{cm}^{-1}}) \times (1.0 \ \cancel{\text{cm}})} = 0.050\ M$$

## Practice

A student would like to determine the amount of copper in a sample of brass. The student creates a calibration curve, shown below, by measuring the absorbances of known concentrations of $Cu(NO_3)_2$. The student then dissolves a 0.212-gram sample of brass in a small amount of concentrated nitric acid. The resulting $Cu^{2+}$ solution is diluted to 25.0 mL. The student makes a second dilution in which 2.00 mL of the $Cu^{2+}$ solution is diluted to 500. mL. She measures an absorbance of 4.375 using a 1.0-cm cuvette. What is the percentage of copper in the original sample?

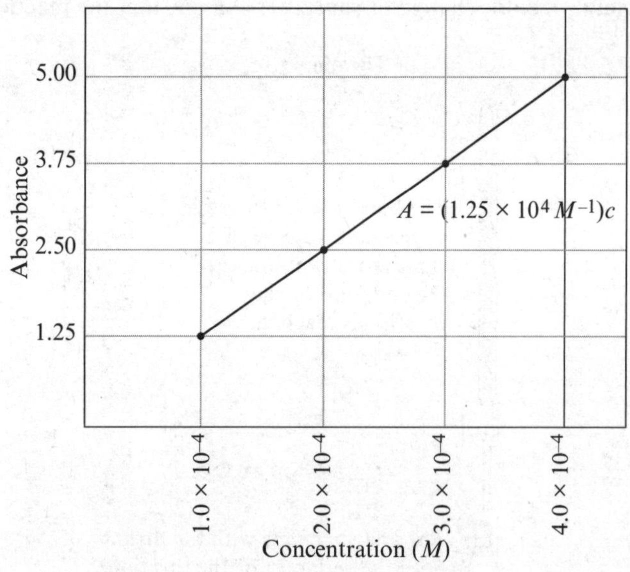

In order to find the concentration of the 500.-mL $Cu^{2+}$ solution, the student must determine the slope of the concentration versus absorbance graph. The slope, which is equal to the product $\varepsilon b$, may be determined by choosing any two points on the graph and finding the difference.

$$\varepsilon b = \frac{\Delta A}{\Delta c} = \frac{5.00 - 1.25}{4.0 \times 10^{-4}\,M - 1.0 \times 10^{-4}\,M} = 1.25 \times 10^{4}\,M^{-1}$$

Since $\varepsilon b$ has been determined, the concentration of $Cu^{2+}$ in the final solution may be found using the measured absorbance.

$$c = \frac{A}{\varepsilon b} = \frac{4.375}{1.25 \times 10^{4}\,M^{-1}} = 3.50 \times 10^{-4}\,M$$

The measured concentration, $M_2$, was the result of diluting 2.00 mL ($V_1$) of the original 25.0-mL solution to a volume of 500. mL ($V_2$). The concentration of the 25.0-mL solution can be found using the dilution equation.

$$M_1 = \frac{M_2 \times V_2}{V_1} = \frac{3.50 \times 10^{-4}\ M \times 500.\ \cancel{\text{mL}}}{2.00\ \cancel{\text{mL}}} = 0.0875\ M$$

The concentration of the 25.0-mL solution is next used to determine the number of moles of copper present in the original sample.

$$0.0250\ \cancel{\text{L}} \times \frac{0.0875\ \text{mol}}{1\ \cancel{\text{L}}} = 0.00219\ \text{mol Cu}$$

The number of moles of copper is converted to grams, and this mass is used to determine the mass percent of the original sample.

$$0.00219\ \cancel{\text{mol Cu}} \times \frac{63.55\ \text{g Cu}}{1\ \cancel{\text{mol Cu}}} = 0.139\ \text{g Cu}$$

$$\text{Percent copper} = \frac{0.139\ \text{g Cu}}{0.212\ \text{g sample}} \times 100\% = 65.6\%\ \text{Cu}$$

## Measuring Concentration Using Titration

**Titration** is a technique by which an unknown concentration of an **analyte** in solution is determined by reaction with an accurately known amount of a standard, or **titrant.** The setup usually involves a known volume of the analyte in an Erlenmeyer flask. The titrant is added using a buret, a specialized piece of glassware that allows measured volumes of titrant to be added. The titrant is added until the endpoint of the titration is reached. The **endpoint** is some indication, usually a color change or sharp pH change, that the reaction is complete.

**Titration**

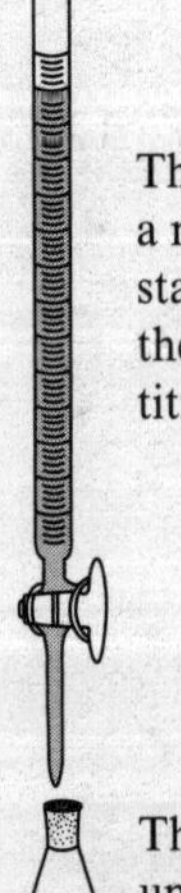

In many cases, an indicator is added to the analyte. An **indicator** is a substance that changes color at or near the equivalence point of the reaction. The **equivalence point** is the point in which stoichiometric amounts of titrant and analyte have reacted. A balanced chemical equation is necessary to determine the stoichiometric ratio of titrant to analyte.

Phenolphthalein is a common indicator that is used to detect the endpoints of titrations of strong acids with strong bases. It changes from colorless in acidic solutions to pink in basic solutions.

## Practice

A 25.0-mL sample of $HNO_3$ is titrated with a standardized solution of 0.460 *M* NaOH. The endpoint of the titration is indicated by a phenolphthalein color change when 13.1 mL of NaOH solution has been added. What is the molarity of the $HNO_3$ solution? The balanced equation for the reaction of NaOH with $HNO_3$ is shown.

$$HNO_3(aq) + NaOH(aq) \rightarrow NaNO_3(aq) + H_2O(l)$$

The balanced chemical equation shows that for each mole of $HNO_3$, one mole of NaOH reacts.

$$13.1\ \cancel{\text{mL NaOH}} \times \frac{1\ \cancel{\text{L NaOH}}}{1000\ \cancel{\text{mL NaOH}}} \times \frac{0.460\ \cancel{\text{mol NaOH}}}{1\ \cancel{\text{L NaOH}}} \times \frac{1\ \text{mol HNO}_3}{1\ \cancel{\text{mol NaOH}}} = 6.03 \times 10^{-3}\ \text{mol HNO}_3$$

The concentration of $HNO_3$ is determined using the number of moles and the original volume.

$$\frac{6.03 \times 10^{-3}\ \text{mol HNO}_3}{25.0\ \cancel{\text{mL}}} \times \frac{1000\ \cancel{\text{mL}}}{1\ \text{L}} = 0.241\ M\ \text{HNO}_3$$

## Separation by Chromatography

**Chromatography** refers to a class of laboratory techniques in which the components of a solution are separated based on differences in the strengths of their intermolecular interactions with a stationary phase and a mobile phase. A **stationary phase** is a solid such as absorbent paper or silica powder, and the **mobile phase** is a liquid or gas that is passed over the stationary phase.

Because different components of a solution interact differently with the stationary and mobile phases, the components travel with the mobile phase through the stationary phase at different speeds and can be separated.

**Paper chromatography** is used to separate very small amounts of a mixture. The bottom of a thin strip of paper is dotted with the mixture to be analyzed. The paper is the polar **stationary phase.** It is sealed in a jar and dipped in a small amount of solvent. As the solvent, or **mobile phase,** begins to creep up the paper, some components of the mixture interact more strongly with the solvent than with the paper. These components move up the paper with the solvent.

**Paper Chromatography**

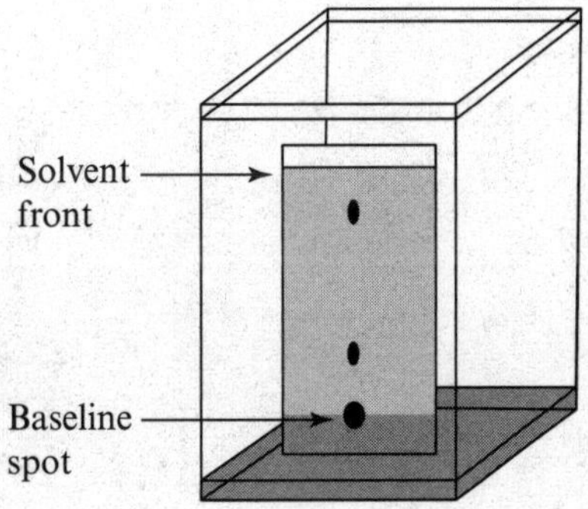

Solvent front

Baseline spot

Different components of a mixture will have different **retention factors** ($R_f$) in a particular solvent system. The $R_f$ for a compound is calculated by dividing the distance travelled by that compound by the distance travelled by the solvent front.

**The Calculation of $R_f$ in Paper Chromatography**

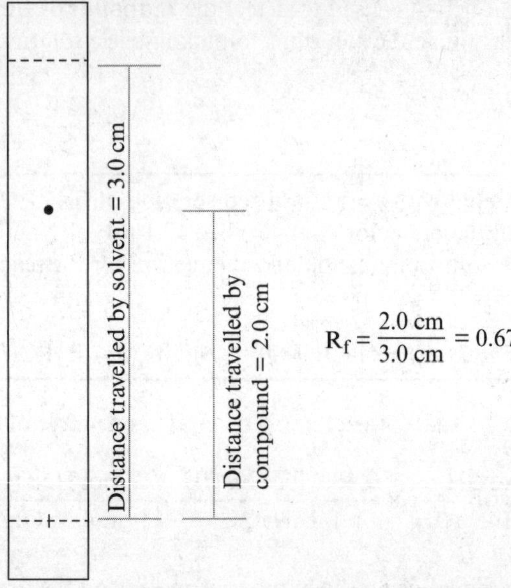


**Column chromatography** is a method for separating the components in a mixture by taking advantage of the different attractions of the components to an adsorbent that is contained in a column. The steps in a column chromatographic separation, represented in the figure below, are as follows:

1. A solid adsorbent, or stationary phase, usually silica powder, is poured into a glass column that contains a plug of glass wool in the tip. The glass wool allows the liquid components to pass through while keeping the silica in.
2. The solid polar silica gel is wetted with a suitable solvent, or mobile phase. Mobile phases tend to be less polar than silica.
3. The sample mixture is loaded onto the surface of the silica in the column.
4. Solvent is used to wash the components of the mixture through the silica. The faster-moving component is the component that interacts more strongly with the mobile phase, and the slower-moving component is the component that interacts more strongly with the stationary phase.
5. As solvent is continually added, the components of the mixture separate.
6. The less polar component elutes first and is collected.
7. The more polar component elutes second and is collected.

**Column Chromatography**

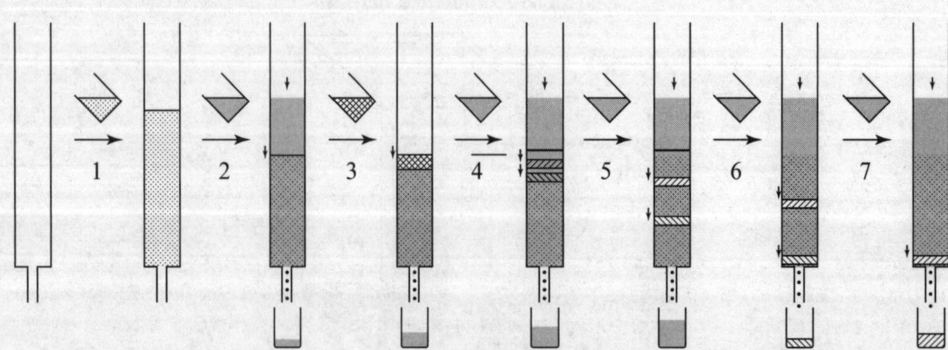

## Practice

A mixture contains a polar component and a nonpolar component. Paper chromatography is performed on the mixture in which a relatively nonpolar solvent is used. The $R_f$ for component A is 0.80, and the $R_f$ for component B is 0.20. Identify components A and B as polar or nonpolar.

Since the $R_f$ of component A indicates that it travelled farther up the paper, it interacts more strongly with the nonpolar solvent than with the paper. Therefore, component A is the nonpolar component, and component B is the polar component.

## Separation by Evaporation and Distillation

A nonvolatile solute can be recovered from a solution by a simple **evaporation,** in which the volatile solvent is allowed to evaporate or is gently boiled away. Sodium chloride can be recovered from a saltwater solution in this way.

If the solvent needs to be recovered, however, or if a solution of two volatile components needs to be separated, distillation is a better choice. **Distillation** is a separation method that is based on differences in vapor pressures and boiling points among the components of the solution.

The mixture is placed in a round-bottom flask equipped with a condenser and a receiving vessel. The mixture in the round-bottom flask is heated to the boiling point of the more volatile component. This volatile component remains in the vapor phase until it is cooled in the condenser. The vapor then forms liquid droplets on the inner surface of the cooled condenser. The liquid, or **distillate,** drips into the receiving vessel, where it is collected.

**Distillation**

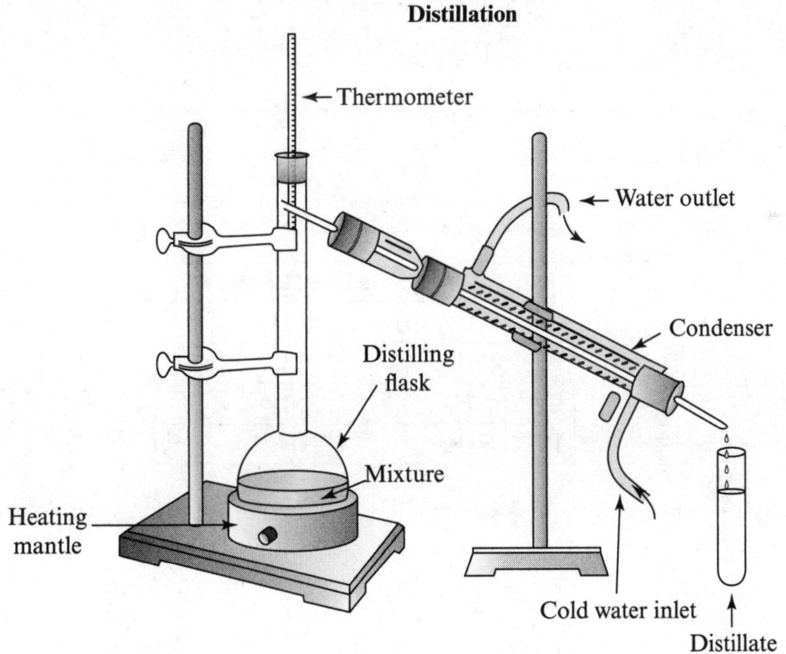


## Practice

A mixture of water and ethyl alcohol ($CH_3CH_2OH$) is distilled. Which component is collected first and why?

Ethyl alcohol is collected first, at a lower temperature. Both ethyl alcohol and water are capable of hydrogen bonding, but ethyl alcohol has a lower boiling point due to the nonpolar portion of the molecule that experiences London dispersion forces only. Ethyl alcohol, therefore, distills before the temperature reaches a temperature at which water boils.

# Review Questions

## Multiple Choice

**1.** Which of the following molecules experiences the strongest intermolecular forces?

A. HCl
B. $H_2O$
C. $H_2O_2$
D. $C_2H_6$

**2.** The figure below shows the boiling points of hydrides in a single group of the periodic table. Which of the following could reflect the identity of the hydrides at points A and B?

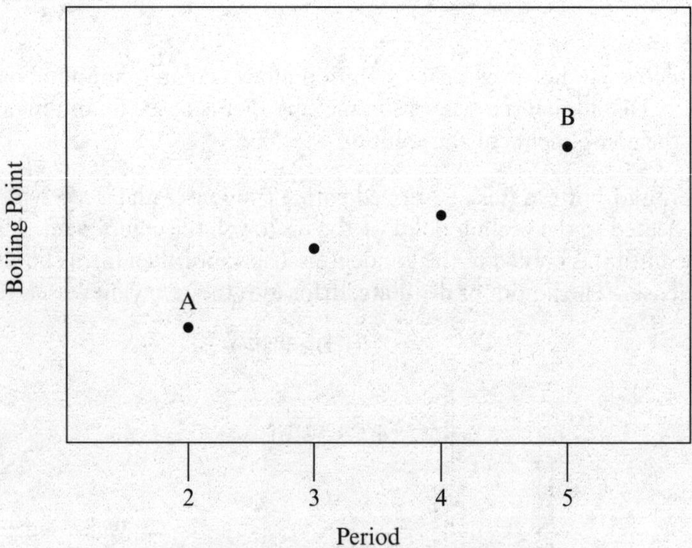


| | Molecule A | Molecule B |
|---|---|---|
| **A.** | $CH_4$ | $SnH_4$ |
| **B.** | $SnH_4$ | $CH_4$ |
| **C.** | $NH_3$ | $SbH_3$ |
| **D.** | $SbH_3$ | $NH_3$ |

**3.** A 30.3-gram sample of gas occupies 33.6 L at standard temperature and pressure. What is the identity of the gas?

A. He
B. Ne
C. Ar
D. $O_2$

**4.** A sample of gas at constant volume is heated from 230 K to 690 K. Which of the following best describes the effect on the gas?

**A.** The final pressure is triple the initial pressure, and the average kinetic energy of the molecules is tripled.
**B.** The final pressure is triple the initial pressure, and the average kinetic energy of the molecules is constant.
**C.** The final pressure is the cube root of the initial pressure, and the average kinetic energy of the molecules is tripled.
**D.** The final pressure is the cube root of the initial pressure, and the average kinetic energy of the molecules is constant.

**5.** The Maxwell distribution for two gases at the same temperature is shown below. Which of the following is true?

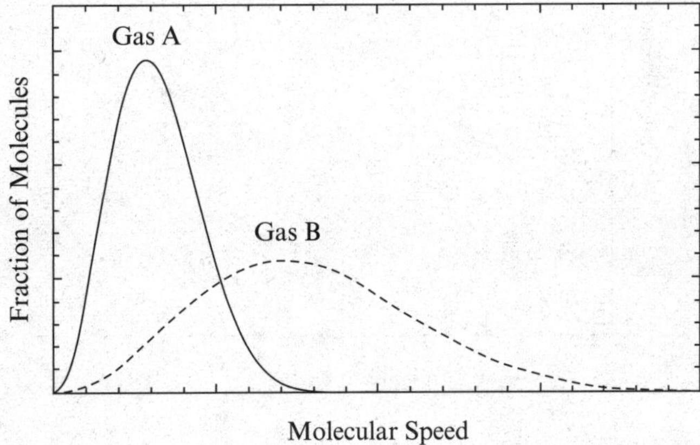


**A.** Gas A has a higher average kinetic energy than gas B.
**B.** Gas B has a higher average kinetic energy than gas A.
**C.** Gas A has a higher molar mass than gas B.
**D.** Gas B has a higher molar mass than gas A.

**6.** Which of the following correctly ranks four gases in order of increasingly ideal behavior at constant temperature and pressure?

**A.** $Ne < N_2 < CO < NH_3$
**B.** $Ne < CO < N_2 < NH_3$
**C.** $NH_3 < N_2 < CO < Ne$
**D.** $NH_3 < CO < N_2 < Ne$

**7.** An analytical chemist needs to make a 0.5000 *M* HCl solution. The chemist has a 12.00 *M* solution available in addition to distilled water, a balance, an Erlenmeyer flask, a variety of graduated pipettes, a 250.0-mL volumetric flask, and a 250.-mL graduated cylinder. Which of the following procedures would result in the correct HCl concentration?

**A.** Place the volumetric flask on the balance and add 4.56 grams of the HCl solution. Dilute to the mark with distilled water.
**B.** Pipette 10.42 mL of the 12.00 *M* HCl into the Erlenmeyer flask. Use the graduated cylinder to add 239.6 mL of distilled water.
**C.** Place the Erlenmeyer flask on the balance and add 4.56 grams of the HCl solution. Use the graduated cylinder to add 245.4 mL of distilled water.
**D.** Pipette 10.42 mL of the 12.00 *M* HCl into a 250.0-mL volumetric flask. Dilute to the mark with distilled water.

**8.** Which of the following compounds would be expected to produce a homogeneous solution when combined?

**A.** $CH_3COOH$ and $C_5H_{12}$
**B.** $LiNO_3$ and $C_6H_6$
**C.** $CH_3COOH$ and $H_2O$
**D.** $C_6H_6$ and $CH_3OH$

**9.** Match the curves relating vapor pressure to temperature for the following compounds.

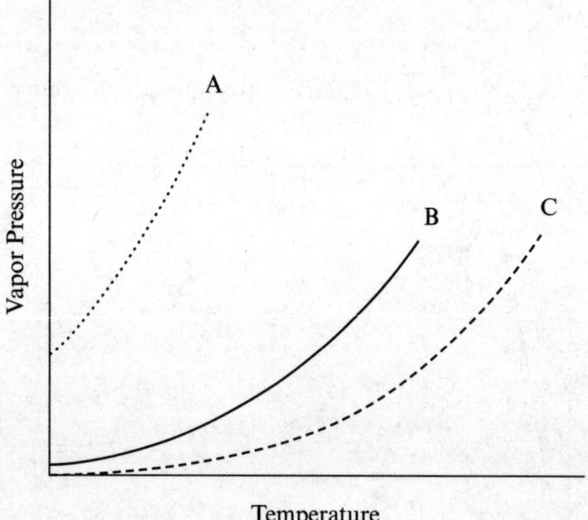


**A.** A = $SiH_4$, B = $H_2S$, C = $NH_3$
**B.** A = $NH_3$, B = $H_2S$, C = $SiH_4$
**C.** A = $SiH_4$, B = $NH_3$, C = $H_2S$
**D.** A = $H_2S$, B = $NH_3$, C = $SiH_4$

*Use the following particulate diagrams to answer questions 10–11.*

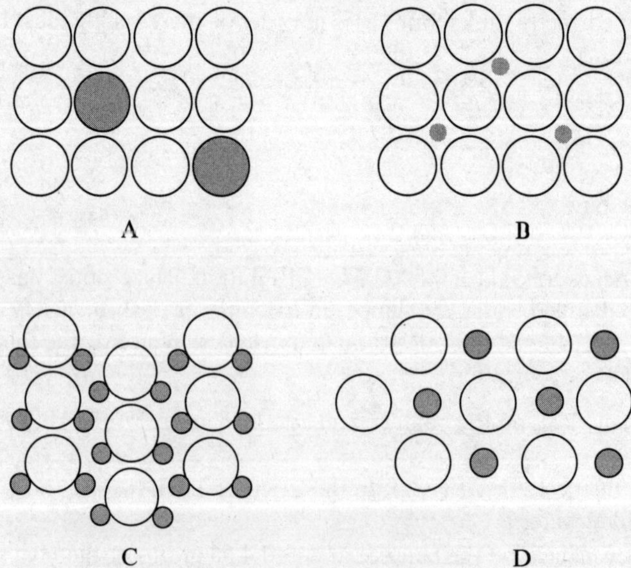

**10.** A jeweler uses an alloy of silver and zinc to solder a silver bracelet closed. The alloy is low-melting, and its density lies between that of silver and zinc. The solid form of this alloy is best represented by which solid?

A. Solid A
B. Solid B
C. Solid C
D. Solid D

**11.** Which statement is most likely to describe the solids represented by the particulate diagrams?

A. Solids A and D are malleable, and solids B and C are brittle.
B. Solids A and B are malleable, and solids C and D are low-melting.
C. Solids A and D are electrically conductive, and solids B and C are electrical insulators.
D. Solids A and B are electrically conductive, and solids C and D are electrical insulators.

**12.** Which of the following polymer structures would be expected to have the highest density?

A.
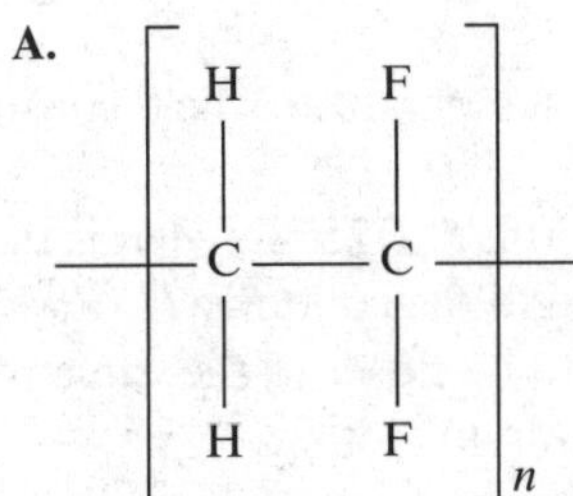


B.
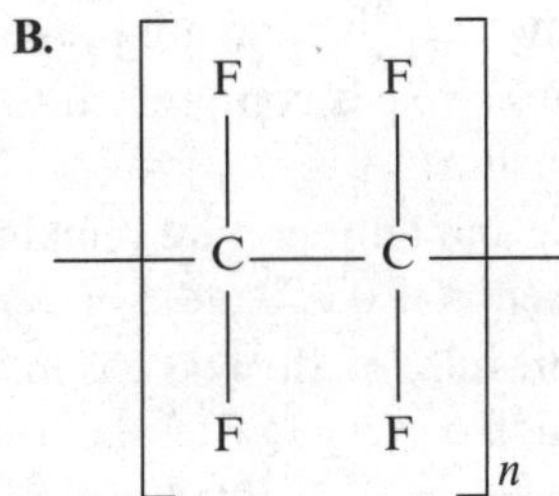


C.
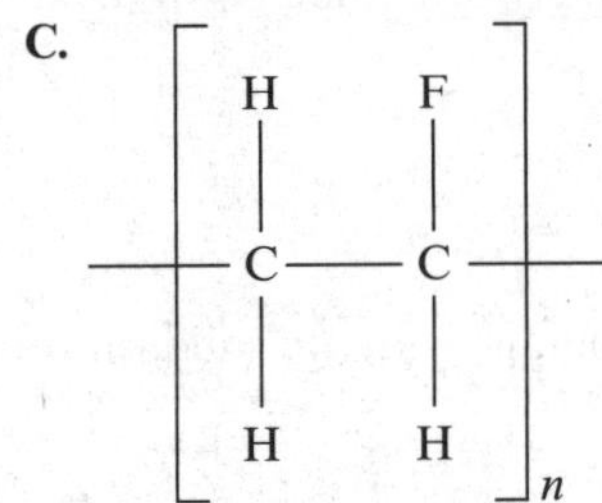


D.
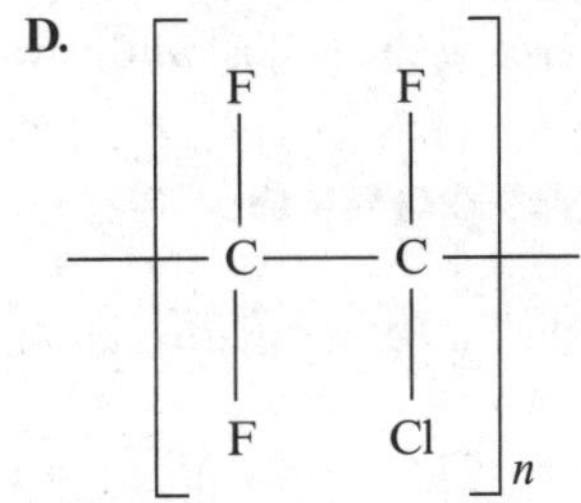


**13.** Which of the following statements is true of graphite?

A. Graphite is a network covalent material that is conductive in its solid form due to delocalized $\pi$ electrons.
B. Graphite is a hard network covalent solid because every atom in the network is $sp^3$ hybridized.
C. The layers of graphite in pencil lead easily slip past one another because they are held together by relatively weak hydrogen bonds.
D. Graphite has an extremely high melting point because it is composed of ionized particles that are arranged in a manner that maximizes coulombic attractions and minimizes coulombic repulsions.

**14.** Which choice does NOT correctly complete the following sentence?

The process of distillation is best described as _______.

A. a method for separating substances such as ethanol and water
B. a method for separating sodium chloride from sand
C. a method for separating two liquids based on differential boiling points
D. a method for separating two liquids based on differential vapor pressures

**15.** Two 20-mL containers of a carbonated beverage are opened to the atmosphere. One container is placed in a 40 °C refrigerator and the other is placed in a warm window. After 8 hours, the refrigerated liquid still retains some carbonation. The container in the warm window has very little detectable carbonation. Which of the following is the best explanation for these observations?

- **A.** The solubility of dissolved gases increases with increased partial pressure of water vapor above the liquid.
- **B.** The solubility of dissolved gases increases with increased pressure of the gas above the liquid.
- **C.** The solubility of dissolved gases decreases with increasing concentration of nongaseous solutes because the solutes change the nature of the intermolecular forces between the solvent and the gas.
- **D.** The solubility of dissolved gases decreases with increasing temperature because the increased strength and frequency of molecular collisions force the gas molecules out of the solution.

## Long Free Response

**1.** A sample of helium, initially at 25 °C and 760. torr, occupies a volume of 2.50 L.

- **a.** How many moles of helium are present?
- **b.** If the helium sample is compressed to 1.50 L and the temperature is held constant, what is the change in pressure in atm?
- **c.** If the volume and temperature remain constant at 1.50 L and 25 °C, what is the resulting partial pressure of xenon if 0.050 mole of xenon gas is added to the container?
- **d.** Is the final pressure of the real gas mixture expected to be closer to the value predicted by the ideal gas law at very high or very low temperatures? Explain your answer.
- **e.** A 10.0-mL portion of distilled water is introduced into the container and the container is allowed to expand until the total pressure is 0.500 atm. The vapor pressure of water is 23.8 torr at 25 °C. What is the mole fraction of the water vapor in the gas mixture?
- **f.** The container is punctured with a tiny pinhole. Which of the three gases (helium, xenon, or water) is expected to escape the fastest and why?

## Short Free Response

**2.** The compounds shown below, methyl cinnamate and cinnamic acid, are solids at room temperature.

Methyl cinnamate

Cinnamic acid

- **a.** One of the compounds is insoluble in water. The other is slightly soluble. Explain which compound is insoluble and which is soluble based on the principles of intermolecular interactions.
- **b.** Which compound has the higher vapor pressure and why?
- **c.** How might infrared spectroscopy and mass spectrometry be used to differentiate between the two compounds?
- **d.** A 50:50 mixture of methyl cinnamate and cinnamic acid might be separated by column chromatography using silica as a stationary phase. Which compound would be expected to elute from the column first and why?

# Answers and Explanations

## Multiple Choice

**1.** **C.** $H_2O_2$, or hydrogen peroxide, like water, is capable of hydrogen bonding, the strongest intermolecular force, choice C. $H_2O_2$, however, is larger than water, so in addition to hydrogen bonding, it experiences larger London dispersion forces. HCl experiences only relatively weak dipole-dipole interactions, and $C_2H_6$ experiences London dispersion attractions only.

**2.** **A.** A group that contains hydrides that experience London dispersion forces will only see a steady increase in boiling points up the periods of the periodic table, as seen in the graph. The boiling points of the group 14 hydrides, $CH_4$, $SiH_4$, $GeH_4$, and $SnH_4$, increase in this way, choice A. The hydride of nitrogen, $NH_3$, however, is capable of hydrogen bonding, so $NH_3$ is a period 2 hydride that is expected to have a boiling point that is anomalously high compared to the rest of the group 15 hydrides, $PH_3$, $AsH_3$, and $SbH_3$.

**3.** **B.** One mole of a gas occupies 22.4 L at STP, so the number of moles can be determined.

$$33.6\text{ L} \times \frac{1\text{ mol}}{22.4\text{ L}} = \frac{3}{2}\text{ mol}$$

The molar mass is the number of grams per mol, so we can determine the molar mass of the unknown gas.

$$\frac{30.6\text{ grams}}{\frac{3}{2}\text{ mol}} = \frac{2(\cancel{30.3}\text{ grams})}{\cancel{3}\text{ mol}} = \frac{2(10.1\text{ grams})}{1\text{ mol}} = 20.2\text{ grams/mol}$$

The gas with a molar mass of 20.2 grams/mol is neon, choice B.

**4.** **A.** Choice A is correct. Average kinetic energy has tripled since the temperature has tripled. The equation relating temperature to pressure is

$$\frac{P_1}{T_1} = \frac{P_2}{T_2}$$
$$P_2 = \frac{P_1T_2}{T_1}$$

Since $T_2 = 3 \times T_1$, the final pressure is triple the initial pressure:

$$P_2 = \frac{3\cancel{T}_1P_1}{\cancel{T}_1}$$
$$P_2 = 3P_1$$

**5.** **C.** Because the two gases have the same temperature, their average kinetic energies are the same. Since gas A has a lower range of molecular speeds than gas B, it has a higher molar mass than gas B, making choice C correct.

**6.** **D.** $NH_3$ participates in hydrogen bonding, so it deviates the most from ideal behavior. CO is polar, so its deviation is greater than $N_2$ and Ne, which are both nonpolar and experience only London dispersion forces. $N_2$ is a larger molecule than Ne and it has more electrons, so it is more polarizable. It also occupies a slightly greater volume, so it deviates more from ideal behavior than Ne. The correct order is, therefore, $NH_3 < CO < N_2 < Ne$, choice D.

**7.** **D.** The correct concentration can be achieved with pipettes for measuring 10.42 mL of the 12.00 $M$ HCl and a volumetric flask for the final dilution, choice D. The balance, graduated cylinder, and Erlenmeyer flask are inappropriate for achieving an accurate or precise dilution in this situation.

**8.** **C.** $CH_3COOH$ and $H_2O$ experience similar intermolecular forces. They both participate in hydrogen bonding, so a solution of these two components should be homogeneous, choice C. All of the other choices are comprised of compounds that experience dissimilar intermolecular forces, so they will form heterogeneous solutions.

**9.** **A.** Choice A correctly identifies the curves: A = $SiH_4$, B = $H_2S$, C = $NH_3$. $SiH_4$ (curve A) is nonpolar and experiences only London dispersion forces. It has the highest vapor pressure. $H_2S$ (curve B) is polar, so its vapor pressure is between that of $SiH_4$ and $NH_3$. $NH_3$ (curve C) has the lowest vapor pressure since it participates in hydrogen bonding, the strongest intermolecular force.

**10.** **A.** Since both silver and zinc are metals and since neither is a small atom such as hydrogen, boron, carbon, or nitrogen, the alloy is substitutional rather than interstitial. Solid A best represents the solid form of this alloy, choice A.

**11.** **D.** Choice D best describes the solids: Solids A and B are electrically conductive, and solids C and D are electrical insulators. Solids A and B represent substitutional and interstitial alloys, respectively. Alloys are metallic and conductive. Solid C represents an insulating molecular solid, and solid D represents an insulating ionic solid. Ionic compounds are only conductive in the molten state or in aqueous solutions.

**12.** **D.** The polymer with the highest density is the one that has the heaviest elements incorporated into the structure. In this case, the polymer with F and Cl substitution, shown in choice D, is the densest.

**13.** **A.** Choice A is the true statement: Graphite is a network covalent material that is conductive in its solid form due to delocalized $\pi$ electrons. The extended $\pi$ system of graphite is responsible for its electrical conductivity. Graphite is a network covalent solid composed of $sp^2$ hybridized carbon atoms.

**14.** **B.** Choice B does not correctly complete the sentence to describe the process of distillation; it is not a method for separating sodium chloride from sand. Sodium chloride and sand are both high-melting solids. Sodium chloride is an ionic compound, while sand is a network covalent compound composed of silicon dioxide. Distillation, on the other hand, is a technique for separating liquid components based on differential vapor pressures and boiling points. It is suitable for the separation of two liquids such as water and ethanol ($CH_3CH_2OH$).

**15.** **D.** The solubility of dissolved gases decreases with increasing temperature, choice D. As the average kinetic energy in a solution of dissolved gas increases, the gas particles escape the solution and enter the vapor phase.

## Long Free Response

**1.** **a.** The number of moles of helium present can be determined using the ideal gas law. Recognizing that 760. torr is equal to 1.00 atm simplifies the task and eliminates the need to convert units.

$$PV = nRT$$

$$n = \frac{PV}{RT} = \frac{1.00\ \cancel{\text{atm}} \times 2.50\ \cancel{\text{L}}}{(0.08206\ \cancel{\text{L}} \cdot \cancel{\text{atm}} \cdot \text{mol}^{-1}\ \cancel{\text{K}^{-1}}) \times (25 + 273\ \cancel{\text{K}})} = 0.102\ \text{mol He}$$

**b.** The new pressure can be determined using Boyle's law.

$$P_1V_1 = P_2V_2$$

$$P_2 = \frac{P_1V_1}{V_2} = \frac{1.00\ \text{atm} \times 2.50\ \cancel{\text{L}}}{1.50\ \cancel{\text{L}}} = 1.67\ \text{atm}$$

The change in pressure is the difference between the final pressure, $P_f$, and initial pressure, $P_i$.

$$\Delta P = P_f - P_i = 1.67\ \text{atm} - 1.00\ \text{atm} = 0.67\ \text{atm}$$

**c.** The total number of moles of gas present after the addition of xenon is

$$n_{total} = 0.102 \text{ mole He} + 0.050 \text{ mole Xe} = 0.152 \text{ mole}$$

The new pressure is

$$P = \frac{nRT}{V} = \frac{0.152 \cancel{\text{mol}} \times (0.08206 \cancel{\text{L}} \cdot \text{atm} \cdot \cancel{\text{mol}^{-1}} \cdot \cancel{\text{K}^{-1}}) \times (25 + 273 \cancel{\text{K}})}{1.50 \cancel{\text{L}}} = 2.48 \text{ atm}$$

The partial pressure of xenon, $P_{Xe}$, is equal to the mole fraction of xenon, $X_{Xe}$, multiplied by the total pressure.

$$P_{Xe} = X_{Xe} \times P_{total} = \left(\frac{n_{Xe}}{n_{total}}\right) \times P_{total}$$

$$P_{Xe} = \left(\frac{0.050 \cancel{\text{mol}}}{0.152 \cancel{\text{mol}}}\right) \times 2.48 \text{ atm} = 0.82 \text{ atm}$$

**d.** The final pressure of the gas mixture is expected to be closer to the value determined by the ideal gas law at very high temperatures. At high temperatures, the gas particles are moving faster and their collisions are energetic enough to overcome the intermolecular attractions that would otherwise cause their behavior to deviate from that predicted by the ideal gas law.

**e.** The vapor pressure of the water is equal to the partial pressure of water vapor in the gas mixture. The pressure must be converted to atmospheres first.

$$23.8 \text{ torr} \times \left(\frac{1 \text{ atm}}{760 \text{ torr}}\right) = 0.0313 \text{ atm}$$

Since the vapor pressure is equal to the partial pressure of water vapor, we can use it to determine the mole fraction of the water vapor.

$$P_{H_2O} = X_{H_2O} \times P_{total}$$

$$X_{H_2O} = \frac{P_{H_2O}}{P_{total}} = \frac{0.0313 \cancel{\text{atm}}}{0.500 \cancel{\text{atm}}} = 0.0626$$

**f.** Of the three gases in the container, helium escapes the fastest. The rate of effusion of a molecule is inversely proportional to the square root of its molar mass.

$$\text{rate} \propto \sqrt{\frac{T}{M}}$$

Since the molar mass of helium is the smallest, 4.00 $g{\cdot}mol^{-1}$ (compared to 18.01 $g{\cdot}mol^{-1}$ for $H_2O$ and 131.29 $g{\cdot}mol^{-1}$ for Xe), its rate of diffusion is the highest.

## Short Free Response

**2.** **a.** Cinnamic acid is a slightly water-soluble compound. It has an O–H bond that is capable of forming hydrogen bonds with water molecules. Methyl cinnamate is a polar molecule, but it is incapable of hydrogen bonding. Methyl cinnamate is water-insoluble.

**b.** The hydrogen bonding between molecules of cinnamic acid results in a lower vapor pressure for cinnamic acid than for methyl cinnamate. Methyl cinnamate experiences dipole-dipole attractions that are weaker than hydrogen bonds, so the methyl cinnamate molecules more easily overcome the intermolecular attractions and escape into the gas phase.

**c.** Infrared spectroscopy gives information about the types of bonds present in molecules. The IR spectrum of cinnamic acid would be expected to have an IR absorption due to the O–H peak, while the spectrum of methyl cinnamate will lack the O–H absorbance. Methyl cinnamate would have an O–$CH_3$ single bond absorbance that would not be expected in the spectrum of cinnamic acid. Mass spectrometry would indicate a larger molecular mass for methyl cinnamate (162 amu) than for cinnamic acid (148 amu).

**d.** Methyl cinnamate is not capable of hydrogen bonding, so it would be expected to elute first from a column packed with silica gel, since silica is highly polar. Cinnamic acid is capable of hydrogen bonding, so it would be expected to interact more strongly with the silica and elute more slowly from the column.

Chapter 4

# Chemical Reactions

**Chemical reactions involve changes in molecular structure.**

During a **physical change,** intermolecular interactions between atoms or compounds are formed or disrupted. This causes a change in properties or phase, but no change takes place to the composition or bonding arrangement of individual molecules. The chemical composition remains unaltered.

By contrast, during a **chemical change,** bonds are broken and formed. This leads to a fundamental change in the identity and chemical composition of a substance.

A combustion reaction, for example, transforms glucose and oxygen to carbon dioxide and water. When the combustion is complete, all of the carbon, oxygen, and hydrogen atoms still exist, but they exist in the form of $CO_2$ and $H_2O$. No glucose remains.

$$C_6H_{12}O_6 + 6\ O_2 \rightarrow 6\ CO_2 + 6\ H_2O$$

A chemical change is accompanied by a change in the energy of the reactants as they are transformed to products. When the products of a reaction are lower in energy than the reactants, energy is released and the reaction is exothermic. When the products are higher in energy, the reaction is endothermic.

## Evidence for Chemical Change

It can be difficult to tell whether a chemical change has occurred, and entire chemical research projects are sometimes devoted to finding out.

There are four general clues that strongly suggest that a chemical change has occurred:

1. *Production of heat or light* usually indicates an exothermic chemical reaction. When glucose is rapidly burned in oxygen, both heat and light are given off. (This evidence is not definitive by itself, though. When a current is passed through the tungsten filament of an incandescent light bulb, both heat and light are emitted, but no chemical reaction takes place.)
2. *Formation or consumption of a gas,* usually observed as a change in pressure or volume, may indicate a chemical reaction has occurred. Formation of bubbles in a liquid generally indicates a chemical change. (Physical changes such as sublimation and vaporization create gases, too, so this evidence should be used cautiously as well.)
3. *Color changes* can also be indicators of a chemical change. The acid-base indicator phenolphthalein changes from colorless to pink when it undergoes a chemical reaction in the presence of a base. (Again, this evidence should be used with care. When yellow dye is added to blue dye, the mixture turns green, but no chemical change has occurred.)
4. *Formation of a* ***precipitate,*** an insoluble solid, can indicate that a chemical reaction has occurred. When a solution containing $Ag^+$ ions is added to a solution containing $Cl^-$ ions, solid AgCl forms and settles to the bottom of the container. (Again, however, a solid can form from a homogeneous solution due to physical changes, too. Honey, an aqueous sugar solution, crystallizes under certain conditions.)

## Practice

Which of the following is likely to be a chemical change?

**A.** When a voltage is applied across a glass tube filled with neon, a bright orange glow is observed.
**B.** When a sample of ammonium dichromate is ignited with a splint, the substance gives off heat, light, and gases as it changes from a crystalline orange solid to a green powdery solid.
**C.** When a flavored drink packet is added to water, the water turns red.

Only choice B, the decomposition of ammonium dichromate, is a chemical change. Several lines of evidence suggest this is the case, including color change, production of heat and light, and formation of gases. Lighting a neon light (choice A) and dissolving a drink packet in water (choice C) are physical changes.

# Balancing Chemical Equations

Balancing chemical equations is an essential skill in chemistry.

Because atoms are conserved in ordinary (non-nuclear) chemical changes, the atoms of each element on the reactant side of the equation must equal the atoms on the product side of the equation. This is achieved with **coefficients,** numbers indicating the number of particles or moles of each species in an equation. In the reaction between nitrogen and hydrogen to make ammonia, the coefficients in the balanced equation are 1, 3, and 2 for $N_2$, $H_2$, and $NH_3$, respectively.

$$N_2 + 3\,H_2 \rightarrow 2\,NH_3$$

A balanced chemical equation can be represented symbolically by a **particle diagram,** which is a picture like the one shown below. A particle diagram represents reacting atoms or molecules as circles or spheres. Notice that there are two nitrogen atoms on the reactant side and two on the product side. Likewise for hydrogen, there are six atoms on each side of the equation.

**Particulate Representation of a Reaction**

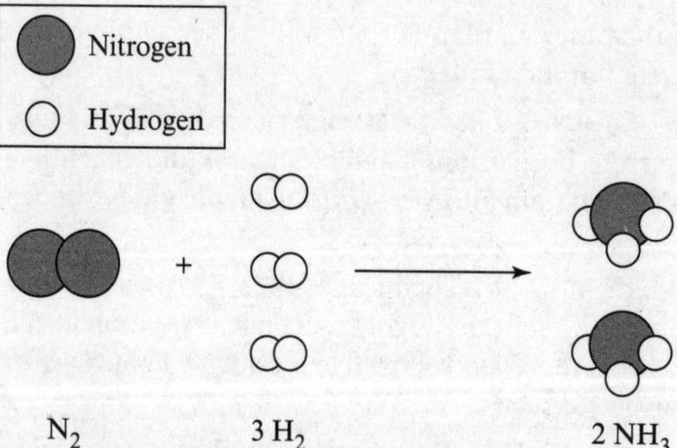


The following steps, when followed consistently, will result in correctly balanced equations.

- Balance elements that appear in only one place on each side of the equation *first.*
- Balance hydrogen, oxygen, and atoms in their free elemental form *last.*
- Balance polyatomic ions, such as $NO_3^-$ and $SO_4^{2-}$, as units.
- Make sure that the sum of the charges of the reactants is equal to the sum of the charges of the products.

Consider the unbalanced equation for the oxidation of aluminum as an example.

$$\text{Unbalanced: } Al + O_2 \rightarrow Al_2O_3$$

When oxygen is present, it is usually easiest to balance last. Since there are two aluminum atoms in the products, a coefficient of 2 will balance the aluminum in the reactants.

$$\text{Still unbalanced: } \mathbf{2}\ Al + O_2 \rightarrow Al_2O_3$$

There are three oxygen atoms on the right and two on the left. A coefficient of 1.5 will equalize the number of oxygen atoms.

$$\text{Still unbalanced: } 2\ Al + \mathbf{1.5}\ O_2 \rightarrow Al_2O_3$$

Finally, each coefficient is multiplied by the factor that will result in the lowest whole-number ratio, 2 in this case.

$$\mathbf{2} \times (2\ Al + 1.5\ O_2 \rightarrow Al_2O_3)$$
$$\text{Balanced: } \mathbf{4}\ Al + \mathbf{3}\ O_2 \rightarrow \mathbf{2}\ Al_2O_3$$

It is always best to double-check that there are the same number of atoms of each element on both sides of the equation. In this case, there are four aluminum atoms on each side of the equation and six oxygen atoms on each side of the equation, so the equation is balanced correctly.

More difficult equations can be balanced systematically, as well. Consider the unbalanced equation for the explosive decomposition of nitroglycerin, $C_3H_5N_3O_9$.

$$C_3H_5N_3O_9 \rightarrow N_2 + H_2O + CO_2 + O_2$$

We can begin by balancing either carbon or nitrogen. Consider carbon.

$$\mathbf{1}\ C_3H_5N_3O_9 \rightarrow N_2 + H_2O + \mathbf{3}\ CO_2 + O_2$$

Next, balance the nitrogen. A fractional coefficient is required to result in three nitrogen atoms on each side of the equation.

$$1\ C_3H_5N_3O_9 \rightarrow \mathbf{1.5}\ N_2 + H_2O + 3\ CO_2 + O_2$$

Now, consider hydrogen.

$$1\ C_3H_5N_3O_9 \rightarrow 1.5\ N_2 + \mathbf{2.5}\ H_2O + 3\ CO_2 + O_2$$

Deal with oxygen last since it appears in multiple species in the products. We can solve for a variable, $n$, equal to the coefficient for $O_2$, the only species that contains only oxygen atoms.

Take an inventory of the number of oxygen atoms on each side. There are nine in the reactants. On the product side, oxygen is present in $H_2O$, $CO_2$, and $O_2$. Set up an equation to balance the oxygen atoms and solve for the coefficient, $n$.

$$9 = (2.5 \times 1) + (3 \times 2) + (n \times 2)$$
$$9 = 2.5 + 6 + 2n$$
$$0.5 = 2n$$
$$0.25 = n$$

$$1\ C_3H_5N_3O_9 \rightarrow 1.5\ N_2 + 2.5\ H_2O + 3\ CO_2 + \mathbf{0.25}\ O_2$$

Notice that the coefficients can be converted to smallest whole numbers when multiplied by 4.

$$\mathbf{4}\ C_3H_5N_3O_9 \rightarrow \mathbf{6}\ N_2 + \mathbf{10}\ H_2O + \mathbf{12}\ CO_2 + O_2$$

## Practice

Balance the following equation using the lowest whole-number coefficients: $C_4H_{10} + O_2 \rightarrow CO_2 + H_2O$.

Carbon should be balanced first. Hydrogen and oxygen atoms should be balanced last.

$$C_4H_{10} + O_2 \rightarrow \mathbf{4}\ CO_2 + H_2O$$

Hydrogen can be balanced second; it's easier to balance than oxygen, which is present in two places on the right side of the equation.

$$C_4H_{10} + O_2 \rightarrow 4\ CO_2 + \mathbf{5}\ H_2O$$

Because oxygen exists as a free element on the left, it should be balanced last. There are 13 oxygen atoms on the right, so the $O_2$ on the left must be multiplied by 6.5 in order to balance.

$$C_4H_{10} + \mathbf{6.5}\ O_2 \rightarrow 4\ CO_2 + 5\ H_2O$$

Multiplication of each coefficient by 2 converts the coefficients to the lowest set of whole numbers.

$$\mathbf{2} \times (C_4H_{10} + 6.5\ O_2 \rightarrow 4\ CO_2 + 5\ H_2O)$$
$$\mathbf{2}\ C_4H_{10} + \mathbf{13}\ O_2 \rightarrow \mathbf{8}\ CO_2 + \mathbf{10}\ H_2O$$

Double-check that there are 8 carbon atoms, 20 hydrogen atoms, and 26 oxygen atoms on each side of the equation.

## Practice

Balance the equation below by filling in the blanks using the smallest whole-number coefficients.

$$\_\ C_5H_{17}N_5O_2 + \_\ O_2 \rightarrow \_\ N_2 + \_\ H_2O + \_\ CO_2$$

Again, a systematic method is to balance carbon first. Balance hydrogen last. Remember that it may be necessary to use fractions.

$$\mathbf{1}\ C_5H_{17}N_5O_2 + \_\ O_2 \rightarrow \mathbf{2.5}\ N_2 + \mathbf{8.5}\ H_2O + \mathbf{5}\ CO_2$$

Set up an equation to determine $n$, the coefficient for $O_2$.

$$(1\times2)+(n\times2)=(8.5\times1)+(5\times2)$$
$$2+2n=8.5+10$$
$$2n=18.5-2$$
$$2n=16.5$$
$$n=8.25$$

$$1\ C_5H_{17}N_5O_2 + \mathbf{8.25}\ O_2 \rightarrow 2.5\ N_2 + 8.5\ H_2O + 5\ CO_2$$

Multiply each coefficient by 4 to obtain the smallest whole-number ratio.

$$\mathbf{4}\ C_5H_{17}N_5O_2 + \mathbf{33}\ O_2 \rightarrow \mathbf{10}\ N_2 + \mathbf{34}\ H_2O + \mathbf{20}\ CO_2$$

## Practice

Practice balancing as many equations as you can find. Some examples include:

$Si + S_8 \rightarrow Si_2S_4$
$As + NaOH \rightarrow Na_3AsO_3 + H_2$
$Mg_3N_2 + H_2O \rightarrow Mg(OH)_2 + NH_3$
$V_2O_5 + Ca \rightarrow CaO + V$
$H_3BO_3 \rightarrow H_4B_6O_{11} + H_2O$
$Ca_3(PO_4)_2 + SiO_2 \rightarrow P_4O_{10} + CaSiO_3$
$Fe_2(C_2O_4)_3 \rightarrow FeC_2O_4 + CO_2$

$$\mathbf{4}\ Si + S_8 \rightarrow \mathbf{2}\ Si_2S_4$$
$$\mathbf{2}\ As + \mathbf{6}\ NaOH \rightarrow \mathbf{2}\ Na_3AsO_3 + \mathbf{3}\ H_2$$
$$Mg_3N_2 + \mathbf{6}\ H_2O \rightarrow \mathbf{3}\ Mg(OH)_2 + \mathbf{2}\ NH_3$$
$$V_2O_5 + \mathbf{5}\ Ca \rightarrow \mathbf{5}\ CaO + \mathbf{2}\ V$$
$$\mathbf{6}\ H_3BO_3 \rightarrow H_4B_6O_{11} + \mathbf{7}\ H_2O$$
$$\mathbf{2}\ Ca_3(PO_4)_2 + \mathbf{6}\ SiO_2 \rightarrow P_4O_{10} + \mathbf{6}\ CaSiO_3$$
$$Fe_2(C_2O_4)_3 \rightarrow \mathbf{2}\ FeC_2O_4 + \mathbf{2}\ CO_2$$

# Classification of Chemical Reactions

Chemical equations can be classified as one or more of six types: synthesis, decomposition, combustion, single replacement, double replacement, and acid-base. In addition to fitting into one or more of these categories, some reactions may be oxidation-reduction (redox) reactions. These will be reviewed in much greater detail in the section on oxidation-reduction reactions later in this chapter.

## Synthesis Reactions

A **synthesis reaction** is one in which reactants combine to form a single product. The reaction between nitrogen and hydrogen on page 132 is classified as a synthesis reaction. Many synthesis reactions occur when elements are heated with oxygen, and many of these reactions are redox, as will be discussed later in this chapter.

Examples of synthesis reactions include:

$$2\ Mg + O_2 \rightarrow 2\ MgO$$
$$C + O_2 \rightarrow CO_2$$
$$2\ Li + H_2 \rightarrow 2\ LiH$$
$$Cl_2 + H_2 \rightarrow 2\ HCl$$

The reactions of metal and nonmetal oxides are important examples of synthesis reactions in AP Chemistry. Metal oxides are often called **basic anhydrides** because they react with water to form hydroxides. *Anhydride* means "without water."

$$Na_2O + H_2O \rightarrow 2\ NaOH$$
$$CaO + H_2O \rightarrow Ca(OH)_2$$

Nonmetal oxides are called **acid anhydrides** because they react with water to form acids.

$$Cl_2O_7 + H_2O \rightarrow 2\ HClO_4 \text{ (perchloric acid)}$$
$$SO_3 + H_2O \rightarrow H_2SO_4 \text{ (sulfuric acid)}$$
$$N_2O_5 + H_2O \rightarrow 2\ HNO_3 \text{ (nitric acid)}$$

Acid anhydrides react with basic anhydrides to form salts.

$$Li_2O + SO_3 \rightarrow Li_2SO_4$$
$$6\ Na_2O + P_4O_{10} \rightarrow 4\ Na_3PO_4$$

## Decomposition Reactions

**Decomposition reactions** can be thought of as the reverse of synthesis reactions. In these reactions, one reactant forms two or more products. In many cases, this is due to heating or electrolysis, in which an electrical current is passed through a solution. The electrolysis of water is classified as a decomposition reaction:

$$2\ H_2O\ (l) \rightarrow 2\ H_2\ (g) + O_2\ (g)$$

Other examples of decompositions include:

$$Cu(OH)_2 \rightarrow CuO + H_2O$$
$$2\ Na_2O_2 \rightarrow 2\ Na_2O + O_2$$
$$2\ KClO_3 \rightarrow 2\ KCl + 3\ O_2$$
$$(NH_4)_2CO_3 \rightarrow 2\ NH_3 + CO_2 + H_2O$$

Sometimes unstable products are formed during the course of a reaction that decompose further to produce water and gases. Three unstable products to be aware of are carbonic acid, $H_2CO_3$, sulfurous acid, $H_2SO_3$, and ammonium hydroxide, $NH_4OH$.

$$H_2CO_3 \rightarrow H_2O + CO_2$$
$$H_2SO_3 \rightarrow H_2O + SO_2$$
$$NH_4OH \rightarrow H_2O + NH_3$$

These reactions may occur during acid-base reactions. For instance, consider the reaction between vinegar (acetic acid) and baking soda (sodium bicarbonate). In the reaction between acetic acid, $H_3C_2O_2H$, and sodium bicarbonate, $NaHCO_3$, the carbonic acid that is generated immediately decomposes to form carbon dioxide and water.

$$H_3C_2O_2H + NaHCO_3 \rightarrow NaH_3C_2O_2 + CO_2 + H_2O$$

## Combustion Reactions

**Combustion reactions** are reactions in which elements or compounds react exothermically with oxygen. Usually "combustion" refers to the burning of hydrocarbon and other organic (carbon- and hydrogen-containing) molecules to yield carbon dioxide and water, but it may also refer to the burning of other combustible materials. A spark is required to initiate these processes.

$$CH_4 + 2\ O_2 \rightarrow CO_2 + 2\ H_2O$$
$$2\ CH_4O + 3\ O_2 \rightarrow 2\ CO_2 + 4\ H_2O$$
$$SiH_4 + 2\ O_2 \rightarrow SiO_2 + 2\ H_2O$$

These reactions give off a great deal of energy, and they are used extensively for energy production. Gasoline, wood, coal, and natural gas are all examples of combustible fuels.

Combustion reactions are also classified as redox reactions.

## Single Replacement Reactions

**Single replacement reactions** are redox reactions in which one element in a compound is replaced by another, more active element. Whether a single replacement is expected to occur can be determined by consulting a table of measured oxidation-reduction potentials. This will be covered in more depth in the section on electrochemistry in Chapter 9.

$$A + BC \rightarrow AB + C$$

**Single Replacement Reaction**

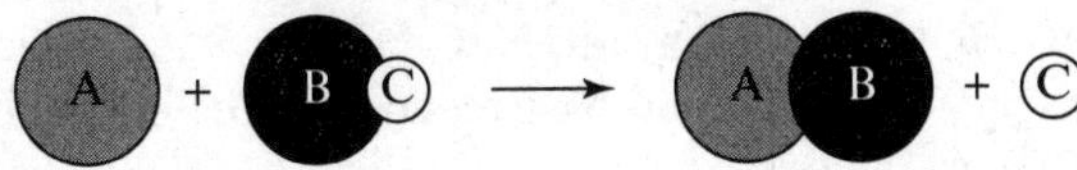


Examples of single replacement reactions include:

$$Cu + 2\ AgNO_3 \rightarrow Cu(NO_3)_2 + 2\ Ag$$
$$Mg + 2\ HCl \rightarrow MgCl_2 + H_2$$

In the first reaction, Cu replaces $Ag^+$ in $AgNO_3$. In the second reaction, Mg replaces $H^+$ in HCl.

Single replacement reactions readily occur between the group 1 metals and water. (Remember, group 1 (IA) metals have an $ns^1$ valence electron configuration, so they readily lose an electron to form a cation with a noble gas configuration.)

When a group 1 metal is added to water, it reacts vigorously to form the metal hydroxide and hydrogen gas. The reaction is exothermic, and enough heat can be given off to cause the hydrogen above the surface of the water to ignite.

Moving down group 1 from lithium to cesium, the elements become more reactive with water. The reactivity can be explained by increasing shielding and decreasing ionization energy down the group.

$$K + H_2O \rightarrow KOH + H_2$$

Fluorine is the most active halogen and iodine is the least active halogen. Group 17 (VIIA) elements have an $ns^2np^7$ valence electron configuration. They achieve a noble gas configuration by gaining one electron. Group 17 elements close to the top of the periodic table are more likely to gain an electron than those at the bottom. Chlorine, which is closer to the top of the periodic table and has a greater electron affinity, replaces bromide as the sodium salt in a single replacement reaction.

$$2\ NaBr + Cl_2 \rightarrow 2\ NaCl + Br_2$$

## Double Replacement Reactions

**Double replacement reactions** are reactions in which cationic and anionic portions of two compounds are exchanged.

$$AB + CD \rightarrow AD + CB$$

**Double Replacement Reaction**

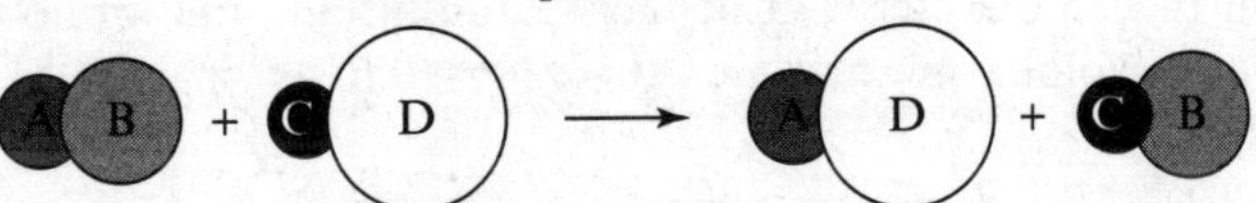

Double replacement reactions occur under any of the following circumstances:

- A precipitate is formed. A precipitate is an insoluble solid that forms when two solutions of ions are mixed. The reaction between $AgNO_3$ and KCl is an example. Alone, each reactant is soluble in water, but when they are mixed, the solid AgCl formed settles to the bottom of the container. AgCl is practically insoluble. In other words, it has a low $K_{sp}$, a concept that will be discussed in Chapter 7, "Chemical Equilibrium."

$$AgNO_3(aq) + KCl(aq) \rightarrow AgCl(s) + KNO_3(aq)$$

- Water, a weak acid, a weak base, or some other weakly-ionizing compound forms. The strong base, NaOH, reacts with a strong acid, $HClO_3$, to produce a salt, $NaClO_3$, and water, which is weakly ionizing. This is also classified as an acid-base reaction. An **acid-base reaction** is one in which an acid reacts with a base to form a solution of salt and water.

$$NaOH(aq) + HClO_3(aq) \rightarrow NaClO_3(aq) + H_2O(l)$$

- An intermediate that can decompose to a gas is formed. Sulfurous acid, $H_2SO_3$, was one of the three compounds mentioned earlier that decomposes to water and gas. (The other two species were $H_2CO_3$ and $NH_4OH$.) It is a transient (or fleeting) product of the reaction of $K_2SO_3$ with HCl, a strong acid. As soon as the intermediate $H_2SO_3$ is formed, it decomposes to $SO_2$ and water.

$$K_2SO_3(aq) + 2\ HCl(aq) \rightarrow 2\ KCl(aq) + H_2SO_3(aq) \rightarrow 2\ KCl(aq) + SO_2(g) + H_2O(l)$$

### Practice

Classify the following reactions as synthesis, decomposition, combustion, single replacement, or double replacement:

**a.** $2\ Na + Cl_2 \rightarrow 2\ NaCl$
**b.** $2\ C_2H_6 + 7\ O_2 \rightarrow 4\ CO_2 + 6\ H_2O$
**c.** $NaCN + HBr \rightarrow NaBr + HCN$
**d.** $2\ HgO \rightarrow 2\ Hg + O_2$
**e.** $Br_2 + 2\ KI \rightarrow I_2 + 2\ KBr$

**(a)** Synthesis. Two reactants (Na and $Cl_2$) come together to make a single product (NaCl).

**(b)** Combustion. The carbon-containing molecule, $C_2H_6$, reacts with oxygen to form carbon dioxide and water.

**(c)** Double replacement. The cations ($Na^+$ and $H^+$) and anions ($CN^-$ and $Br^-$) are exchanged.

**(d)** Decomposition. One reactant (HgO) breaks down into two products (Hg and $O_2$).

**(e)** Single replacement. The iodide ion in KI replaces bromide.

# Ionic Solubility Rules

The solubility of ionic substances varies from compound to compound. Some substances, such as AgCl, are minimally soluble in water, while others, such as KI, are much more soluble.

Solubility will be explored further in Chapter 7, "Chemical Equilibrium," but for now, it is important to note that ionic compounds that are poorly soluble in water are often referred to as "insoluble," despite the fact that they may dissociate to a slight extent.

Other ionic compounds **dissociate,** or break up into ions, almost completely in an aqueous solution. These are considered to be highly soluble. Compounds that behave in this way are strong electrolytes. A dissociation equation shows the ionic compound breaking up into ions when it is added to water.

$$NaCl(s) \rightarrow Na^+(aq) + Cl^-(aq)$$

The AP Chemistry curriculum requires memorization of only a few solubility rules:

- Sodium salts are soluble in water. Examples include NaF, $Na_2SO_4$, and NaCN.
- Potassium salts are soluble in water. Examples include KCl, $K_2CO_3$, and $K_3PO_4$.
- Ammonium salts are soluble in water. Examples include $NH_4Cl$ and $NH_4CH_3CO_2$.
- Nitrate salts are soluble in water. Examples include $AgNO_3$, $Cu(NO_3)_2$, and $Pb(NO_3)_2$.

### Practice

Which of the following compounds are expected to conduct an electric current in an aqueous solution: $PbCl_2$, $NH_4Cl$, $HgNO_3$?

The two soluble salts listed that are expected to conduct an electric current in an aqueous solution are $NH_4Cl$ and $HgNO_3$.

## Net Ionic Equations

**Net ionic equations** show only the species involved in a reaction. **Spectator ions,** which are ions that do not participate in the reaction, are not shown. Net ionic equations are a very useful way of representing double replacement, single replacement, and acid-base reactions in aqueous solutions.

Consider the double replacement reaction between aqueous potassium iodide and aqueous lead nitrate. An insoluble solid precipitate, lead iodide, is formed. The overall equation is first written and balanced. State symbols (*aq* and *s* in this example) are included here to emphasize the nature of the precipitation reaction, but they are not required for credit on the AP Chemistry exam.

$$Pb(NO_3)_2(aq) + 2\ KI(aq) \rightarrow PbI_2(s) + 2\ KNO_3(aq)$$

Next, a **total ionic equation** is written. Any strong electrolytes, which include soluble salts, strong acids, and strong bases, are written as dissociated ions in the total ionic equation. Remember that all nitrates and potassium salts are soluble.

$$Pb^{2+}(aq) + 2\ NO_3^-(aq) + 2\ K^+(aq) + 2\ I^-(aq) \rightarrow PbI_2(s) + 2\ K^+(aq) + 2\ NO_3^-(aq)$$

Finally, the spectator ions, which appear in both the reactants and the products, are cancelled. The species left over make up the net ionic equation.

$$\text{Total ionic equation: } Pb^{2+}(aq) + \cancel{2\ NO_3^-(aq)} + \cancel{2\ K^+(aq)} + 2\ I^-(aq) \rightarrow PbI_2(s) + \cancel{2\ K^+(aq)} + \cancel{2\ NO_3^-(aq)}$$

$$\text{Net ionic equation: } Pb^{2+}(aq) + 2\ I^-(aq) \rightarrow PbI_2(s)$$

### Practice

Write balanced net ionic equations for the reactions described below.

**a.** Solid aluminum reacts with aqueous hydrochloric acid to form aqueous aluminum chloride and hydrogen gas.

**b.** When aqueous sodium hydroxide is added to an aqueous solution of magnesium sulfate, magnesium hydroxide precipitate is formed.

**(a)** The reaction between aluminum and hydrochloric acid is a single replacement reaction.

Overall balanced equation:

$$2\ Al(s) + 6\ HCl(aq) \rightarrow 2\ AlCl_3(aq) + 3\ H_2(g)$$

Total ionic equation:

$$2\,Al(s) + 6\,H^+(aq) + \cancel{6\,Cl^-(aq)} \rightarrow 2\,Al^{3+}(aq) + \cancel{6\,Cl^-(aq)} + 3\,H_2(g)$$

Net ionic equation:

$$2\,Al(s) + 6\,H^+(aq) \rightarrow 2\,Al^{3+}(aq) + 3\,H_2(g)$$

Alternative net ionic equation without state symbols:

$$2\,Al + 6\,H^+ \rightarrow 2\,Al^{3+} + 3\,H_2$$

**(b)** The reaction between sodium hydroxide and magnesium sulfate is a double replacement reaction.

Overall balanced equation:

$$MgSO_4(aq) + 2\,NaOH(aq) \rightarrow Mg(OH)_2(s) + Na_2SO_4(aq)$$

Total ionic equation:

$$Mg^{2+}(aq) + \cancel{SO_4^{2-}(aq)} + \cancel{2\,Na^+(aq)} + 2\,OH^-(aq) \rightarrow Mg(OH)_2(s) + \cancel{2\,Na^+(aq)} + \cancel{SO_4^{2-}(aq)}$$

Net ionic equation:

$$Mg^{2+}(aq) + 2\,OH^-(aq) \rightarrow Mg(OH)_2(s)$$

Alternative net ionic equation without state symbols:

$$Mg^{2+} + 2\,OH^- \rightarrow Mg(OH)_2$$

# Acids and Bases

Strong acids and strong bases dissociate completely into ions in water. Strong acids and bases are strong electrolytes. Weak acids and bases and weakly-soluble hydroxides are weak electrolytes.

## Acids

Six **strong acids** should be memorized. These are HCl, HBr, HI, $HClO_4$, $H_2SO_4$, and $HNO_3$.

Other strong, oxygen-containing ternary acids (made up of three elements) can be recognized by the fact that they are composed of at least two more oxygen atoms than hydrogen atoms. Both $HClO_3$ and $HBrO_4$ are strong acids. They each have at least two more oxygen atoms than hydrogen atoms.

A **weak acid** is any acid that partially ionizes in an aqueous solution. Some weak acids are HF, $H_2CO_3$, $H_3PO_4$, and $HNO_2$, among many others.

In addition, all **carboxylic acids,** which contain the functional group $-CO_2H$, are weak acids. Acetic acid, $CH_3CO_2H$, is an example of a carboxylic acid.

## Bases

Seven **strong bases** should be memorized. They are the group 1 hydroxides, LiOH, NaOH, KOH, RbOH, and CsOH. Two hydroxides from group 2, $Ba(OH)_2$ and $Sr(OH)_2$, also dissociate completely and are considered to be strong bases.

Bases that are not strong are either weakly-soluble hydroxides or weak bases.

Weakly-soluble hydroxides include compounds such as $Mg(OH)_2$ or $Al(OH)_3$.

Ammonia, $NH_3$, and the amines are common weak bases. **Amines** have the general formula $NR_3$, where R is either hydrogen or a hydrocarbon group. Examples of amines are $CH_3NH_2$, $(CH_3)_2NH$, and $(CH_3)_3N$.

## Reactions of Acids and Bases

Chemists employ several definitions of acids and bases, but the most important one for AP Chemistry is the Brønsted-Lowry definition. Familiarity with the older Arrhenius acid-base theory facilitates understanding of Brønsted-Lowry acids and bases.

### Arrhenius Acids and Bases

An **Arrhenius acid** is a species that releases a proton. The word "proton" is synonymous with "hydrogen ion," or "$H^+$." All of the acids discussed above (HCl, $H_2SO_4$, $H_2CO_3$, $CH_3CO_2H$, etc.) are Arrhenius acids. The ionization of an acid in water produces a **hydrogen ion,** $H^+$, and the anion of the acid.

$$HCl(aq) \rightarrow H^+(aq) + Cl^-(aq)$$

**Arrhenius bases** are species that release a hydroxide ion, $OH^-$, upon dissociation. Strong soluble bases such as NaOH, KOH, and $Ba(OH)_2$ are Arrhenius bases. The dissociation of basic hydroxide produces a hydroxide ion, $OH^-$.

$$NaOH(aq) \rightarrow Na^+(aq) + OH^-(aq)$$

### Brønsted-Lowry Acids and Bases

A **Brønsted-Lowry acid** is a proton donor, exactly like an Arrhenius acid. The anion that is produced upon the release of the proton is the **conjugate base** of the acid. It is called the conjugate base because it could accept a proton under appropriate conditions.

In the ionization reaction of HCl in water shown below, $Cl^-$ is the conjugate base of HCl and $H_3O^+$ (hydronium) is the **conjugate acid** of $H_2O$. In other words, $HCl/Cl^-$ are a conjugate acid/base pair, and $H_3O^+/H_2O$ are a conjugate acid/base pair.

*A conjugate acid/base pair differs by one $H^+$.*

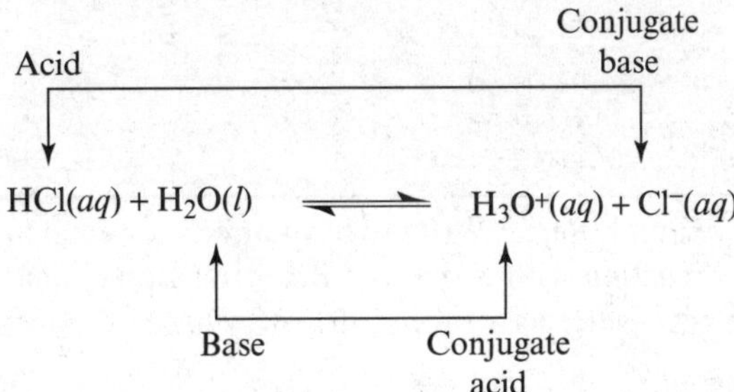

A **Brønsted-Lowry base** is a proton acceptor. A Brønsted-Lowry base such as ammonia, $NH_3$, reacts with water by accepting a proton. The resulting ammonium ion, $NH_4^+$, is the conjugate acid of $NH_3$, because it could donate a proton under appropriate conditions. The conjugate base of water, $OH^-$, is produced in the reaction.

**Conjugate Acid/Conjugate Base Relationships**

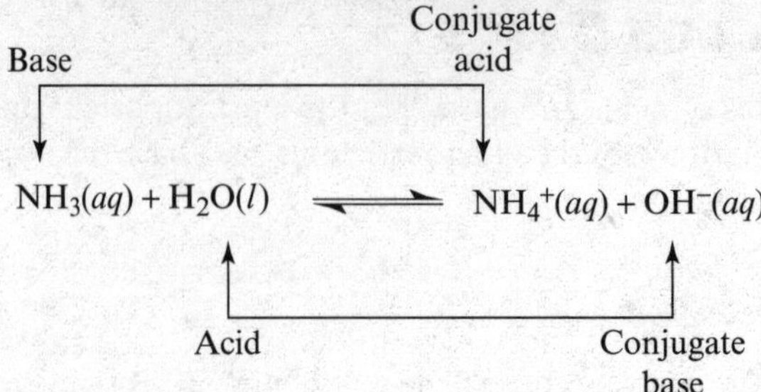


Water is behaving as a base in the reaction with HCl and as an acid in the reaction with $NH_3$. This illustrates an important point: Water is an amphoteric species. An **amphoteric** species can act as both an acid and a base.

Water is also an **amphiprotic** species, which is a species that can either donate or accept a proton.

**TIP: Since all amphiprotic species are amphoteric, but not all amphoteric species are amphiprotic, it is "safest" to use *amphoteric*.**

## Practice

In the acid/base reactions shown below, identify the Brønsted-Lowry acid, the Brønsted-Lowry base, the conjugate acid, and the conjugate base.

**a.** $C_5H_5N(aq) + H_2O(l) \rightarrow C_5H_5NH^+(aq) + OH^-(l)$
**b.** $HI(aq) + H_2O(l) \rightarrow H_3O^+(aq) + I^-(aq)$

**(a)** $C_5H_5N$ is the Brønsted-Lowry base, and $H_2O$ is the Brønsted-Lowry acid. $C_5H_5NH^+$ is the conjugate acid, and $OH^-$ is the conjugate base.

**(b)** HI is the Brønsted-Lowry acid, and $H_2O$ is the Brønsted-Lowry base. $H_3O^+$ is the conjugate acid, and $I^-$ is the conjugate base.

## Practice

Which of the following species are amphoteric? Write two equations for each amphoteric species. One equation should show the species reacting with water as a Brønsted-Lowry acid, and the other equation should show the species reacting with water as a Brønsted-Lowry base.

$$HCO_3^-,\ SO_4^{2-},\ HPO_4^{2-}$$

$HCO_3^-$ and $HPO_4^{2-}$ are both amphoteric species. $SO_4^{2-}$ can accept a proton, but it does not have a proton to donate, so it is not amphoteric.

The reactions of $HCO_3^-$ and $HPO_4^{2-}$, in which the species are behaving as acids, are

$$HCO_3^- + H_2O \rightarrow CO_3^{2-} + H_3O^+$$
$$HPO_4^{2-} + H_2O \rightarrow PO_4^{3-} + H_3O^+$$

The reactions of $HCO_3^-$ and $HPO_4^{2-}$, in which the species are behaving as bases, are

$$HCO_3^- + H_2O \rightarrow H_2CO_3 + OH^-$$
$$HPO_4^{2-} + H_2O \rightarrow H_2PO_4^- + OH^-$$

## Conjugate Acid/Base Strength

The stronger an acid, the weaker its conjugate base.

The stronger a base, the weaker its conjugate acid.

HCl is a very strong acid, and its conjugate base, $Cl^-$, is a very weak base.

Acetic acid, $CH_3CO_2H$, is a weaker acid than HCl. Its conjugate base, $CH_3CO_2^-$, is a stronger base than $Cl^-$. Notice that this refers to relative strength. Acetate anion, $CH_3CO_2^-$, is not a strong base. It is merely a *stronger* base than $Cl^-$.

Ammonia, $NH_3$, is a stronger base than acetate, $CH_3CO_2^-$. The conjugate acid of ammonia, $NH_4^+$, is a weaker acid than acetic acid, $CH_3CO_2H$, the conjugate acid of acetate.

Further review of acid and base strength and the influence of molecular structure on acid/base strength will be covered in Chapter 8, "Acids and Bases."

### Practice

---

Determine the weaker acid in each of the following pairs:

**a.** $H_2O$ or $NH_4^+$
**b.** $(CH_3)_2NH$ or $(CH_3)_2NH_2^+$

---

To determine the weaker acid, look for the stronger conjugate base in each pair.

**(a)** In this pair, the conjugate bases are $OH^-$ and $NH_3$. The hydroxide ion is a much stronger base than ammonia, a weak base. Water, the conjugate acid of hydroxide, is a weaker acid than $NH_4^+$, the conjugate acid of ammonia.

**(b)** In this pair, the conjugate bases are $(CH_3)_2N^-$ and $(CH_3)_2NH$, respectively. The stronger conjugate base is $(CH_3)_2N^-$. When a weakly basic amine such as $(CH_3)_2NH$ loses a proton, the resulting anionic species is strongly basic. The conjugate base of the second species is $(CH_3)_2NH$. It is a weak base, so its conjugate acid is the stronger acid. The first species, $(CH_3)_2NH$, is the weaker acid.

### Practice

---

Determine the weaker base in each of the following pairs:

**a.** $H_2O$ or $HSO_3^-$
**b.** $Cl^-$ or $CN^-$

---

To determine the weaker base, look for the stronger conjugate acid in each pair.

**(a)** In this pair, the conjugate acids are $H_3O^+$ and $H_2SO_3$. Hydronium, $H_3O^+$, is a strong acid, and $H_2SO_3$ is a weak acid. Therefore, water is a weaker base than $HSO_3^-$.

**(b)** In this pair, the conjugate acids are HCl and HCN. HCl is a strong acid and HCN is a weak acid. Therefore, chloride, $Cl^-$, is a weaker base than $CN^-$.

# Oxidation-Reduction (Redox) Reactions

**Oxidation-reduction reactions,** also known as **redox reactions,** involve the transfer of electron(s) from a species that is oxidized to a species that is reduced.

An easy mnemonic is "oil rig." **O**xidation **i**s **l**oss of electrons. **R**eduction **i**s **g**ain of electrons.

It is important to remember that since electrons are negatively charged, the species that is reduced becomes more negative when electrons are gained and the species that is oxidized becomes more positive when electrons are lost.

Although the terms "oxidizing agent" and "reducing agent" are not tested on the AP Chemistry exam, these phrases are important for anyone working with laboratory chemicals. An oxidizing agent is a chemical that is easily reduced, and a reducing agent is easily oxidized. This terminology is included on the material safety data sheet (MSDS) for many compounds such as peroxides, permanganates, and nitrates.

## Assigning Oxidation Numbers

A prerequisite skill for identifying which species in a reaction is oxidized and which is reduced is the ability to assign oxidation numbers to each element in a compound.

The following rules are used for assigning oxidation numbers:

- A free, neutral element such as Fe, $H_2$, or $S_8$ always has an oxidation number of zero.
- The oxidation number of a monatomic ion is equal to the charge on that ion. The oxidation number of $Na^+$ is +1 and the oxidation number of $S^{2-}$ is –2.
- In compounds, hydrogen generally has an oxidation number of +1 when it is combined with nonmetals and –1 when it is combined with metals. The oxidation number of hydrogen in HCl is +1 and the oxidation number of hydrogen in KH is –1.
- Oxygen generally has an oxidation number of –2 except in peroxides, in which it has an oxidation number of –1. Oxygen has an oxidation number of –2 in $H_2O$, but it has an oxidation number of –1 in $H_2O_2$. (A **peroxide** is a compound containing an O–O bond. Hydrogen peroxide, $H_2O_2$, is the simplest peroxide: H–O–O–H.)
- The sum of the oxidation numbers of the element in a polyatomic ion is equal to the charge of that ion. The sum of the oxidation numbers of the atoms in $MnO_4^-$ is –1.
- The sum of the oxidation numbers of the elements in a neutral compound such as $CO_2$ or $Fe_2O_3$ is zero.

The species that is reduced in a redox reaction can be identified by its decrease in oxidation number. The species that is oxidized has an increase in oxidation number.

In determining which species are oxidized and reduced in a redox reaction, it is generally easiest to evaluate the net ionic form of the equation, since spectator ions are not oxidized or reduced.

### Practice

Assign oxidation numbers to each element in the following:

**a.** $Na_2SO_4$
**b.** $Cr_2O_7^{2-}$
**c.** $KMnO_4$

**(a)** For $Na_2SO_4$, each sodium atom has an oxidation number of +1 and each oxygen atom has an oxidation number of –2. That leaves the oxidation number of sulfur, which can be assigned the variable *x*. Since the sum of the oxidation numbers is zero, an equation can be written.

$$2(+1)+x+4(-2)=0$$
$$2+x-8=0$$
$$x=6$$

The oxidation number of sulfur in $Na_2SO_4$ is +6.

**(b)** For $Cr_2O_7^{2-}$, oxygen again has an oxidation number of –2. Let *x* be the oxidation number of chromium. The overall charge on the ion is –2, so the sum of the oxidation numbers is –2.

$$2x+7(-2)=-2$$
$$2x-14=-2$$
$$2x=12$$
$$x=6$$

The oxidation number of chromium in $Cr_2O_7^{2-}$ is +6.

**(c)** For $KMnO_4$, the oxidation number of potassium is +1 and the oxidation number of oxygen is –2. Let *x* be the oxidation number of manganese. Again, the overall charge of the compound is zero.

$$1(+1)+x+4(-2)=0$$
$$1+x-8=0$$
$$x=7$$

The oxidation number of manganese in $KMnO_4$ is +7.

## Practice

Determine which element is oxidized and which element is reduced in the following reaction:

$$2\ Zn + 2\ HCl \rightarrow 2\ ZnCl + H_2$$

This equation is easiest to deal with in net ionic equation form, since spectator ions are not oxidized or reduced.

Total ionic equation:

$$2\ Zn + 2\ H^+ + \cancel{2\ Cl^-} \rightarrow 2\ Zn^{2+} + \cancel{2\ Cl^-} + H_2$$

Net ionic equation:

$$2\ Zn + 2\ H^+ \rightarrow 2\ Zn^{2+} + H_2$$

In the reactants, zinc has an oxidation number of zero, and in the products it has an oxidation number of +2. The oxidation number has increased. Zinc has lost electrons and has been oxidized.

In the reactants, hydrogen has an oxidation number of +1, and in the products it has an oxidation number of zero. The oxidation number has decreased. Hydrogen has gained electrons and has been reduced.

# Balancing Redox Equations

Balancing redox equations is most easily accomplished using a systematic technique called the **half reaction method.**

Once a reaction has been identified as redox, the overall reaction is divided into two half reactions: an oxidation half reaction and a reduction half reaction. The half reactions are then balanced and then added back together.

Often a redox reaction occurs in acidic or basic solution. All balancing is first performed as if the solution were acidic. Then, if the solution is basic, the acidic form of the balanced equation is modified to suit the basic conditions.

Steps for balancing redox reactions:

1. Write the net ionic form of the unbalanced equation.
2. Using oxidation numbers, divide the equation into oxidation and reduction half reactions.
3. Balance all atoms other than hydrogen and oxygen.
4. Balance oxygen by adding one $H_2O$ for every missing oxygen.
5. Balance hydrogen by adding one $H^+$ for every missing hydrogen.
6. Balance charge by adding the appropriate number of electrons to the side that has a more positive overall charge.
7. Multiply the half reactions by a factor that will equalize the number of electrons.
8. Add the half reactions back together. The electrons on both sides must cancel.
9. If the reaction is in basic solution, add enough $OH^-$ to both sides of the equation to neutralize all the $H^+$. Simplify the number of water molecules.
10. Check to make sure that the number of each type of atom is equal in the products and the reactants. The overall charge must be the same on both sides of the equation.

Consider the reaction in basic solution between sulfite and chromate. A chromium(III) hydroxide precipitate and a sulfate solution are formed.

1. First, write the unbalanced equation. State symbols are not required for credit and are excluded here for clarity.

$$SO_3^{2-} + CrO_4^{2-} \rightarrow SO_4^{2-} + Cr(OH)_3$$

2. Divide the reaction into oxidation and reduction half reactions. The oxidation number of sulfur increases from +4 to +6, and the oxidation number of chromium decreases from +6 to +3. Sulfur is oxidized and chromium is reduced.

$$\text{oxidation: } SO_3^{2-} \rightarrow SO_4^{2-} \qquad \text{reduction: } CrO_4^{2-} \rightarrow Cr(OH)_3$$

3. Balance all atoms other than hydrogen and oxygen. In this case, sulfur and chromium are balanced, so no coefficients are needed.

$$SO_3^{2-} \rightarrow SO_4^{2-} \qquad CrO_4^{2-} \rightarrow Cr(OH)_3$$

4. Balance oxygen by adding $H_2O$.

$$SO_3^{2-} + \mathbf{H_2O} \rightarrow SO_4^{2-} \qquad CrO_4^{2-} \rightarrow Cr(OH)_3 + \mathbf{H_2O}$$

5. Balance hydrogen by adding $H^+$.

$$SO_3^{2-} + H_2O \rightarrow SO_4^{2-} + \mathbf{2\ H^+} \qquad CrO_4^{2-} + \mathbf{5\ H^+} \rightarrow Cr(OH)_3 + H_2O$$

6. Balance charge by adding electrons.

$$SO_3^{2-} + H_2O \rightarrow SO_4^{2-} + 2\ H^+ + \mathbf{2}\ \boldsymbol{e^-} \qquad CrO_4^{2-} + 5\ H^+ + \mathbf{3}\ \boldsymbol{e^-} \rightarrow Cr(OH)_3 + H_2O$$

7. Multiply the half reactions by a factor that will equalize the number of electrons.

$$3[SO_3^{2-} + H_2O \rightarrow SO_4^{2-} + 2\,H^+ + 2\,e^-] \qquad 2[CrO_4^{2-} + 5\,H^+ + 3\,e^- \rightarrow Cr(OH)_3 + H_2O]$$

$$3\,SO_3^{2-} + 3\,H_2O \rightarrow 3\,SO_4^{2-} + 6\,H^+ + 6\,e^- \qquad 2\,CrO_4^{2-} + 10\,H^+ + 6\,e^- \rightarrow 2\,Cr(OH)_3 + 2\,H_2O$$

8. Add the half reactions back together. The electrons on both sides must cancel. The resulting equation is balanced in acidic solution.

$$3\,SO_3^{2-} + \cancel{3}^1\,H_2O + 2\,CrO_4^{2-} + \cancel{10}^4\,H^+ + \cancel{6e^-} \rightarrow 3\,SO_4^{2-} + \cancel{6H^+} + \cancel{6e^-} + 2\,Cr(OH)_3 + \cancel{2\,H_2O}$$

$$3\,SO_3^{2-} + H_2O + 2\,CrO_4^{2-} + 4\,H^+ \rightarrow 3\,SO_4^{2-} + 2\,Cr(OH)_3$$

9. If the reaction is in basic solution, add enough $OH^-$ to both sides of the equation to neutralize all of the $H^+$. Simplify the number of water molecules.

$$3\,SO_3^{2-} + H_2O + 2\,CrO_4^{2-} + 4\,H^+ + \mathbf{4\,OH^-} \rightarrow 3\,SO_4^{2-} + 2\,Cr(OH)_3 + \mathbf{4\,OH^-}$$

$$3\,SO_3^{2-} + H_2O + 2\,CrO_4^{2-} + 4\,H_2O \rightarrow 3\,SO_4^{2-} + 2\,Cr(OH)_3 + 4\,OH^-$$

$$3\,SO_3^{2-} + 2\,CrO_4^{2-} + 5\,H_2O \rightarrow 3\,SO_4^{2-} + 2\,Cr(OH)_3 + 4\,OH^-$$

10. *Always* double-check to ensure that the number of each type of atom is equal in the products and the reactants. The overall charge must be the same on both sides of the equation.

## Practice

Practice balancing the following reactions in the indicated solutions.

**a.** Acidic solution: $Al + Fe^{2+} \rightarrow Al^{3+} + Fe$
**b.** Acidic solution: $BrO_3^- + N_2H_4 \rightarrow Br^- + N_2$
**c.** Basic solution: $H_2O_2 + ClO_2 \rightarrow ClO_2^- + O_2$
**d.** Basic solution: $Al + MnO_4^- \rightarrow MnO_2 + Al(OH)_4^-$

**(a)** Aluminum is oxidized from zero to +3, and iron is reduced from +2 to zero.

$$2\,Al + 3\,Fe^{2+} \rightarrow 2\,Al^{3+} + 3\,Fe$$

**(b)** Nitrogen is oxidized from –2 to zero, and bromine is reduced from +5 to –1.

$$2\,BrO_3^- + 3\,N_2H_4 \rightarrow 2\,Br^- + 6\,H_2O + 3\,N_2$$

**(c)** Hint: $H_2O_2$ is hydrogen peroxide, so oxygen has an oxidation number of –1 instead of –2.

The oxygen in peroxide is oxidized. It goes from an oxidation state of –1 in $H_2O_2$ to an oxidation state of zero in $O_2$. Chlorine is reduced from +4 to +3.

$$H_2O_2 + 2\,OH^- + 2\,ClO_2 \rightarrow 2\,ClO_2^- + 2\,H_2O + O_2$$

**(d)** Aluminum is oxidized from zero to +3, and manganese is reduced from +7 to +4.

$$Al + 2\,H_2O + MnO_4^- \rightarrow MnO_2 + Al(OH)_4^-$$

# Reaction Stoichiometry

**Reaction stoichiometry** refers to the relationship between the quantities of reactants and products in chemical reactions.

# The Stoichiometry Roadmap

Dimensional analysis, as we have seen, is the use of unit factors to convert a quantity with certain units to an equal quantity with different units. It is used extensively in stoichiometric calculations.

In reaction stoichiometry, the coefficients in the balanced chemical equation are used as conversion factors.

Consider the coefficients in the balanced equation for the decomposition of potassium chlorate upon heating.

$$2\ KClO_3(s) \rightarrow 2\ KCl(s) + 3\ O_2(g)$$

The coefficients have multiple meanings.

1. They are a molecular ratio. Two molecules of $KClO_3$ decompose to produce two molecules of KCl and three molecules of $O_2$.
2. They are also a mole ratio. Two moles of $KClO_3$ decompose to produce two moles of KCl and three moles of $O_2$.

The following is an incomplete list of the equalities that can be obtained from this balanced equation:

$$2 \text{ molecules } KClO_3 \text{ reacted} = 2 \text{ molecules KCl formed}$$
$$2 \text{ molecules KCl formed} = 3 \text{ molecules } O_2 \text{ formed}$$
$$2 \text{ moles } KClO_3 \text{ reacted} = 3 \text{ moles of } O_2 \text{ formed}$$

The number of moles of $O_2$ formed when 8.75 moles of $KClO_3$ decompose, according to the equation above, can be determined using the relationship between the number of moles of $KClO_3$ and the number of moles of $O_2$ in the balanced equation.

$$8.75 \cancel{\text{mol } KClO_3} \times \frac{3 \text{ mol } O_2}{2 \cancel{\text{mol } KClO_3}} = 13.1 \text{ mol } O_2$$

In real situations, moles cannot be measured directly. Mass and volume are quantities that can be measured. They must be converted to moles using molar masses, concentrations, densities, or gas law relationships.

The stoichiometry roadmap below summarizes the relationships between moles, masses, volumes, and particles of generic substances A and B that participate in a chemical reaction.

**Stoichiometry Roadmap**

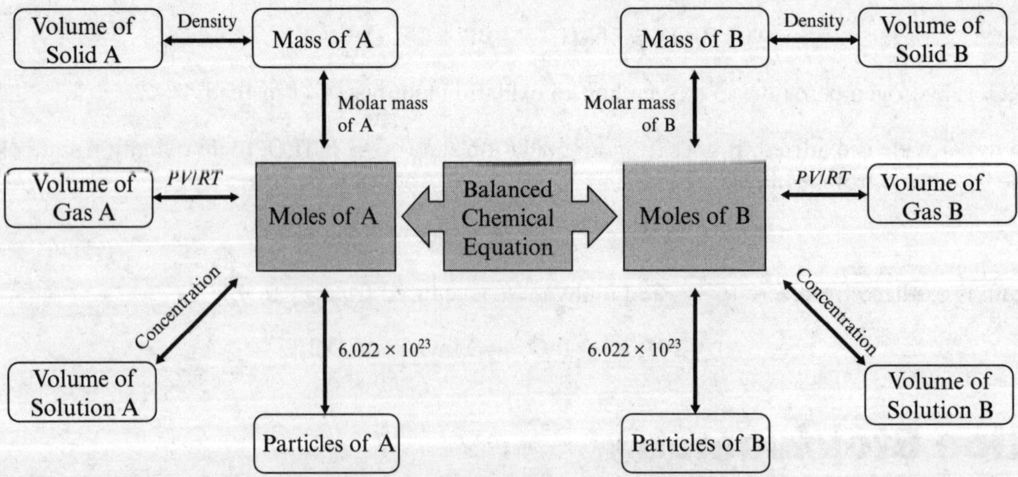

## Practice

How many liters of $O_2$ are produced at 27 °C and 0.975 atm when 12.3 grams of $KClO_3$ decompose, according to the equation below?

$$2\ KClO_3(s) \rightarrow 2\ KCl(s) + 3\ O_2(g)$$

Begin by converting the mass of $KClO_3$ to moles using the molar mass. The number of moles of $KClO_3$ is related to the number of moles of $O_2$ by the stoichiometric coefficients in the balanced equation.

$$12.3\ \cancel{g\ KClO_3} \times \frac{1\ \cancel{mol\ KClO_3}}{122.55\ \cancel{g\ KClO_3}} \times \frac{3\ mol\ O_2}{2\ \cancel{mol\ KClO_3}} = 0.151\ mol\ O_2\ (0.15055...)$$

The number of moles of $O_2$ can then be converted to liters using the rearranged ideal gas law.

$$V = \frac{nRT}{P} = \frac{(0.15055\ \cancel{mol})(0.08206\ L \cdot \cancel{atm} \cdot \cancel{mol^{-1}} \cdot \cancel{K^{-1}})(27 + 273\ \cancel{K})}{0.975\ \cancel{atm}} = 3.80\ L$$

## Practice

What mass of $CO_2$ results from the combustion of 5.0 mL of pure benzene, $C_6H_6$, with excess oxygen (enough oxygen for complete combustion)? The density of benzene is 0.877 g·mL$^{-1}$.

It is first necessary to balance the combustion equation.

$$2\ C_6H_6 + 15\ O_2 \rightarrow 12\ CO_2 + 6\ H_2O$$

The volume of benzene must be converted to moles, and then the moles of benzene can be used to find the number of moles of $CO_2$. The moles of $CO_2$ are multiplied by the molar mass of $CO_2$ to give the mass produced in the reaction.

$$5.0\ \cancel{mL\ C_6H_6} \times \frac{0.877\ \cancel{g\ C_6H_6}}{1\ \cancel{mL\ C_6H_6}} \times \frac{1\ \cancel{mol\ C_6H_6}}{78.12\ \cancel{g\ C_6H_6}} \times \frac{12\ \cancel{mol\ CO_2}}{2\ \cancel{mol\ C_6H_6}} \times \frac{44.01\ g\ CO_2}{1\ \cancel{mol\ CO_2}} = 15\ g\ CO_2$$

# Gas Stoichiometry

A convenient shortcut is to remember that the volume of a gas at a particular temperature and pressure is proportional to the number of moles. With this in mind, gas-phase reaction stoichiometry can be simplified by using volumes in place of moles whenever the gases under investigation are at the same temperature and pressure.

When hydrogen and oxygen gases are ignited, they form water.

$$2\ H_2(g) + O_2(g) \rightarrow 2\ H_2O(g)$$

How many liters of oxygen at STP must be available to completely react with 3.0 liters of hydrogen at STP?

The balanced equation shows that 2.0 liters of hydrogen react with exactly 1.0 liter of oxygen at a given temperature and pressure.

$$3.0\ \cancel{L\ H_2} \times \frac{1\ L\ O_2}{2\ \cancel{L\ H_2}} = 1.5\ L\ O_2$$

## Practice

Acetylene gas, $C_2H_2$, undergoes combustion in oxygen to form carbon dioxide and water. How many liters of oxygen must be present at STP to react completely with 4.0 liters of acetylene?

The first step in solving this problem is to write and balance the equation for the combustion of acetylene.

$$2\ C_2H_2(g) + 5\ O_2(g) \rightarrow 4\ CO_2(g) + 2\ H_2O(g)$$

Next, the coefficients in the balanced equation are used as a volume ratio to determine the stoichiometric amount of oxygen needed.

$$4.0\ \text{L}\,\cancel{C_2H_2} \times \frac{5\ \text{L}\ O_2}{2\ \text{L}\,\cancel{C_2H_2}} = 10.\ \text{L}\ O_2$$

# Limiting Reagent

The **limiting reagent** (or limiting reactant) is the substance that runs out first in a chemical reaction. It determines the maximum amount of product that can be produced, or **theoretical yield,** of a reaction.

To find the limiting reagent in a chemical reaction, first calculate the theoretical yield of one of the products using the amounts of each reactant given. Any reactant that is "in excess" is not the limiting reagent. *Whenever a question provides the amounts of two or more starting materials, the procedure for determining limiting reagent and theoretical yield must be followed.*

Steps for determining limiting reagent and theoretical yield:

1. Write the balanced chemical equation.
2. Choose one of the products. Use reaction stoichiometry to convert the amount of one of the reactants to moles of the chosen product.
3. Repeat step 2 with each reactant in turn. Always convert to moles of the same product.
4. The smallest number of moles of the chosen product resulting from steps 2 and 3 is the theoretical yield of that product.
5. The reactant that gave the theoretical yield is the limiting reagent.

Suppose a question asks for identification of the limiting reagent and the theoretical yield in grams of $P_4O_{10}$ in the reaction between 12.3 grams of $KClO_3$ and 12.3 grams of $P_4$. We can use this example to illustrate the steps for determining limiting reagent and theoretical yield.

1. The first step is to write the balanced equation.

$$3\ P_4 + 10\ KClO_3 \rightarrow 3\ P_4O_{10} + 10\ KCl$$

2. The second step is to convert each reactant to moles of product. Although either product will work in this step, the theoretical yield of $P_4O_{10}$ is required. It is best to choose this product in order to minimize the effort. Notice that although the numbers of moles of $P_4O_{10}$ can only be reported with three significant figures, it is better to use the unrounded value in subsequent calculations.

$$12.3\ \text{g}\,\cancel{KClO_3} \times \frac{1\ \text{mol}\,\cancel{KClO_3}}{122.55\ \text{g}\,\cancel{KClO_3}} \times \frac{3\ \text{mol}\ P_4O_{10}}{10\ \text{mol}\,\cancel{KClO_3}} = 0.0301\ \text{mol}\ P_4O_{10}$$

(0.030110)

**3.** The amount of $P_4$ is also converted to number of moles of product.

$$12.3\ \cancel{g\ P_4} \times \frac{1\ \cancel{mol\ P_4}}{123.88\ \cancel{g\ P_4}} \times \frac{3\ mol\ P_4O_{10}}{3\ \cancel{mol\ P_4}} = 0.0993\ mol\ P_4O_{10}$$

(0.0992896)

**4.** Since 0.030110 is less than 0.0992896, the theoretical yield of $P_4O_{10}$ is 0.0301 mole. The question asked for the theoretical yield in grams, so a simple conversion is required.

$$0.030110\ \cancel{mol\ P_4O_{10}} \times \frac{283.88\ g\ P_4O_{10}}{1\ \cancel{mol\ P_4O_{10}}} = 8.55\ g\ P_4O_{10}$$

**5.** The limiting reagent is $KClO_3$ since it results in the smaller theoretical yield.

Often students are asked to determine the amount of the excess reactant that is left over. In this case, the theoretical yield of product is used to determine the number of moles of excess reagent that is consumed in the reaction. The number of moles that reacted is subtracted from the number of moles of excess reagent that were present at the start of the reaction. The difference is the amount that is left over.

The excess reagent in the example is $P_4$. We can find the number of grams of $P_4$ consumed using the theoretical yield.

$$0.030110\ \cancel{mol\ P_4O_{10}} \times \frac{3\ \cancel{mol\ P_4}}{3\ \cancel{mol\ P_4O_{10}}} \times \frac{123.88\ g\ P_4}{1\ \cancel{mol\ P_4}} = 3.73\ g\ P_4$$

To find the amount left over, we subtract the amount consumed from the initial amount.

$$12.3\ g - 3.73\ g = 8.6\ g\ P_4$$

## Practice

100. mL of 0.600 $M$ $BaCl_2$ is mixed with 100. mL of 0.250 $M$ $K_3AsO_4$. A precipitation reaction occurs, and the solid $Ba_3(AsO_4)_2$ that forms is filtered, dried completely, and weighed. What mass of precipitate is formed? The molar mass of $Ba_3(AsO_4)_2$ is 689.82 $g{\cdot}mol^{-1}$.

The first step is to write the balanced equation.

$$3\ BaCl_2 + 2\ K_3AsO_4 \rightarrow Ba_3(AsO_4)_2 + 6\ KCl$$

Next, each reagent is converted to moles of $Ba_3(AsO_4)_2$.

$$100.\ \cancel{mL\ BaCl_2} \times \frac{1\ \cancel{L\ BaCl_2}}{1000\ \cancel{mL\ BaCl_2}} \times \frac{0.600\ \cancel{mol\ BaCl_2}}{1\ \cancel{L\ BaCl_2}} \times \frac{1\ mol\ Ba_3(AsO_4)_2}{3\ \cancel{mol\ BaCl_2}} = 0.0200\ mol\ Ba_3(AsO_4)_2$$

$$100.\ \cancel{mL\ K_3AsO_4} \times \frac{1\ \cancel{L\ K_3AsO_4}}{1000\ \cancel{mL\ K_3AsO_4}} \times \frac{0.250\ \cancel{mol\ K_3AsO_4}}{1\ \cancel{L\ K_3AsO_4}} \times \frac{1\ mol\ Ba_3(AsO_4)_2}{2\ \cancel{mol\ K_3AsO_4}} = 0.0125\ mol\ Ba_3(AsO_4)_2$$

Since the smaller number of moles of $Ba_3(AsO_4)_2$ is produced by complete reaction of the $K_3AsO_4$, the theoretical yield of $Ba_3(AsO_4)_2$ is 0.0125 mole. The limiting reagent is $K_3AsO_4$.

To determine the mass of precipitate formed, the number of moles capable of forming based on the limiting reagent is converted to grams using the molar mass.

$$0.0125\ \cancel{mol\ Ba_3(AsO_4)_2} \times \frac{689.82\ g\ Ba_3(AsO_4)_2}{1\ \cancel{mol\ Ba_3(AsO_4)_2}} = 8.62\ g\ Ba_3(AsO_4)_2$$

## Practice

In the preceding question, which reactant is in excess and what is the molarity of this reactant when the reaction is complete?

Since $K_3AsO_4$ is the limiting reagent, $BaCl_2$ is in excess. In order to find the molarity of this reactant at the end of the reaction, we need to know how many moles are left over and the volume of the combined solution.

The amount of $BaCl_2$ left over is equal to the difference between the initial amount and the amount that was consumed.

$$\text{Initial moles of BaCl}_2 = 100.\ \cancel{\text{mL BaCl}_2} \times \frac{1\ \cancel{\text{L BaCl}_2}}{1000\ \cancel{\text{mL BaCl}_2}} \times \frac{0.600\ \text{mol BaCl}_2}{1\ \cancel{\text{L BaCl}_2}} = 0.0600\ \text{mol BaCl}_2$$

The amount of $BaCl_2$ that was consumed can be determined from the number of moles of product that were produced.

$$\text{Moles of BaCl}_2\ \text{reacted} = 0.0125\ \cancel{\text{mol Ba}_3\text{(AsO}_4)_2} \times \frac{3\ \text{mol BaCl}_2}{1\ \text{mol}\ \cancel{\text{Ba}_3\text{(AsO}_4)_2}} = 0.0375\ \text{mol BaCl}_2$$

$$\text{Moles of BaCl}_2\ \text{left over} = 0.0600\ \text{mol} - 0.0375\ \text{mol} = 0.0225\ \text{mol BaCl}_2$$

The final concentration of $BaCl_2$ is the number of moles divided by the total volume, 200. mL of combined solution.

$$\frac{0.0225\ \text{mol BaCl}_2}{200.\ \cancel{\text{mL}}} \times \frac{1000\ \cancel{\text{mL}}}{1\ \text{L}} = 0.113\ M\ \text{BaCl}_2$$

# Percent Yield

**Percent yield** is a measure of the efficiency of a reaction. It indicates how much of the desired product of a reaction was obtained compared to how much was expected. Often, less product is obtained than anticipated. This can be due to incomplete reaction, competing side reactions, impure starting materials, or difficult purification of products. *A yield greater than 100% is not possible.*

$$\text{Percent yield} = \frac{\text{Actual yield of product}}{\text{Theoretical yield of product}} \times 100\%$$

As long as the units of actual yield and theoretical yield are the same, either grams or moles may be used.

When 56.1 grams of 1-butene, $C_4H_8$, are allowed to react with 240. grams of bromine, $Br_2$, the reaction yields 200. grams of product, $C_4H_8Br_2$. In order to determine the percent yield, we must first find the theoretical yield. Since the amounts of both starting materials are given, we must treat this as a limiting reagent problem.

$$C_4H_8 + Br_2 \rightarrow C_4H_8Br_2$$

$$56.1\ \cancel{\text{g C}_4\text{H}_8} \times \frac{1\ \cancel{\text{mol C}_4\text{H}_8}}{56.12\ \cancel{\text{g C}_4\text{H}_8}} \times \frac{1\ \text{mol C}_4\text{H}_8\text{Br}_2}{1\ \cancel{\text{mol C}_4\text{H}_8}} = 1.00\ \text{mol C}_4\text{H}_8\text{Br}_2$$

$$240.0\ \cancel{\text{g Br}_2} \times \frac{1\ \cancel{\text{mol Br}_2}}{159.8\ \cancel{\text{g Br}_2}} \times \frac{1\ \text{mol C}_4\text{H}_8\text{Br}_2}{1\ \cancel{\text{mol Br}_2}} = 1.50\ \text{mol C}_4\text{H}_8\text{Br}_2$$

Since fewer moles of product are expected to form based on the initial amount of 1-butene, the theoretical yield of $C_4H_8Br_2$ is 1.00 mole. The actual yield, 200.0 grams of $C_4H_8Br_2$, must be divided by the theoretical yield of $C_4H_8Br_2$ in grams.

$$\text{Theoretical yield} = 1.00\ \cancel{\text{mol } C_4H_8Br_2} \times \frac{215.92\ \text{g } C_4H_8Br_2}{1\ \cancel{\text{mol } C_4H_8Br_2}} = 216\ \text{g } C_4H_8Br_2$$

$$\text{Percent yield} = \frac{200.\ \cancel{\text{g } C_4H_8Br_2}}{216\ \cancel{\text{g } C_4H_8Br_2}} \times 100\% = 92.6\%$$

### Practice

The reaction below can be used to produce hydrazine, $N_2H_4$. A group of chemists would like to prepare 10.0 kg of hydrazine. If the expected yield of hydrazine is 70.0%, how many moles of chloramine, $NH_2Cl$, must the chemists use?

$$NH_2Cl + NH_3 + NaOH \rightarrow N_2H_4 + H_2O + NaOH$$

The solution to this problem begins with the desired mass of hydrazine. The actual yield of $N_2H_4$ must be 10.0 kg. The percent yield equation can be rearranged to calculate the theoretical yield that must be planned.

$$\text{Theoretical yield} = \frac{\text{Actual yield}}{\text{Percent yield}} \times 100\% = \frac{(10.0\ \text{kg } N_2H_4)}{70.0\%} \times 100\% = 14.3\ \text{kg } N_2H_4\ (14.2857)$$

The theoretical yield is used to calculate the number of moles of chloramine that must be used. The full, unrounded calculated output is needed.

$$14.2857\ \cancel{\text{kg } N_2H_4} \times \frac{1000\ \cancel{\text{g } N_2H_4}}{1\ \cancel{\text{kg } N_2H_4}} \times \frac{1\ \cancel{\text{mol } N_2H_4}}{32.06\ \cancel{\text{g } N_2H_4}} \times \frac{1\ \cancel{\text{mol } NH_2Cl}}{1\ \cancel{\text{mol } N_2H_4}} \times \frac{51.48\ \cancel{\text{g } NH_2Cl}}{1\ \cancel{\text{mol } NH_2Cl}} \times \frac{1\ \text{kg } NH_2Cl}{1000\ \cancel{\text{g } NH_2Cl}}$$
$$= 22.9\ \text{kg } NH_2Cl$$

## Redox Titrations

Titration is a technique for determining an unknown concentration of an analyte by reaction with an accurately known amount of a standard, or titrant. A **redox titration** is a titration that relies on a redox reaction between the analyte and the standard.

For example, hydrogen peroxide, $H_2O_2$, is an active ingredient in consumer bottles of antiseptic solution. It decomposes over time due to exposure to light, heat, and trace amounts of transition metals.

Redox titration of $H_2O_2$ with permanganate, $MnO_4^-$, is a method for determining the concentration of hydrogen peroxide in a solution. The oxygen atoms in hydrogen peroxide are oxidized from a –1 oxidation state to an oxidation state of zero. The manganese atoms in the permanganate ions are reduced from a purple +7 oxidation state to a colorless +2 oxidation state. The permanganate serves as both the titrant and the indicator.

$$H_2O_2 \rightarrow O_2 + 2\ H^+ + 2\ e^- \quad \text{(oxidation)}$$
$$MnO_4^- + 8\ H^+ + 5\ e^- \rightarrow Mn^{2+} + 4\ H_2O \quad \text{(reduction)}$$

In order to balance the overall equation, the hydrogen peroxide oxidation must be multiplied by a factor of 5 and the permanganate reduction must be multiplied by a factor of 2 before the reactions can be added.

$$5(H_2O_2 \rightarrow O_2 + 2\ H^+ + 2\ e^-) \qquad 2(MnO_4^- + 8\ H^+ + 5\ e^- \rightarrow Mn^{2+} + 4\ H_2O)$$
$$5\ H_2O_2 \rightarrow 5\ O_2 + 10\ H^+ + 10\ e^- \qquad 2\ MnO_4^- + 16\ H^+ + 10\ e^- \rightarrow 2\ Mn^{2+} + 8\ H_2O$$

Add the half reactions back together. The electrons on both sides must cancel.

$$5\,H_2O_2 + 2\,MnO_4^- + \cancel{16}^{6}\,H^+ + \cancel{10e^-} \rightarrow 5\,O_2 + \cancel{10H^+} + \cancel{10e^-} + 2\,Mn^{2+} + 8\,H_2O$$

$$5\,H_2O_2 + 2\,MnO_4^- + 6\,H^+ \rightarrow 5\,O_2 + 2\,Mn^{2+} + 8\,H_2O$$

The balanced equation shows that for every five moles of hydrogen peroxide present, two moles of permanganate react.

In the titration, an exact volume of hydrogen peroxide solution is diluted with a small volume of water. The volume of water added does not affect the result. The hydrogen peroxide solution is acidified with a small amount of sulfuric acid. A precisely known concentration of potassium permanganate solution is then added drop-wise to the hydrogen peroxide via a buret. The endpoint of the titration is reached when the hydrogen peroxide solution turns light pink and remains pink. At this point, no peroxide is left to react with additional $Mn^{+7}$, so the pink color lingers.

## Practice

---

Titration of a dilute, acidified, 2.00-mL aliquot of a commercial hydrogen peroxide solution requires 13.4 mL of 0.050 *M* $KMnO_4$. Assume the commercial solution has a density of 1.00 g·mL$^{-1}$.

**a.** What is the molarity of the $H_2O_2$ solution?
**b.** What is the mass percent of the $H_2O_2$ solution?

---

The volume of titrant required is used to determine the number of moles of $H_2O_2$ in the aliquot.

$$13.4\ \cancel{mL\ KMnO_4} \times \frac{1\ \cancel{L\ KMnO_4}}{1000\ \cancel{mL\ KMnO_4}} \times \frac{0.050\ \cancel{mol\ KMnO_4}}{1\ \cancel{L\ KMnO_4}} \times \frac{5\ mol\ H_2O_2}{2\ \cancel{mol\ KMnO_4}} = 0.0017\ mol\ H_2O_2$$

(0.001675)

**(a)** The molarity is determined by dividing the moles of $H_2O_2$ by the volume of the commercial sample used. (If we had been asked to report the number of moles of peroxide, we would report only two significant figures. The value is used before rounding in the following calculation, however.)

$$\text{Molarity of } H_2O_2 = \frac{0.001675\ mol\ H_2O_2}{2.00\ \cancel{mL\ H_2O_2}} \times \frac{1000\ \cancel{mL\ H_2O_2}}{1\ L\ H_2O_2} = 0.84\ M\ H_2O_2$$

**(b)** The mass percent is the percentage of the total mass of the solution that is $H_2O_2$.

$$\text{mass of } H_2O_2 = 0.001675\ mol\ H_2O_2 \times \frac{34.02\ g\ H_2O_2}{1\ mol\ H_2O_2} = 0.057\ g\ H_2O_2$$

(0.05698)

Since the density of the solution is assumed to be 1.00 g·mL$^{-1}$, the mass of the solution is 2.00 grams. Again, the answer is limited to two significant figures by the molarity of $KMnO_4$, but in order to avoid propagation of error, the entire calculated mass of $H_2O_2$ is used in the calculation of mass percent.

$$\text{percent by mass } H_2O_2 = \frac{\text{mass } H_2O_2}{\text{mass solution}} \times 100\% = \frac{0.05698\ g\ H_2O_2}{2.00\ g\ solution} \times 100\% = 2.8\%$$

# Review Questions

## Multiple Choice

**1.** Three 0.5-mL samples of different unknown clear, colorless solutions are contained in vials labeled A, B, and C. Twenty drops of an aqueous solution of ammonium chloride are added to each vial. Based on the following data, which statement is most likely to be true?

| Vial | A | B | C |
|---|---|---|---|
| **Observation** | A gas is evolved, which turns damp litmus paper blue. | No detectable change occurs. | A white precipitate forms. |

**A.** A chemical change occurs in vial A, while the processes in vials B and C are physical.
**B.** Chemical changes probably only occur in vials A and C.
**C.** Chemical changes occur in all three vials.
**D.** A chemical change only occurs in vial C.

*Use the following particulate diagram to answer questions 2 and 3. The reaction takes place in an aqueous solution, but water is not shown.*

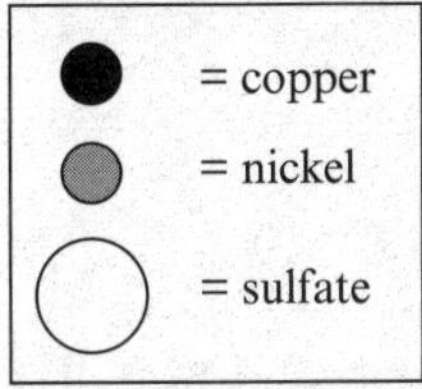


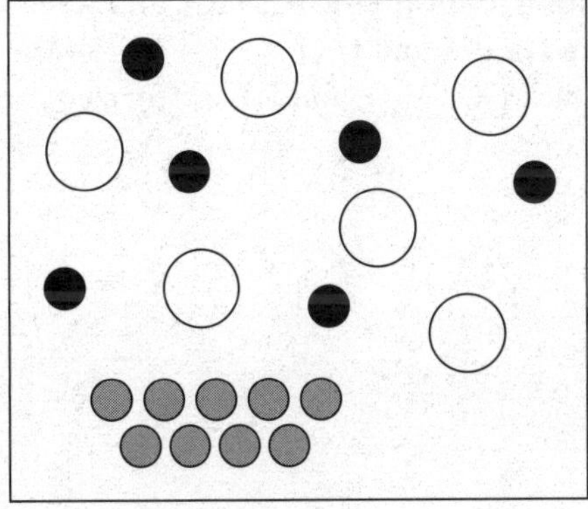

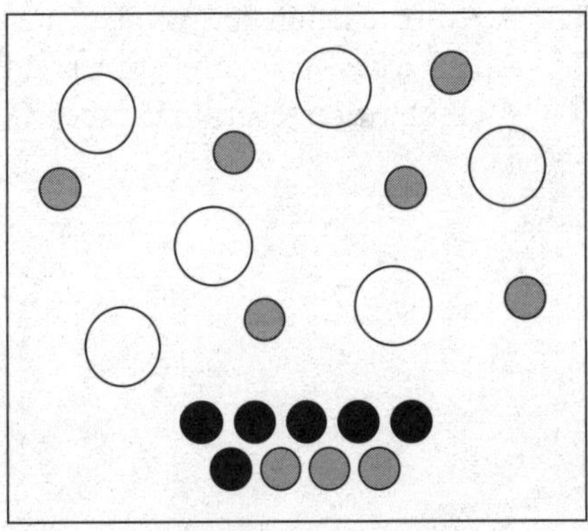

**2.** Which type of chemical reaction is most likely represented by the particulate diagram above?

**A.** Synthesis
**B.** Decomposition
**C.** Redox
**D.** Precipitation

**3.** Which of the following statements is true?

**A.** Copper ions are the limiting reagent.
**B.** Nickel ions are the limiting reagent.
**C.** Solid copper is the limiting reagent.
**D.** Solid nickel is the limiting reagent.

*Refer to the unbalanced redox reaction between tin and periodate ions in acidic solution to answer questions 4–6.*

$$Sn^{2+} + IO_4^- + H^+ \rightarrow Sn^{4+} + I^- + H_2O$$

**4.** Which choice shows the oxidation number of iodine in $IO_4^-$?

**A.** –5
**B.** –1
**C.** +5
**D.** +7

**5.** Which statement regarding this reaction is true?

**A.** Tin is being reduced.
**B.** Iodine is being reduced.
**C.** Oxygen is being reduced.
**D.** Hydrogen is being reduced.

**6.** How many $Sn^{2+}$ ions are required when the equation is balanced with the smallest possible whole-number coefficients?

**A.** 1
**B.** 2
**C.** 4
**D.** 8

**7.** For which of the following reactions is a net ionic equation least appropriate?

**A.** A double replacement reaction between $MgSO_4$ and NaOH, in which a precipitate of $Mg(OH)_2$ forms
**B.** A redox reaction between NaClO and KI in an HCl solution to form $I_2$, KCl, NaCl, and water
**C.** An acid-base reaction between $H_3PO_4$ and NaOH to form $Na_3PO_4$ and $H_2O$
**D.** A combustion reaction between $C_2H_3F$ and oxygen, in which $CO_2$, $H_2O$, and HF are formed

**8.** The reactants represented in the diagram below are placed in a vessel and a reaction occurs.

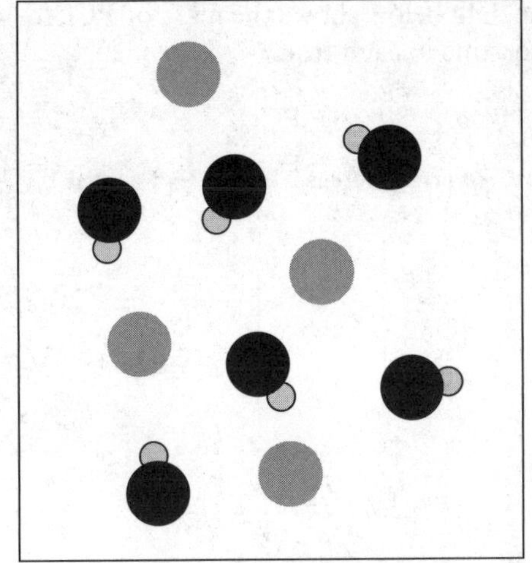

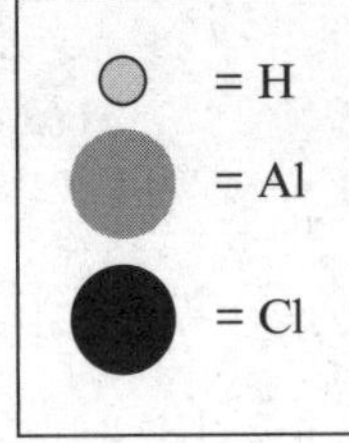


Which of the following best represents the contents of the vessel after the reaction proceeds as completely as possible?

**A.**

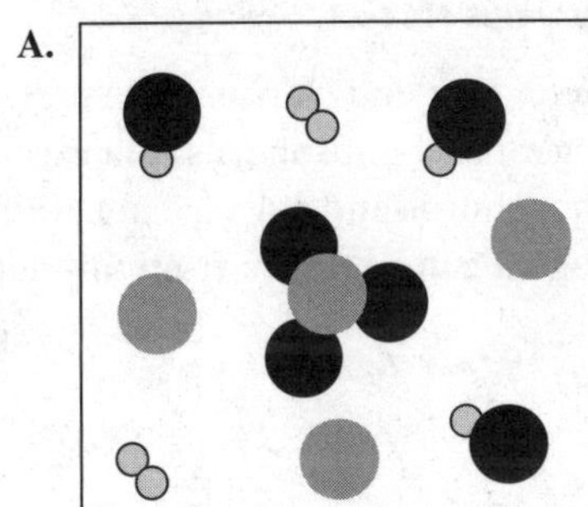

**B.**

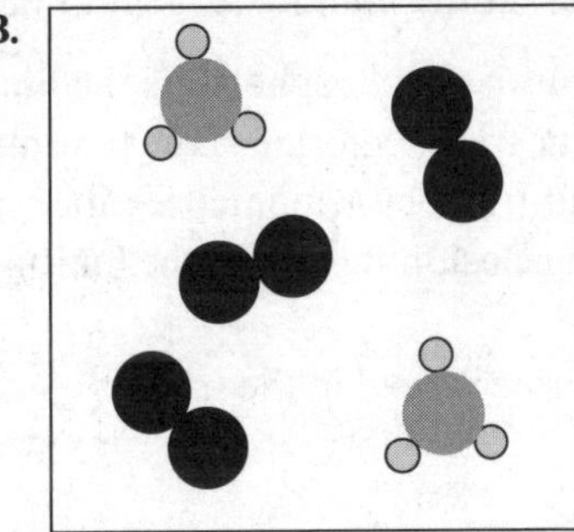

**C.**

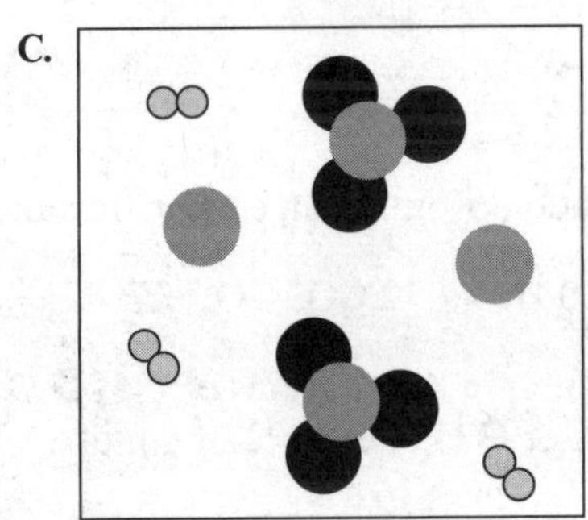

**D.**

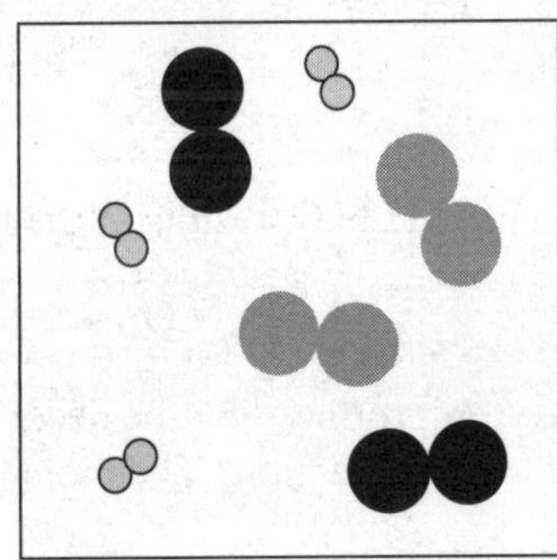

**9.** A student prepares 1.0 *M* aqueous solutions of each of the following salts: $HCO_2Na$, $NH_4Cl$, $NaNH_2$, and $CaI_2$. Which choice shows the correct ranking of the solutions from most acidic to most basic?

| | Most Acidic | | | Most Basic |
|---|---|---|---|---|
| **A.** | $HCO_2Na$ | $NH_4Cl$ | $NaNH_2$ | $CaI_2$ |
| **B.** | $HCO_2Na$ | $NH_4Cl$ | $CaI_2$ | $NaNH_2$ |
| **C.** | $NH_4Cl$ | $CaI_2$ | $HCO_2Na$ | $NaNH_2$ |
| **D.** | $NH_4Cl$ | $HCO_2Na$ | $NaNH_2$ | $CaI_2$ |

*Use the following information to answer questions 10 and 11.*

Phosphorus trichloride, $PCl_3$, a reagent used in organic chemistry, is synthesized by the reaction of white phosphorus, $P_4$, with chlorine gas, $Cl_2$. The data below shows the mass of $PCl_3$ formed when varying amounts of $P_4$ are used. The mass of chlorine is the same in each trial.

$$\_\_P_4 + \_\_Cl_2 \rightarrow \_\_PCl_3$$

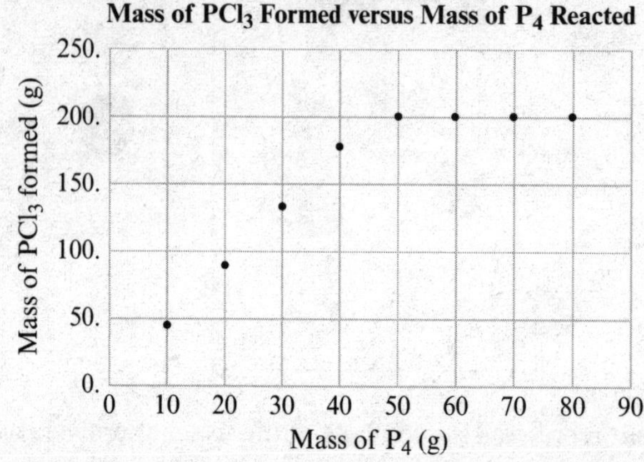


**10.** Which statement about the limiting reagent in these trials is correct?

**A.** In the first four trials, phosphorus is the limiting reagent and chlorine is in excess.
**B.** In the first four trials, chlorine is the limiting reagent and phosphorus is in excess.
**C.** In the last four trials, stoichiometric ratios of phosphorus and chlorine are available to react.
**D.** No reliable conclusion regarding the limiting reagent can be drawn from this data set.

**11.** What mass of chlorine is used in the trials?

**A.** 39 grams
**B.** 77 grams
**C.** 103 grams
**D.** 155 grams

**12.** The decomposition of $C_3H_5N_3O_9$ results in the formation of nitrogen, carbon dioxide, oxygen, and water.

$$4\ C_3H_5N_3O_9 \rightarrow 6\ N_2 + 10\ H_2O + 12\ CO_2 + O_2$$

If the volume of oxygen produced by the decomposition of 454.2 grams of $C_3H_5N_3O_9$ is 8.4 L at STP, what is the percent yield for this reaction? The molar mass of $C_3H_5N_3O_9$ is 227.1 g·mol$^{-1}$.

**A.** 38%
**B.** 67%
**C.** 75%
**D.** 100%

*Questions 13–14 refer to the information below.*

A 0.20-gram pellet of calcium metal was added to 10.0 mL of a solution of 0.30 *M* $HNO_3$.

**13.** Which of the following best represents the balanced net ionic equation for the resulting reaction?

- **A.** $2\ Ca + 2\ HNO_3 \rightarrow 2\ CaNO_3 + H_2$
- **B.** $2\ Ca + 2\ H^+ \rightarrow 2\ Ca^+ + H_2$
- **C.** $Ca + 2\ HNO_3 \rightarrow Ca(NO_3)_2 + H_2$
- **D.** $Ca + 2\ H^+ \rightarrow Ca^{2+} + H_2$

**14.** When the reaction reaches completion, which choice best reflects the correct quantity of the reactant that is left over?

- **A.** 0.14 grams of calcium are left over
- **B.** 0.06 grams of calcium are left over
- **C.** 0.0015 moles of nitric acid are left over
- **D.** 0.0035 moles of nitric acid are left over

**15.** The reaction between 3.0 liters of $SO_2(g)$ and 1.0 liter of $O_2$ gas in a piston produces the maximum amount of $SO_3(g)$. At constant temperature and pressure, which statement describes the change in the volume of the piston?

- **A.** The volume of the piston increased by 2.0 liters.
- **B.** The volume of the piston increased by 1.0 liter.
- **C.** The volume of the piston decreased by 2.0 liters.
- **D.** The volume of the piston decreased by 1.0 liter.

## Long Free Response

**1.** A study is conducted in which varying volumes of 0.500 *M* sodium hypochlorite, NaOCl, solution are added to varying volumes of a sodium thiosulfate solution such that the final volume was kept constant at 50.0 mL. The sodium thiosulfate is 0.500 *M* $Na_2S_2O_3$ in 0.200 *M* NaOH. The products of the reaction are sodium sulfate, sodium chloride, and water.

- **a.** Write and balance the net ionic equation for the reaction between $OCl^-$, $S_2O_3^{2-}$, and $OH^-$ to form $SO_4^{2-}$, $Cl^-$, and $H_2O$.
- **b.** Identify the atom that is oxidized in the reaction. What are the initial and final oxidation states?
- **c.** What advantage does the net ionic equation have over the total equation?

**d.** A plot of the temperature changes for the trials is constructed. Explain the shape of the plot in terms of limiting reagent before and after trial 9.

| Trial # | Volume of NaOCl (mL) | Volume of $Na_2S_2O_3$ (mL) |
|---|---|---|
| 1 | 0 | 50 |
| 2 | 5 | 45 |
| 3 | 10 | 40 |
| 4 | 15 | 35 |
| 5 | 20 | 30 |
| 6 | 25 | 25 |
| 7 | 30 | 20 |
| 8 | 35 | 15 |
| 9 | 40 | 10 |
| 10 | 45 | 5 |
| 11 | 50 | 0 |

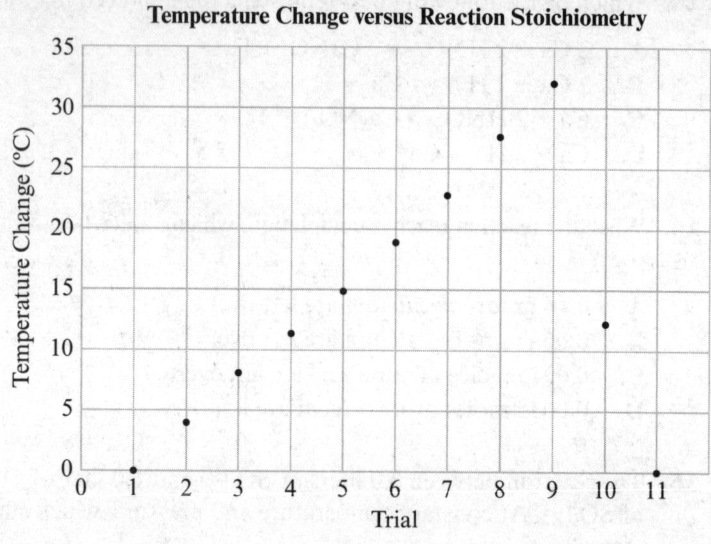


**e.** What conclusion about reagent ratio and theoretical yield can be drawn from the maximum temperature change in trial 9?

**f.** The final volume of the mixture is kept constant throughout the trials. Why is it important to maintain a constant volume throughout the experiment?

## Short Free Response

**2.** A solution of potassium permanganate, $KMnO_4$, can be standardized with sodium oxalate, $Na_2C_2O_4$. When $MnO_4^-$ is reacted with $C_2O_4^{2-}$ in acidic solution, the following reaction results.

$$MnO_4^- + C_2O_4^{2-} \rightarrow Mn^{2+} + CO_2$$

**a.** Identify the atom that is oxidized and the atom that is reduced in the reaction. Justify your response by referring to the starting and ending oxidation states for the oxidized and reduced atoms.

**b.** Write the balanced redox equation.

**c.** A 25.8-mL volume of $MnO_4^-$ solution was required to react completely with 0.723 grams of sodium oxalate, $Na_2C_2O_4$, in acidic solution.

**i.** How many moles of sodium oxalate reacted?

**ii.** How many moles of $MnO_4^-$ were required to reach the endpoint?

**iii.** What is the molarity of the $MnO_4^-$ solution?

# Answers and Explanations

## Multiple Choice

1. **B.** Chemical changes probably only occur in vials A and C, choice B. The gas formation in vial A and the precipitate formation in vial C are indicative of chemical changes. The absence of any change in vial B is evidence that no chemical change is occurring, but further testing would be required to prove this definitively.

2. **C.** This reaction is a redox reaction, choice C. The copper ions start out in the solution. The copper ions are found in a 1:1 ratio with sulfate ions, which have a –2 charge, so the copper ions must be $Cu^{2+}$. The nickel is shown as a solid crystal at the bottom of the container. On the product side, some of the nickel has been replaced by copper, which has been reduced. The nickel that has been replaced has entered the solution as $Ni^{2+}$. The net ionic equation for the reaction shown is

$$Cu^{2+} + Ni \rightarrow Cu + Ni^{2+}$$

3. **A.** The $Cu^{2+}$ is completely consumed to form products, but some $Ni^{0}$ remains. The copper ions, therefore, are the limiting reagent, choice A.

4. **D.** Iodine has a +7 oxidation state in the periodate ion, choice D. This molecule is not a peroxide, so oxygen has an oxidation number of –2. The oxidation numbers of iodine plus oxygen must add up to the overall charge on the periodate ion, –1.

$$x + 4(-2) = -1$$
$$x - 8 = -1$$
$$x = 7$$

5. **B.** Iodine is being reduced, choice B. It gains electrons in going from a +7 oxidation state in the periodate ion to a –1 oxidation state as iodide. Tin is being oxidized, while the oxidation states of hydrogen and oxygen remain unchanged.

6. **C.** The coefficient for tin is 4, choice C. The balanced redox equation as determined by the half reaction method is shown.

$$4\ Sn^{2+} + IO_4^- + 8\ H^+ \rightarrow 4\ Sn^{4+} + I^- + 4\ H_2O$$

7. **D.** The presence of spectator ions, which do not participate in the chemistry of a reaction, necessitates the use of net ionic equations. The only reaction in which there are no spectator ions is the last equation, the combustion of $C_2H_3F$, choice D. $Na^+$ and $SO_4^{2-}$ are spectators in the double replacement reaction, $K^+$ and $Na^+$ are spectators in the redox reaction, and $Na^+$ is a spectator in the acid-base reaction.

8. **C.** Choices B and D can be eliminated on the grounds that the numbers of reactants and products are not equal. Choice A can be eliminated because it shows unreacted Al and HCl. The correct choice is C, in which the reactants undergo the following reaction completely, leaving behind two unreacted Al atoms:

$$4\ Al + 6\ HCl \rightarrow 2\ AlCl_3 + 3\ H_2$$

9. **C.** The correct ranking of the solutions from most acidic to most basic is given in choice C. $NH_4^+$ ions are slightly acidic, aqueous $CaI_2$ is neutral, $HCO_2^-$ is weakly basic, and $NH_2^-$ is strongly basic. It is best to approach this problem by first recognizing that $Na^+$, $Cl^-$, $Ca^{2+}$, and $I^-$ do not affect the pH of the solution since they are the conjugates of strong acids (**H**Cl and **H**I) and bases (**Na**OH and **Ca**$(OH)_2$). $NH_4^+$ is the conjugate acid of ammonia, a weak base ($NH_3$). It forms a weakly acidic solution. $CaI_2$ is composed of $Ca^{2+}$, the cation of a strong base, $Ca(OH)_2$, and $I^-$, the conjugate base of a strong acid, HI, so it is a neutral salt. $HCO_2^-$ is the conjugate base of a weak acid, $HCO_2H$. It forms a weakly basic solution. $NH_2^-$ is the conjugate base of ammonia, which is already weakly basic and is an amphoteric species. The deprotonated form of ammonia is strongly basic, and this solute is the most basic of the four species.

**10.** **A.** The amount of $PCl_3$ formed levels off when the stoichiometric ratio of $P_4$ to $Cl_2$ is reached. Before this point, the limiting reagent is $P_4$, choice A, because as more $P_4$ is added, more product is formed. After this point, the amount of product formed does not increase with increasing quantities of $P_4$. In these cases, chlorine is the limiting reagent.

**11.** **D.** The first step in any stoichiometry problem is to balance the chemical equation.

$$P_4 + 6\,Cl_2 \rightarrow 4\,PCl_3$$

The maximum amount of chlorine available is used in the final four trials to form the maximum amount of product, 200. grams. The number of grams of chlorine can be calculated from the number of grams of product that are formed.

$$200.\ \cancel{g\ PCl_3} \times \frac{1\ \cancel{mol\ PCl_3}}{137.32\ \cancel{g\ PCl_3}} \times \frac{6\ \cancel{mol\ Cl_2}}{4\ \cancel{mol\ PCl_3}} \times \frac{70.90\ g\ Cl_2}{1\ \cancel{mol\ Cl_2}} = 155\ g$$

Without a calculator, we can make an estimation. The molar mass of chlorine is approximately 70, or $\frac{140}{2}$.

$$200 \times \frac{1}{\cancel{140}} \times \frac{3}{2} \times \frac{\cancel{140}}{2} = \frac{300}{2} = 150$$

The closest choice is 155 grams, choice D.

**12.** **C.** Theoretical yield and actual yield are needed in order to determine percent yield. The theoretical yield of oxygen is determined using the mass of the reagent.

$$454.2\ \cancel{g\ C_3H_5N_3O_9} \times \frac{1\ \cancel{mol\ C_3H_5N_3O_9}}{227.1\ \cancel{g\ C_3H_5N_3O_9}} \times \frac{1\ \cancel{mol\ O_2}}{4\ \cancel{mol\ C_3H_5N_3O_9}} \times \frac{22.4\ L\ O_2}{1\ \cancel{mol\ O_2}} = 11.2\ L\ O_2$$

The actual yield of $O_2$ is 8.4 L.

$$\frac{8.4\ \cancel{L\ O_2}}{11.2\ \cancel{L\ O_2}} \times 100\% = 75\%, \text{ choice C}$$

**13.** **D.** Calcium forms a 2+ ion since it is a group 2 alkaline earth metal. Nitrate is a spectator ion that is not included in the net ionic equation. Therefore, choice D is the correct answer.

**14.** **A.** From the balanced net ionic equation, we can determine the theoretical yield of $H_2$ that would be expected from complete reaction of each of the two reactants.

$$0.20\ \cancel{g\ Ca} \times \frac{1\ \cancel{mol\ Ca}}{40.08\ \cancel{g\ Ca}} \times \frac{1\ mol\ H_2}{1\ \cancel{mol\ Ca}} = 0.0050\ mol\ H_2$$

$$0.0100\ \cancel{L\ HNO_3} \times \frac{0.30\ \cancel{mol\ HNO_3}}{1\ \cancel{L\ HNO_3}} \times \frac{1\ mol\ H_2}{2\ \cancel{mol\ HNO_3}} = 0.0015\ mol\ H_2$$

Because $HNO_3$ yields the smaller number of moles of $H_2$, $HNO_3$ is the limiting reagent, and 0.0015 moles of $H_2$ is the theoretical yield. We can use the theoretical yield of $H_2$ to determine the number of grams of calcium that reacted.

$$0.0015\ \cancel{mol\ H_2} \times \frac{1\ \cancel{mol\ Ca}}{1\ \cancel{mol\ H_2}} \times \frac{40\ g\ Ca}{1\ \cancel{mol\ Ca}} = 0.060\ g\ Ca$$

The amount of calcium left over is the difference between the amount that was present initially and the amount that reacted.

$$0.20 \text{ g Ca} - 0.060 \text{ g Ca} = 0.14 \text{ g Ca, choice A}$$

**15.** **D.** The chemical equation must first be written and balanced.

$$2\ SO_2 + O_2 \rightarrow 2\ SO_3$$

Next, determine the limiting reagent.

$$3.0\ \text{L}\ \cancel{SO_2} \times \frac{2\ \text{L}\ SO_3}{2\ \text{L}\ \cancel{SO_2}} = 3.0\ \text{L}\ SO_3$$

$$1.0\ \text{L}\ \cancel{O_2} \times \frac{2\ \text{L}\ SO_3}{1\ \text{L}\ \cancel{O_2}} = 2.0\ \text{L}\ SO_3$$

The maximum amount of $SO_3$ that can be produced is 2.0 L. $O_2$ is the limiting reagent, so the entire volume of $O_2$ reacts. The amount of $SO_2$ that reacts must be calculated.

$$2.0\ \text{L}\ \cancel{SO_3} \times \frac{2\ \text{L}\ SO_2}{2\ \text{L}\ \cancel{SO_3}} = 2.0\ \text{L}\ SO_2$$

One methodical approach to determining how many liters are present at the end of the process is to set up a table to take an inventory of the species present before and after the reaction.

| Species | $SO_2$ | $O_2$ | $SO_3$ |
|---|---|---|---|
| **Before Reaction** | 3.0 L | 1.0 L | 0 L |
| **Change** | –2.0 L | –1.0 L | +2.0 L |
| **After Reaction** | 1.0 L | 0 L | 2.0 L |

Before the reaction, there are 4.0 L of gas ($SO_2 + O_2$) in the piston; after the reaction, there are 3.0 L of gas ($SO_2 + SO_3$). There is a decrease of 1.0 L, choice D.

## Long Free Response

**1.** **a.** The change in oxidation states of the atoms in the species shows that this is a redox reaction in basic solution.

First, write the unbalanced net ionic equation. Recognizing that $Na^+$ is always a spectator ion in an aqueous solution will expedite the task.

$$OCl^- + S_2O_3^{2-} + OH^- \rightarrow SO_4^{2-} + Cl^- + H_2O$$

Divide the reaction into oxidation and reduction half reactions. Balance all atoms other than hydrogen and oxygen.

oxidation: $S_2O_3^{2-} \rightarrow \mathbf{2}\ SO_4^{2-}$ reduction: $OCl^- \rightarrow Cl^-$

Balance oxygen by adding $H_2O$.

$S_2O_3^{2-} + \mathbf{5\ H_2O} \rightarrow 2\ SO_4^{2-}$ $OCl^- \rightarrow Cl^- + \mathbf{H_2O}$

Balance hydrogen by adding $H^+$.

$S_2O_3^{2-} + 5\ H_2O \rightarrow 2\ SO_4^{2-} + \mathbf{10\ H^+}$ $OCl^- + \mathbf{2\ H^+} \rightarrow Cl^- + H_2O$

Balance charge by adding electrons.

$$S_2O_3^{2-} + 5\,H_2O \rightarrow 2\,SO_4^{2-} + 10\,H^+ + \mathbf{8\,e^-} \qquad OCl^- + 2\,H^+ + \mathbf{2\,e^-} \rightarrow Cl^- + H_2O$$

Multiply the half reactions by a factor that will equalize the number of electrons.

$$\mathbf{1}(S_2O_3^{2-} + 5\,H_2O \rightarrow 2\,SO_4^{2-} + 10\,H^+ + 8\,e^-) \qquad \mathbf{4}(OCl^- + 2\,H^+ + 2\,e^- \rightarrow Cl^- + H_2O)$$

$$S_2O_3^{2-} + 5\,H_2O \rightarrow 2\,SO_4^{2-} + 10\,H^+ + 8\,e^- \qquad 4\,OCl^- + 8\,H^+ + 8\,e^- \rightarrow 4\,Cl^- + 4\,H_2O$$

Add the half reactions back together. The electrons on both sides must cancel.

$$S_2O_3^{2-} + 5\,H_2O + 4\,OCl^- + \cancel{8\,H^+} + \cancel{8\,e^-} \rightarrow 2\,SO_4^{2-} + {}^{2}\cancel{10}\,H^+ + \cancel{8\,e^-} + 4\,Cl^- + 4\,H_2O$$

$$S_2O_3^{2-} + 5\,H_2O + 4\,OCl^- \rightarrow 2\,SO_4^{2-} + 2\,H^+ + 4\,Cl^- + 4\,H_2O$$

The reaction is in basic solution. Add enough $OH^-$ to both sides of the equation to neutralize all the $H^+$. Simplify the number of water molecules.

$$S_2O_3^{2-} + 5\,H_2O + 4\,OCl^- + 2\,OH^- \rightarrow 2\,SO_4^{2-} + 2\,H^+ + 4\,Cl^- + 4\,H_2O + 2\,OH^-$$

$$S_2O_3^{2-} + \cancel{5\,H_2O} + 4\,OCl^- + 2\,OH^- \rightarrow 2\,SO_4^{2-} + 4\,Cl^- + {}^{1}\cancel{6}\,H_2O$$

$$\mathbf{S_2O_3^{2-} + 4\,OCl^- + 2\,OH^- \rightarrow 2\,SO_4^{2-} + 4\,Cl^- + H_2O}$$

**b.** The oxidation number of sulfur goes from +2 to +6. Sulfur is oxidized.

**c.** The net ionic equation has the advantage of omitting spectator ions that do not undergo any reaction.

**d.** Before trial 9, the heat released increases as the ratio of $S_2O_3^{2-}$ to $OCl^-$ increases. In these trials, $OCl^-$ is the limiting reagent since adding more $OCl^-$ from one trial to the next increases the amount of heat released. After trial 9, $S_2O_3^{2-}$ is the limiting reagent. At this point, there is not enough $S_2O_3^{2-}$ to react with the larger quantity of $OCl^-$.

**e.** The maximum temperature change in trial 9 indicates that a perfect stoichiometric ratio of $OCl^-$ and $S_2O_3^{2-}$ has been added. The concentrations of both reactants are the same, so the 4:1 volume ratio of trial 9 corresponds to a 4:1 mole ratio of $OCl^-$ to $S_2O_3^{2-}$. This means that the maximum theoretical yield is expected.

**f.** It is important to maintain a constant final volume in this experiment so that no correction must be made for heat transferred to varying amounts of the solvent. If more solvent were available in a trial, the same amount of heat would be released in the reaction, but more heat would be transferred to the solvent. This would result in a smaller heat transfer to the thermometer to register as a temperature change relative to a trial with a smaller volume of solvent.

## Short Free Response

**2. a.** The oxidation number of Mn in $MnO_4^-$ is

$$x + 4(-2) = -1$$
$$x = +7$$

The oxidation number of carbon in $C_2O_4^{2-}$ is

$$2x + 4(-2) = -2$$
$$2x = 6$$
$$x = +3$$

The oxidation number of carbon in $CO_2$ is

$$x + 2(-2) = 0$$
$$x = +4$$

Mn is reduced from a +7 to +2 oxidation state.

C is oxidized from a +3 to +4 oxidation state.

**b.** The unbalanced net ionic equation is

$$MnO_4^- + C_2O_4^{2-} \rightarrow Mn^{2+} + CO_2$$

Divide the reaction into oxidation and reduction half reactions. Balance all atoms other than hydrogen and oxygen.

oxidation: $C_2O_4^{2-} \rightarrow \mathbf{2}\ CO_2$ reduction: $MnO_4^- \rightarrow Mn^{2+}$

Balance oxygen by adding $H_2O$.

$C_2O_4^{2-} \rightarrow 2\ CO_2$ $MnO_4^- \rightarrow Mn^{2+} + \mathbf{4\ H_2O}$

Balance hydrogen by adding $H^+$.

$C_2O_4^{2-} \rightarrow 2\ CO_2$ $\mathbf{8\ H^+} + MnO_4^- \rightarrow Mn^{2+} + 4\ H_2O$

Balance charge by adding electrons.

$C_2O_4^{2-} \rightarrow 2\ CO_2 + \mathbf{2}\ \boldsymbol{e^-}$ $8\ H^+ + MnO_4^- + \mathbf{5}\ \boldsymbol{e^-} \rightarrow Mn^{2+} + 4\ H_2O$

Multiply the half reactions by a factor that will equalize the number of electrons.

$\mathbf{5}(C_2O_4^{2-} \rightarrow 2\ CO_2 + 2\ e^-)$ $\mathbf{2}(8\ H^+ + MnO_4^- + 5\ e^- \rightarrow Mn^{2+} + 4\ H_2O)$

$5\ C_2O_4^{2-} \rightarrow 10\ CO_2 + 10\ e^-$ $16\ H^+ + 2\ MnO_4^- + 10\ e^- \rightarrow 2\ Mn^{2+} + 8\ H_2O$

Add the half reactions back together. The electrons on both sides must cancel to result in the final net ionic equation.

$$5\ C_2O_4^{2-} + 16\ H^+ + 2\ MnO_4^-\ \cancel{10\ e^-} \rightarrow 10\ CO_2 + \cancel{10\ e^-} + 2\ Mn^{2+} + 8\ H_2O$$

$$5\ C_2O_4^{2-} + 16\ H^+ + 2\ MnO_4^- \rightarrow 10\ CO_2 + 2\ Mn^{2+} + 8\ H_2O$$

**c.** **i.** The molar mass of sodium oxalate can be used to determine the number of moles that reacted.

$$0.723\ g\ \cancel{Na_2C_2O_4} \times \frac{1\ mol\ Na_2C_2O_4}{134\ g\ \cancel{Na_2C_2O_4}} = 0.00540\ mol\ Na_2C_2O_4\ (0.0053955...)$$

**ii.** The number of moles of permanganate that were required to reach the endpoint can be determined stoichiometrically.

$$0.0053955\ mol\ C_2O_4^{2-} \times \frac{2\ mol\ MnO_4^-}{5\ mol\ C_2O_4^{2-}} = 0.00216\ mol\ MnO_4^-\ (0.0021582...)$$

**iii.** The molarity of the permanganate solution is determined by dividing the number of moles by the volume.

$$\frac{0.0021582\ mol\ MnO_4^-}{0.0258\ L} = 0.0837\ M\ MnO_4^-$$

Chapter 5

# Kinetics

**Chemical kinetics is the study of the rates of chemical reactions.**

Chemical reactions occur at a wide variety of speeds. The explosive reaction of gunpowder in fireworks occurs almost instantly, while the transformation of diamond to graphite occurs on a geologic timescale. **Kinetics** is the study of reaction speed and the factors that influence it.

## Rates of Reactions

In chemistry, the **rate of a reaction** is expressed as a change in the amounts of reactants or products per unit time.

As an aid to visualization, imagine a catering service is preparing sandwiches at a rate of one sandwich per minute according to the following "equation."

$$2 \text{ slices of bread} + 3 \text{ slices of cheese} \rightarrow 1 \text{ sandwich}$$

The rate of sandwich production is expressed as a change in quantity per unit time.

$$\text{rate of sandwich production} = \frac{\Delta \text{ sandwiches}}{\Delta \text{ time}} = \frac{1 \text{ sandwich}}{\text{minute}}$$

The rate of sandwich production can be related to the rate of disappearance of bread and cheese slices. Since the amounts of bread and cheese decrease as sandwiches are produced, Δ bread and Δ cheese are negative.

As one sandwich is produced, two slices of bread disappear. Alternatively stated, the rate of sandwich production is one-half the rate of bread disappearance.

$$\text{rate of sandwich production} = -\frac{1}{2}\left(\frac{\Delta \text{ bread}}{\Delta \text{ time}}\right)$$

As one sandwich is produced, three slices of cheese disappear. The rate of cheese disappearance is one-third the rate of sandwich production.

$$\text{rate of sandwich production} = -\frac{1}{3}\left(\frac{\Delta \text{ cheese}}{\Delta \text{ time}}\right)$$

By this analogy, for the general reaction

$$a\text{A} + b\text{B} \rightarrow c\text{C} + d\text{D}$$

the rate expression in terms of concentrations of reactants and products is:

$$\text{rate of reaction} = -\frac{1}{a}\left(\frac{\Delta[\text{A}]}{\Delta t}\right) = -\frac{1}{b}\left(\frac{\Delta[\text{B}]}{\Delta t}\right) = \frac{1}{c}\left(\frac{\Delta[\text{C}]}{\Delta t}\right) = \frac{1}{d}\left(\frac{\Delta[\text{D}]}{\Delta t}\right)$$

## Practice

The equation for a hypothetical reaction is shown below. Using the tabulated data provided, determine the coefficients *b*, *c*, and *d*.

$$3\text{ A} + b\text{ B} \rightarrow c\text{ C} + d\text{ D}$$

| Change | Rate of Change ($M \cdot s^{-1}$) |
|---|---|
| $\frac{\Delta[A]}{\Delta t}$ | –0.750 |
| $\frac{\Delta[B]}{\Delta t}$ | –1.00 |
| $\frac{\Delta[C]}{\Delta t}$ | 0.500 |
| $\frac{\Delta[D]}{\Delta t}$ | 0.250 |

Since the coefficient of A is known, the rate of change of the concentration of A can be used to determine the rate of the reaction.

$$\text{rate of reaction} = -\frac{1}{a}\left(\frac{\Delta[A]}{\Delta t}\right) = -\frac{1}{3}\left(-0.750\ M \cdot s^{-1}\right) = 0.250\ M \cdot s^{-1}$$

Once the rate of reaction is known, it can be used to determine the coefficients *b*, *c*, and *d*.

$$0.250\ M \cdot s^{-1} = -\frac{1}{b}\left(-1.00\ M \cdot s^{-1}\right)$$
$$b = 4$$
$$0.250\ M \cdot s^{-1} = \frac{1}{c}\left(0.500\ M \cdot s^{-1}\right)$$
$$c = 2$$
$$0.250\ M \cdot s^{-1} = \frac{1}{d}\left(0.250\ M \cdot s^{-1}\right)$$
$$d = 1$$

# Instantaneous Rate

At any time in the reaction, the **instantaneous rate** with respect to a reactant or product is given by the slope of the tangent line to the concentration versus time curve.

Consider the following hypothetical reaction between reactants A and B.

$$\text{A} + \text{B} \rightarrow \text{C} + \text{D}$$

Let's calculate the instantaneous rate of disappearance of B at 1 second and at 5 seconds of reaction time.

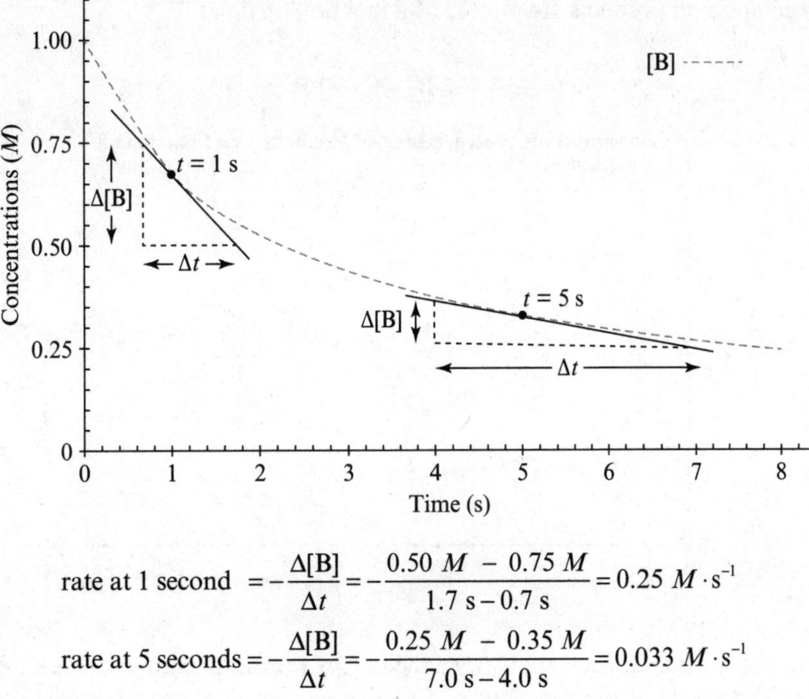


$$\text{rate at 1 second} = -\frac{\Delta[B]}{\Delta t} = -\frac{0.50\ M - 0.75\ M}{1.7\ s - 0.7\ s} = 0.25\ M \cdot s^{-1}$$

$$\text{rate at 5 seconds} = -\frac{\Delta[B]}{\Delta t} = -\frac{0.25\ M - 0.35\ M}{7.0\ s - 4.0\ s} = 0.033\ M \cdot s^{-1}$$

Notice also that the rate of disappearance of B decreases with time.

## Practice

The product of a reaction is monitored and its concentration is plotted as a function of time. At which of the marked points, X or Y, is the instantaneous rate of the reaction the fastest?

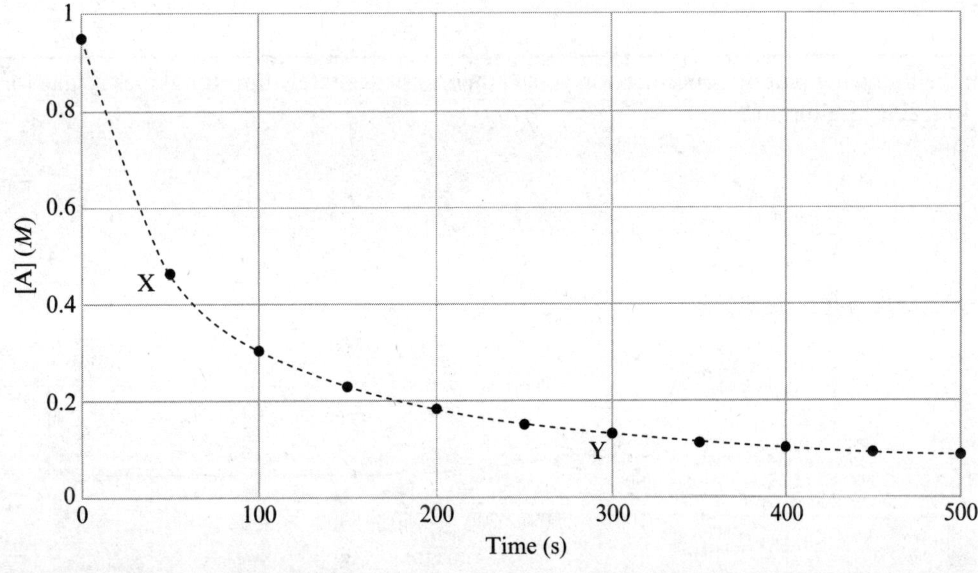


The rate of the disappearance is faster at point X than at point Y. The slope of the tangent line at point X, which is a measure of the instantaneous rate of the reaction, is greater at point X than it is at point Y.

## Equilibrium Rates

Now, imagine another hypothetical reaction between reactants A and B in an aqueous solution; the concentrations of reactants and products are plotted as a function of time.

$$A + 2\,B \rightarrow C + 2\,D$$

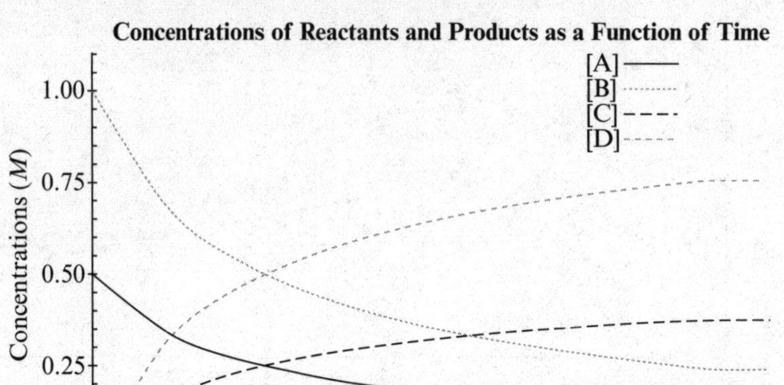


The rate of disappearance of B at any instant is twice the rate of the disappearance of A, and the rate of appearance of D is twice the rate of the appearance of C. The rates of disappearance of reactants and the rates of appearance of products are decreasing with time. In other words, as concentrations of reactants decrease, the rate of the reaction also decreases.

When the concentrations of reactants and products stop changing, after 8 seconds in this case, **equilibrium** has been established. At equilibrium, the rates of the forward and reverse reactions are no longer changing. Chemical equilibrium is reviewed in Chapter 7, "Chemical Equilibrium." It is important to note that at equilibrium, the forward and reverse reactions are still occurring, but the concentrations of reactants and products are constant.

### Practice

Based on the following plot of concentration versus time, approximately how long does it take for the reaction to reach equilibrium?

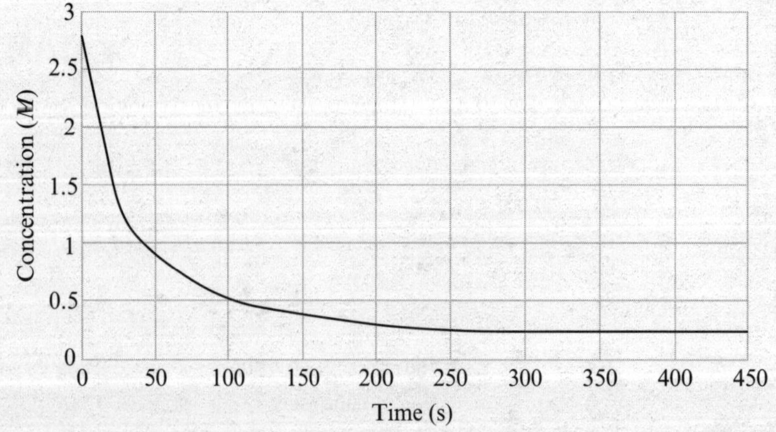


The concentration of the reactant stops changing at about 250 seconds, so at 250 seconds the reaction has reached equilibrium.

# Measuring the Rates of Reactions

Methods for determining the rate of a chemical reaction experimentally involve measuring the change in some observable quantity over time.

Some quantities that can be monitored to deduce the rate of a reaction include:

- *Volume of a gas produced.* A reaction that produces a gas as a product can be set up in an Erlenmeyer flask with a stopper and an empty syringe. As the reaction proceeds, the plunger of the syringe is pushed out by the pressure of the gas produced, and the volume of gas in the syringe is recorded at regular time intervals during the reaction.

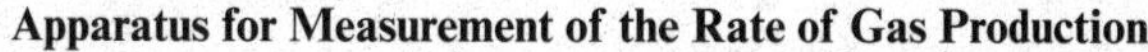

**Apparatus for Measurement of the Rate of Gas Production**

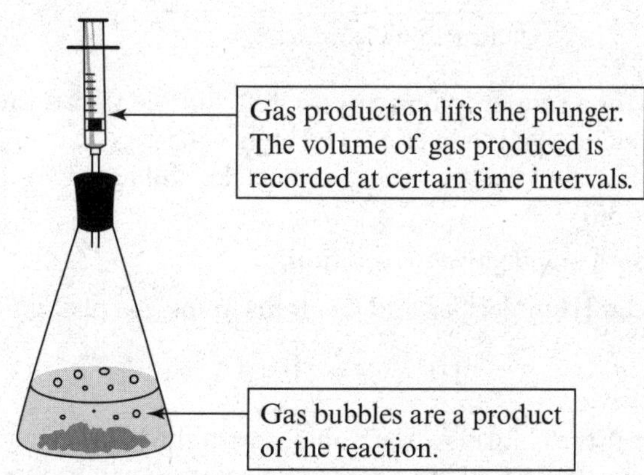


- *Change in pressure as gases are produced or consumed.* In a sealed, rigid container, the pressure will increase as gaseous products are produced or decrease as gaseous reactants are consumed. The total pressure in the container can be measured with a sensor.
- *Change in the mass of a reaction as a solid or liquid is converted to gas.* A reaction can be carried out in a container on a balance. The mass can be recorded at time intervals to monitor the rate at which gaseous products are released to the atmosphere.
- *Change in concentration as determined by spectrophotometry.* An increase in the concentration of a colored product or a decrease in the concentration of a colored reactant can be monitored in a spectrophotometer. The values of the concentrations are determined using Beer's law analysis of a known calibration curve.
- *Change in pH as a reaction proceeds.* Reactions in which acids or bases are formed or consumed will undergo a change in pH as the reaction proceeds. A pH meter can be used to monitor the change. The pH scale will be reviewed in Chapter 8, "Acids and Bases."

# Collision Theory

In order for two molecules to react with one another to form a product, the molecules must collide. Collisions between particles, however, do not guarantee that a reaction will occur. In order for a collision to successfully result in a reaction, two minimum conditions must be met:

1. The energy of the collision between the particles must be greater than or equal to the activation energy in order for the reaction to occur. The **activation energy, $E_a$,** is a potential energy barrier that the reactants must overcome to achieve the necessary **transition state,** the high-energy arrangement of atoms in which bond breaking and/or bond formation occurs. The transition state is sometimes referred to as the **activated complex.** The lower the activation energy of a reaction, the faster the rate of the reaction.

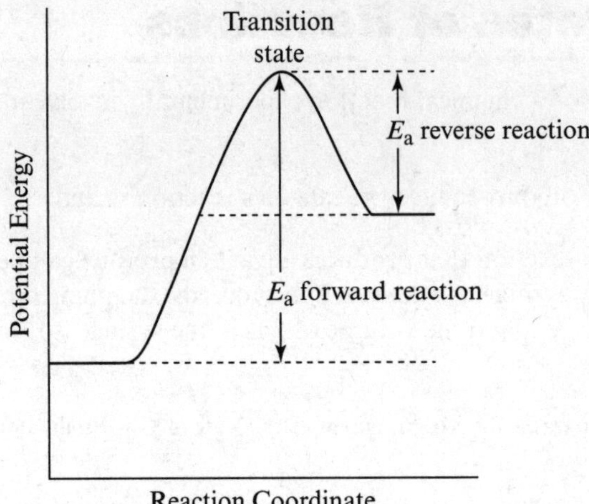


Just as it takes more energy for a runner to jump over a high hurdle than a low one, it takes more energy for reactants to overcome a higher activation barrier. The higher the kinetic energy of the reacting particles, the higher the likelihood that a collision between the two particles will have enough energy to overcome the transition barrier for a reaction.

2. The reactants must collide with an effective orientation.

   Consider the reaction between HI molecules and Cl atoms in the gas phase.

   $$HI + Cl \rightarrow HCl + I$$

   The reaction can only occur if the chlorine atom collides with the hydrogen atom of the HI molecule. If the iodine is oriented toward the colliding chlorine atom, the collision will be ineffective and no reaction will occur.

**Effective Orientation of Reactants during Collisions**

Correct orientation during collision results in reaction.

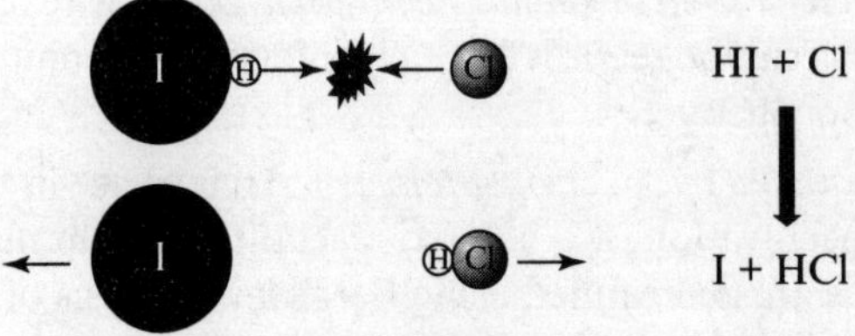


Incorrect orientation during collision results in no reaction.

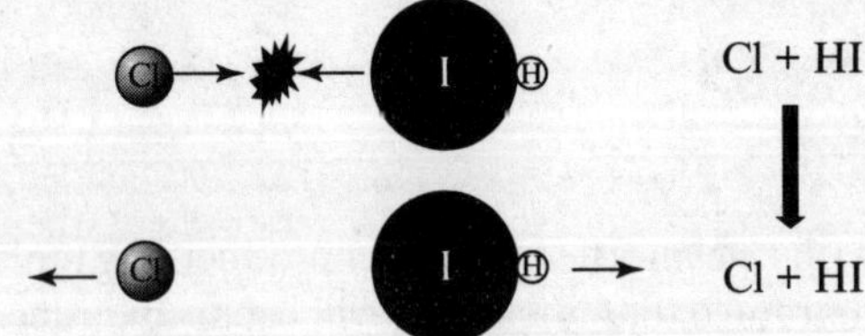


## Practice

Explain in terms of collision theory how an increase in temperature leads to an increase in the rate of a reaction.

The higher the temperature of a reaction mixture, the higher the average kinetic energy of the particles. The higher the kinetic energy, the greater the frequency of collisions that have the correct molecular orientation and that are sufficiently energetic to overcome the activation barrier for the reaction. At higher temperature, the reaction rate will increase.

## The Transition State

The **transition state,** or **activated complex,** is the highest-energy configuration that a group of atoms must achieve in order to break or form bonds. Along the reaction coordinate, it is the state that is achieved by colliding molecules with sufficient activation energy and correct orientation. At this point, bonds are at the process of breaking or forming, so they are longer and weaker than regular covalent bonds.

$$CH_3Br + OH^- \rightarrow CH_3OH + Br^-$$

In the exothermic reaction with methyl bromide, $CH_3Br$, hydroxide, $OH^-$, collides with the carbon atom on the side opposite the bromine. When the collision occurs with sufficient energy to overcome the activation barrier, the C–Br bond breaks as the C–O bond forms. The dotted bonds in the activated complex indicate partially formed or partially broken bonds.

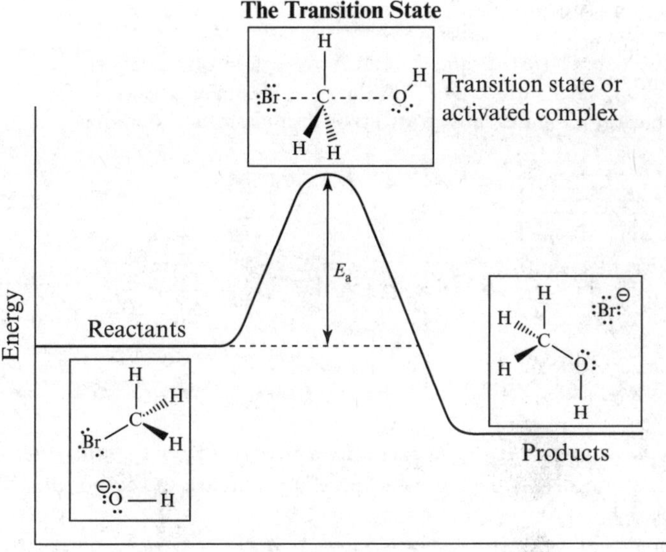


## Factors That Influence Rates of Reactions

Reactions can occur at an enormous variety of rates. The rate of a given reaction can vary, depending on the conditions. Often, the explanations for these factors invoke collision theory.

Consider, for example, a glow stick. This novelty item consists of a thin glass tube filled with one reagent inside a flexible plastic casing that contains a second reagent. When the tube is bent, the inner stick breaks and the reagents mix. The resulting reaction mixture glows until the reaction is complete. At room temperature, the reaction is complete in several hours and the stick stops glowing. In the freezer, the stick may continue to glow for days, though. Temperature influences the rate of the reaction.

There are several experimental factors that have important effects on the measured rates of chemical reactions.

- *The physical state of the reactants has an effect on the rate of a reaction.* For example, when aqueous solutions of $Pb(NO_3)_2$ and KI are mixed, a precipitation reaction will occur, but if solid $Pb(NO_3)_2$ and KI are brought together, no reaction takes place because the ions are unable to collide in their solid crystalline forms. In aqueous solutions, the reactants are dissociated into free ions and are capable of colliding.

**The Effect of Physical States on Reaction Rate**

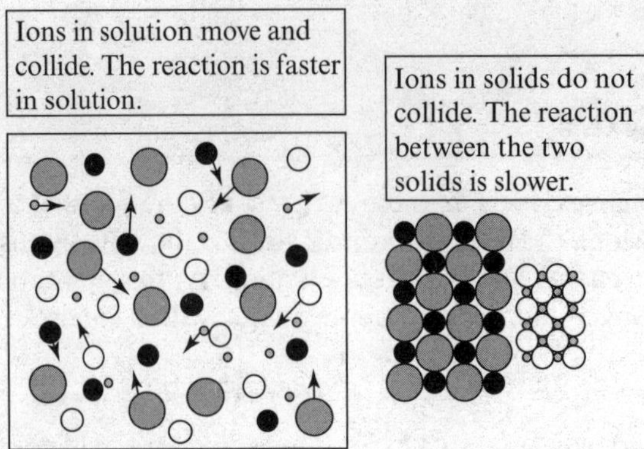


Physical differences in the surface areas of solid reactants have an effect on reaction rates. A stick of chalk that has been crushed has more surface area and reacts much more quickly with aqueous acid than a stick of chalk that is intact. This is because more of the chalk particles in the crushed chalk are exposed to the colliding acid molecules.

**Surface Area**

Two cubes have the same volume, but the cube on the right has a greater surface area. For the same reason, a crushed reactant has more surface area than a single, intact solid.

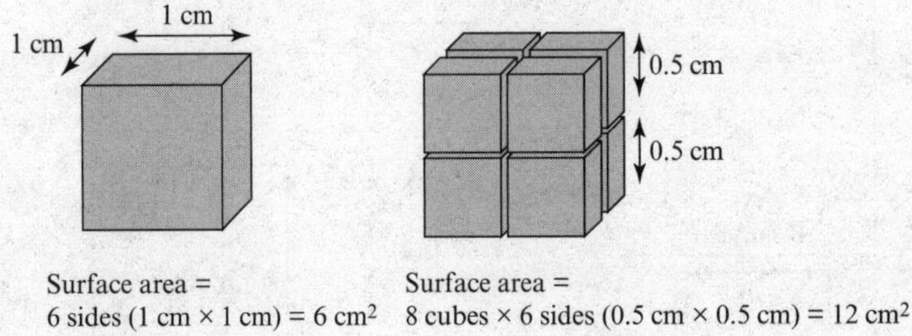


Surface area = 6 sides (1 cm × 1 cm) = 6 $cm^2$

Surface area = 8 cubes × 6 sides (0.5 cm × 0.5 cm) = 12 $cm^2$

- *The concentrations or partial pressures of the reactants have an impact on the rates of nonzero order reactions.* For reactions in which reactant particles must collide, an increase in the number of particles will increase the frequency of collisions.
- *The temperature at which a reaction is carried out affects the rate of the reaction.* At higher temperatures, reactions proceed faster than at lower temperatures. An increase in temperature is an increase in the average kinetic energy of the molecules. The higher the average kinetic energy, the higher the frequency and energy of the collisions. *For reactions at or near room temperature, a 10 °C increase in temperature will, in many cases, double the reaction rate.*

**The Effect of Temperature on Reaction Rate**

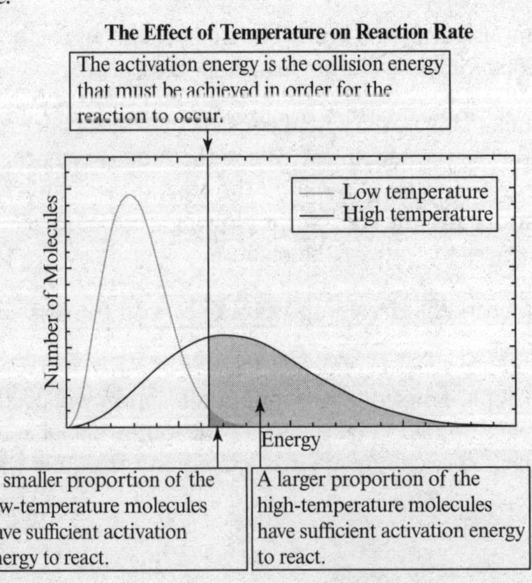

- *The presence of a catalyst increases the rate of a reaction.* A catalyst is a substance that provides an alternate route, or mechanism, leading to an alternative transition state for a reaction. This alternative transition state is lower in energy than the activated complex of the uncatalyzed reaction. In other words, the activation energy of the catalyzed reaction is lower than that of the uncatalyzed reaction.

**The Effect of Catalyst on Reaction Rate**

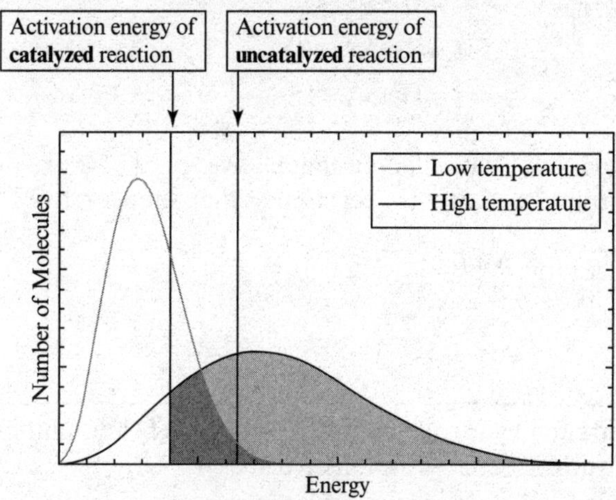


# The Rate Law

A **rate law** (also called the **differential rate law** or **rate expression**) is a mathematical expression of the relationship between the rate of a reaction and the concentration of the reactants.

For the general reaction

$$a\,\mathrm{A} + b\,\mathrm{B} + c\,\mathrm{C} \rightarrow \text{products}$$

where A, B, and C are reactants and *a*, *b*, and *c* are stoichiometric coefficients, the rate law has the form

$$\text{rate} = k[\mathrm{A}]^x[\mathrm{B}]^y[\mathrm{C}]^z$$

The constant, *k*, is called the **rate constant** for a reaction. It is an experimentally determined factor that is characteristic of a particular reaction *at a certain temperature*. The exponents, *x*, *y*, and *z*, are the **orders** of the reaction with respect to each reactant. The **overall reaction order** is given by the sum of the exponents.

The reaction between chlorine dioxide and hydroxide ion is shown to illustrate:

$$2\,ClO_2 + 2\,OH^- \rightarrow ClO_3^- + ClO_2^- + H_2O$$

The rate law for this reaction at 25 °C is

$$\text{rate} = (230\ M^{-2}\cdot s^{-1})[ClO_2]^2[OH^-]$$

The exponents for $[ClO_2]$ and $[OH^-]$, 2 and 1, respectively, indicate that the reaction is second order with respect to $ClO_2$ and first order with respect to $OH^-$. Since the sum of the exponents is 3, the reaction is third order overall. Sometimes the rate of a reaction is independent of the concentration of one (or more) of the reactants. In this case, the order of a reaction with respect to that reactant is zero.

*The orders are unrelated to the coefficients in the overall balanced equation since the reaction mechanism may not occur in a single step.* (The orders of reactions are related to the stoichiometric coefficients when the reaction occurs via a single step mechanism, as will be discussed in "Reaction Mechanisms," page 187.)

Notice that $k = 230\ M^{-2}\cdot s^{-1}$. Again, the numerical value of the rate constant has been determined experimentally for this reaction at this temperature. The units of $k$ are chosen so that the rate has units of $M\cdot s^{-1}$. Considering *only the units,*

$$\text{rate} \propto k \cdot M^3$$
$$M \cdot s^{-1} = k \cdot M^3$$
$$(M \cdot s^{-1})M^{-3} = k$$
$$M^{-2} \cdot s^{-1} = k$$

The value of the rate constant provides information about the relative speed of a reaction at a specific temperature. The larger the rate constant, the faster the reaction at that temperature.

*Remember, the rate constant is temperature-specific.*

## Practice

A reaction between chlorine dioxide and hydroxide ion at 25 °C, shown below, initially has a chlorine dioxide concentration of 0.020 *M* and a hydroxide concentration of 0.400 *M*.

$$2\ ClO_2 + 2\ OH^- \rightarrow ClO_3^- + ClO_2^- + H_2O$$

The rate law is

$$\text{rate} = (230\ M^{-2}\cdot s^{-1})[ClO_2]^2[OH^-]$$

What is the initial rate of the reaction?

The initial rate can be found by solving the rate law expression using the initial concentrations.

$$\text{rate} = (230\ M^{-2} \cdot s^{-1})[ClO_2]^2[OH^-]$$
$$\text{rate} = (230\ \cancel{M^{-2}} \cdot s^{-1})(0.020\ \cancel{M})^2(0.400 M) = 0.037\ M \cdot s^{-1}$$

## Practice

A reaction between molecules X, Y, and Z has the following rate law.

$$\text{rate} = k[X][Y]^3$$

What will happen to the reaction rate if the concentrations of X and Z are doubled and the concentration of Y is halved?

Because the reaction is zero-order in Z, we can disregard Z ($[Z]^0 = 1$).

Let $X_1$ and $Y_1$ be the initial concentrations and let $X_2$ and $Y_2$ be the final concentrations of X and Y. We can establish relationships between the initial concentrations and the final concentrations based on the wording of the problem.

$$[X_2] = 2[X_1]$$
$$[Y_2] = \frac{1}{2}[Y_1]$$

We would like to know the ratio of the rate of the final reaction to the initial reaction, so we can set up an equation and substitute the values of $X_1$ and $Y_1$ for $X_2$ and $Y_2$, respectively.

$$\frac{\text{rate}_2}{\text{rate}_1} = \frac{k[X_2][Y_2]^3}{k[X_1][Y_1]^3}$$

$$\frac{\text{rate}_2}{\text{rate}_1} = \frac{k(2[X_1])\left(\frac{1}{2}[Y_1]\right)^3}{k[X_1][Y_1]^3} = 2\left(\frac{1}{2}\right)^3 = \frac{1}{4}$$

When the concentrations of X and Z are doubled and the concentration of Y is halved, the rate of the reaction decreases to $\frac{1}{4}$ the initial rate.

## Rate Law Determination

There are several methods for using experimental data to determine the rate law for a process. These are:

- the **method of initial rates**
- **integrated rate equations**
- use of proposed **reaction mechanisms**

## Method of Initial Rates

The **method of initial rates** utilizes the initial concentrations and initial rate of a reaction to deduce its rate law and the magnitude of its rate constant.

Chemical reactions proceed both in the forward and in the reverse direction. When the *initial* rates of reactions are measured, products have not appeared yet, and the complication of the reverse reaction can be avoided.

For example, the initial rates of reaction and initial concentrations of reactants A, B, and C are measured experimentally and tabulated.

$$A + 3\,B + 2\,C \rightarrow \text{products}$$

| Trial Number | Initial Concentration of A ($M$) | Initial Concentration of B ($M$) | Initial Concentration of C ($M$) | Initial Rate of Reaction ($M{\cdot}s^{-1}$) |
|---|---|---|---|---|
| 1 | 0.010 | 0.012 | 0.010 | $1.2 \times 10^{-5}$ |
| 2 | 0.010 | 0.024 | 0.010 | $2.4 \times 10^{-5}$ |
| 3 | 0.020 | 0.012 | 0.010 | $4.8 \times 10^{-5}$ |
| 4 | 0.010 | 0.012 | 0.020 | $1.2 \times 10^{-5}$ |

First, the reaction orders are determined. Let the orders with respect to A, B, and C be $x$, $y$, and $z$, respectively.

Begin by choosing two sets of data in which the concentration of only one of the reactants changes.

Between trials 1 and 3, only the concentration of A changes, so these trials can be used to determine $x$, the order with respect to A. A ratio is set up between the rate laws, and since the concentration of A is greater in trial 3 than it is in trial 1, it is much easier to put trial 3 in the numerator and trial 1 in the denominator.

$$\frac{\text{rate}_3}{\text{rate}_1} = \frac{k[A]_3^x[B]_3^y[C]_3^z}{k[A]_1^x[B]_1^y[C]_1^z}$$

$$\frac{4.8\times10^{-5}}{1.2\times10^{-5}} = \frac{\cancel{k}(0.020)^x\,\cancel{(0.012)^y}\,\cancel{(0.010)^z}}{\cancel{k}(0.010)^x\,\cancel{(0.012)^y}\,\cancel{(0.010)^z}}$$

$$4 = 2^x$$

$$x = 2$$

The reaction is second order with respect to A.

Only the concentration of B changes between trials 1 and 2, so these trials can be chosen to determine $y$, the order with respect to B.

$$\frac{\text{rate}_2}{\text{rate}_1} = \frac{k[\text{A}]_2^x[\text{B}]_2^y[\text{C}]_2^z}{k[\text{A}]_1^x[\text{B}]_1^y[\text{C}]_1^z}$$

$$\frac{2.4\times10^{-5}}{1.2\times10^{-5}} = \frac{\cancel{k}\,\cancel{(0.010)^x}\,(0.024)^y\,\cancel{(0.010)^z}}{\cancel{k}\,\cancel{(0.010)^x}\,(0.012)^y\,\cancel{(0.010)^z}}$$

$$2 = 2^y$$

$$y = 1$$

The reaction is first order with respect to B.

Finally, trials 1 and 4 can be used to determine the order with respect to C.

$$\frac{\text{rate}_4}{\text{rate}_1} = \frac{k[\text{A}]_4^x[\text{B}]_4^y[\text{C}]_4^z}{k[\text{A}]_1^x[\text{B}]_1^y[\text{C}]_1^z}$$

$$\frac{1.2\times10^{-5}}{1.2\times10^{-5}} = \frac{\cancel{k}\,\cancel{(0.010)^x}\,\cancel{(0.012)^y}\,(0.020)^z}{\cancel{k}\,\cancel{(0.010)^x}\,\cancel{(0.012)^y}\,(0.010)^z}$$

$$1 = 2^z$$

$$z = 0$$

The order with respect to C is zero, so the rate is independent of the concentration of C.

The rate law, therefore, has the general form

$$\text{rate} = k[\text{A}]^2[\text{B}]$$

Any of the four trials can be chosen to find the rate constant. The data from the selected trial is used to solve the rate expression for $k$. The data from trial 1 is shown as an example, but the same result would be obtained from any of the trial data.

$$1.2\times10^{-5}\,M\cdot\text{s}^{-1} = k(0.010M)^2(0.012M)$$

$$k = \frac{1.2\times10^{-5}\,\cancel{M}\cdot s^{-1}}{(0.010M)^2(0.012\cancel{M})} = 10.\ M^{-2}\cdot\text{s}^{-1}$$

**TIP: For problems in which the concentrations of one of the reagents is different for all trials, solve for the order of that reagent last.**

## Practice

The initial rates and concentrations for the reaction between $C_5H_5N$ and $CH_3I$ at 25 °C are shown below.

$$C_5H_5N + CH_3I \rightarrow [C_5H_5NCH_3]^+ + I^-$$

**a.** What is the rate law for this reaction?
**b.** What is the value of the rate constant?
**c.** What is the rate of the reaction in trial 3 when the concentration of $C_5H_5N$ has decreased to $7.60 \times 10^{-4}$ $M$?

| Trial Number | Initial Concentration of $C_5H_5N$ ($M$) | Initial Concentration of $CH_3I$ ($M$) | Initial Rate of Reaction ($M{\cdot}s^{-1}$) |
|---|---|---|---|
| 1 | $3.70 \times 10^{-4}$ | $3.70 \times 10^{-4}$ | $1.03 \times 10^{-5}$ |
| 2 | $1.11 \times 10^{-3}$ | $1.11 \times 10^{-3}$ | $9.24 \times 10^{-5}$ |
| 3 | $1.11 \times 10^{-3}$ | $2.22 \times 10^{-3}$ | $1.85 \times 10^{-4}$ |

**(a)** The general form of the rate expression for this reaction is

$$\text{rate} = k[C_5H_5N]^x[CH_3I]^y$$

Since the concentration of $C_5H_5N$ is the same in trials 2 and 3, these two trials can be used to determine the order with respect to $CH_3I$.

$$\frac{\text{rate}_3}{\text{rate}_2} = \frac{k[C_5H_5N]_3^x[CH_3I]_3^y}{k[C_5H_5N]_2^x[CH_3I]_2^y}$$

$$\frac{1.85\times10^{-4}\ \cancel{M\cdot s^{-1}}}{9.24\times10^{-5}\ \cancel{M\cdot s^{-1}}} = \frac{\cancel{k}\ \cancel{(1.11\times10^{-3}\ M)^x}\ (2.22\times10^{-3}\ \cancel{M})^y}{\cancel{k}\ \cancel{(1.11\times10^{-3}\ M)^x}\ (1.11\times10^{-3}\ \cancel{M})^y}$$

$$2 = 2^y$$

$$y = 1$$

At this point, either of the other two combinations of trials (1 and 2 or 1 and 3) can be selected. Either choice will provide the correct order of the reaction with respect to $C_5H_5N$.

$$\frac{\text{rate}_3}{\text{rate}_1} = \frac{k[C_5H_5N]_3^x[CH_3I]_3^y}{k[C_5H_5N]_1^x[CH_3I]_1^y}$$

$$\frac{1.85\times10^{-4}\ \cancel{M\cdot s^{-1}}}{1.03\times10^{-5}\ \cancel{M\cdot s^{-1}}} = \frac{\cancel{k}(1.11\times10^{-3}\ \cancel{M})^x(2.22\times10^{-3}\ \cancel{M})^1}{\cancel{k}(3.70\times10^{-4}\ \cancel{M})^x(3.70\times10^{-4}\ \cancel{M})^1}$$

$$3 = 3^x$$

$$x = 1$$

The rate expression is

$$\text{rate} = k[C_5H_5N][CH_3I]$$

**(b)** The value of $k$ can be determined using the data from any of the three trials. Trial 1 is used as an example here.

$$1.03\times10^{-5}\ M\cdot s^{-1} = k\left(3.70\times10^{-4}\ M\right)\left(3.70\times10^{-4}\ M\right)$$

$$k = \frac{1.03\times10^{-5}\ \cancel{M}\cdot s^{-1}}{\left(3.70\times10^{-4}\ \cancel{M}\right)\left(3.70\times10^{-4}\ M\right)} = 75.2\ M^{-1}\cdot s^{-1}$$

**(c)** From the one-to-one stoichiometry of the reaction, the rates of consumption of $C_5H_5N$ and $CH_3I$ are equal.

$$-\frac{\Delta[C_5H_5N]}{\cancel{\Delta t}} = -\frac{\Delta[CH_3I]}{\cancel{\Delta t}}$$

The change in concentration of $C_5H_5N$ is determined.

$$\Delta[C_5H_5N] = [C_5H_5N]_{\text{final}} - [C_5H_5N]_{\text{initial}} = 7.60 \times 10^{-4}\ M - 1.11 \times 10^{-3}\ M = -3.50 \times 10^{-4}\ M$$

This change, $\Delta[C_5H_5N]$, is equal to that of $\Delta[CH_3I]$.

$$\Delta[CH_3I] = [CH_3I]_{final} - [CH_3I]_{initial} = -3.50 \times 10^{-4}\ M$$
$$\Delta[CH_3I] = [CH_3I]_{final} - 2.22 \times 10^{-3}\ M$$
$$[CH_3I]_{final} = -3.50 \times 10^{-4}\ M - (-2.22 \times 10^{-3}\ M) = 1.87 \times 10^{-3}\ M$$

Now the two final concentrations can be used in the rate law to determine the new rate.

$$\text{rate} = k[C_5H_5N][CH_3I] = \left(75.2\ \cancel{M^{-1}} \cdot s^{-1}\right)\left(7.60 \times 10^{-4}\ \cancel{M}\right)\left(1.87 \times 10^{-3}\ M\right) = 1.07 \times 10^{-4}\ M \cdot s^{-1}$$

## Integrated Rate Laws

The integrated rate equation is used to determine how much of a certain reactant remains after a specific time, $t$, has elapsed.

The integrated rate equations for zero order, first order, and second order reactions are shown on the *AP Chemistry Equations and Constants* sheet, but they are not labeled as such. It is the student's responsibility to know which equation corresponds to which reaction order.

### Zero Order Reactions

For the zero order reaction

$$A \rightarrow \text{products}$$

the rate expression is

$$\text{rate} = k[A]^0 = k$$

The units of the rate constant must be $M{\cdot}s^{-1}$, just like the rate itself.

This rate is independent of the concentration of A. A plot of the concentration, [A], versus time, $t$, gives a straight line.

**A Plot of [A] versus *t* for a Zero Order Decay**

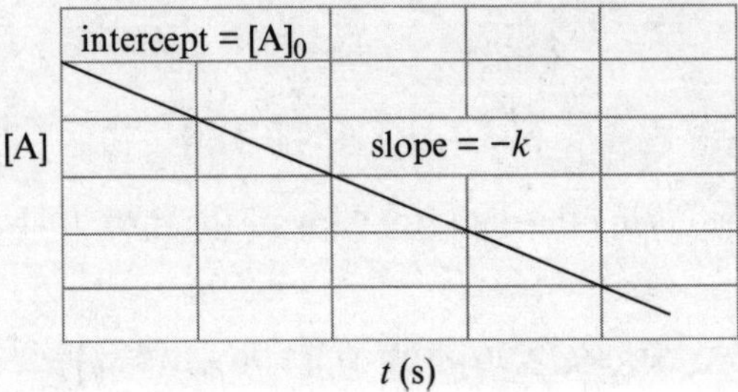


The absolute value of the slope of the plot gives the rate constant, $k$, and the intercept gives the initial concentration, $[A]_0$, of the reactant before any reaction time has elapsed.

The integrated rate law of a zero order process relates the concentration of A at time $t$, $[A]_t$, to the initial concentration, $[A]_0$.

$$[A]_t - [A]_0 = -kt$$

### First Order Reactions and Nuclear Decay

For the first order reaction

$$A \rightarrow \text{products}$$

the rate is

$$\text{rate} = k[\text{A}]$$

A plot of [A] versus time, *t*, gives a curved decay rather than a straight line. Unlike the rate of a zero order reaction, the rate of a first order reaction is concentration-dependent: It decreases as the reaction progresses. The unit of a first order rate constant is $s^{-1}$.

**A Plot of [A] versus *t* for a First Order Decay**

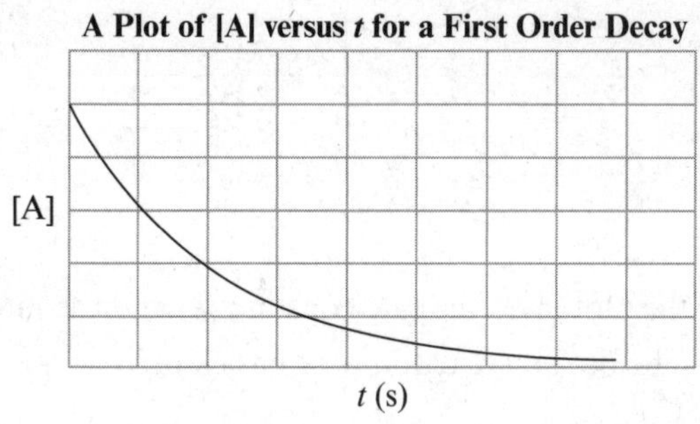


The plot of the natural logarithm of [A] versus time gives a straight line.

**A Plot of ln [A] versus *t* for a First Order Decay**

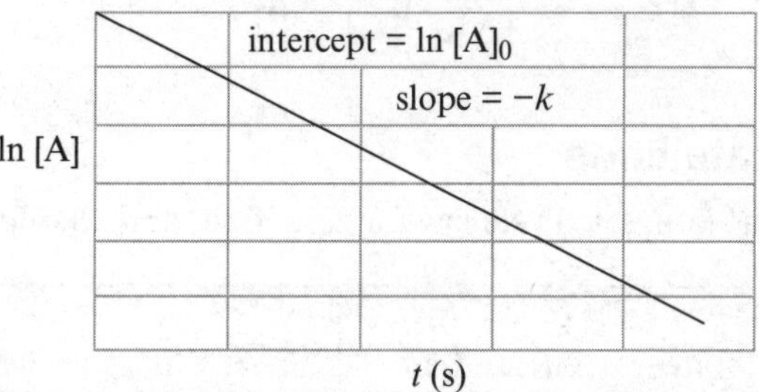


The absolute value of the slope of the plot gives the rate constant, *k*, and the intercept gives the natural log of the initial reactant concentration, ln $[\text{A}]_0$, before any reaction time has elapsed.

The integrated rate law of a first order process is given on the *AP Chemistry Equations and Constants* sheet.

$$\ln[\text{A}]_t - \ln[\text{A}]_0 = -kt$$

The integrated rate law for a first order reaction determines the reaction **half-life,** $t_{1/2}$, the time it takes for a quantity of a reactant to decrease by half. *The half-life of a first order reaction is relevant for nuclear decay, as nuclear decay is always a first order process.* The half-life equation shown on the *AP Chemistry Equations and Constants* sheet is for a first order reaction.

$$\frac{0.693}{k} = t_{1/2}$$

## Second Order Reactions

For the second order reaction

$$\text{A} \rightarrow \text{products}$$

the rate expression is

$$\text{rate} = k[\text{A}]^2$$

Again, the rate of a second order reaction decreases as the reaction progresses; it is concentration-dependent. The units of the rate constant are $M^{-1}\cdot s^{-1}$.

The plot of $\frac{1}{[A]}$ versus time for a second order process gives a straight line.

**A Plot of ln $[A]^{-1}$ versus $t$ for a Second Order Decay**

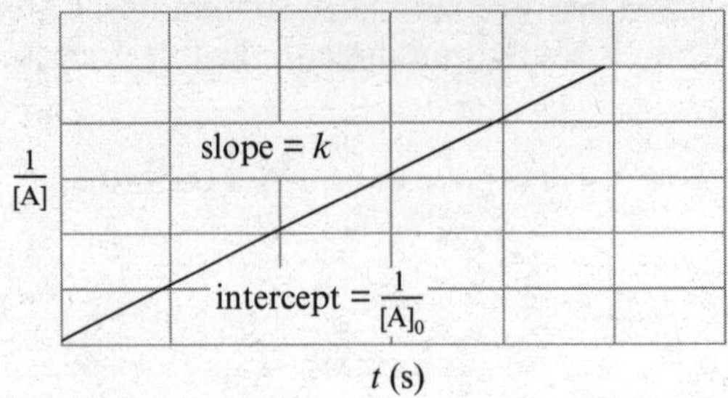


The absolute value of the slope of the plot gives the rate constant, $k$, and the intercept gives the reciprocal of the initial reactant concentration, $\frac{1}{[A]_0}$, before any reaction time has elapsed.

The integrated rate law of a second order process is given on the *AP Chemistry Equations and Constants* sheet, but it is not labeled as such, so students will need to recognize it.

$$\frac{1}{[A]_t} - \frac{1}{[A]_0} = kt$$

## Summary of Integrated Rate Laws

The following table summarizes the integrated rate law for zero, first, and second order reaction rates.

| Reaction Order | Differential Rate Law | Integrated Rate Law | Linear Plot | Half-Life* | Units of Rate Constant |
|---|---|---|---|---|---|
| Zero | rate = $k$ | $[A]_t - [A]_0 = -kt$ | [A] versus $t$ | $t_{½} = \frac{[A]_0}{2k}$ | $M \cdot s^{-1}$ |
| First | rate = $k[A]$ | $\ln [A]_t - \ln [A]_0 = -kt$ | ln [A] versus $t$ | $t_{½} = \frac{0.693}{k}$ | $s^{-1}$ |
| Second | rate = $k[A]^2$ | $\frac{1}{[A]_t} - \frac{1}{[A]_0} = kt$ | $\frac{1}{[A]_t}$ versus $t$ | $t_{½} = \frac{1}{k[A]_0}$ | $M^{-1} \cdot s^{-1}$ |

***TIP: Don't be concerned with memorization of the zero and second order half-life equations. It is only important to notice that zero and second order half-lives are concentration-*dependent* and that the half-life of a first order reaction is concentration-*independent*.**

## Practice

The decomposition of a reactant, X, occurs by the following reaction.

$$X \rightarrow \text{products}$$

The reaction was monitored over the course of 10.0 hours at constant temperature and the data obtained is shown in the table below. The information is plotted in the graphs that follow.

**a.** What is the order of the reaction with respect to X? Justify your response.

**b.** Write the rate law for the reaction and calculate the specific rate constant with appropriate units.

**c.** Calculate the concentration of X after 14.0 hours have elapsed.
**d.** Determine the half-life of X.

| Time (hours) | [X] | ln [X] | $[X]^{-1}$ |
|---|---|---|---|
| 0.0 | 0.368 | –1.00 | 2.72 |
| 1.0 | 0.247 | –1.40 | 4.06 |
| 2.0 | 0.165 | –1.80 | 6.05 |
| 3.0 | 0.111 | –2.20 | 9.03 |
| 5.0 | 0.0498 | –3.00 | 20.1 |
| 7.0 | 0.0224 | –3.80 | 44.7 |
| 10.0 | 0.00674 | –5.00 | 148 |

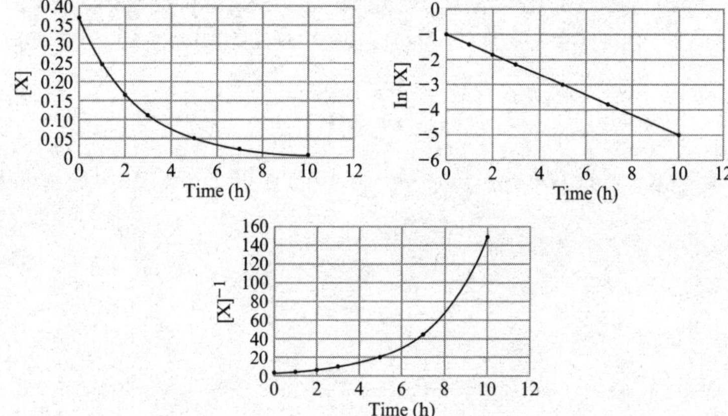


---

**(a)** The reaction is first order with respect to X. The plot of ln [X] versus time produces a straight line.

**(b)** The general form of the rate law for a reaction that is first order in X is

$$\text{rate} = k[\text{X}]$$

The specific rate constant is given by the absolute value of the slope of the plot of ln [X] versus time. Any two points can be chosen to determine the slope of the graph.

$$k = \left|\frac{-5.00-(-1.00)}{10.0-0.0}\right| = 0.400\ \text{h}^{-1}$$

Since rate is reported in units of concentration per unit time, the rate constant must have units of $\text{h}^{-1}$.

**(c)** The concentration of X after 14.0 hours have elapsed can be calculated using the equation for a first order reaction.

$$\ln\,[\text{X}]_{14\,\text{h}} - \ln\,[\text{X}]_0 = -kt$$
$$\ln\,[\text{X}]_{14\,\text{h}} - \ln\,(0.368\ M) = -(0.400\ \text{h}^{-1})(14.0\ \text{h})$$
$$\ln\,[\text{X}]_{14\,\text{h}} = -5.60 - 1.00$$
$$[\text{X}]_{14\,\text{h}} = e^{-6.60} = 0.0014\ M$$

**(d)** The half-life of X is determined using the rate constant.

$$t_{1/2} = \frac{0.693}{k} = \frac{0.693}{0.400\ \text{h}^{-1}} = 1.73\ \text{h}$$

## Reactions Between Multiple Reactants

The integrated rate laws discussed so far have been for reactions involving single reactants. Often more than one reactant is involved, and this complicates the measurement of the rate law. One way to handle the problem is by creating conditions in which one reagent is so much less concentrated than the others that it experiences a substantially greater change in concentration as the reaction proceeds. The concentration of the other reactant changes so little that it is approximately constant.

For visualization purposes, imagine a "reaction" occurs between a grain of rice and a small glass bead. Now, imagine that there is a 2-liter container filled with rice grains (an estimated 126,000 grains) and 126 glass beads. The number of rice grains is 1000 times greater than the number of beads.

$$\text{Rice grain} + \text{Glass bead} \rightarrow \text{Imaginary product}$$
$$R + G \rightarrow P$$

When 63 of the glass beads have reacted, the "concentration" of beads has decreased by half. A decrease of 63 rice grains, however, still leaves approximately 126,000 grains of rice (rounded to three significant figures). The rate law for this reaction is

$$\text{rate} = k[R]^x[G]^y$$

Since the concentration of R appears to be constant, the product of $[R]^x$ and $k$ appears constant as well. We refer to this product as $k'$.

$$k' = k[R]^x$$

The rate law can be rewritten

$$\text{rate} = k'\,[G]^y$$

Now, the concentration of G can be measured experimentally as a function of time and the order with respect to G can be determined using the integrated rate law method, as usual. The slope of the plot of [G], ln [G], or $\frac{1}{[G]}$ versus time gives $k'$ rather than $k$.

## Practice

Bromophenol blue is an acid-base indicator that is blue in basic solution. Upon prolonged exposure to alkaline conditions, the blue molecule fades to a colorless form.

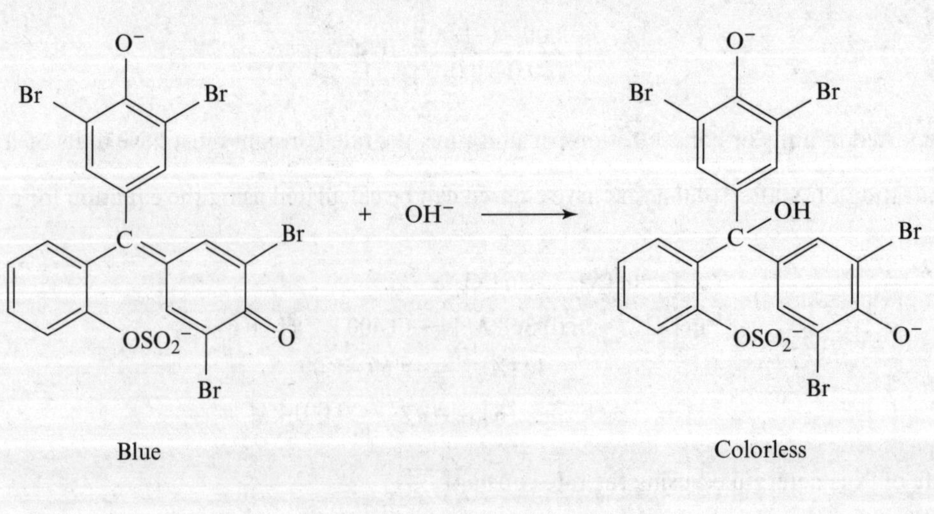

$$\text{Blue} + OH^- \longrightarrow \text{Colorless}$$

Blue

Colorless

Two solutions of $1.00 \times 10^{-3}$ *M* bromophenol blue (BB) in 1.00 and 0.500 *M* hydroxide ($OH^-$), respectively, are monitored spectrophotometrically. The concentrations at various times are tabulated below.

**1.00 M $OH^-$ Solution**

| Time (min) | [BB] (*M*) |
|---|---|
| 0 | $1.00 \times 10^{-3}$ |
| 1 | $9.33 \times 10^{-4}$ |
| 3 | $8.12 \times 10^{-4}$ |
| 5 | $7.07 \times 10^{-4}$ |
| 10 | $5.00 \times 10^{-4}$ |
| 20 | $2.50 \times 10^{-4}$ |

**0.500 M $OH^-$ Solution**

| Time (min) | [BB] (*M*) |
|---|---|
| 0 | $1.00 \times 10^{-3}$ |
| 1 | $9.66 \times 10^{-4}$ |
| 3 | $9.01 \times 10^{-4}$ |
| 5 | $8.41 \times 10^{-4}$ |
| 10 | $7.07 \times 10^{-4}$ |
| 20 | $4.99 \times 10^{-4}$ |

Determine the rate law for the fading of bromophenol blue in hydroxide by

**a.** finding the order with respect to BB (Practice graphing with your calculator for this exercise.)
**b.** finding the apparent rate constant, $k'$, for the fading in 1.00 *M* $OH^-$
**c.** finding the apparent rate constant, $k'$, for the fading in 0.500 *M* $OH^-$
**d.** using the apparent rate constants to determine the order with respect to $OH^-$

---

**(a)** The order of the reaction with respect to BB can be determined using either set of data. Three plots are generated from the chosen data set. The concentration, [BB], versus time, ln [BB] versus time, and $[BB]^{-1}$ versus time are plotted. The order is determined by the plot that gives a straight line.

Using the 1.00 *M* $OH^-$ data, the straight line plot is obtained when ln [BB] is plotted versus time. The reaction is, therefore, first order in [BB].

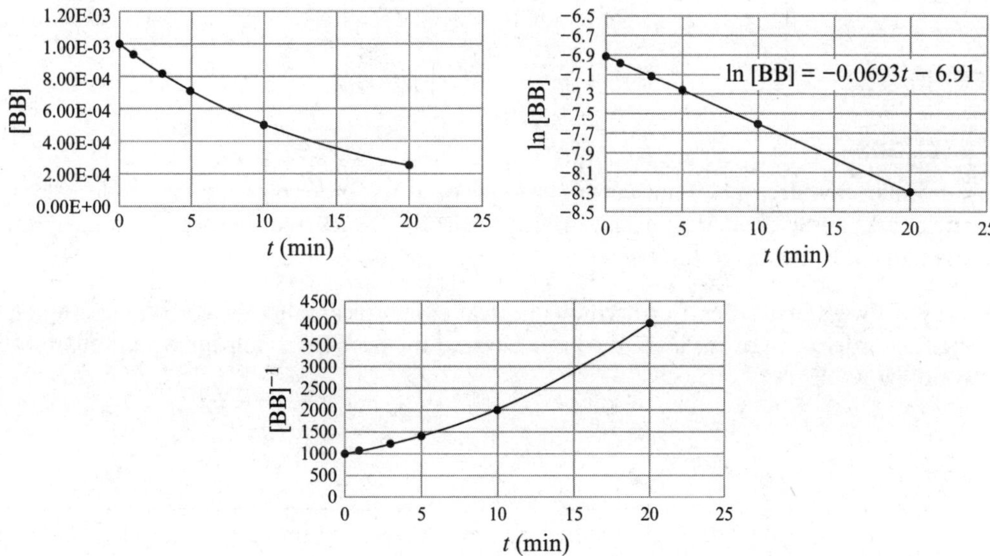


**(b)** The apparent rate constant for the fading of BB in 1.00 *M* $OH^-$ is the absolute value of the slope of the plot of ln [BB] versus time.

$$\text{rate} = k'[\text{BB}]$$

where the slope of the line is given in the equation of the line on the plot.

$$k' = k[\text{OH}^-]^x = |\text{slope}| = 0.0693 \text{ min}^{-1}$$

**(c)** First, ln[BB] versus time in the 0.500 M $OH^-$ solution must be plotted. The apparent rate constant for the fading of BB in 0.500 *M* $OH^-$ is the absolute value of the slope of the plot.

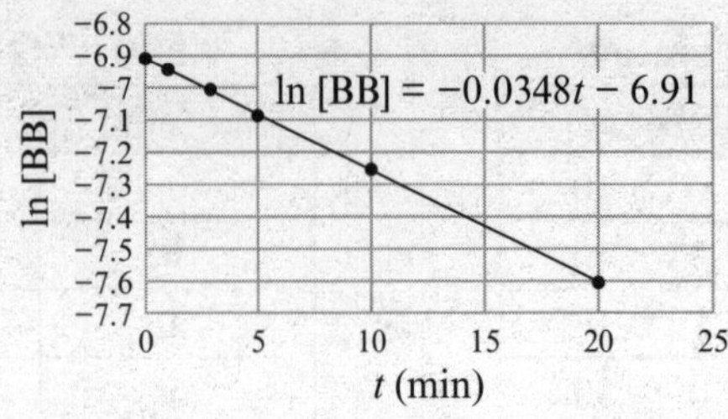


$$k' = k[OH^-]^x = |\text{slope}| = 0.0348\ \text{min}^{-1}$$

**(d)** There are two equations for $k'$ at two different concentrations of $OH^-$.

For $[OH^-] = 1.00\ M$,

$$k'_{1.00\ M} = k\left[OH^-\right]^x = 0.0693\ \text{min}^{-1} = k[1.00]^x$$

For $[OH^-] = 0.500\ M$,

$$k'_{0.500\ M} = k\left[OH^-\right]^x = 0.0348\ \text{min}^{-1} = k[0.500]^x$$

The ratio of the rate constants can be used to determine the order, $x$, with respect to $OH^-$.

$$\frac{k'_{1.00\ M}}{k'_{0.500\ M}} = \frac{0.0693\ \cancel{\text{min}^{-1}}}{0.0348\ \cancel{\text{min}^{-1}}} = \frac{\cancel{k}[1.00]^x}{\cancel{k}[0.500]^x}$$

$$1.99 = 2^x$$

$$x = 1$$

## Nuclear Reactions

**Nuclear decay,** or **radioactive decay,** is an important example of a first order process. In a nuclear decay, subatomic particles are emitted from the nucleus of an atom with a concurrent release of a large amount of energy and a change in the identity of the element.

Radioactive decay is always first order. Furthermore, nuclear reactions are unaffected by temperature, pressure, or catalysts. The first order form of the integrated rate law and the first order half-life equation are sufficient for solving nuclear decay problems.

## Practice

The half-life of iodine-131 is 8.02 days. Determine the percentage of a sample of iodine-131 that will remain after 21.2 days.

The rate constant for this first order process can be calculated from the half-life.

$$t_{1/2} = \frac{0.693}{k}$$

$$k = \frac{0.693}{8.02\ \text{days}} = 0.0864\ \text{day}^{-1}$$

Let $A_0$ represent the initial amount of iodine-131. Before any time has elapsed, the sample is assumed to be 100% iodine-131, so $A_0 = 1$. Let A represent the fraction that remains after 21.2 days. The first order integrated rate law equation can be solved for A.

$$\ln A - \ln A_0 = -kt$$

$$\ln A - \ln 1 = -\left(0.0864\ \cancel{\text{day}^{-1}}\right)\left(21.2\ \cancel{\text{day}}\right)$$

$$\ln A = -1.83$$

$$A = e^{-1.83} = 0.160$$

The fraction that remains after 21.2 days is 0.160 or 16.0%.

# Reaction Mechanisms

Most chemical reactions occur by a process that involves more than one step. A **reaction mechanism** is a step-by-step sequence of bond breaking, bond formation, and other transformations that result in an overall reaction. These steps are called **elementary reactions.** The stoichiometric coefficients for an elementary reaction determine the rate law *of that step*. (Remember, the rate law for a reaction made up of many steps cannot be deduced from the stoichiometric coefficients of the balanced equation.)

An elementary step can be **unimolecular, bimolecular,** or **termolecular.** Termolecular steps, those in which three reactants simultaneously collide and react, are extremely unlikely.

| Molecularity | Phenomenon | Possible Elementary Steps | Possible Rate Laws for the Elementary Steps |
|---|---|---|---|
| Unimolecular | One reactant decomposes | $A \rightarrow$ products | Rate = $k$[A] |
| Bimolecular | Two reactants collide | $A + A \rightarrow$ products<br>$A + B \rightarrow$ products | Rate = $k[A]^2$<br>Rate = $k$[A][B] |
| Termolecular | Three reactants collide simultaneously | $A + A + A \rightarrow$ products<br>$A + A + B \rightarrow$ products<br>$A + B + C \rightarrow$ products | Rate = $k[A]^3$<br>Rate = $k[A]^2[B]$<br>Rate = $k$[A][B][C] |

Imagine a hypothetical reaction between reactants A and B to form a product, E.

$$2\,A + B \rightarrow 2\,E$$

It is highly unlikely that this reaction occurs in a single termolecular step. The reaction between A and B below might occur in the following series of elementary steps.

$$A + A \rightarrow C \quad \text{(bimolecular)}$$

$$B + C \rightarrow D \quad \text{(bimolecular)}$$

$$D \rightarrow E + E \quad \text{(unimolecular)}$$

These elementary steps can be "added" together to give the overall equation.

$$\begin{array}{c} A + A \rightarrow \cancel{C} \\ B + \cancel{C} \rightarrow \cancel{D} \\ \cancel{D} \rightarrow E + E \\ \hline 2A + B \rightarrow 2E \end{array}$$

Species C and D, which are produced in earlier elementary steps and then consumed later in the mechanism, are called **reaction intermediates.** They are neither reactants nor products; rather, they are only temporarily present.

## Practice

The reaction between reactants A and B

$$A + B \rightarrow C + D$$

has a rate law that has been determined experimentally to be

$$\text{rate} = k[A]^2[B]$$

Does the reaction between A and B take place in a single step? Explain your reasoning.

The stoichiometric coefficients of a reaction determine the rate law for that step, so if this reaction took place in a single bimolecular step, the rate law would be expected to be first order in A and first order in B. This reaction cannot occur in a single step and remain consistent with a rate law that is second order in A and first order in B.

## The Rate-Determining Step

The rate of an overall reaction is determined by the rate of the slowest step in the process. The slowest step, for this reason, is called the **rate-determining step.**

Visualize an assembly line in which three workers are assembling bicycles. Worker 1 attaches the fork and handlebars to the frame of the bicycle, worker 2 assembles the wheels and gears, and worker 3 attaches the seat. Suppose that the task of worker 2 is the most difficult and takes the longest amount of time. The bicycles can be produced only as fast as worker 2 can perform the task. In other words, the task of worker 2 is the rate-determining step.

In a chemical reaction, the rate-determining step has the highest activation energy of any step in the mechanism. The reaction profile for a hypothetical multistep process is shown below. This process is a three-step reaction with three transition states (activated complexes) and two intermediates.

**Reaction Profile for a Three-Step Reaction**

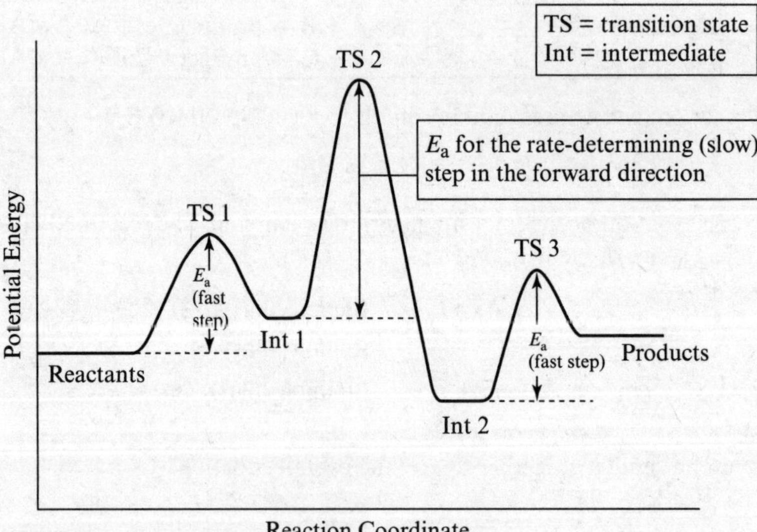


For a hypothetical reaction mechanism, the rate law is determined by the slow step.

$$A + A \rightarrow C \quad \text{(slow)}$$
$$B + C \rightarrow D \quad \text{(fast)}$$
$$D \rightarrow E + E \quad \text{(fast)}$$

The rate law for the overall transformation is based on the stoichiometry of the slow step.

$$\text{rate} = k[A]^2$$

The mechanisms for reactions that occur in multiple steps can be very difficult to elucidate experimentally. A rate law is determined for an overall process using either the method of initial rates or measurement of change in amounts of reactants and products with time. A proposed mechanism must be consistent with the observed rate law.

## Practice

$$2\ NO + Br_2 \rightarrow 2\ NOBr$$

The rate law for the reaction shown above between NO and $Br_2$ is determined experimentally to be

$$\text{rate} = k[NO][Br_2]$$

Which mechanism, A or B, is consistent with this rate law?

| | | |
|---|---|---|
| Mechanism A: | $NO + NO \rightarrow N_2O_2$ | (slow) |
| | $N_2O_2 + Br_2 \rightarrow NOBr + NOBr$ | (fast) |
| Mechanism B: | $NO + Br_2 \rightarrow NOBr_2$ | (slow) |
| | $NO + NOBr_2 \rightarrow NOBr + NOBr$ | (fast) |

Mechanism B is consistent with the observed rate law. The rate law based on the slow step of mechanism B is the one that is observed. The rate law based on the slow step of mechanism A, on the other hand, is

$$\text{rate} = k[NO]^2$$

# Steady State Analysis

In some cases, the rate-determining step is not the first step. Sometimes, the elementary reactions that precede the slow step are fast equilibria. **Equilibrium,** as will be reviewed in Chapter 7, "Chemical Equilibrium," describes the condition in which the rate of a reaction in the forward direction is equal to the rate of the reverse reaction. It is designated with an **equilibrium arrow,** $\rightleftharpoons$.

A reaction between reactants A and C to form D,

$$2A + C \rightarrow D$$

may proceed by the following mechanism in which the first step is a fast equilibrium reaction and intermediate B is transiently formed.

$$A + A \underset{k_{-1}}{\overset{k_1}{\rightleftharpoons}} B \quad \text{(fast)}$$

$$B + C \xrightarrow{k_2} D \quad \text{(slow)}$$

The rate constant for the first step in the forward direction is written $k_1$, and the rate constant for the first step in the reverse direction is $k_{-1}$. The rate law based on the slow step is

$$\text{rate} = k_2[B][C]$$

but there is a problem. Intermediates do not generally appear in the rate law. This issue is easily solved by recognizing that the forward rate of the first reaction is equal to the reverse rate.

$$\text{rate}_1 = \text{rate}_{-1}$$

$$k_1[A]^2 = k_{-1}[B]$$

Now, [B] can be calculated and substituted into the rate law that was written based on the slow step.

$$[\mathrm{B}] = \frac{k_1[\mathrm{A}]^2}{k_{-1}}$$

$$\text{rate} = k_2\left(\frac{k_1[\mathrm{A}]^2}{k_{-1}}\right)[\mathrm{C}]$$

The product of the constants $k_2 \cdot k_1 \cdot \frac{1}{k_{-1}}$ is the apparent rate constant, $k$, observed in rate experiments. The overall rate law is

$$\text{rate} = k[\mathrm{A}]^2[\mathrm{C}]$$

## Practice

Chlorine reacts with chloroform to generate carbon tetrachloride and hydrogen chloride.

$$Cl_2 + CHCl_3 \rightarrow CCl_4 + HCl$$

The proposed mechanism follows. Based on this mechanism, determine the rate law for the reaction.

$$Cl_2 \underset{k_{-1}}{\overset{k_1}{\rightleftharpoons}} 2\,Cl \qquad \text{(fast, equilibrium)}$$

$$Cl + CHCl_3 \xrightarrow{k_2} HCl + CCl_3 \qquad \text{(slow)}$$

$$CCl_3 + Cl \xrightarrow{k_3} CCl_4 \qquad \text{(fast)}$$

The overall rate of the reaction is based on the slow step.

$$\text{rate} = k_2[\mathrm{Cl}][\mathrm{CHCl_3}]$$

Since the chlorine atom is an intermediate, it should not be included in the rate law. For the first step in the mechanism, the rates of the forward and reverse steps are equal since the reaction is at equilibrium.

$$\text{rate}_1 = \text{rate}_{-1}$$

$$k_1[\mathrm{Cl_2}] = k_{-1}[\mathrm{Cl}]^2$$

Rearrangement allows us to solve for [Cl] in terms of $[Cl_2]$ and the apparent rate constant.

$$[\mathrm{Cl}] = \left(\frac{k_1[\mathrm{Cl_2}]}{k_{-1}}\right)^{\frac{1}{2}}$$

$$\text{rate} = k_2\left(\frac{k_1[\mathrm{Cl_2}]}{k_{-1}}\right)^{\frac{1}{2}}[\mathrm{CHCl_3}]$$

$$\text{rate} = k_2\left(\frac{k_1}{k_{-1}}\right)^{\frac{1}{2}}[\mathrm{Cl_2}]^{\frac{1}{2}}[\mathrm{CHCl_3}]$$

$$\text{rate} = k[\mathrm{Cl_2}]^{\frac{1}{2}}[\mathrm{CHCl_3}]$$

# Temperature and Reaction Rate

As temperature increases, reaction rate increases because more reactant molecules collide more frequently with the necessary orientation and with sufficient energy to overcome the activation barrier. In other words, the rate constant, $k$, is temperature-dependent. Near room temperature, for every 10 K increase in temperature, the rate of a reaction approximately doubles.

The **Arrhenius equation** is an expression that relates the rate constant for an elementary reaction to the temperature. Calculations involving the Arrhenius equation have been excluded from the AP Chemistry curriculum, but an understanding of the implications and graphical results is explicitly required. The Arrhenius equation has the form

$$k = Ae^{-\frac{E_a}{RT}}$$

As temperature increases, the fraction of molecules with the minimum critical activation energy increases exponentially as a function of $e^{-\frac{E_a}{RT}}$, where $E_a$ is the activation energy, $R$ is the gas constant, 8.314 J/(mol·K), and $T$ is the temperature in Kelvin. The **frequency factor,** $A$, describes the frequency with which collisions occur with the proper orientation for a reaction.

*In other words, the Arrhenius equation relates the rate constant to the two facets of collision theory (page 171), activation energy and frequency of successful collisions, as a function of temperature.*

Taking the natural log of both sides of the equation gives an equation with the form $y = mx + b$.

$$\ln k = -\left(\frac{E_a}{R}\right)\left(\frac{1}{T}\right) + \ln A$$

When the rate constants for a reaction are measured at different temperatures, a plot of ln $k$ versus $\frac{1}{T}$ results in a straight line with a slope of $-\frac{E_a}{R}$.

**A Plot of ln [A] versus $T^{-1}$ for a Reaction**

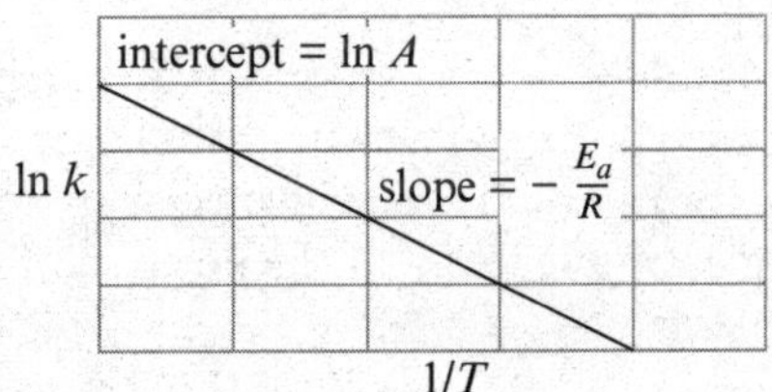


## Practice

The plot below shows Arrhenius plots for two reactions, X and Y.

**a.** Which reaction has a higher activation energy?
**b.** Which reaction has a higher frequency of successful collisions?

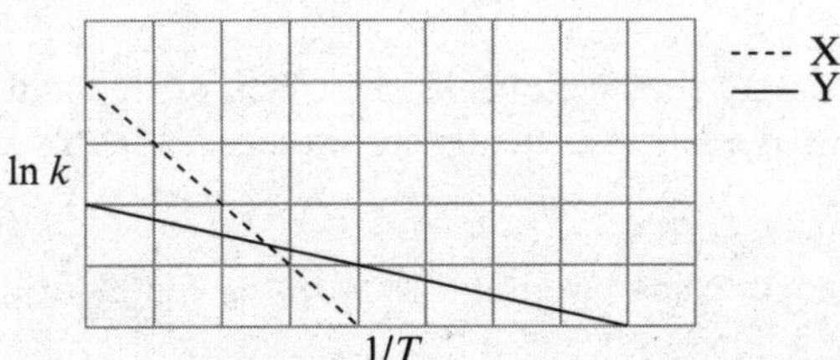

**(a)** Reaction X has a higher activation energy. The steeper slope of plot X shows that the magnitude of $\left| -\frac{E_a}{R} \right|$ is larger than it is for Y.

**(b)** Reaction X has the higher frequency factor. Because the intercept is equal to ln *A*, the frequency factor is *e* raised to the power of the intercept.

$$\text{intercept} = \ln A$$
$$A = e^{\text{intercept}}$$

The intercept of the plot for reaction X is greater than that for reaction Y, so X has the higher frequency factor.

## Catalysts

A catalyst is a substance that increases the rate of a reaction by providing an alternative transition state with a lower activation energy than that of the uncatalyzed reaction. The effect of a catalyst on the activation energy of a reaction step is shown below. The decrease in activation energy causes the rates of both the forward and reverse reactions to increase.

**The General Effect of a Catalyst on Activation Energy**

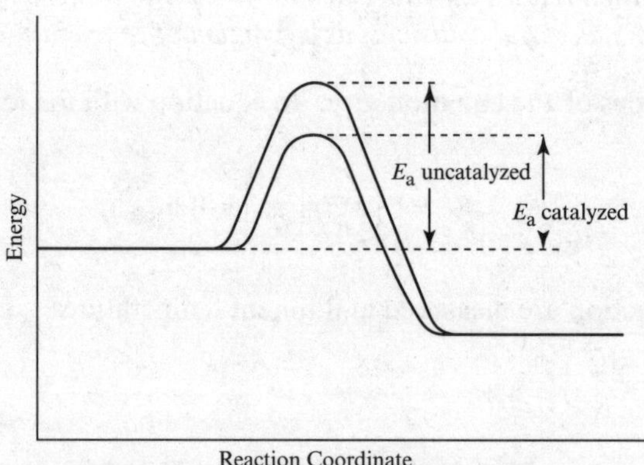


In a reaction mechanism, a catalyst is a substance that is used in an early step and regenerated in a later step. Since a catalyst is regenerated in the mechanism, it can react again and again. For this reason, only a small (*substoichiometric,* or less than stoichiometric) amount is usually necessary to increase the rate of a reaction. A catalyst will not change the thermodynamics of a reaction or the position of equilibrium (how far a reaction will proceed before the forward and reverse rates equalize).

For an overall reaction, a catalyst is shown above the reaction arrow.

$$A + B \xrightarrow{\text{catalyst}} AB$$

The mechanism for this reaction, in which C is the catalyst, might be

$$A + C \rightarrow AC$$
$$AC + B \rightarrow AB + C$$

In this example, the catalyst, C, is the substance that is used in the first step and regenerated in the second step, and the intermediate, AC, is generated and subsequently consumed. It is common for the catalyst to appear in the rate law for a reaction.

The energy of a catalyzed transformation can be compared with the energy of an uncatalyzed reaction. Notice that the ‡ symbol in the figure that follows indicates a transition state.

**The Effect of a Catalyst on a Reaction Mechanism**

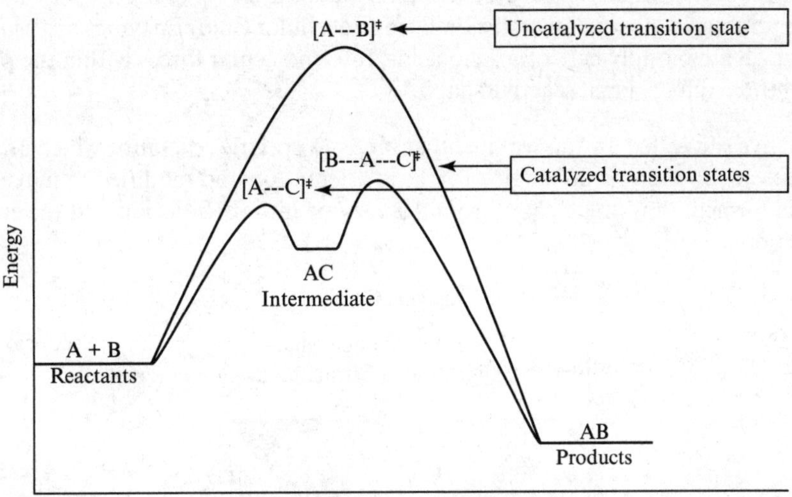


The activation energies for the formation of the intermediate and the product are lower than that of the uncatalyzed process. The decreased activation barriers increase the rate of the reaction.

The two major catalyst classifications are homogeneous and heterogeneous catalysts.

- A **homogeneous catalyst** is in the same phase or physical state as the reactants. Examples of homogeneous catalysts include acids, bases, and enzymes in solution.
- A **heterogeneous catalyst** is in a different phase than the reactants. Catalytic converters in cars, in which solid catalysts speed the conversion of unburned gaseous fuel, carbon monoxide, and nitrogen oxides to carbon dioxide, water, and nitrogen, are heterogeneous catalysts.

## Acid-Base Catalysis

**Acid-base catalysis** can be homogeneous or heterogeneous catalysis. In an acid- or base-catalyzed reaction, the rate is increased by transfer of a proton, $H^+$, to or from a reactant. An example is the acid-catalyzed hydration of an alkene.

$$C_2H_4 + H_2O\ (+\ H^+) \rightarrow C_2H_6O\ (+\ H^+)$$

**Acid-Catalyzed Hydration of an Alkene**

$$H_2C{=}CH_2 + H_2O \xrightarrow{H^+} H{-}CH_2{-}CH_2{-}\ddot{\underset{\cdot\cdot}{O}}{-}H$$

The rate of the reaction of $C_2H_4$ with water is increased when the $C_2H_4$ is activated by a proton, $H^+$, in the first step of the mechanism. This results in a positively charged intermediate that reacts quickly with water in the second elementary step. Regeneration of the proton in the third step leads to the product, $C_2H_6O$. Notice that there is no *net* consumption or production of $H^+$ in the reaction. (Since $H^+$ is both consumed in the first step and regenerated in the last step, it is canceled from the overall equation.)

$$C_2H_4 + \cancel{H^+} \rightarrow \cancel{C_2H_5^+}$$
$$\cancel{C_2H_5^+} + H_2O \rightarrow \cancel{C_2H_7O^+}$$
$$\cancel{C_2H_7O^+} \rightarrow C_2H_6O + \cancel{H^+}$$

## Enzyme Catalysis

**Enzymes** are catalysts that occur in biological systems. Enzyme catalysis is homogeneous catalysis, which usually occurs under mild conditions in an aqueous solution like intracellular fluid. Enzymes are usually composed mostly of proteins, which are biopolymers of amino acids. Intermolecular forces within the protein hold the large molecule in a defined three-dimensional reactive shape.

An enzyme has a reactive site called an "active site" that binds to specific reactants, which are called **substrates.** The reacting substrates are bound in a favorable geometric orientation and modified to make reaction faster. Once the products are formed, they are released from the enzyme into the solution and the enzyme is free to catalyze another reaction.

**Enzyme Catalysis**

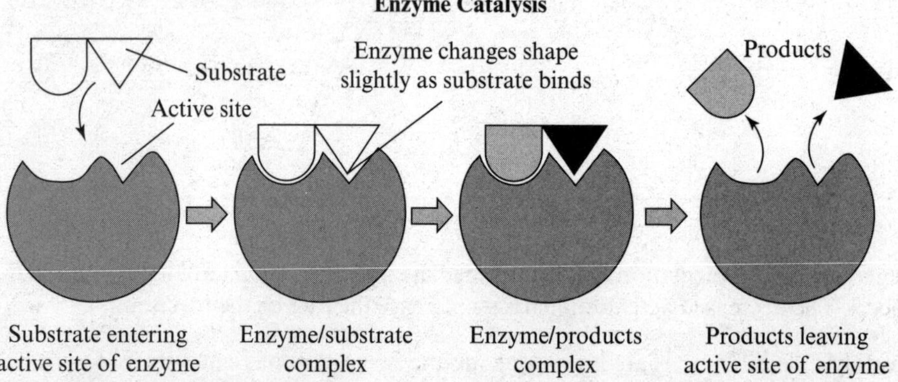


## Heterogeneous Catalysis

A heterogeneous catalyst is in a different phase or physical state than the reactants. There are many practical examples of heterogeneous catalysts, such as the catalytic converters in cars and the polymerization of propene (also known as propylene) to form polypropylene.

The heterogeneous catalytic cycle consists of the following steps:

1. **Adsorption** refers to the binding of reactant molecules to the surface of the catalyst and activation of the reactant.
2. The reaction then occurs while the activated reactant is adsorbed; the activation energy is lower than that of the uncatalyzed process.
3. The product molecule undergoes **desorption,** or release from the surface of the catalyst.

An example of heterogeneous catalysis is used in the production of ammonia from hydrogen and nitrogen gases. This reaction is carried out at very high temperatures in the presence of a solid iron catalyst. The hydrogen and nitrogen adsorb to the iron surface and their bonds break. New bonds between hydrogen and nitrogen form on the surface and the resulting ammonia is released.

**Heterogeneous Catalysis**

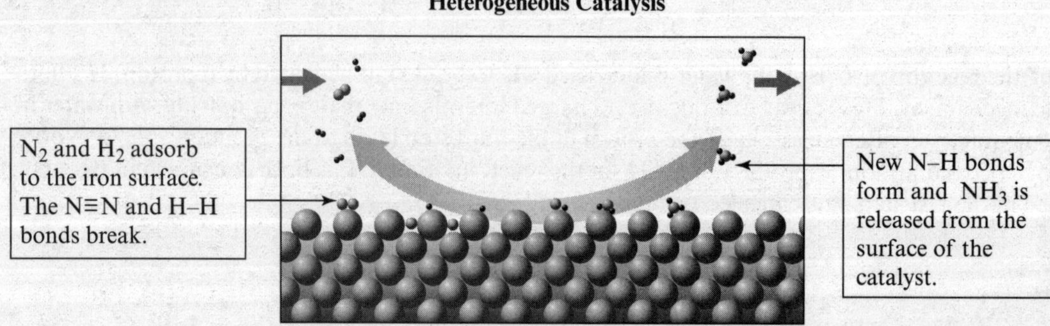

# Review Questions

## Multiple Choice

**1.** Dinitrogen pentoxide decomposes to form dinitrogen tetroxide and oxygen gas.

$$2\ N_2O_5 \rightarrow 2\ N_2O_4 + O_2$$

In an experiment at 45 °C, the concentration of $N_2O_5$ decreases from 0.0200 *M* to 0.0071 *M* in 0.10 second. What is the average rate of formation of $O_2$ during this time period?

**A.** $1.29 \times 10^{-2}\ M \cdot s^{-1}$
**B.** $6.45 \times 10^{-2}\ M \cdot s^{-1}$
**C.** $1.29 \times 10^{-1}\ M \cdot s^{-1}$
**D.** $2.58 \times 10^{-1}\ M \cdot s^{-1}$

**2.** When the kinetics of the reaction $2\ HI \rightarrow H_2 + I_2$ are studied using the method of initial rates, the following data is obtained. What is the rate law for the reaction?

| Trial | $[HI]_0$ (*M*) | Initial Rate of Formation of $I_2$ ($M \cdot s^{-1}$) |
|---|---|---|
| 1 | 0.010 | $4.0 \times 10^{-6}$ |
| 2 | 0.020 | $1.6 \times 10^{-5}$ |

**A.** rate = $k$
**B.** rate = $k[HI]$
**C.** rate = $k[HI]^2$
**D.** rate = $k[HI]^{0.5}$

**3.** Chlorine and fluorine react to produce chlorine trifluoride.

$$Cl_2(g) + 3\ F_2(g) \rightarrow 2\ ClF_3(g)$$

The reaction was monitored over a period of time and the rate of reaction was determined to be 0.018 $M \cdot s^{-1}$. Which choice correctly describes the change in concentration of each of the species with time?

| | $\frac{\Delta[F_2]}{\Delta t} M \cdot s^{-1}$ | $\frac{\Delta[ClF_3]}{\Delta t} M \cdot s^{-1}$ |
|---|---|---|
| **A.** | –0.054 | 0.036 |
| **B.** | –0.006 | 0.009 |
| **C.** | 0.054 | –0.036 |
| **D.** | 0.006 | –0.009 |

**4.** An increase in which of the following is NOT expected to increase the rate of a reaction?

**A.** Reactant concentration
**B.** Surface area of a reactant
**C.** Activation energy of the reaction
**D.** Temperature of the reaction

*The nuclear decay of sulfur-35 was monitored and plotted below. Use this data to answer questions 5–7.*

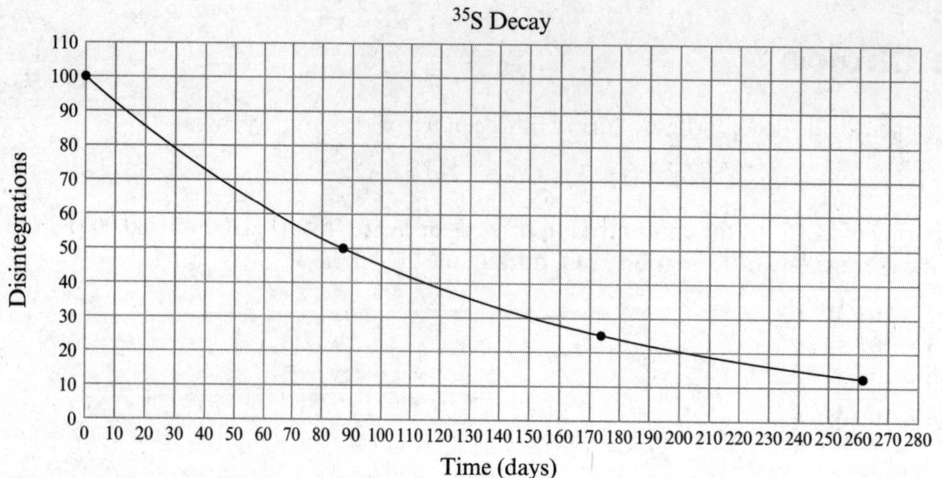


**5.** What is the approximate half-life for the decay of sulfur-35?

**A.** 44 days
**B.** 87 days
**C.** 130 days
**D.** 260 days

**6.** What is the rate constant for the decay of sulfur-35?

**A.** 0.004 $day^{-1}$
**B.** 0.008 $day^{-1}$
**C.** 0.04 $day^{-1}$
**D.** 0.08 $day^{-1}$

**7.** A pure sample of sulfur-35 has a mass of 2.00 grams. How much will remain after 348 days?

**A.** 0.125 gram
**B.** 0.250 gram
**C.** 0.500 gram
**D.** 1.00 gram

**8.** The rate of a chemical reaction at 320 K is found to be four times faster than the rate of the same reaction at temperature, *T*. What is the likely value of *T*?

**A.** 0 K
**B.** 273 K
**C.** 300 K
**D.** 350 K

*Questions 9 and 10 refer to the reaction described below.*

The reaction between 2-chloro-2-methylpropane, $(CH_3)_3CCl$, and hydroxide to form 2-methyl-2-propanol, $(CH_3)_3COH$, occurs according to the mechanism shown below.

Mechanism:

$$(CH_3)_3CCl \rightarrow (CH_3)_3C^+ + Cl^- \quad \text{(slow)}$$
$$(CH_3)_3C^+ + OH^- \rightarrow (CH_3)_3COH \quad \text{(fast)}$$

**9.** What is the rate law for this reaction?

A. $k[(CH_3)_3CCl][OH^-]$
B. $k[(CH_3)_3CCl]$
C. $k[(CH_3)_3C^+][OH^-]$
D. $k[(CH_3)_3C^+]$

**10.** Which of the species is an intermediate in the reaction?

A. $(CH_3)_3CCl$
B. $Cl^-$
C. $OH^-$
D. $C(CH_3)_3C^+$

**11.** A proposed mechanism for the reaction of nitrogen monoxide with bromine is:

$$2\,NO \underset{k_{-1}}{\overset{k_1}{\rightleftharpoons}} N_2O_2 \qquad \text{(fast)}$$

$$N_2O_2 + Br_2 \xrightarrow{k_2} 2\,NOBr \qquad \text{(slow)}$$

Which rate law is consistent with this mechanism?

A. $k[NO]^2[Br_2]$
B. $k[NO][Br_2]$
C. $k[NO]^2$
D. $k[Br_2]$

**12.** The concentration of a reactant, A, in a zero order reaction is plotted as a function of time. Which of the following graphs represents the results?

A.

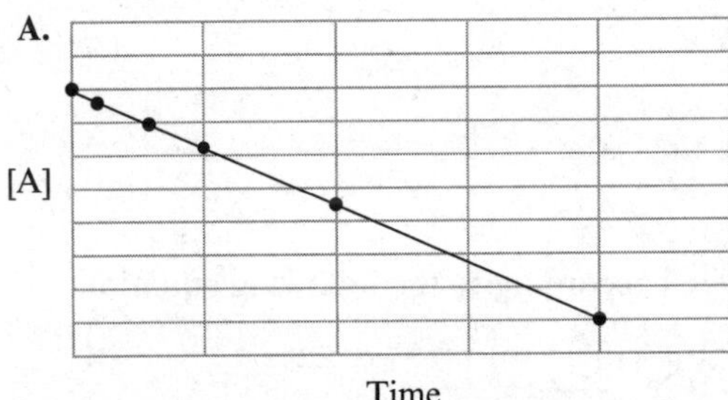


B.

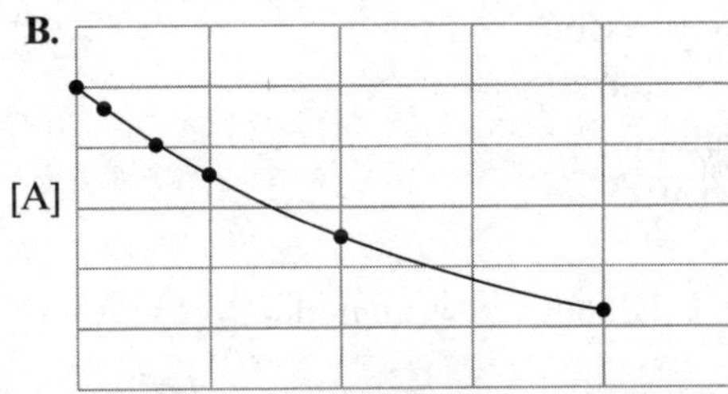


C.

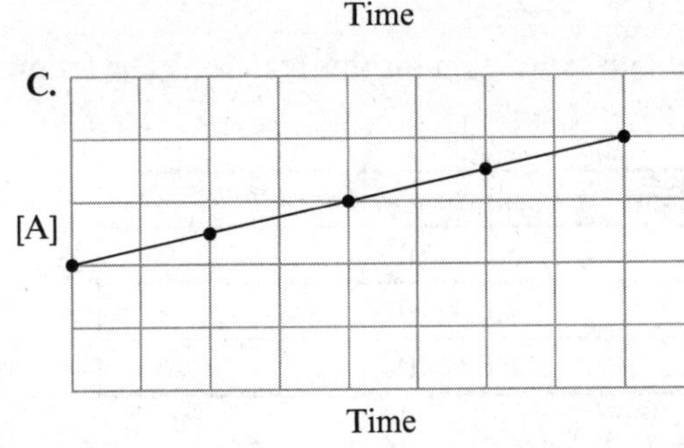


D.

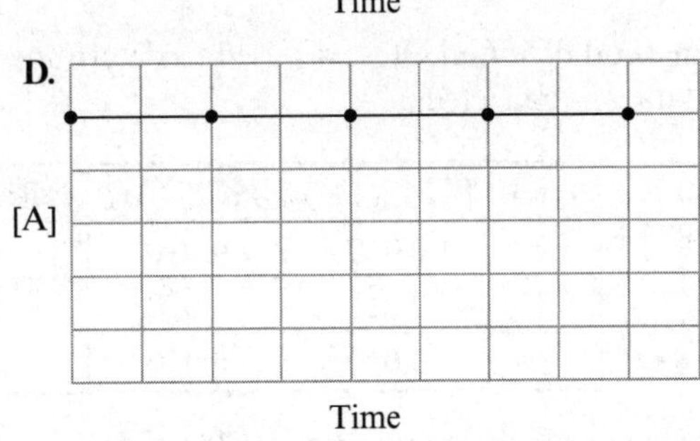

**13.** Which is true of the reaction represented by the energy profile below?

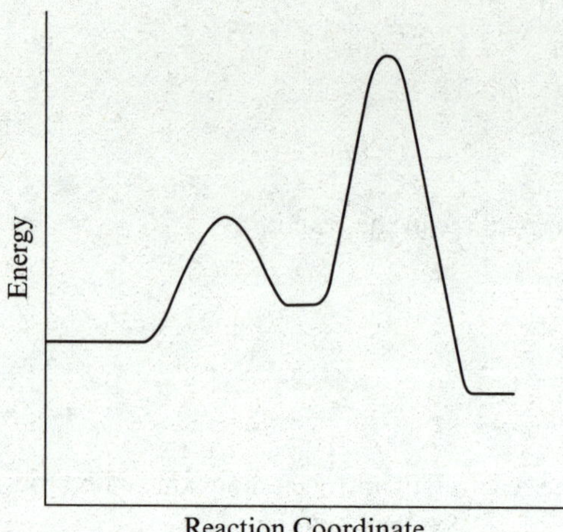


**A.** The reaction mechanism has a single transition state and two intermediates.
**B.** The reaction mechanism has two catalysts and two intermediates.
**C.** The reaction mechanism has two catalysts and no intermediates.
**D.** The reaction mechanism has two transition states and one intermediate.

**14.** The reaction mechanism below takes place in an aqueous solution. The role of $H_3O^+$ in the reaction is best described as that of a(n) _________.

$$C_7H_6O + H_3O^+ \rightarrow C_7H_7O^+ + H_2O$$
$$C_7H_7O^+ + C_8H_8O \rightarrow C_{15}H_{15}O_2^+$$
$$C_{15}H_{15}O_2^+ + H_2O \rightarrow C_{15}H_{14}O_2 + H_3O^+$$

**A.** heterogeneous catalyst
**B.** acid catalyst
**C.** reactant
**D.** intermediate

**15.** Propene, $C_3H_6$, undergoes an addition reaction with HCl according to the following equation.

$$C_3H_6 + HCl \rightarrow C_3H_7Cl$$

The method of initial rates was used to determine the rate expression for this reaction. The following initial rate data was collected.

| Trial | $[C_3H_6]_0$ ($M$) | $[HCl]_0$ ($M$) | Initial Rate of Formation of $C_3H_7Cl$ ($M \cdot s^{-1}$) |
|---|---|---|---|
| 1 | 0.010 | 0.010 | $5.0 \times 10^{-8}$ |
| 2 | 0.010 | 0.020 | $4.0 \times 10^{-7}$ |
| 3 | 0.020 | 0.010 | $1.0 \times 10^{-7}$ |

The rate law for the addition of HCl to $C_3H_6$ is rate = _________.

**A.** $k[C_3H_6]^3[HCl]$
**B.** $k[C_3H_6][HCl]$
**C.** $k[C_3H_6][HCl]^3$
**D.** $k[C_3H_6][HCl]^2$

## Long Free Response

**1.** At 2000 K, cyclopropane rearranges to form propene with a specific rate constant of $1.24 \times 10^8\ s^{-1}$.

Cyclopropane → Propene

**a.** A student claims that this reaction is relatively slow at 2000 K. Briefly describe why you agree or disagree.

**b.** What is the order of the reaction with respect to cyclopropane? How do you know?

**c.** If the initial mass of cyclopropane is 2.00 grams and no propene is initially present, how long will it take for the mass of propene to reach 1.00 gram?

**d.** The rate constant of the uncatalyzed reaction was measured at various temperatures and used to generate the plot shown below.

**i.** Compare the rate of the reaction at high temperature with the rate at low temperature, referring to the plot in your response.

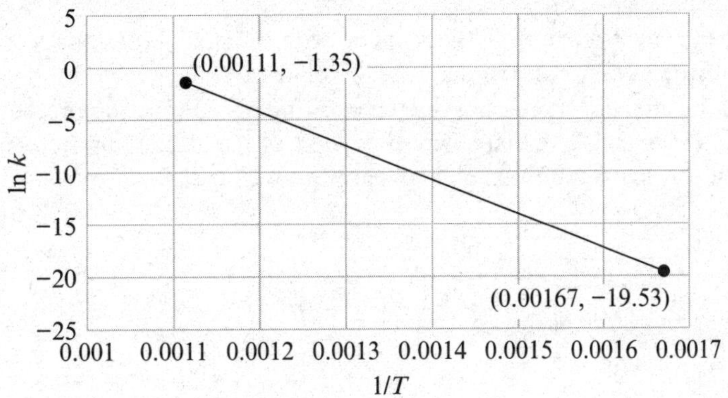


**ii.** What is the activation energy of the uncatalyzed formation of propene from cyclopropane? Include units in your response. (Temperatures are in Kelvin.)

**e.** At 500 K, the rate of this reaction increases in the presence of boron trichloride, $BCl_3$, but $BCl_3$ is not consumed.

**i.** Explain the likely role of $BCl_3$ in the reaction.

**ii.** On the plot below, sketch two energy profiles, one for the reaction in the presence of $BCl_3$ and one for the reaction in the absence of $BCl_3$ at 500 K. Label products, reactants, and activation energies. Clearly indicate which reaction is taking place in the presence of $BCl_3$. The reaction is exothermic.

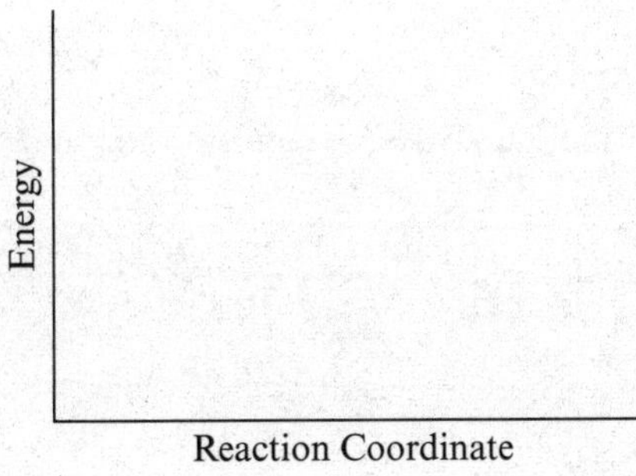

## Short Free Response

**2.** The following reaction occurs at 1000 K. It is a second order process. When equilibrium has been established, the concentrations of products are greater than the concentration of the reactants.

$$2\ A(g) \rightleftharpoons 2\ B(g) + C(g)$$

**a.** Using the axes below, sketch a graph showing how the concentrations of products and reactants change with time when pure A is placed in a piston with a fixed volume. Be sure to indicate which concentration each curve represents.

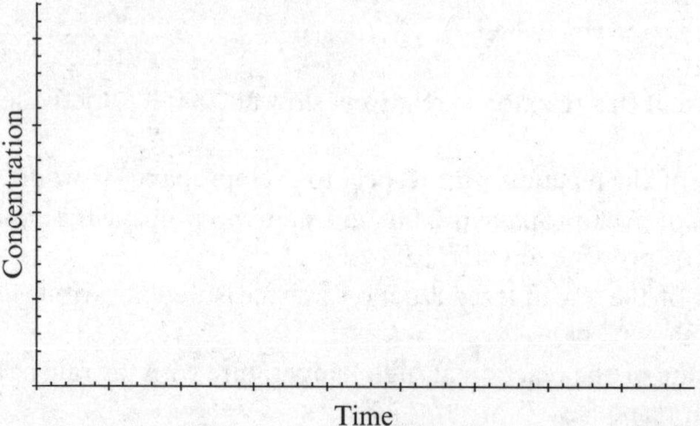


**b.** What would be the effect of running the reaction with the same initial mass of A, but increasing the volume of the piston at constant temperature? Explain your answer.

**c.** What would be the effect of increasing the temperature of the reaction on the rate at constant volume? On the plot below, sketch a curve that represents energy of the reactant molecules at a higher temperature. Label the curves "low temperature" and "high temperature."

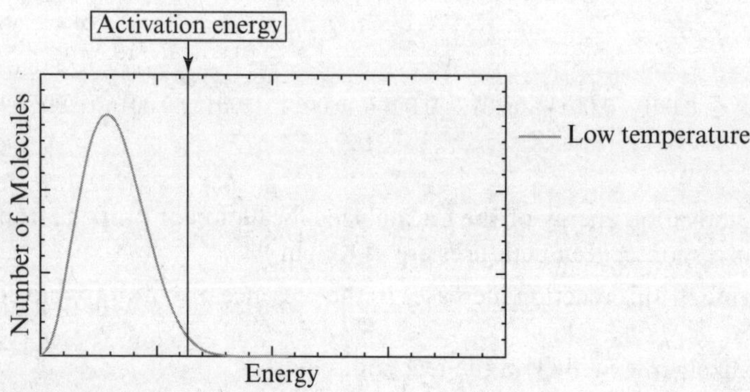


# Answers and Explanations

## Multiple Choice

**1. B.** The relationship between the rates of $N_2O_5$ consumption and $O_2$ production are given by the expression

$$\text{rate} = \frac{\Delta[O_2]}{\Delta t} = -\frac{1}{2}\left(\frac{\Delta[N_2O_5]}{\Delta t}\right) = -\frac{1}{2}\left(\frac{0.0071\ M\ -\ 0.0200\ M}{0.10\ \text{s}}\right) = \frac{0.0129}{0.20} = 6.45\times10^{-2}\ M\cdot\text{s}^{-1}\text{, choice B}$$

**2.** **C.** The general form of the rate law is

$$\text{rate} = k[\text{HI}]^x$$

Setting up a ratio between the rate laws allows us to solve for *x*.

$$\frac{\text{rate}_2}{\text{rate}_1} = \frac{\cancel{k}}{\cancel{k}}\left(\frac{[\text{HI}]_{0_2}}{[\text{HI}]_{0_1}}\right)^x$$

$$\frac{1.6\times10^{-5}}{4.0\times10^{-6}} = \left(\frac{0.020}{0.010}\right)^x$$

$$4 = 2^x$$

$$x = 2$$

Therefore, the rate law for the reaction is rate = $k[\text{HI}]^2$, choice C.

**3.** **A.** The relationship between the rate and the change in concentration of $F_2$ is

$$\text{rate} = -\frac{1}{3}\frac{\Delta[\text{F}_2]}{\Delta t} = 0.018M\cdot\text{s}^{-1}$$

$$\frac{\Delta[\text{F}_2]}{\Delta t} = -0.054M\cdot\text{s}^{-1}$$

The relationship between the rate and the concentration of $ClF_3$ is

$$\text{rate} = \frac{1}{2}\frac{\Delta[\text{ClF}_3]}{\Delta t} = 0.018M\cdot\text{s}^{-1}$$

$$\frac{\Delta[\text{ClF}_3]}{\Delta t} = 0.036M\cdot\text{s}^{-1}$$

Therefore, choice A is correct.

**4.** **C.** An increase in the activation energy of a reaction, choice C, would slow the reaction rather than speed it up. At a given temperature, fewer molecules would have sufficient energy to overcome a higher activation energy barrier than a lower activation barrier.

**5.** **B.** Half-life is the amount of time it takes for the initial number of disintegrations to decrease by half. The time it takes for the initial number of disintegrations, 100, to decrease to 50 is between 85 and 90 days according to the graph. The best choice is 87 days, choice B.

**6.** **B.** This reaction is first order because all nuclear decay processes occur with first order kinetics. The relationship between half-life for a first order decay and the rate constant is given on the *AP Chemistry Equations and Constants* sheet.

$$t_{1/2} = \frac{0.693}{k}$$

$$k = \frac{0.693}{t_{1/2}} = \frac{0.693}{87\text{ days}} \approx \frac{0.7}{90} \approx 0.008\text{ day}^{-1}\text{, choice B}$$

**7.** **A.** The easiest way to approach this problem is to recognize that 348 days corresponds to four half-lives (87 days × 4 = 348 days). After the first half-life, the number of grams of sulfur-35 has decreased to 1.00 gram. After two half-lives, the amount has decreased to 0.500 gram. After three half-lives, the amount has decreased to 0.250 gram, and after four half-lives, the amount has decreased to 0.125 gram, choice A.

**8.** **C.** For many chemical reactions near room temperature, a 10-Kelvin increase in temperature results in a twofold increase in the rate of a reaction. Since the reaction rate has quadrupled, this means that a temperature of 320 K must be 20 Kelvins higher than the previous temperature. Therefore, the likely value of *T* is 300 K, choice C.

**9.** **B.** The rate of a reaction can be determined by the molecularity of the slow step, or rate-determining step. The first step of the reaction is the rate-determining step, so the rate law is written based on this step: $k[(CH_3)_3CCl]$, choice B.

**10.** **D.** A reaction intermediate is a species that is generated in an earlier step of a reaction mechanism and consumed in a later step. An intermediate does not appear in the overall equation. $(CH_3)_3CCl$ and $OH^-$ are reactants and $Cl^-$ is a product. $C(CH_3)_3C^+$, choice D, does not appear in the overall equation, and is, therefore, an intermediate.

**11.** **A.** The rate law based on the slow step is

$$\text{rate} = k_2[N_2O_2][Br_2]$$

Since $N_2O_2$ is an intermediate, however, it is not included in the rate law.

The rate of the first step in the forward direction is equal to the rate in the reverse direction.

$$k_1[NO]^2 = k_{-1}[N_2O_2]$$

Solving for the intermediate results in

$$[N_2O_2] = \frac{k_1[NO]^2}{k_{-1}}$$

Substitution into the rate law gives

$$\text{rate} = \frac{k_1[NO]^2[Br_2]}{k_{-1}}$$

$$\text{rate} = k[NO]^2[Br_2], \text{ choice A}$$

**12.** **A.** A plot of concentration versus time for a zero order reaction results in a straight line with a negative slope, choice A. The concentration of A decreases steadily, but the rate of the reaction does not change.

**13.** **D.** The energy profile shows two transition states and one intermediate, choice D, as labeled in the plot below. The transition states are at the energy maxima, and the intermediate is at an energy minimum between the two transition states.

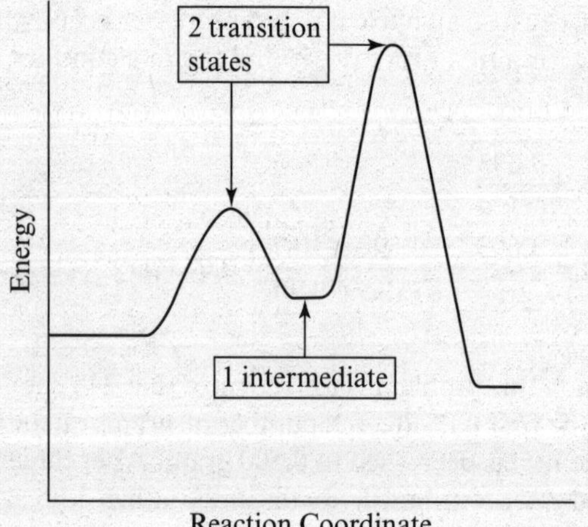

**14.** B. The $H_3O^+$ reacts in the first step and is regenerated in a subsequent step, so it is a catalyst in this mechanism. It activates the reactant by donating a proton, so it is specifically an acid catalyst, choice B.

**15.** C. The overall rate of the reaction is given by

$$\text{rate} = k[C_3H_6]^x[HCl]^y$$

Select two trials (trials 1 and 3) in which HCl does not change in order to solve for $x$.

$$\frac{\text{rate}_3}{\text{rate}_1} = \frac{\cancel{k}(0.020)^x \cancel{(0.010)^y}}{\cancel{k}(0.010)^x \cancel{(0.010)^y}} = \frac{1.0\times10^{-7}}{5.0\times10^{-8}}$$

$$2^x = 2$$

$$x = 1$$

The order with respect to HCl can be determined using trials 1 and 2.

$$\frac{\text{rate}_2}{\text{rate}_1} = \frac{\cancel{k}\cancel{(0.010)^x}(0.020)^y}{\cancel{k}\cancel{(0.010)^x}(0.010)^y} = \frac{4.0\times10^{-7}}{5.0\times10^{-8}}$$

$$2^y = 8$$

$$y = 3$$

The rate law for the reaction is

$$\text{rate} = k[C_3H_6][HCl]^3\text{, choice C}$$

## Long Free Response

**1.** **a.** Disagree with the student. The reaction is fast, as indicated by the very large rate constant on the order of $10^8$.

**b.** The reaction is first order with respect to cyclopropane. The units associated with the rate constant, $s^{-1}$, must be multiplied by $M$ to result in units of rate, $M \cdot s^{-1}$.

**c.** The time it takes for the concentration to decrease by half is the half-life of a reaction. Since the molar masses are identical for each gram of cyclopropane that is consumed, 1 gram of propene is produced. The time for the cyclopropane to decrease from 2 grams to 1 gram is the half-life. This reaction is first order, so the half-life is given by

$$t_{1/2} = \frac{0.693}{k} = \frac{0.693}{1.24\times10^8\ s^{-1}} = 5.59\times10^{-9}\ s$$

**d.** **i.** The rate constant for the reaction at higher temperatures is lower than the rate constant of the reaction at lower temperatures. At $T^{-1}$ of 0.00167 and 0.00111, the temperatures are 599 and 901 K, respectively. The rate constant, $k$, at 599 K is $e^{-19.53}$, which is less than the rate constant at 901 K, $e^{-1.35}$.

**ii.** The activation energy of a reaction can be determined from the slope of the plot of ln $k$ versus $T^{-1}$.

$$\text{slope} = -\frac{E_a}{R}$$

$$E_a = -\text{slope}\times R = -\left(\frac{-19.53-(-1.35)}{0.00167\ \cancel{K^{-1}} - 0.00111\ \cancel{K^{-1}}}\right)\times 8.314\ J\cdot mol^{-1}\cdot \cancel{K^{-1}}$$

$$E_a = 2.70\times10^5\ J\cdot mol^{-1}$$

e. i. Because $BCl_3$ increases the rate of the reaction without being consumed, it is a catalyst. It changes the reaction mechanism to one with lower activation energy than the uncatalyzed process.

ii. Your answer should resemble the sketch below. Notice that the activation energy for the reaction in the presence of $BCl_3$ is lower than that of the reaction in the absence of $BCl_3$. Since the reaction is exothermic, product potential energy is lower than reactant potential energy.

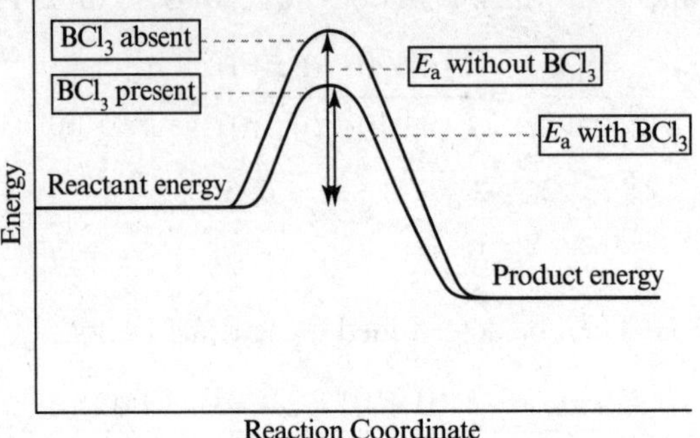


## Short Free Response

2. a. The sketch should clearly show that A disappears at approximately the same rate that B appears. The rate of appearance of C is half the rate of appearance of B.

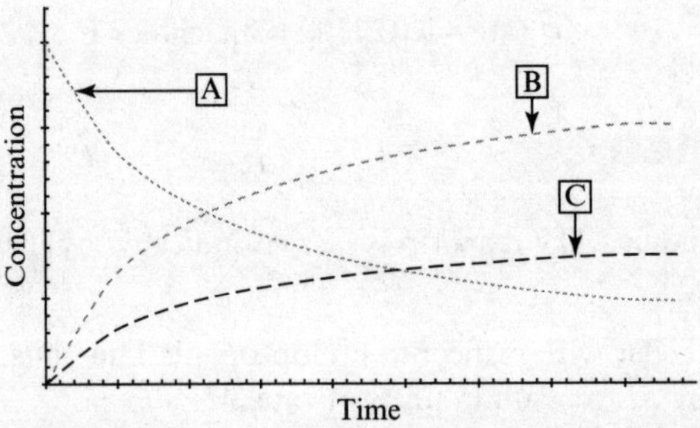


b. Because the volume of the vessel is increased, the concentration of A is decreased. A decreased concentration of reactant leads to a decrease in the rate of the reaction when the rate is not zero order.

c. At a higher temperature, a larger number of molecules will have sufficient kinetic energy to overcome the activation energy barrier and react, while at the lower temperature, a smaller number of molecules will have sufficient kinetic energy to do so. This results in an increased reaction rate at high temperature.

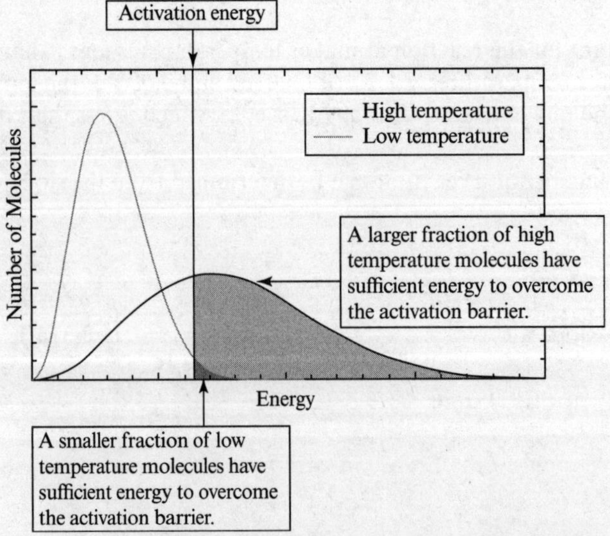

# Chapter 6

# Thermodynamics

**Thermodynamics is the study of the energetics of chemical and physical changes.**

**Energy** is the capacity to transfer heat or do work. **Thermodynamics** studies the transfer of energy during physical and chemical processes.

## The First Law of Thermodynamics

According to the **first law of thermodynamics,** the amount of matter and energy in the universe is fixed. In ordinary chemical and physical processes, energy is neither created nor destroyed.

### Heat

**Heat, *q*,** is thermal energy that is transferred from a warmer substance to a cooler substance. When two samples at different temperatures are in contact with one another, the particles at the interface between the two samples collide. If one of the samples has a higher average kinetic energy than the other, the collisions between the particles of the two samples result in the transfer of energy from the higher-energy substance to the lower-energy substance.

The kinetic energy of the particles of the warmer substance will continue to be transferred to the particles of the cooler substance until the particles of both samples have the same average kinetic energy. This condition is referred to as **thermal equilibrium.** Two objects in contact with one another are at thermal equilibrium when they have achieved the same temperature. When thermal equilibrium has been reached, no further heat transfer will take place. If the two substances are thermally isolated from the rest of the universe, then the heat lost by one substance will be exactly equal to the heat gained by the other.

At constant pressure, a transfer of heat to or from a system is equal to the **enthalpy change, $\Delta H$,** of the system. The heat at constant pressure is denoted $q_p$.

$$\Delta H = q_p$$

The transfer of heat to a system is an **endothermic** process. The final energy of the system is greater than the initial energy of the system, and $\Delta H$ is positive.

The release of heat by a system to the surroundings is an **exothermic** process. The final energy of the system is less than the initial energy of the system, and $\Delta H$ is positive.

#### Practice

Determine whether the following processes are endothermic or exothermic from the perspective of the underlined system.

**a.** Water condenses on the outside of a glass containing an iced liquid.
**b.** A bar of chocolate melts in a saucepan on a stove.
**c.** Your fingers get cold on a cold day.

**(a)** As the water condenses, it is losing heat. This is an exothermic process.

**(b)** As the chocolate bar melts, it is gaining heat. This is an endothermic process.

**(c)** As your fingers get cold, they are losing heat. This is an exothermic process.

## Work

**Work, *w*,** is the product of a force acting on an object or a system over a distance; the term encompasses energy transfers that are not heat transfers. In AP Chemistry, discussion is limited to pressure-volume work. Pressure-volume work only occurs when there is a change in the volume of a gas against an external pressure.

When a system expands against its surroundings, the system is doing work on the surroundings. When the system contracts, the surroundings are doing work on the system.

The equation for work is

$$w = -P\Delta V$$

where $P$ is the external pressure and $\Delta V$ is the change in volume, $V_f - V_i$. This gives work in units of L·atm, which can be converted to joules:

$$1 \text{ L·atm} = 101.3 \text{ J}$$

In the event that a work calculation is required on the AP Chemistry exam, either L·atm will be appropriate units *or* this conversion factor is likely to be included in the question. Be sure to read carefully.

### Practice

A piston expands from a volume of 2.50 L to a volume of 3.10 L against a constant external pressure of 1.0 atm.

**a.** How much work is performed? Express your answer in joules (101.3 J = 1 L·atm).
**b.** Is work done by the system or on the system?

**(a)** Using the equation for work,

$$w = -P\Delta V = -(1.00 \text{ atm})(3.10 \text{ L} - 2.50 \text{ L}) = -0.60 \text{ L·atm}$$

$$-0.60 \cancel{\text{L·atm}} \times \frac{101.3 \text{ J}}{1 \cancel{\text{L·atm}}} = -61 \text{ J}$$

**(b)** The expansion of the piston indicates that work is done by the system. The value of $w$ will always be negative when work is done by a system.

## Internal Energy

The **internal energy, *E*,** of a system is the sum of the kinetic and potential energies of the particles that make up the system. **Kinetic energy** is the energy of the motion of the particles in a system. **Potential energy** is the energy a system possesses as a consequence of the arrangement of electrons and nuclei in molecules.

Energy can be transferred between system and surroundings. A **system** is a sample or a region under investigation, and the **surroundings** are everything else. Both the system and the surroundings taken together make up the **thermodynamic universe.**

According to the first law of thermodynamics, if energy is lost by a system, the amount of energy lost must be equal to the amount of energy gained by the surroundings.

$$\Delta E_{sys} = -\Delta E_{surr}$$

When $\Delta E_{sys}$ is negative, energy is flowing out of the system. When $\Delta E_{sys}$ is positive, energy is flowing into the system.

The change in the internal energy of a system is the sum of the heat transferred and the work done.

$$\Delta E = q + w$$

The heat transferred to a system is equal in magnitude and opposite in sign to the heat transferred from the surroundings.

$$q_{sys} = -q_{surr}$$

Likewise, the work done on a system is equal in magnitude and opposite in sign to the work done by the surroundings.

$$w_{sys} = -w_{surr}$$

The following conventions apply to the signs of $q$ and $w$:

- When heat is transferred from the surroundings to a system, $q > 0$.
- When heat is transferred from a system to the surroundings, $q < 0$.
- When work is done by the surroundings on a system, $w > 0$.
- When work is done by the system on the surroundings, $w < 0$.

**Transfer of Heat and Work between System and Surroundings**

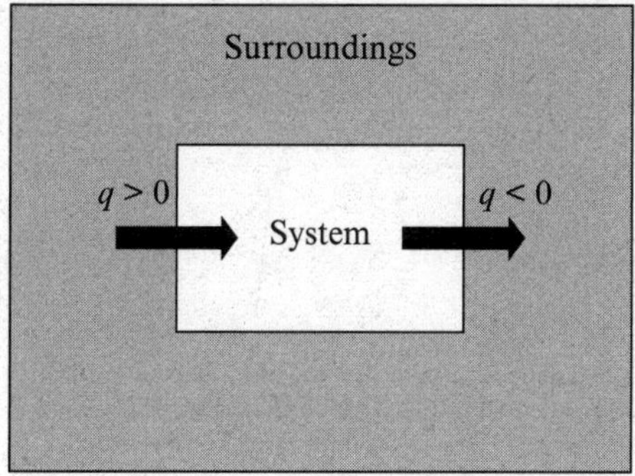


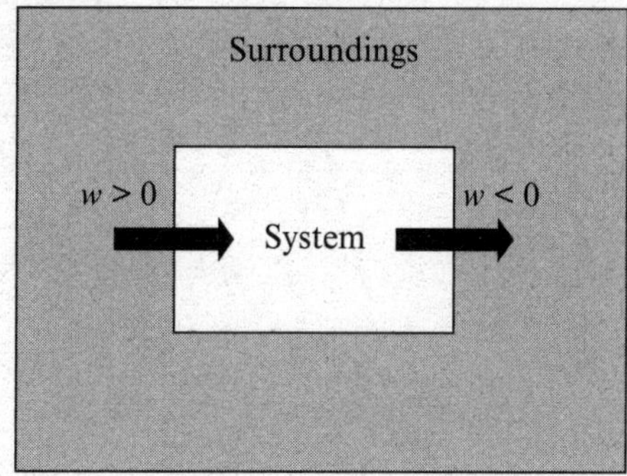


Solids and liquids do not expand or contract significantly as pressure changes, so for processes involving only solids and liquids, $\Delta V = 0$ and $w = 0$. For reactions in which the number of moles of gaseous reactants is equal to the number of moles of gaseous products at constant pressure, $\Delta V = 0$ and $w = 0$.

For reactions in which the number of moles of gaseous reactants is greater than the number of moles of gaseous products at constant pressure, $\Delta V < 0$ and $w > 0$.

For reactions in which the number of moles of gaseous reactants is less than the number of moles of gaseous products at constant pressure, $\Delta V > 0$ and $w < 0$.

## Practice

Ammonium nitrate, an agricultural fertilizer, decomposes explosively according to the following equation.

$$2\ NH_4NO_3(s) \rightarrow 2\ N_2(g) + 4\ H_2O(g) + O_2(g)$$

Predict the signs of **(a)** $w$, **(b)** $q$, and **(c)** $\Delta E_{sys}$, where the system is the reaction mixture. Explain your reasoning.

**(a)** The number of moles of gaseous products ($n = 2 + 4 + 1 = 7$ moles) is greater than the number of moles of gaseous reactants ($n = 0$). The change in volume is positive, $V_f - V_i > 0$, so $w < 0$.

**(b)** Prediction of the sign of $q$ requires an understanding that explosions tend to give off heat to the surroundings, so $q < 0$.

**(c)** Because $w < 0$ and $q < 0$, $\Delta E_{sys}$, the change in internal energy of the system, which is equal to the sum of $q$ and $w$, is also less than zero.

# Potential Energy Diagrams

Endothermic and exothermic reactions are often described in **potential energy diagrams.** Energy is plotted on the vertical axis, and reaction coordinate is shown on the horizontal axis. **Reaction coordinate** can be interpreted as progress of a reaction or as time, usually with reactants on the left and products on the right.

For endothermic reactions, the energy of the products is higher than the energy of the reactants. Energy is absorbed over the course of the reaction.

For exothermic reactions, energy of the reactants is greater than the energy of the products. Energy is lost to the surroundings over the course of the reaction.

**Potential Energy Diagrams for Endothermic and Exothermic Reactions**

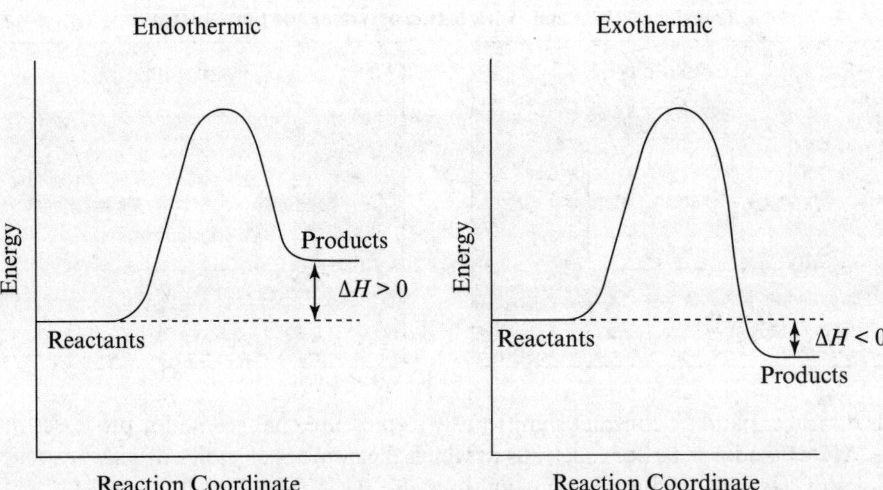


## Practice

A beaker at room temperature contains a small amount of sodium hydroxide. When a dilute aqueous solution of hydrochloric acid is added to the beaker, the outside of the beaker begins to feel very warm to the touch.

**a.** Sketch a potential energy diagram that shows the relative energy of the reactants and the products.
**b.** Is the reaction endothermic or exothermic? Justify your response.

**(a)** Your sketch should have the following features:

- Vertical axis labeled *Energy* or *Potential Energy*
- Horizontal axis labeled *Reaction Coordinate*, *Reaction Progress*, or *Time*
- Reactants labeled $HCl$ and $NaOH$
- Products labeled $NaCl$ and $H_2O$
- Energies of reactants and products labeled
- Energy of reactants is higher than energy of products

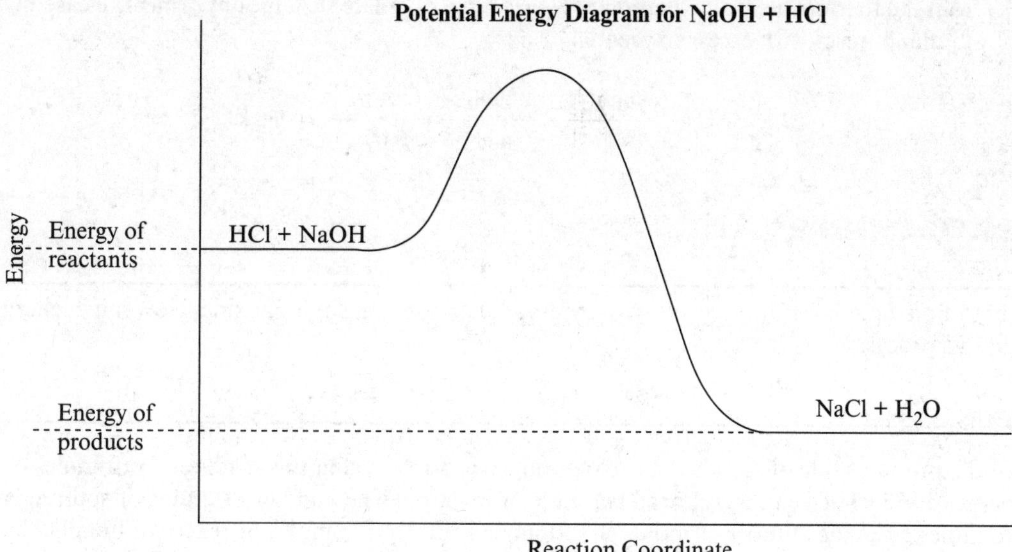


**(b)** The reaction is exothermic. Because the beaker became warm to the touch, we know that heat was given off as reactants were converted to products. The energy of the products is lower than the energy of the reactants.

# Thermochemical Equations

A **thermochemical equation** is a balanced chemical equation that includes the value of $\Delta H$ for the reaction.

For exothermic and endothermic reactions, thermochemical equations take the forms

reactants → products + **heat** (exothermic)

reactants + **heat** → products (endothermic)

Consider $\Delta H$ for the exothermic reaction of aluminum and oxygen.

$$2\ \mathrm{Al}(s) + \frac{3}{2}\mathrm{O_2}(s) \rightarrow \mathrm{Al_2O_3}(s) + 1776\ \mathrm{kJ/mol_{rxn}}$$

In thermochemical equations, the stoichiometric coefficients refer to the number of moles of reactants consumed and products formed to produce the reported enthalpy change.

The $\mathrm{mol_{rxn}}$ refers to a "mole of this reaction as it is written." Because everything is read in terms of moles, the use of fractional coefficients is allowed when balancing the chemical equation. An enthalpy change of $\Delta H = -1776$ kJ is a consequence of the heat transferred during the reaction of *two moles* of Al with *one-and-a-half moles* of $O_2$ to form *one mole* of $Al_2O_3$.

$$\Delta H = \frac{q}{n} = \frac{\text{heat transferred (kJ)}}{\text{moles of reaction (mol}_{\text{rxn}})}$$

The negative change in enthalpy indicates a release of heat. This is detected as an increase in the temperature of the surroundings.

The enthalpy change, $\Delta H$, can be used as a ratio in dimensional analysis. It can be used to quantify the heat transfer for a given quantity of a component involved in a reaction.

For example, from the thermochemical equation above, we can calculate the amount of heat released when 2.0 grams of aluminum react with excess oxygen.

$$2.0\ \cancel{\text{g Al}} \times \frac{1\ \cancel{\text{mol Al}}}{26.98\ \cancel{\text{g Al}}} \times \frac{1\ \cancel{\text{mol}}_{\text{rxn}}}{2\ \cancel{\text{mol Al}}} \times \frac{-1776\ \text{kJ}}{1\ \cancel{\text{mol}}_{\text{rxn}}} = -66\ \text{kJ}$$

## Practice

A 46-gram sample of solid sodium reacts with excess oxygen to form sodium oxide. How much energy is released in this process?

$$4\ \text{Na}(s) + \text{O}_2(g) \rightarrow 2\ \text{Na}_2\text{O}(s) + 832\ \text{kJ}$$

Energy is on the products side of the chemical equation, so it is released in this reaction. Four moles of sodium are consumed and 832 kJ of heat are released per mole of reaction. The number of grams of sodium is first converted to moles; then the number of moles of sodium is converted to moles of reaction. Finally, the number of moles of reaction is multiplied by the value of $\Delta H$:

$$46.0\ \cancel{\text{g Na}} \times \frac{1\ \cancel{\text{mol Na}}}{22.99\ \cancel{\text{g Na}}} \times \frac{1\ \cancel{\text{mol}}_{\text{rxn}}}{4\ \cancel{\text{mol Na}}} \times \frac{-832\ \text{kJ}}{1\ \cancel{\text{mol}}_{\text{rxn}}} = -416\ \text{kJ}$$

Thus, 416 kJ of energy are released when 46 grams of sodium are reacted with excess oxygen.

## Practice

The amount of heat liberated when 0.0720 mole of sodium reacts with excess water is 13.2 kJ. What is the enthalpy change per $\text{mol}_{\text{rxn}}$ for this process?

$$2\ \text{Na}(s) + 2\ \text{H}_2\text{O}(l) \rightarrow \text{H}_2(g) + 2\ \text{NaOH}(aq) + \text{heat}$$

The enthalpy change that occurs when 2 moles of sodium react (according to the balanced equation) can be determined from the heat released when 0.0720 mole of sodium reacts.

$$\Delta H_{\text{rxn}} = \frac{q}{n} = \frac{-13.2\ \text{kJ}}{0.0720\ \text{mol Na}} \times \frac{2\ \text{mol Na}}{1\ \text{mol}_{\text{rxn}}} = -367\ \text{kJ/mol}_{\text{rxn}}$$

# Heat Capacity and Temperature Changes

The **heat capacity,** ***c*****,** of a substance is the amount of heat required to raise the temperature of a quantity of the substance by 1 K or 1 °C.

$$c = \frac{\text{heat absorbed}}{\text{temperature increase}}$$

The amount of heat required to raise the temperature of *one gram* of the substance by 1 K (or 1 °C) is called the **specific heat capacity.** It is reported in units of J/(g·K) or J/(g·°C).

The amount of heat required to raise the temperature of *one mole* of the substance by 1 K (or 1 °C) is called the **molar heat capacity.** It is reported in units of J/(mol·K) or J/(mol·°C).

The higher the specific heat capacity of a substance, the greater the heat required to raise the temperature of the substance. For example, liquid water has a specific heat capacity of 4.18 J/(g·K). Copper has a specific heat capacity of 0.385 J/(g·K). It requires less heat to raise the temperature of 1 gram of copper than it does to achieve the same temperature increase in 1 gram of water.

The amount of heat absorbed or released in order to change the temperature of a mass, *m*, of a particular substance of specific heat capacity *c* is

$$q = mc\Delta T$$

where $\Delta T$ is the change in temperature, $T_{final} - T_{initial}$. This equation is given on the *AP Chemistry Equations and Constants* sheet.

### Practice

What quantity of heat is needed to raise the temperature of 10.0 g of water from 25 °C to 50 °C? The specific heat capacity of water is 4.18 J/(g·°C).

The equation for heat is used.

$$q = mc\Delta T = (10.0\ \cancel{g})\left[4.18\ \text{J}/(\cancel{g}\cdot\cancel{°C})\right](50.0\ \cancel{°C} - 25.0\ \cancel{°C}) = 1050\ \text{J}$$

## Heat Transfer between Substances in Thermal Contact

When two substances are in thermal contact, the hotter body will transfer heat to the cooler body until both substances have reached the same temperature (thermal equilibrium). The heat lost by the higher-temperature substance, $q_{lost}$, is equal in magnitude and opposite in sign to the heat gained by the lower-temperature substance, $q_{gained}$.

$$q_{lost} = -q_{gained}$$

Suppose a student would like to measure the specific heat of aluminum. A 5.00-gram block of aluminum at 100.0 °C is placed in 20.0 grams of water at 30.0 °C. The temperature of the water increases until it reaches a maximum of 33.6 °C.

The heat transferred from the aluminum to the water is equal in magnitude to the heat gained by the water. Although both substances are at different initial temperatures, at thermal equilibrium their final temperature, $T_f$, is the same.

$$q_{aluminum} = -q_{water}$$

Substituting $mc\Delta T$ for the heat transferred between the water and the aluminum allows us to solve for $c_{Al}$, the specific heat capacity of aluminum. The specific heat of water is 4.18 J/(g·°C).

$$m_{Al}c_{Al}\Delta T_{Al} = -m_{H_2O}c_{H_2O}\Delta T_{H_2O}$$

$$c_{Al} = \frac{-m_{H_2O}c_{H_2O}\Delta T_{H_2O}}{m_{Al}\Delta T_{Al}}$$

$$c_{Al} = \frac{-(20.0\ \cancel{g})\left[(4.18\ \text{J}/(\cancel{g\cdot°C})\right](33.6\ \cancel{°C} - 30.0\ \cancel{°C})}{(5.0\ \text{g})(33.6\ °\text{C} - 100.0\ °\text{C})}$$

$$c_{Al} = 0.91\ \text{J}/(\text{g}\cdot°\text{C})$$

## Practice

A 100.-gram block of silver at 20.0 °C is placed in contact with a 10.0-gram block of iron at 60.0 °C. Assuming that no heat is lost to the surroundings, what is the final temperature of the two metals? The specific heat of silver is 0.240 J/(g·°C) and the specific heat of iron is 0.450 J/(g·°C).

The heat lost by the iron is equal in magnitude and opposite in sign to the heat gained by the silver. The final temperature, $T_f$, is the same for both metals.

$$q_{\text{lost}} = -q_{\text{gained}}$$
$$m_{\text{Fe}}c_{\text{Fe}}\Delta T_{\text{Fe}} = -m_{\text{Ag}}c_{\text{Ag}}\Delta T_{\text{Ag}}$$
$$(10.0\text{ g})[0.450\text{ J/(g}\cdot{}^\circ\text{C)}](T_f - 60.0\ {}^\circ\text{C}) = -(100.\text{ g})[0.240\text{ J/(g}\cdot{}^\circ\text{C)}](T_f - 20.0\ {}^\circ\text{C})$$
$$T_f(4.50\text{ J/}^\circ\text{C}) - 270\text{ J} = -T_f(24.0\text{ J/}^\circ\text{C}) + 576\text{ J}$$
$$T_f(24.5\text{ J/}^\circ\text{C}) = 846\text{ J}$$
$$T_f = 34.5\ {}^\circ\text{C}$$

# Calorimetry

**Calorimetry** is an experimental technique used to measure the amount of heat transferred to or from a system.

In most AP Chemistry classrooms, a **coffee cup calorimeter** is used. Coffee cup calorimetry is used to measure heat transferred by reactions in aqueous solutions. A reaction is carried out in a well-insulated disposable foam cup equipped with a stirrer and a thermometer. The insulation is important so that heat transfer to or from the surroundings is minimized.

**Coffee Cup Calorimeter**

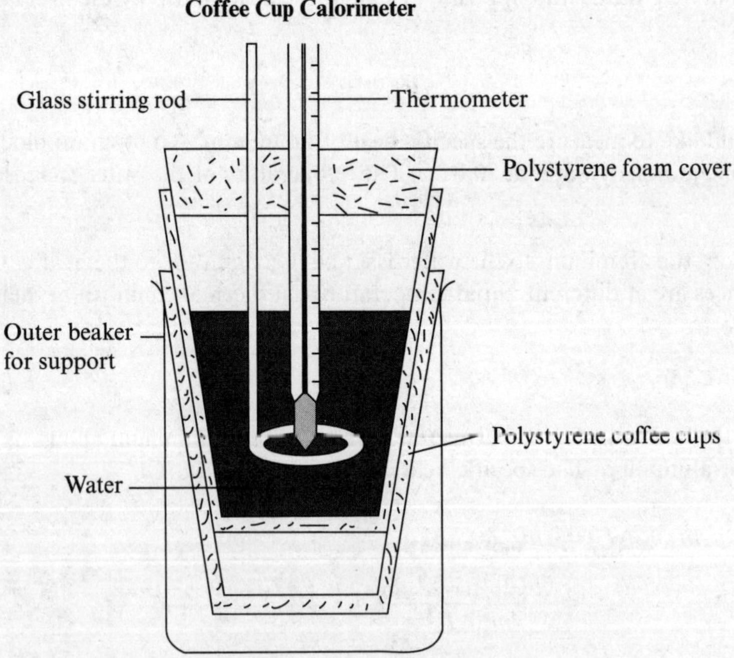


Coffee cup calorimetry is carried out at atmospheric pressure; pressure is constant in this kind of calorimetry experiment, so the change in enthalpy of the system is equal to the heat lost or gained in the reaction under investigation. (Remember $\Delta H = q_p$.)

The system is the reactants and products of the reaction, and the surroundings include the calorimeter, the thermometer, *and the solvent*.

When the temperature of the solvent increases, it is because energy is being released by the system to the surroundings as reactants form products in an exothermic process.

When the temperature of the solvent decreases, the decrease is due to the uptake of energy from the surroundings by the system as reactants form products in an endothermic process.

Since a coffee cup calorimeter is not a perfectly insulated apparatus, some heat will be transferred to or from the surroundings. The value of $\Delta H$ calculated in a calorimetry experiment can differ from the actual value when a significant amount of heat transfer between the system and surroundings occurs.

## Practice

Two aqueous solutions, 1.00 $M$ HCl and 1.00 $M$ NaOH, were at an initial temperature of 30.0 °C.

100. mL of the HCl solution was mixed with 100. mL of the NaOH in a coffee cup calorimeter.

At the end of the experiment, the maximum temperature reached by the combined solutions was 40.0 °C.

Assume that the densities and specific heats of the two solutions were the same as that for water, 1.00 g/mL and 4.18 J/(g·°C), respectively.

**a.** What is the enthalpy of reaction for the neutralization of HCl with NaOH? Report the value in units of $J/mol_{rxn}$.
**b.** If the calorimeter is poorly insulated and a significant amount of heat transfer with the surroundings occurs, will the enthalpy of neutralization calculated be too high or too low? Explain your response.

**(a)** The heat given off in the reaction is equal to the heat gained by the solution. Since the reaction occurs at constant pressure (atmospheric pressure), this heat is equal to the enthalpy change for the reaction. Because the density of the solution is 1.00 g/mL, 100. mL of solution has a mass of 100. grams.

$$q_{rxn} = -q_{soln} = -m_{soln}c_{soln}\Delta T_{soln}$$

$$q_{rxn} = -(100.\ \cancel{g} + 100.\ \cancel{g})\left[4.18\ \text{J}/(\cancel{g}\cdot\cancel{°C})\right](40.0\ \cancel{°C} - 30.0\ \cancel{°C})$$

$$q_{rxn} = -8360\ \text{J}$$

The amount of heat given off per mole of HCl or NaOH reacted is reported as the enthalpy change per mole reaction.

$$HCl(aq) + NaOH(aq) \rightarrow H_2O(l) + NaCl(aq)$$

The same number of moles of HCl and NaOH are present.

$$100.\ \cancel{\text{mL HCl}} \times \frac{1\ \cancel{L}}{1000\ \cancel{\text{mL}}} \times \frac{1.00\ \cancel{\text{mol HCl}}}{1\ \cancel{\text{L HCl}}} \times \frac{1\ \text{mol}_{rxn}}{1\ \cancel{\text{mol HCl}}} = 0.100\ \text{mol}_{rxn}$$

The enthalpy of reaction is the amount of heat given off per mole of reaction.

$$\Delta H_{rxn} = \frac{q}{n} = \frac{-8360\ \text{J}}{0.100\ \text{mol}} = -8.36\times10^4\ \text{J/mol}_{rxn}$$

**(b)** If a significant amount of heat is lost to the surroundings, the temperature of the solution will not increase as much as expected. This means that $\Delta T$ will appear to be smaller than the true value, and the calculated magnitude of $q_{rxn}$ will be too low. Since the reaction is exothermic and the value is negative, however, the calculated value of the enthalpy change will be too high (less negative than the true value).

# Energy of Phase Changes

Phase changes, including freezing, melting, boiling, condensation, sublimation, and deposition, involve the transfer of energy to or from a substance as a result of the breaking or formation of intermolecular attractions.

Melting, vaporization, boiling, and sublimation are endothermic processes. Energy is required to disrupt the intermolecular forces in a substance.

The reverse is true for freezing, condensation, and deposition. When molecules come together due to intermolecular attractions, energy is released. These processes are exothermic.

A **heating curve** represents the transfer of energy that occurs during phase transitions. The heating curve below shows the change in temperature as heat energy is added to water. It should be noted that the temperature of a substance remains constant during the phase changes, and it changes when energy is being added or removed to a pure solid, liquid, or gas.

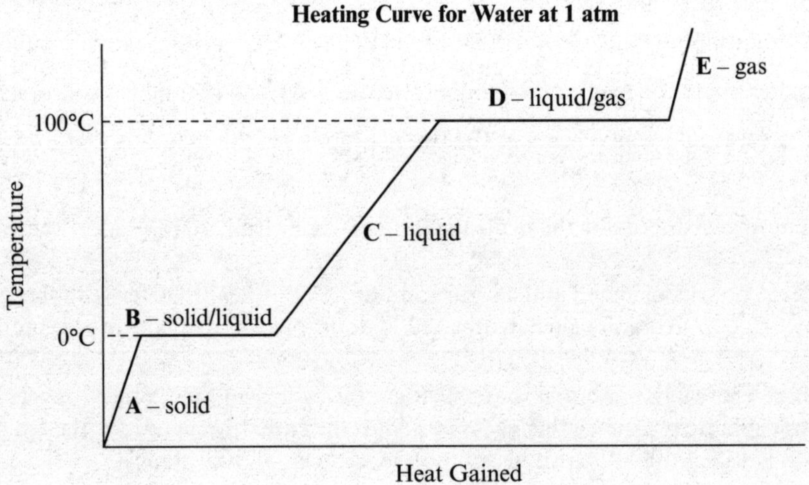


**A – solid:** As heat is added to ice at temperatures below 0 °C, the closely packed molecules begin to vibrate faster and the temperature rises.

**B – solid/liquid:** When the ice reaches 0 °C, the melting point, the temperature stops rising, even though heat is still being added. All the added energy at this temperature is used to disrupt the intermolecular forces and convert the solid to liquid.

**C – liquid:** Once the ice has been completely melted, the temperature of the liquid water rises again as more heat is added.

**D – liquid/gas:** A second and longer plateau occurs at the liquid-gas transition temperature, 100 °C. Boiling requires more heat than melting because not only do intermolecular forces need to be overcome, but the molecules must be separated completely from each other in space.

**E – gas:** Finally, once the water is converted completely to vapor, the temperature begins to rise again.

A heating curve can be used to determine the energy that must be transferred to result in a temperature or change or a phase change.

When energy is added to or removed from a solid, liquid, or gas to raise or lower its temperature, the amount of heat transferred can be determined using

$$q = mc\Delta T$$

Notice that the specific heat capacity of a substance depends on the physical state (solid, liquid, or gas), as shown for water in the table below.

| Form of Water | Specific heat capacity |
|---|---|
| solid | $c_{ice} = 2.1$ J/(g·°C) |
| liquid | $c_{liquid} = 4.18$ J/(g·°C) |
| gas | $c_{steam} = 2.0$ J/(g·°C) |

When a substance freezes or melts, the temperature remains constant throughout the phase change. The amount of heat transferred is determined by the number of moles of the substance, $n$, multiplied by the **enthalpy of fusion, $\Delta H_{fus}$,** which is the heat absorbed during the conversion of *one mole* of a substance from a solid to a liquid at its freezing temperature.

$$q = n\Delta H_{fus}$$

Sometimes the enthalpy of fusion is given as the energy transferred during the conversion of *one gram* of a substance from a solid to a liquid at its freezing temperature. In this case, the heat transfer is determined by the product of the mass of the substance, $m$, and the enthalpy of fusion.

$$q = m\Delta H_{fus}$$

When a substance boils or condenses, the amount of heat transferred is determined by the number of moles multiplied by the **enthalpy of vaporization, $\Delta H_{vap}$.** The enthalpy of vaporization is the energy required to convert *one mole* of a substance from liquid to gas at its boiling temperature.

$$q = n\Delta H_{vap}$$

The enthalpy of vaporization may also be expressed as the energy required to convert *one gram* of a substance from liquid to gas. In this case, the heat transfer is determined by the product of the mass of the substance, $m$, and the enthalpy of vaporization.

$$q = m\Delta H_{vap}$$

**Heating Curve for Water at 1 atm**

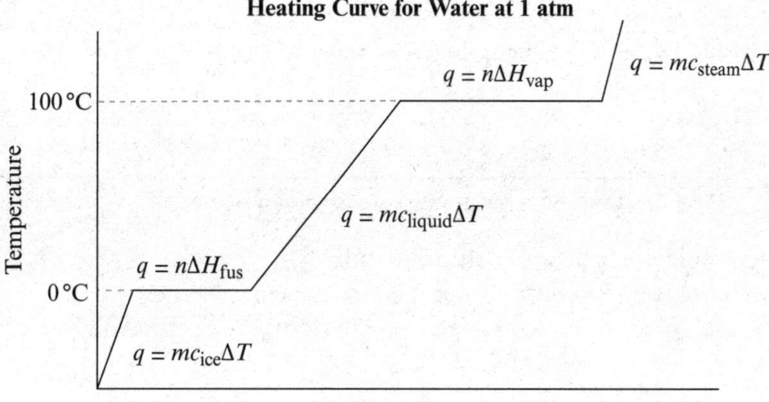

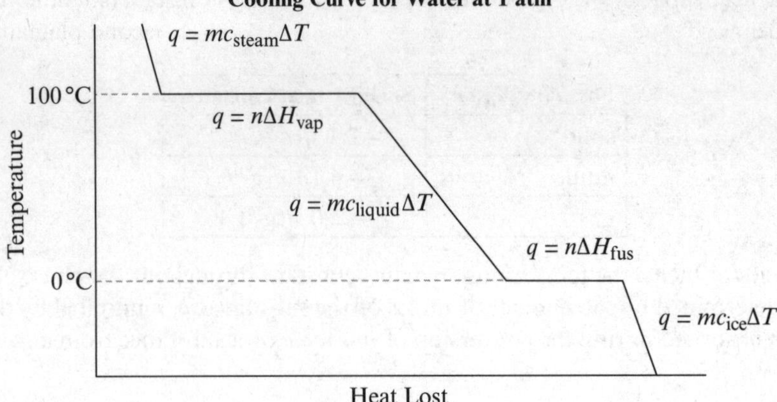


Suppose a block of ice with a mass of 25.0 grams is initially at a temperature of –15 °C. The amount of energy that is required to convert the ice to steam at a temperature of 102 °C can be calculated from the known heat of fusion of ice, heat of vaporization of steam, and the specific heat capacities of ice, liquid water, and steam: $\Delta H_{fus}$ = 6.01 kJ/mol, $\Delta H_{vap}$ = 40.7 kJ/mol, $c_{ice}$ = 2.1 J/(g·°C), $c_{liq}$ = 4.18 J/(g·°C), and $c_{steam}$ = 2.0 J/(g·°C).

The heat required for this process is the sum of the amounts of heat needed to raise the temperature of the block of ice from –15 °C to 0 °C, melt the ice at 0 °C, raise the temperature of the liquid water from 0 °C to 100 °C, convert the liquid to steam at 100. °C, and then raise the temperature of the steam from 100. to 102 °C.

$$q = mc_{ice}\Delta T + n\Delta H_{fus} + mc_{liq}\Delta T + n\Delta H_{vap} + mc_{steam}\Delta T$$

$$q = [(25.0\ \cancel{g})[2.1\ \text{J}/(\cancel{g}\cdot\cancel{°C})](0\ \cancel{°C} - (-15\ \cancel{°C}))] + \left[\left(25.0\ \cancel{g} \times \frac{1\ \cancel{mol}}{18.02\ \cancel{g}}\right)(6.01\ \cancel{kJ}\cdot\cancel{mol^{-1}})\left(\frac{1000\ \text{J}}{1\ \cancel{kJ}}\right)\right] +$$

$$[(25.0\ \cancel{g})[4.18\ \text{J}/(\cancel{g}\cdot\cancel{°C})](100\ \cancel{°C} - 0\cancel{°C})] + \left[\left(25.0\ \cancel{g} \times \frac{1\ \cancel{mol}}{18.02\ \cancel{g}}\right)(40.7\ \cancel{kJ}\cdot\cancel{mol^{-1}})\left(\frac{1000\ \text{J}}{1\ \cancel{kJ}}\right)\right] +$$

$$[(25.0\ \cancel{g})[2.0\ \text{J}/(\cancel{g}\cdot\cancel{°C})](102\ \cancel{°C} - 100\ \cancel{°C})]$$

$$q = 76{,}100\ \text{J}$$

A total of 76,100 J (or 76.1 kJ) of energy are required to convert 25.0 grams of ice at –15 °C to steam at 102 °C.

## Practice

The heating curve for a pure substance at a given pressure is shown below.

**a.** What are the melting and boiling points of the substance?
**b.** At which points on the heating curve are chemical bonds broken?
**c.** Is the specific heat capacity for the substance greater in its liquid state or its solid state?

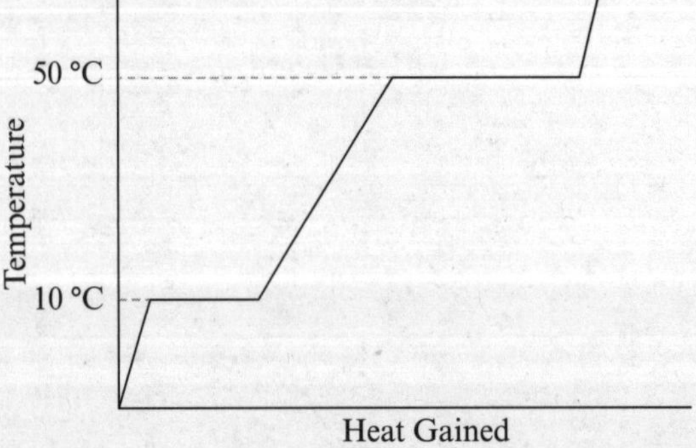

**(a)** The melting point of the substance is 10 °C. The first plateau in the curve represents the conversion of solid to liquid. This occurs at the melting temperature. The boiling point is 50 °C. The second plateau represents the conversion of liquid to gas at the boiling point.

**(b)** The heating of the substance involves a physical process. The average kinetic energy of the particles is increasing, and during melting and boiling, intermolecular forces are being disrupted, but at no point are chemical bonds broken.

**(c)** The slope of the heating curve of the liquid state is less steep, which means it requires more heat to raise the temperature of the substance in the liquid state by 1 °C than it does to raise the temperature of the substance in the solid state by 1 °C. The specific heat capacity of the liquid form is greater than the specific heat capacity of the solid substance.

## Practice

A block of ice with a mass of 10.0 grams and a temperature of –10.0 °C is added to 100.0 mL of water at 40.0 °C. What is the final temperature of the mixture? ($\Delta H_{fus}$ = 6.01 kJ/mol, $c_{ice}$ = 2.1 J/(g·°C), and $c_{liq}$ = 4.18 J/(g·°C))

This type of problem can be treated exactly like a calorimeter problem. The heat gained by the ice as it warms, melts to a liquid, and continues to warm until thermal equilibrium is reached is equal to the heat lost by the water until both are at the same temperature, $T_{final}$. The number of grams of water is equal to the volume of the water multiplied by the density of water, 1.00 g/mL. The number of moles of ice is the mass divided by the molar mass, 18.02 g/mol.

The energy lost by the 40.0 °C water is equal in magnitude to the energy that is gained by the ice.

$$q_{lost} = -q_{gained}$$
$$m_{liq}c_{liq}\Delta T = -(m_{ice}c_{ice}\Delta T + n_{ice}\Delta H + m_{liq}c_{liq}\Delta T)$$

$$\left[\left(100.\ \cancel{mL} \times \frac{1\ \cancel{g}}{1\ \cancel{mL}}\right)(4.18\ \cancel{J}/(\cancel{g\cdot{}^\circ C}))(T_f - 40.0\ ^\circ C)\right] = -\left[(10.0\ \cancel{g})(2.1\ \cancel{J}/(\cancel{g\cdot{}^\circ C}))(0^\circ C - (-10.0^\circ C))\right]$$
$$-\left[\left(10.0\ \cancel{g} \times \frac{1\ \cancel{mol}}{18.02\ \cancel{g}}\right)(6.01\ \cancel{kJ}/\cancel{mol})\left(\frac{1000\ \cancel{J}}{1\ \cancel{kJ}}\right)\right]$$
$$-\left[(10.0\ \cancel{g})(4.18\ \cancel{J}/(\cancel{g\cdot{}^\circ C}))(T_f - 0\ ^\circ C)\right]$$
$$T_f = 28.7\ ^\circ C$$

# Thermodynamics of Dissolution

Formation of a solution may be exothermic or endothermic.

During an exothermic dissolution, the temperature of the solvent increases as dissolving species release heat ($\Delta H < 0$).

During an endothermic dissolution, the temperature of the solvent decreases as dissolving species gain heat ($\Delta H > 0$).

Disruption of interactions (either ionic bonds or intermolecular forces) between particles in a dissolution requires an input of energy. Formation of interactions releases energy.

When a solution is formed, three processes contribute to the overall change in enthalpy:

1. Intermolecular forces or ionic bonds between solute particles are disrupted, requiring an input of heat ($\Delta H_1$).
2. Intermolecular forces between solvent particles are disrupted, requiring an input of heat ($\Delta H_2$).
3. Interactions between solute and solvent particles are formed, releasing heat ($\Delta H_3$).

The overall enthalpy change, $\Delta H_{solution}$, for the dissolution process is the sum of the enthalpy changes for the three processes.

$$\Delta H_{solution} = \Delta H_1 + \Delta H_2 + \Delta H_3$$

**The Enthalpy of Solution Formation**

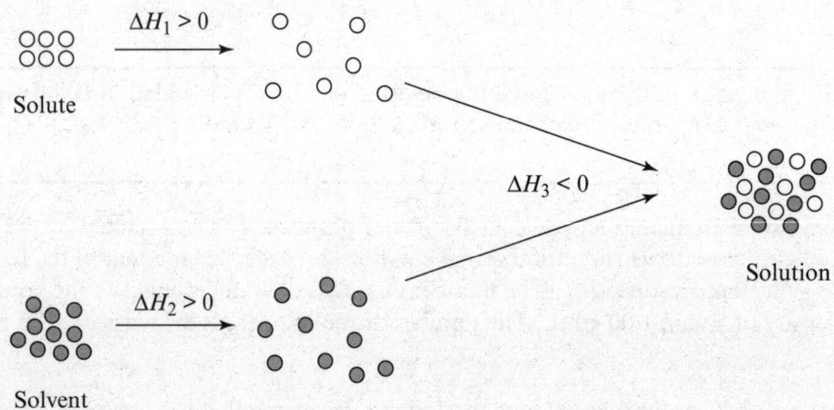


The enthalpy change associated with disrupting intermolecular forces between solute particles, in the case of an ionic solute, is the lattice energy, $\Delta H_L$. Recall that this is the energy required to separate the ionic compound into separate gas-phase ions.

$$\Delta H_1 = \Delta H_L$$

Usually, dilute solutions of ionic compounds are considered in AP Chemistry. *For a dilute solution, the enthalpy change associated with interruption of solvent-solvent interactions, $\Delta H_2$, is negligible and can be ignored.*

$$\Delta H_2 \approx 0$$

The enthalpy change associated with formation of interactions between solute and solvent particles is the enthalpy of hydration, $\Delta H_{hydration}$.

$$\Delta H_3 = \Delta H_{hydration}$$

Substituting these values into the equation for the enthalpy of solution gives the following equation.

$$\Delta H_{solution} = \Delta H_L + \Delta H_{hydration}$$

## Practice

The lattice energy of $CaCl_2$ is 2270 kJ/mol. Given the following data for the enthalpies of hydration of gaseous ions, will the temperature of an aqueous solution increase or decrease as $CaCl_2$ is dissolved? Justify your response with a calculation.

| Ion | $\Delta H_{hydration}$ (kJ/mol) |
|---|---|
| $Ca^{2+}$ | –1653 |
| $Cl^-$ | –338 |

The lattice energy, 2270 kJ/mol, is the enthalpy change for the dissociation of one mole of $CaCl_2$ into its constituent ions.

$$CaCl_2(s) \rightarrow Ca^{2+}(g) + 2\ Cl^-(g)$$

The overall enthalpy of hydration for a mole of $CaCl_2$ is the sum of the $\Delta H_{hydration}$ values for each mole of ions. One mole of $Ca^{2+}$ and two moles of $Cl^-$ are hydrated for each mole of $CaCl_2$ dissolved.

$$\Delta H_{hydration} = -1653 \text{ kJ/mol} + 2(-338 \text{ kJ/mol}) = -2329 \text{ kJ/mol}$$

The overall enthalpy change for the dissolution is

$$\Delta H_{solution} = \Delta H_L + \Delta H_{hydration}$$
$$\Delta H_{solution} = 2270 \text{ kJ/mol} - 2329 \text{ kJ/mol} = -59 \text{ kJ/mol}$$

The negative enthalpy change means that this is an exothermic process. The temperature of an aqueous solution will increase as $CaCl_2$ is dissolved.

# Estimation of Reaction Enthalpy

In addition to calorimetry experiments, there are several ways to estimate reaction enthalpy: using bond energies of reactants and products, using Hess's law, and using the standard heats of formation.

## Bond Energies

Energy is required to break bonds, and energy is released when bonds are formed. In other words, bond breaking is endothermic and bond formation is exothermic.

The average amount of energy required to break a particular type of bond is the **bond enthalpy** or **bond energy.** The higher the bond energy, the stronger the bond.

Bond energies can be used to estimate the enthalpy change of a reaction in which all reactants and products are in the gas phase. The energies of the bonds in the products are subtracted from the energies of the bonds in the reactants to estimate the enthalpy of reaction.

$$\Delta H_{rxn} = \Sigma \text{ bond energies of reactants} - \Sigma \text{ bond energies of products}$$
$$\Delta H_{rxn} = \Sigma \text{ energy of bonds broken} - \Sigma \text{ energy of bonds formed}$$

For example, suppose a chemist would like to know the enthalpy of reaction for the addition in the gas phase of HBr to ethylene, $C_2H_4$, to form bromoethane, $CH_3CH_2Br$. A table of bond energies provides the following data.

| Bond | Bond Energy (kJ/mol) |
|---|---|
| C–C | 348 |
| C=C | 612 |
| C–H | 412 |
| Br–H | 363 |
| C–B | 276 |

The Lewis structures for reactants and products show the number of each kind of bond present.

**Bromoethane Formation**

$$H_2C{=}CH_2 + H{-}Br \longrightarrow H{-}CH_2{-}CH_2{-}Br$$

The reactants contain one C=C bond, four C–H bonds, and one H–Br bond. The product contains one C–C single bond, five C–H bonds, and one C–Br bond. The enthalpy change is calculated using the tabulated values.

$$\Delta H_{rxn} = \Sigma\, BE_{reactants} - \Sigma\, BE_{products}$$

$$\Delta H_{rxn} = [(1 \times BE_{C=C}) + (4 \times BE_{C-H}) + (1 \times BE_{Br-H})] - [(1 \times BE_{C-C}) + (5 \times BE_{C-H}) + (1 \times BE_{C-Br})]$$

$$\Delta H_{rxn} = [(1 \times 612\text{ kJ/mol}) + (4 \times 412\text{ kJ/mol}) + (1 \times 363\text{ kJ/mol})] - [(1 \times 348\text{ kJ/mol}) + (5 \times 412\text{ kJ/mol}) + (1 \times 276\text{ kJ/mol})]$$

$$\Delta H_{rxn} = -61\text{ kJ/mol}$$

Notice that this technique works equally well when only the bonds that are broken and formed are considered. The broken bonds are the C=C bond and the H–Br bond. The bonds formed are the C–C, C–Br, and C–H bonds.

$$\Delta H_{rxn} = (\Sigma\text{ Energy of bonds broken}) - (\Sigma\text{ Energy of bonds formed})$$

$$\Delta H_{rxn} = [(1 \times 612\text{ kJ/mol}) + (1 \times 363\text{ kJ/mol})] - [(1 \times 348\text{ kJ/mol}) + (1 \times 412\text{ kJ/mol}) + (1 \times 276\text{ kJ/mol})]$$

$$\Delta H_{rxn} = -61\text{ kJ/mol}$$

## Practice

Use the tabulated bond energies below to find the enthalpy of reaction for the addition of nitrogen gas to fluorine gas to produce nitrogen trifluoride gas.

$$N_2 + 3\,F_2 \rightarrow 2\,NF_3$$

| Bond | Bond Energy (kJ/mol) |
|---|---|
| N≡N | 944 |
| F–F | 158 |
| N–F | 270 |

The Lewis structures of reactants and products should be determined first.

$$:N{\equiv}N: + 3\ :\ddot{F}{-}\ddot{F}: \longrightarrow 2\ :N(F)_3$$

There is one N–N triple bond and there are three F–F single bonds in the reactants. The product contains six N–F single bonds. The enthalpy of reaction is determined as follows.

$$\Delta H_{rxn} = [(1 \times 944\text{ kJ/mol}) + (3 \times 158\text{ kJ/mol})] - (6 \times 270\text{ kJ/mol})$$

$$\Delta H_{rxn} = -202\text{ kJ/mol}$$

## Hess's Law

A second method for calculating the enthalpy of a reaction employs Hess's law. **Hess's law** states that the enthalpy change for a process is equal to the sum of the enthalpy changes for reaction steps that add up to the process, *regardless of whether those steps were actually involved in the reaction mechanism.*

When using Hess's law to determine reaction enthalpy, the steps for which the enthalpy changes are known must be rearranged so that they add up to the overall process.

In order to use Hess's law to find overall reaction enthalpy, the enthalpies of the contributing reactions must be handled as follows:

- If a reaction is reversed, the sign of $\Delta H$ is reversed.
- If the stoichiometric coefficients must be multiplied by an integer, $\Delta H$ is also multiplied by that integer.

For instance, consider the decomposition of sulfur trioxide, $SO_3$.

$$2\,SO_3(g) \rightarrow 2\,SO_2(g) + O_2(g) \qquad \Delta H = ?$$

The following information is given.

$$S(s) + \frac{3}{2}O_2(g) \rightarrow SO_3(g) \qquad \Delta H = -395.7 \text{ kJ/mol}$$
$$S(s) + O_2(g) \rightarrow SO_2(g) \qquad \Delta H = -296.8 \text{ kJ/mol}$$

These two equations must be manipulated so that they add up to the overall transformation of interest. In this case, since $SO_3(g)$ is a reactant with a stoichiometric coefficient of 2 in the overall equation, the first equation must be reversed, and the enthalpy change must be multiplied by –2.

In the overall reaction, $SO_2(g)$ is a product with a stoichiometric coefficient of 2. In order for the two reactions to add up to the overall balanced equation, the second reaction and its enthalpy change must be multiplied by 2.

The sum of the two transformed enthalpy changes is $\Delta H$ for the overall reaction.

$$2\,SO_3(g) \rightarrow \cancel{2\,S(s)} + \frac{\cancel{6}^{\,2}}{2}O_2(g) \qquad \Delta H = -2(-395.7 \text{ kJ/mol})$$
$$\cancel{2\,S(s)} + \cancel{\frac{4}{2}O_2(g)} \rightarrow 2\,SO_2(g) \qquad \Delta H = 2(-296.8 \text{ kJ/mol})$$
$$\overline{2\,SO_3(g) \rightarrow 2\,SO_2(g) + O_2(g) \qquad \Delta H = 197.8 \text{ kJ/mol}}$$

### Practice

Determine the enthalpy of reaction for the reaction of HCl with $F_2$.

$$2\,HCl(g) + F_2(g) \rightarrow 2\,HF(l) + Cl_2(g) \qquad \Delta H = ?$$

Use the following chemical equations and associated enthalpy changes.

$$4\,HCl(g) + O_2(g) \rightarrow 2\,H_2O(l) + 2\,Cl_2(g) \qquad \Delta H = -202.4 \text{ kJ/mol}$$
$$H_2O(l) \rightarrow H_2(g) + \frac{1}{2}O_2(g) \qquad \Delta H = 285.8 \text{ kJ/mol}$$
$$HF(l) \rightarrow \frac{1}{2}H_2(g) + \frac{1}{2}F_2(g) \qquad \Delta H = 299.8 \text{ kJ/mol}$$

The first equation is the only one in which HCl appears. It is divided by 2 to result in 2 moles of HCl(*g*).

The third equation is the only one in which HF appears. Reversal and multiplication of the third equation by 2 results in the correct number of HF moles in the products.

Since $H_2O$, $H_2$, and $O_2$ do not appear in the overall reaction, the second equation is left as it is written in the problem so that $H_2O$, $H_2$, and $O_2$ cancel in the overall equation.

$$2\,\text{HCl}(g) + \tfrac{1}{2}\cancel{\text{O}_2(g)} \rightarrow \cancel{\text{H}_2\text{O}(l)} + \text{Cl}_2(g) \qquad \Delta H = (-202.4\ \text{kJ/mol})\tfrac{1}{2}$$

$$\cancel{\text{H}_2\text{O}(l)} \rightarrow \cancel{\text{H}_2(g)} + \tfrac{1}{2}\cancel{\text{O}_2(g)} \qquad \Delta H = 285.8\ \text{kJ/mol}$$

$$\cancel{\text{H}_2(g)} + \text{F}_2(g) \rightarrow 2\,\text{HF}(l) \qquad \Delta H = (-299.8\ \text{kJ/mol})2$$

$$2\,\text{HCl}(g) + \text{F}_2(g) \rightarrow 2\,\text{HF}(l) + \text{Cl}_2(g) \qquad \Delta H = -415.0\ \text{kJ/mol}$$

# Enthalpy of Formation

The third method for determining the enthalpy change for a reaction involves the use of standard molar heats of formation of the reactants.

The **standard molar heat of formation, $\Delta H^\circ_f$**, of a substance is the enthalpy change that occurs when one mole of a substance is formed from its constituent elements in their standard states. The **standard state** of an element is its form under thermodynamic standard conditions.

The standard molar heat of formation of a pure element in its standard state, such as $Cl_2(g)$, He(*g*), $P_4(s)$, or Zn(*s*), is zero.

In thermodynamics, standard conditions are defined as 1 atmosphere pressure or 1 atm of partial pressures in a gas mixture or 1 *M* solution. (Often the standard temperature reported in thermodynamic tables is 25 °C, unlike standard temperature for gas law calculations, which is 0 °C.)

In order to calculate the enthalpy of a reaction, the sum of the standard molar enthalpies of formation of the reactants is subtracted from the sum of the standard molar enthalpies of formation of the products.

$$\Delta H^\circ_{\text{rxn}} = \Sigma n\Delta H^\circ_{\text{f(products)}} - \Sigma n\Delta H^\circ_{\text{f(reactants)}}$$

The enthalpy of reaction for the combustion of propyne, $C_3H_4$, for example, can be calculated using the tabulated standard molar heats of formation. Notice that $O_2(g)$ is an element in its standard state, so its $\Delta H^\circ_f$ is zero.

$$\text{C}_3\text{H}_4(g) + 4\,\text{O}_2(g) \rightarrow 3\,\text{CO}_2(g) + 2\,\text{H}_2\text{O}(l)$$

| Substance | $\Delta H^\circ_f$ (kJ/mol) |
|---|---|
| $C_3H_4(g)$ | 184.9 |
| $CO_2(g)$ | –393.5 |
| $H_2O(l)$ | –285.8 |

The enthalpies of formation are multiplied by the number of moles, *n*.

$$\Delta H^\circ_{\text{rxn}} = \Sigma n\Delta H^\circ_{\text{f(products)}} - \Sigma n\Delta H^\circ_{\text{f(reactants)}}$$

$$\Delta H^\circ_{\text{rxn}} = [3\Delta H^\circ_{\text{f(CO}_2)} + 2\Delta H^\circ_{\text{f(H}_2\text{O)}}] - [1\Delta H^\circ_{\text{f(C}_3\text{H}_4)} + 4\Delta H^\circ_{\text{f(O}_2)}]$$

$$\Delta H^\circ_{\text{rxn}} = [(3(-393.5\ \text{kJ/mol})) + (2(-285.8\ \text{kJ/mol}))] - [(1(184.9\ \text{kJ/mol})) + (4(0\ \text{kJ/mol}))]$$

$$\Delta H^\circ_{\text{rxn}} = -1937\ \text{kJ/mol}$$

## Practice

Determine the amount of energy absorbed or released when 30. moles of solid iron are produced from iron(III) oxide by the following reaction.

$$Fe_2O_3(s) + 3\,CO(g) \rightarrow 3\,CO_2(g) + 2\,Fe(s)$$

| Substance | $\Delta H^\circ_f$ (kJ/mol) |
|---|---|
| $Fe_2O_3(s)$ | –824.2 |
| $CO_2(g)$ | –393.5 |
| $CO(g)$ | –110.5 |

The overall enthalpy change for the reaction can be determined using the standard molar enthalpies of formation.

$$\Delta H^\circ_{rxn} = \Sigma n\Delta H^\circ_{f(products)} - \Sigma n\Delta H^\circ_{f(reactants)}$$

$$\Delta H^\circ_{rxn} = [(3(-393.5\text{ kJ/mol})) + (2(0\text{ kJ/mol}))] - [(1(-824.2\text{ kJ/mol})) + (3(-110.5\text{ kJ/mol}))]$$

$$\Delta H^\circ_{rxn} = -25\text{ kJ/mol}$$

The enthalpy change "per mole of reaction" is the enthalpy change when two moles of iron are formed (using the stoichiometric coefficient). The amount of heat released when 30. moles of iron are formed is determined using the stoichiometry of the reaction.

$$30.\ \cancel{\text{mol Fe}} \times \frac{1\ \cancel{\text{mol}_{rxn}}}{2\ \cancel{\text{mol Fe}}} \times \frac{-25\text{ kJ}}{1\ \cancel{\text{mol}_{rxn}}} = -370\text{ kJ}$$

The amount of heat released for every 30. moles of iron formed is 372 kJ. Reported to two significant figures, this is 370 kJ.

Notice that the standard molar enthalpy change is not rounded until the end of the calculation. When the calculated value is used, the amount of heat released is 370 kJ rather than 380 kJ (rounded up from 375).

# Review Questions

## Multiple Choice

**1.** Which of these processes is exothermic in terms of the underlined system?

**A.** Sweat evaporates from skin.
**B.** Frost forms on cold surfaces in the winter.
**C.** Ice melts on a warm countertop.
**D.** Solid iodine crystals undergo sublimation to form a gas.

**2.** The temperature of a 12.0-gram cube of metal increases from 15.0 °C to 35.0 °C when 216 J of energy are added. Based on the following heat capacities, determine the identity of the metal.

| Element | Heat Capacity (J/(g·°C)) |
|---|---|
| Aluminum | 0.90 |
| Nickel | 0.44 |
| Tin | 0.21 |
| Lead | 0.16 |

**A.** Aluminum
**B.** Nickel
**C.** Tin
**D.** Lead

**3.** A 25.0-mL sample of a 0.500 *M* solution of aqueous sodium hydroxide is added to a 25.0-mL sample of a 0.500 *M* solution of aqueous hydrochloric acid in a coffee cup calorimeter. Both solutions are initially at 22.5 °C. After mixing, the temperature of the combined solution reaches a maximum of 26.0 °C. Which statement is true?

**A.** The negative value of $\Delta H_{rxn}$ that can be calculated from this experiment is more negative than the true value.
**B.** The positive value of $\Delta H_{rxn}$ that can be calculated from this experiment is more positive than the true value.
**C.** The negative value of $\Delta H_{rxn}$ that can be calculated from this experiment is less negative than the true value.
**D.** The positive value of $\Delta H_{rxn}$ that can be calculated from this experiment is less positive than the true value.

**4.** Which of the following objects would require input of the most heat in order to raise the temperature of the object by 1 °C?

| | Material | Heat Capacity (J/(g·°C) | Mass (g) |
|---|---|---|---|
| **A.** | Stainless steel spoon | 0.5 | 50 |
| **B.** | Aluminum foil square | 0.9 | 10 |
| **C.** | Ceramic tea cup | 0.7 | 300 |
| **D.** | Silicone spatula | 1.3 | 100 |

*Use the following equations to answer questions 5–6.*

The combustion of methane occurs by the following reaction.

$$CH_4(g) + 2\,O_2(g) \rightarrow CO_2(g) + 2\,H_2O(g) \qquad \Delta H = -804 \text{ kJ/mol}$$

| Reaction | $\Delta H$ (kJ/mol) |
|---|---|
| $2\,O(g) \rightarrow O_2(g)$ | $v$ |
| $2\,H(g) + O(g) \rightarrow H_2O(g)$ | $w$ |
| $C(graphite) + 2\,O(g) \rightarrow CO_2(g)$ | $x$ |
| $C(graphite) + 2\,H_2(g) \rightarrow CH_4(g)$ | $y$ |
| $2\,H(g) \rightarrow H_2(g)$ | $z$ |

**5.** Which expression represents the heat of combustion of methane?

**A.** $v + w + x + y + z$
**B.** $-v + w + x - y + z$
**C.** $2v + w + x - y + 2z$
**D.** $-2v + 2w + x - y - 2z$

**6.** Which of the values shown are greater than zero?

**A.** $w$, $x$, and $y$ only
**B.** $v$ and $z$ only
**C.** $v$, $w$, $x$, $y$, and $z$
**D.** None of the values are greater than zero.

*Use the following equation for the oxidation of iron to answer questions 7–8.*

$$4\,Fe + 3\,O_2 \rightarrow 2\,Fe_2O_3$$
$$\Delta H_{rxn} = -826 \text{ kJ/mol}_{rxn}$$

**7.** What is the enthalpy change when a 27.9-gram sample of iron reacts with 11.2 L of $O_2$ at STP?

**A.** –103 kJ
**B.** –207 kJ
**C.** –275 kJ
**D.** –413 kJ

**8.** Which of the following sketches best represents the potential energy profile for the reaction of iron with oxygen to form $Fe_2O_3$?

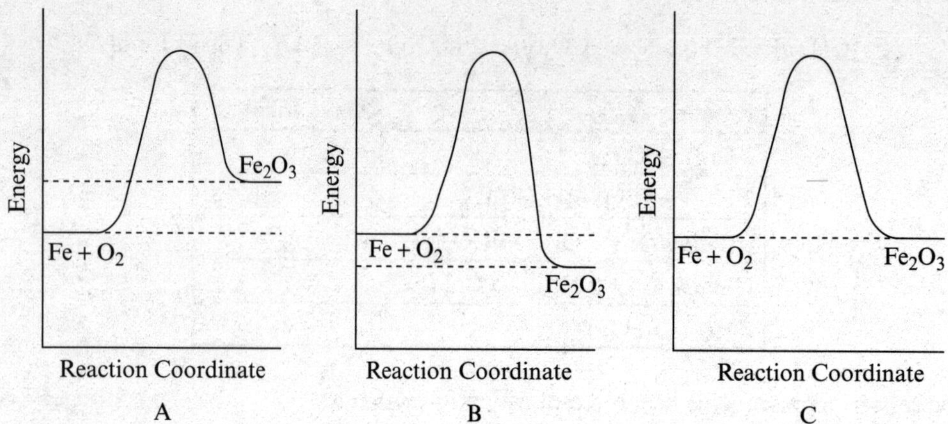


**A.** Energy profile A
**B.** Energy profile B
**C.** Energy profile C
**D.** The energy profile cannot be predicted based on the information given.

**9.** The sketch below represents the steps involved in the dissolution of a solid solute in a liquid solvent. For which (if any) of the steps is $\Delta H < 0$?

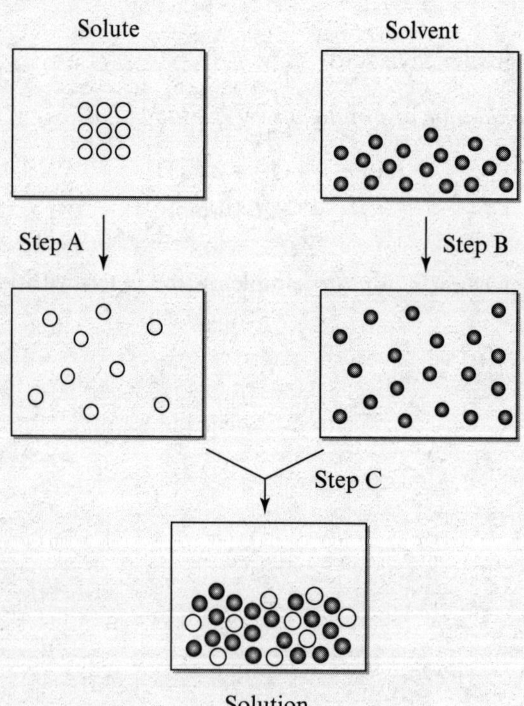


**A.** Step A
**B.** Step B
**C.** Step C
**D.** $\Delta H < 0$ for all three steps

**10.** Which of the following is true of carbon-nitrogen bonds?

**A.** Breaking a C=N double bond requires a greater input of energy than breaking a C–N single bond.
**B.** Breaking a C=N double bond requires a smaller input of energy than breaking a C–N single bond.
**C.** Breaking a C=N double bond releases a larger amount of energy than breaking a C–N single bond.
**D.** Breaking a C=N double bond releases a smaller amount of energy than breaking a C–N single bond.

*Questions 11–12 refer to the heating curve shown below.*

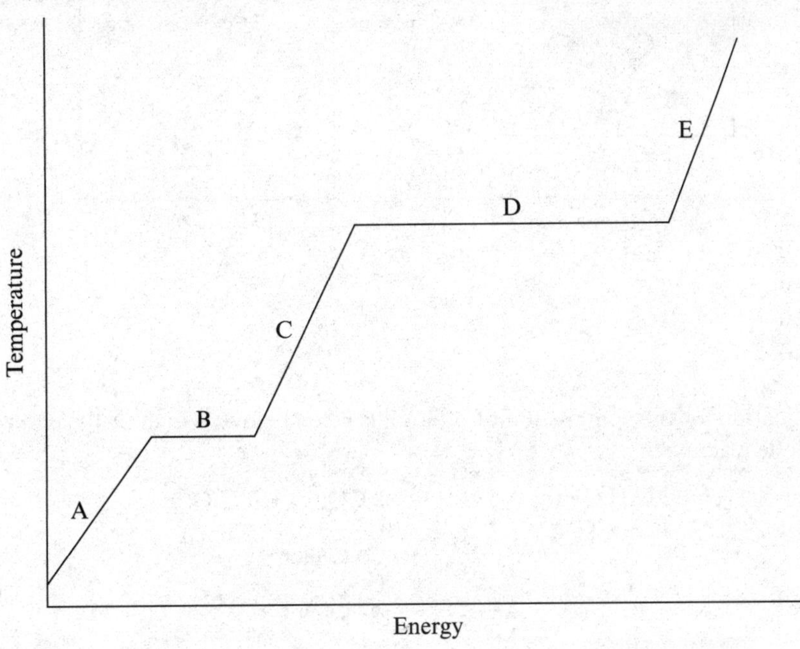


**11.** For which region on the heating curve does the heat transfer equal $mc_{\text{liq}}\Delta T$?

**A.** Region A
**B.** Region B
**C.** Region C
**D.** Region E

**12.** The heat of vaporization for a substance is the amount of energy that must be supplied to convert a substance from the liquid to the vapor phase. The length of which region of the curve is given by $n\Delta H°_{\text{vap}}$?

**A.** Region A
**B.** Region B
**C.** Region C
**D.** Region D

**13.** A 10-gram piece of metal is heated in an oven and then dropped into a 5-liter bucket of cool water. Which of the following best describes the result?

**A.** The water becomes warmer and the metal becomes cooler until both reach the same temperature, and the final temperature is closer to the initial temperature of the water.
**B.** The water becomes warmer and the metal becomes cooler until both reach the same temperature, and the final temperature is closer to the initial temperature of the metal.
**C.** The water becomes warmer and the metal becomes cooler until both reach the same temperature, and the final temperature is exactly halfway between the initial temperature of the metal and the initial temperature of the water.
**D.** The water stays at the same temperature, and the metal cools until it reaches the temperature of the water.

**14.** The following is a representation of a system undergoing a change at constant temperature. Select the answer choice that best describes the heat and work involved in this process.

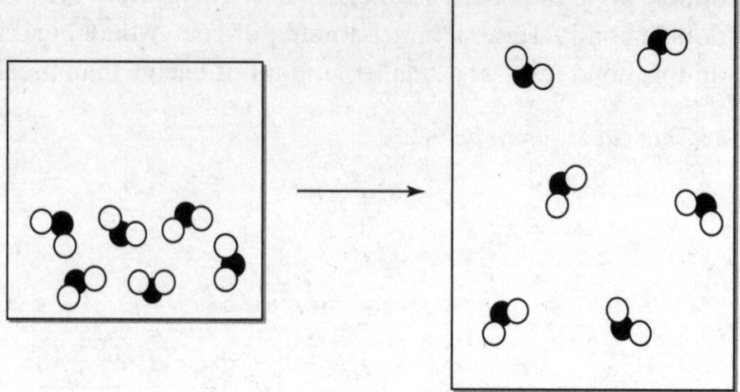

A. $q > 0, w > 0$
B. $q > 0, w < 0$
C. $q < 0, w < 0$
D. $q < 0, w > 0$

**15.** The chemical equation for the combustion of ethanol is given below. Which of the choices correctly reflects the enthalpy of the reaction?

$$C_2H_5OH(g) + 3\ O_2(g) \rightarrow 2\ CO_2(g) + 3\ H_2O(g)$$

$$\Delta H^\circ_{f(C_2H_5OH)} = -278 \text{ kJ/mol}$$

$$\Delta H^\circ_{f(CO_2)} = -394 \text{ kJ/mol}$$

$$\Delta H^\circ_{f(H_2O)} = -242 \text{ kJ/mol}$$

A. 1236 kJ/mol
B. 1792 kJ/mol
C. –1236 kJ/mol
D. –1792 kJ/mol

# Long Free Response

**1.** Consider the reaction of graphite with oxygen to form carbon dioxide.

$$C(gr) + \frac{1}{2}O_2(g) \rightarrow CO(g) \qquad \Delta H = ?$$

a. Is the work done on or by the system positive, negative, or approximately zero? Explain your response.
b. The following two known reaction enthalpies can be used to elucidate the enthalpy of reaction for the oxidation of graphite to carbon dioxide. Show how the reactions may be manipulated to add up to the overall equation.

$$C(gr) + O_2(g) \rightarrow CO_2(g) \qquad \Delta H = -393.5 \text{ kJ/mol}$$

$$CO(g) + \frac{1}{2}O_2(g) \rightarrow CO_2(g) \qquad \Delta H = -283.0 \text{ kJ/mol}$$

c. Calculate $\Delta H_{rxn}$.
d. If 5.0 L of $O_2(g)$ is available at 30.0 °C and 1.02 atm, how many moles of $O_2$ are there?
e. If 2.00 grams of graphite are available to react with the $O_2$ in part d, which is the limiting reagent?
f. What is $\Delta H$ for this reaction if the limiting reagent is consumed completely?

## Short Free Response

**2.** A student constructed a coffee cup calorimeter in order to determine the enthalpy change for the reaction of ammonia, $NH_3(aq)$, and hydrochloric acid, $HCl(aq)$, to form $NH_4Cl(aq)$.

When 50.0 mL of 2.00 *M* aqueous $NH_3$ were added to 50.0 mL of 2.00 *M* aqueous HCl in the calorimeter, the temperature reached 33.1 °C. Both solutions were initially at 23.0 °C.

Assume that the density of all solutions is 1.00 g/mL and the heat capacity of all solutions is the same as that of water, $c = 4.18$ J/(g·°C).

**a.** Write the net ionic equation for the reaction.

**b.** What is the enthalpy change in kilojoules per mole of reaction? Assume that the heat transferred between the system and the surroundings is negligible.

**c.** Would the calculated value for the enthalpy change be higher or lower than the actual value if a significant amount of heat transfer occurred between the solution and the calorimeter? Discuss your response.

**d.** The student uses chemical literature resources to find the standard molar enthalpy of reaction for an aqueous acid-base reaction.

$$H^+(aq) + OH^-(aq) \rightarrow H_2O(l) \qquad \Delta H^\circ = -54 \text{ kJ/mol}$$

Using this information and the information in parts a and b, determine the enthalpy change for the reaction below.

$$NH_3(aq) + H_2O(l) \rightarrow NH_4^+(aq) + OH^-(aq)$$

# Answers and Explanations

## Multiple Choice

**1. B.** The formation of frost, choice B, is an example of sublimation, a phase change in which water vapor in the air transitions directly to the gas phase without passing through the liquid phase. This is the only choice in which the energy of the product (ice) is lower than the energy of the reactant (water vapor), so this is the only exothermic process.

**2. A.** The heat capacity of the block of metal can be found by rearranging the equation for heat transfer, $q = mc\Delta T$.

$$c = \frac{q}{m\Delta T} = \frac{216 \text{ J}}{(12.0 \text{ g})(20 \text{ °C})} = 0.90 \text{ J/(g·°C)}$$

This is the heat capacity of aluminum, choice A. Without a calculator, one suggestion for estimating this value is to start by expanding the numerator and the denominator.

$$\frac{\cancel{2}(108)}{(12)(\cancel{2})(10)} = \frac{\cancel{10}(10.8)}{(12)(\cancel{10})} \approx \frac{11}{12}$$

The numerical heat capacity value of 0.90 is the closest choice to $\frac{11}{12}$.

**3. C.** The increase in temperature shows that the reaction is exothermic. An exothermic reaction has a negative enthalpy change, $\Delta H < 0$. Because some of the heat given off by the reaction is lost to the calorimeter and the surroundings, the temperature change that is measured is artificially low. This results in a less negative enthalpy change than the actual $\Delta H$ for the reaction, choice C, as calculated using

$$\Delta H_{rxn} = \frac{q_{rxn}}{n} = \frac{q_{soln}}{n} = \frac{-mc\Delta T}{n}$$

**4.** **C.** The heat required to raise the temperature of each object by 1 °C can be found using the equation $q = mc\Delta T$, where $m$ is the mass of the object, $c$ is the specific heat capacity of the material, and $\Delta T$ is the change in temperature, or 1 °C. The largest amount of heat is required for the ceramic tea cup, choice C.

$$q = mc\Delta T = (300\ \cancel{g})[0.7\ \text{J}/(\cancel{g \cdot {}^\circ C})](1\ \cancel{^\circ C}) = 210\ \text{J}$$

**5.** **D.** The reactions must be arranged so that their sum results in the overall chemical equation. If the reaction is reversed, its enthalpy sign is reversed. If the coefficients are multiplied by a factor, its enthalpy is also multiplied by that factor. It is simplest to begin with the reaction that contains $CH_4$.

| | |
|---|---|
| $CH_4(g) \rightarrow \cancel{C(graphite)} + \cancel{2\,H_2(g)}$ | $-y$ |
| $\cancel{C(graphite)} + \cancel{2\,O(g)} \rightarrow CO_2(g)$ | $x$ |
| $2[O_2(g) \rightarrow \cancel{2\,O(g)}]$ | $-2v$ |
| $2[\cancel{H_2(g)} \rightarrow \cancel{2\,H(g)}]$ | $-2z$ |
| $2[\cancel{2\,H(g)} + \cancel{O(g)} \rightarrow H_2O(g)]$ | $2w$ |
| $CH_4(g) + 2\,O_2(g) \rightarrow CO_2(g) + 2\,H_2O(g)$ | $\Delta H = -2v + 2w + x - y - 2z$, choice D |

**6.** **D.** As the reactions are written, all have enthalpy changes that are less than zero, so choice D is correct. Energy is released when bonds are formed, corresponding to negative values of $\Delta H$.

**7.** **B.** The equation for the reaction must be balanced first.

$$4\,Fe + 3\,O_2 \rightarrow 2\,Fe_2O_3$$

The problem gives the amount of heat released by the formation of two moles of $Fe_2O_3$, as indicated by the negative value of the enthalpy change. The next step in the solution is determining the number of moles of $Fe_2O_3$ that are generated based on the amounts of starting materials, 27.9 grams of iron and 11.2 L of $O_2$ at STP.

$$27.9\ \cancel{g\,Fe} \times \frac{1\ \cancel{mol\,Fe}}{55.85\ \cancel{g\,Fe}} \times \frac{2\ mol\ Fe_2O_3}{4\ \cancel{mol\,Fe}} = 0.250\ mol\ Fe_2O_3$$

$$11.2\ \cancel{L\,O_2} \times \frac{1\ \cancel{mol\,O_2}}{22.4\ \cancel{L\,O_2}} \times \frac{2\ mol\ Fe_2O_3}{3\ \cancel{mol\,O_2}} = 0.333\ mol\ Fe_2O_3$$

Iron is the limiting reagent. The amount of heat released is calculated from the theoretical yield of $Fe_2O_3$.

$$0.250\ \cancel{mol\,Fe_2O_3} \times \frac{-826\ kJ}{2\ \cancel{mol\,Fe_2O_3}} = -103\ \text{kJ released, choice B}$$

**8.** **B.** Energy profile B represents an exothermic reaction. The negative enthalpy change indicates a product energy that is lower than the energy of the reactants, so heat has been released over the course of the reaction.

**9.** **C.** Step C is the only step in which intermolecular attractions are causing particles to move closer to one another. As particles approach one another as the result of intermolecular attractions, potential energy decreases and the process is exothermic ($\Delta H < 0$).

**10.** **A.** Bond breaking requires an input of energy. Multiple bonds are stronger than single bonds, so breaking a C=N double bond requires a greater input of energy than breaking a C–N single bond, choice A.

**11.** **C.** Region C shows the rise in temperature that occurs as energy is added to the liquid phase of the substance. The heat transferred to or from the substance in the liquid phase is given by

$$q = mc_{\text{liquid}}\Delta T.$$

**12.** **D.** Region D shows the plateau in temperature that occurs as added energy causes the liquid phase of the substance to be converted to gas. The amount of heat required to vaporize a given number of moles of the liquid is given by

$$q = n\Delta H_{vap}$$

**13.** **A.** When two materials of different temperatures are placed in sustained contact with one another, the temperature of the warm material will decrease and the temperature of the cool material will increase until the materials are at the same temperature, or thermal equilibrium, choice A. Water has a higher heat capacity than most metals, so a quantity of energy transferred changes its temperature less than the same energy transfer for a metal.

**14.** **B.** It is apparent that work is being done on the surroundings by the system and that $w < 0$ because the volume of the sample is increasing at constant temperature. Energy is required to disrupt the intermolecular forces between the molecules as the substance is converted from liquid to gas, so heat is absorbed by the system, $q > 0$. Therefore, choice B is correct.

**15.** **C.** The enthalpy of reaction is found using the following equation.

$$\Delta H^\circ_{rxn} = \Sigma n\Delta H^\circ_{f(products)} - \Sigma n\Delta H^\circ_{f(reactants)}$$

$$\Delta H^\circ_{rxn} = [2\Delta H^\circ_{f(CO_2)} + 3\Delta H^\circ_{f(H_2O)}] - [1\Delta H^\circ_{f(C_2H_5OH)} + 3\Delta H^\circ_{f(O_2)}]$$

$$\Delta H^\circ_{rxn} = [(2(-394\text{ kJ/mol})) + (3(-242\text{ kJ/mol}))] - [(1(-278\text{ kJ/mol})) + (3(0\text{ kJ/mol}))]$$

$$\Delta H^\circ_{rxn} = -1236\text{ kJ/mol, choice C}$$

## Long Free Response

**1.** **a.** Work is done by the system, since the number of moles of gaseous products is greater than the number of moles of gaseous reactants. Therefore, $w < 0$.

**b.** Graphite, C(*gr*), is a reactant in the overall equation, so the first step should be used as written. The second equation, however, needs to be reversed so that CO(*g*) is a product. When the equation is reversed, the sign of the enthalpy change for the second equation is multiplied by –1.

$$C(gr) + \frac{\cancel{2}^{1}}{2}O_2(g) \rightarrow \cancel{CO_2(g)}$$

$$\cancel{CO_2(g)} \rightarrow CO(g) + \frac{1}{\cancel{2}}\cancel{O_2(g)}$$

$$C(gr) + \frac{1}{2}O_2(g) \rightarrow CO(g)$$

**c.** The sum of the two steps results in the equation that represents the overall reaction between graphite and oxygen to form carbon monoxide. So, the sum of the enthalpy changes gives the enthalpy of this reaction.

$$\Delta H = -393.5\text{ kJ/mol}$$

$$+\ \Delta H = -(-283.0\text{ kJ/mol})$$

$$\Delta H_{rxn} = -110.5\text{ kJ/mol}$$

**d.** To find the number of moles of $O_2$, we need to use the ideal gas law. Note that the temperature must be converted to Kelvin.

$$\begin{aligned} T &= 273.15\text{ K} + 30.0 = 303.2\text{ K} \\ PV &= nRT \\ n &= \frac{PV}{RT} \\ &= \frac{(1\ \cancel{\text{atm}})(5.0\ \cancel{\text{L}})}{\left(0.08206\ \frac{\cancel{\text{L}}\cdot\cancel{\text{atm}}}{\text{mol}\cdot\cancel{\text{K}}}\right)(303.2\ \cancel{\text{K}})} \\ &= 0.20\text{ mol } O_2 \end{aligned}$$

**e.** To determine the limiting reagent, the number of moles of C(*gr*) must be determined first.

$$2.00\ \cancel{\text{g C}} \times \frac{1\text{ mol C}}{12.01\ \cancel{\text{g C}}} = 0.167\text{ mol C}$$

Next, we determine the number of moles of product that could be produced by each of the available reactants.

$$0.20\ \cancel{\text{mol } O_2} \times \frac{1\text{ mol CO}}{\frac{1}{2}\ \cancel{\text{mol } O_2}} = 0.40\text{ mol CO}$$

$$0.167\ \cancel{\text{mol C}} \times \frac{1\text{ mol CO}}{1\ \cancel{\text{mol C}}} = 0.167\text{ mol CO}$$

Because the possible number of moles of CO formed is smaller for the graphite than for the oxygen, graphite is the limiting reagent.

**f.** We can determine the $\Delta H$ for the reaction using the theoretical yield of CO and the $\Delta H_{rxn}$ calculated in part c.

$$0.167\ \cancel{\text{mol CO}} \times \frac{1\ \cancel{\text{mol}_{rxn}}}{1\ \cancel{\text{mol CO}}} \times \frac{-110.5\text{ kJ}}{1\ \cancel{\text{mol}_{rxn}}} = -18.5\text{ kJ}$$

# Short Free Response

**2. a.** Ammonia is a weak base. It reacts with a strong acid to form an ammonium salt. Ammonium salts are soluble in aqueous solutions. The overall equation is

$$NH_3(aq) + HCl(aq) \rightarrow NH_4Cl(aq)$$

The total ionic equation is

$$NH_3(aq) + H^+(aq) + \cancel{Cl^-(aq)} \rightarrow NH_4^+(aq) + \cancel{Cl^-(aq)}$$

The net ionic equation is

$$NH_3(aq) + H^+(aq) \rightarrow NH_4^+(aq)$$

**b.** The total mass of the solution is given by

$$50.0 \text{ mL} + 50.0 \text{ mL} = 100.0 \text{ mL of the final solution}$$

$$100.0 \cancel{\text{mL}} \times \frac{1.00 \text{ g}}{1 \cancel{\text{mL}}} = 100.0 \text{ g}$$

The number of moles of either reactant is

$$50.0 \cancel{\text{mL}} \times \frac{1 \cancel{\text{L}}}{1000 \cancel{\text{mL}}} \times \frac{2.00 \text{ mol}}{1 \cancel{\text{L}}} = 0.100 \text{ mol}$$

The heat, $q$, given off by this reaction divided by the number of moles, $n$, of either reactant is equal to the enthalpy change for the process.

$$\Delta H_{rxn} = \frac{q_{rxn}}{n} = \frac{-mc\Delta T}{n} = \frac{-(100.0 \cancel{\text{g}})[4.18 \cancel{\text{J}}/(\cancel{\text{g}} \cdot \cancel{^\circ\text{C}})](33.1 \cancel{^\circ\text{C}} - 23.0 \cancel{^\circ\text{C}})}{0.100 \text{ mol}} \times \frac{1 \text{ kJ}}{1000 \cancel{\text{J}}} = -42.2 \text{ kJ/mol}$$

**c.** The calculated value of the enthalpy change is less negative (higher) than the actual value because some of the heat released in the reaction is absorbed by the calorimeter and does not result in a temperature change for the solution.

**d.** The equation for the reaction of $NH_3(aq)$ with $H^+(aq)$ can be added to the reverse of the equation for the reaction of $OH^-(aq)$ with $H^+(aq)$ to result in an equation representing the overall reaction in question. The sum of the enthalpies results in the enthalpy change for the overall reaction.

| | |
|---|---|
| $NH_3(aq) + \cancel{H^+(aq)} \rightarrow NH_4^+(aq)$ | $\Delta H = -42$ kJ/mol |
| $H_2O(l) \rightarrow \cancel{H^+(aq)} + OH^-(aq)$ | $\Delta H = (-1)(-54$ kJ/mol$)$ |
| $NH_3(aq) + H_2O(l) \rightarrow NH_4^+(aq) + OH^-(aq)$ | $\Delta H = 12$ kJ/mol |

# Chapter 7

# Chemical Equilibrium

**Chemical reactions are reversible and proceed to equilibrium, the condition in which the rate of the forward reaction and the rate of the reverse reaction are equal.**

Although some chemical reactions proceed almost to completion, in most chemical reactions, reactants are not completely converted to products.

In addition, reactions are **reversible.** The forward and the reverse reactions occur simultaneously, though they may occur at different rates over the course of the reaction.

An example of a reversible process is the freezing and melting of water. A sample of water will freeze when the temperature drops below 0 °C at one atmosphere, and it will melt when the temperature rises above 0 °C. An equilibrium arrow denotes reversibility.

$$H_2O(l) \rightleftharpoons H_2O(s)$$

**Chemical equilibrium** refers to the state in which a reversible reaction occurs with equal rates in the forward and reverse directions. *The reaction has not stopped; the molecules are continually reacting, but the concentrations of reactants and products are no longer changing.*

Consider the following reversible reaction.

$$A + B \rightleftharpoons C$$

Suppose reactants A and B are combined in a reaction vessel. None of the product, C, is initially present. The reaction begins to proceed with a faster rate in the forward direction than in the reverse direction.

$$A + B \rightarrow C$$

As C forms, it converts back to A and B in the reverse direction, but until equilibrium is established, the rate of this reverse reaction is slower than the rate of the forward reaction.

$$C \rightarrow A + B$$

As equilibrium is approached, the rate of the forward reaction slows and the rate of the reverse reaction increases until equilibrium is reached. At equilibrium, the rates of forward reaction and reverse reaction are equal and the concentrations have stopped changing.

**The Disappearance of A and Appearance of C as Equilibrium Is Established**

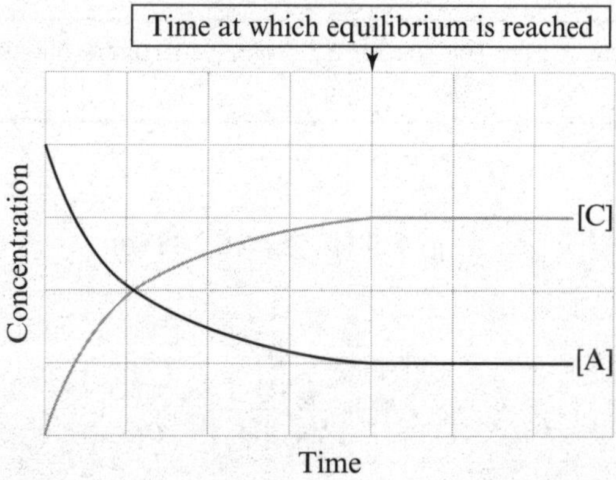

# The Equilibrium Constant

Consider the following equilibrium reaction.

$$aA + bB \rightleftharpoons cC + dD$$

At equilibrium, the concentrations are no longer changing, so the ratio of products to reactants is constant.

The **equilibrium constant, $K_{eq}$,** is the constant for this reaction that describes the equilibrium concentration ratio. Sometimes, the equilibrium constant is expressed as $K_c$. The subscript, c, in $K_c$ denotes a *concentration* equilibrium constant. Brackets around the reactants and products in the equilibrium expression denote concentrations. The equilibrium constant is determined experimentally for a reaction at a particular temperature.

$$K_{eq} = \frac{[C]^c[D]^d}{[A]^a[B]^b}$$

or

$$K_c = \frac{[C]^c[D]^d}{[A]^a[B]^b}$$

The concentrations of the reactants and products at a particular temperature are described by the mass action expression. Note that in the mass action expression, $\frac{[C]^c[D]^d}{[A]^a[B]^b}$, each concentration is raised to the power of its coefficient in the balanced chemical equation.

- For a given reaction, only a change in temperature can change the equilibrium constant.
- The equilibrium constant is independent of the initial concentrations of reactants and products.
- For a reaction in which $K > 1$, products are favored over reactants at equilibrium.
- For a reaction in which $K < 1$, reactants are favored over products at equilibrium.
- The larger the value of $K$, the larger the equilibrium concentrations of products.
- The smaller the value of $K$, the larger the equilibrium concentrations of reactants.
- *Only dissolved aqueous substances and gases appear in the mass action expression.*

No solids or liquids contribute to the equilibrium expression. For the purposes of AP Chemistry, the equilibrium constant is unitless. Furthermore, no calculations involving dissolved species in equilibrium with gaseous species will be assessed.

## Practice

Write the mass action expression, $K_c$, for the reaction between nitrogen gas and hydrogen gas to form gaseous ammonia.

The balanced chemical equation is

$$N_2(g) + 3\ H_2(g) \rightleftharpoons 2\ NH_3(g)$$

The mass action expression is

$$K_c = \frac{[NH_3]^2}{[N_2][H_2]^3}$$

## RICE Tables

Often the equilibrium constant is used along with a RICE table to determine equilibrium concentrations of the species in the reaction. A **RICE table** is a tool that can be used to determine equilibrium concentrations. RICE stands for Reaction, Initial concentration, Change in concentration, and Equilibrium concentration.

For example, suppose 3.00 moles of $A_2$ are added to 3.00 moles of $B_2$ in a 1.00-L container. The reactants $A_2$ and $B_2$ form the product AB. The equilibrium constant at this temperature, $K_c$, is 25.0.

$$A_2 + B_2 \rightleftharpoons 2\,AB$$

$$K_c = \frac{[AB]^2}{[A_2][B_2]}$$

At the beginning of the reaction, the Initial molarities of the reactants, $[A_2]$ and $[B_2]$, are known to be 3.00 *M*. There is no product, so initially [AB] is zero.

Let $x$ equal the magnitude of the decrease in the molarities of $A_2$ and $B_2$. In other words, it is the Change in $A_2$ and $B_2$.

According to the stoichiometry of the reaction, the concentration of AB increases by $+2x$. In other words, the concentration of AB increases twice as much as the concentration of $A_2$ or $B_2$ decreases.

The Equilibrium concentrations of $A_2$ and $B_2$ are $3.00 - x$, and the equilibrium concentration of AB is equal to $2x$.

| Reaction | $A_2$ | $B_2$ | 2 AB |
|---|---|---|---|
| Initial Concentrations (*M*) | 3.00 | 3.00 | 0 |
| Change | $-x$ | $-x$ | $+2x$ |
| Equilibrium Concentrations (*M*) | $3.00 - x$ | $3.00 - x$ | $2x$ |

The equilibrium concentrations are substituted into the mass action expression along with the equilibrium constant; the equation is then solved for $x$.

$$K_c = \frac{[AB]^2}{[A_2][B_2]}$$

$$25.0 = \frac{(2x)^2}{(3.00 - x)^2}$$

In this example, $x$ is determined by taking the square root of both sides of the equation.

$$\sqrt{25.0} = \sqrt{\frac{(2x)^2}{(3.00 - x)^2}}$$

$$5.00 = \frac{2x}{3.00 - x}$$

$$15.0 - 5.00x = 2x$$

$$15.0 = 7.00x$$

$$x = 2.14 \qquad (2.142857143)$$

The unrounded value of $x$ is used to calculate $[A_2]$, $[B_2]$, and [AB].

$$[A_2] = [B_2] = 3.00\ M - x = (3.00 - 2.142857143)\ M = 0.86\ M$$

$$[AB] = 2x = 2(2.142857143\ M) = 4.29\ M$$

The equilibrium expression is not always as easily solved as in the example above. In some problems, it is necessary to use the quadratic formula or perform even more labor-intensive operations in order to solve for $x$. This is extremely unlikely to ever be the case on the AP Chemistry exam. It is much more likely that problems resembling the following practice question will appear.

## Practice

A 5.00-mol sample of $SO_3(g)$ is placed in a rigid, evacuated 10.0-L container at 1000. K. When the reaction reaches equilibrium, the amount of $SO_3(g)$ has decreased to 3.18 moles. What is $K_c$ for the reaction at this temperature?

$$2\ SO_3(g) \rightleftharpoons O_2(g) + 2\ SO_2(g)$$

Because the volume of the container is 10.0 liters, the initial concentration of $SO_3(g)$ is

$$[SO_3]_{initial} = \frac{5.00\text{ mol}}{10.0\text{ L}} = 0.500\ M$$

The final concentration of $SO_3(g)$ is

$$[SO_3]_{final} = \frac{3.18\text{ mol}}{10.0\text{ L}} = 0.318\ M$$

The RICE table can be constructed to determine the equilibrium concentrations of the products.

| **Reaction** | **2 $SO_3$** | **$O_2$** | **2 $SO_2$** |
|---|---|---|---|
| **Initial Concentrations ($M$)** | 0.500 | 0 | 0 |
| **Change** | $-2x$ | $+x$ | $+2x$ |
| **Equilibrium Concentrations ($M$)** | 0.318 | $x$ | $2x$ |

There is enough information about the change in $[SO_3]$ to solve for $x$.

$$0.500 - 2x = 0.318$$
$$0.182 = 2x$$
$$x = 0.0910$$

The equilibrium concentrations are as follows.

$[SO_3] = 0.500 - 2x = 0.500 - 2(0.0910) = 0.318\ M$

$[O_2] = x = 0.0910\ M$

$[SO_2] = 2x = 2(0.0910) = 0.182\ M$

The equilibrium constant, $K_c$, is determined using the equilibrium concentrations.

$$K_c = \frac{[O_2][SO_2]^2}{[SO_3]^2} = \frac{(0.0910)(0.182)^2}{(0.318)^2} = 0.0298$$

# Other Forms of the Equilibrium Constant

Other forms of the equilibrium constant are also useful. For reactions in the gas phase, it is sometimes appropriate to determine an equilibrium constant, $K_p$, as a function of the partial pressures of the reactants and products. A RICE table involving $K_p$ should contain the partial pressures, $P$, for the reactants and products.

$$a\,A(g) + b\,B(g) \rightleftharpoons c\,C(g) + d\,D(g)$$

$$K_p = \frac{P_C^{\,c} P_D^{\,d}}{P_A^{\,a} P_B^{\,b}}$$

Other forms of the equilibrium constant include the solubility product, $K_{sp}$, the acid dissociation constant, $K_a$, and the base dissociation constant, $K_b$. $K_a$ and $K_b$ will be covered in more detail later in this chapter and in Chapter 8, "Acids and Bases."

## Practice

At 2117 K, the reaction of carbon dioxide with graphite proceeds with an equilibrium constant, $K_p$, of $8.60 \times 10^4$. Under equilibrium conditions in which the partial pressure of $CO(g)$ is 4.15 atm, what is the partial pressure of $CO_2(g)$?

$$C(s) + CO_2(g) \rightleftharpoons 2\,CO(g)$$

Graphite is a solid, so it is ignored, and the mass action expression for this reaction can be written and rearranged to solve for the partial pressure of $CO_2$.

$$K_p = \frac{P_{CO}^{\;2}}{P_{CO_2}}$$

$$P_{CO_2} = \frac{P_{CO}^{\;2}}{K_p} = \frac{(4.15)^2}{8.60 \times 10^4} = 2.00 \times 10^{-4} \text{ atm}$$

# Properties of the Equilibrium Constant

When equations representing chemical reactions are reversed, multiplied, or added, their equilibrium constants must be modified to reflect the new equations.

When the equation of a reaction is reversed, the new equilibrium constant is the reciprocal of the equilibrium constant for the original direction.

$$A + B \rightleftharpoons AB \qquad \text{where } K_{\text{forward}} = \frac{[AB]}{[A][B]}$$

$$AB \rightleftharpoons A + B \qquad K_{\text{reverse}} = \frac{[A][B]}{[AB]} = K_{\text{forward}}^{-1}$$

When an equation for a reaction is multiplied by a factor, its equilibrium constant is raised to the power of that factor.

$$A + B \rightleftharpoons AB \qquad \text{where } K_1 = \frac{[AB]}{[A][B]}$$

$$2\,A + 2\,B \rightleftharpoons 2\,AB \qquad K_2 = \frac{[AB]^2}{[A]^2[B]^2} = K_1^2$$

When two equations representing two chemical reactions are added, the equilibrium constant for the resulting equation is the product of the equilibrium constants of the constituent reactions.

$$A + B \rightleftharpoons AB \qquad \text{where } K_1 = \frac{[AB]}{[A][B]}$$

$$AB \rightleftharpoons C + D \qquad \text{where } K_2 = \frac{[C][D]}{[AB]}$$

$$A + B \rightleftharpoons C + D \qquad K = \frac{[C][D]}{[A][B]} = K_1 \times K_2$$

### Practice

Calculate the equilibrium constant for reaction 1, illustrated by the following equation:

$$\text{reaction 1: } A_3 \rightleftharpoons A + A_2$$

using the information given for reactions 2 and 3, illustrated by the following equations:

$$\text{reaction 2: } AB + A \rightleftharpoons A_2B \qquad K_2 = 4.5$$
$$\text{reaction 3: } A_3 + AB \rightleftharpoons A_2B + A_2 \qquad K_3 = 3.1$$

Equation 1 is the result of the reverse of equation 2 added to equation 3. The reciprocal of the equilibrium constant for reaction 2 is multiplied by the equilibrium constant for reaction 3.

$$\cancel{A_2B} \rightleftharpoons \cancel{AB} + A \qquad K_2^{-1} = 4.5^{-1}$$

$$A_3 + \cancel{AB} \rightleftharpoons \cancel{A_2B} + A_2 \qquad K_3 = 3.1$$

$$A_3 \rightleftharpoons A + A_2 \qquad K_1 = K_2^{-1} \times K_3$$

$$= \left(\frac{1}{4.5}\right) \times 3.1 = 0.69$$

## Measuring the Equilibrium Constant

Equilibrium constants are determined experimentally by measurement of the relative amounts of reactants and products at equilibrium.

A common laboratory exercise is the measurement of the equilibrium constant for the reaction between iron(III) ions, $Fe^{3+}$, and thiocyanate ions, $SCN^-$. The product that is formed, $FeSCN^{2+}$, is red, and its concentration can be determined by spectrophotometry.

$$Fe^{3+}(aq) + SCN^-(aq) \rightleftharpoons FeSCN^{2+}(aq)$$

$$K_c = \frac{[FeSCN^{2+}]}{[Fe^{3+}][SCN^-]}$$

### Practice

A student generated a standard calibration curve for the concentration of $FeSCN^{2+}$ using a spectrophotometer. The student then mixed 5.0 mL of $2.00 \times 10^{-3}$ *M* $Fe(NO_3)_3$ with 5.0 mL of $2.00 \times 10^{-3}$ *M* KSCN.

When the color stopped changing, the student determined that the concentration of $FeSCN^{2+}$ was $1.3 \times 10^{-4}$ *M*. What is the equilibrium constant, $K_c$, for this reaction?

The initial concentrations of $Fe^{3+}$ and $SCN^-$, immediately after mixing the two solutions and before any reaction has occurred, can be determined using the dilution formula.

Before the solutions were mixed, the concentration of $Fe^{3+}$ was $2.00 \times 10^{-3}$ *M*. The volume before mixing was 5.0 mL. The final volume of the solution after mixing is 10.0 mL.

$$M_{\text{initial}} \times V_{\text{initial}} = M_{\text{final}} \times V_{\text{final}}$$
$$(2.00 \times 10^{-3}\ M)(5.0\ \cancel{\text{mL}}) = M_{\text{final}}(10.0\ \cancel{\text{mL}})$$
$$M_{\text{final}} = 1.0 \times 10^{-3}\ M$$

The initial concentrations of $Fe^{3+}$ and $SCN^-$ are equal, $1.0 \times 10^{-3}$ *M*. We also know the final concentration of $FeSCN^{2+}$, $1.3 \times 10^{-4}$ *M*. A RICE table is used to determine the equilibrium concentrations.

| Reaction | $Fe^{3+}$ | $SCN^-$ | $FeSCN^{2+}$ |
|---|---|---|---|
| Initial Concentrations (*M*) | $1.0 \times 10^{-3}$ | $1.0 \times 10^{-3}$ | 0 |
| Change | $-1.3 \times 10^{-4}$ | $-1.3 \times 10^{-4}$ | $+1.3 \times 10^{-4}$ |
| Equilibrium Concentrations (*M*) | $8.7 \times 10^{-4}$ | $8.7 \times 10^{-4}$ | $1.3 \times 10^{-4}$ |

The equilibrium concentrations are used to determine the equilibrium constant.

$$K_c = \frac{[FeSCN^{2+}]}{[Fe^{3+}][SCN^-]} = \frac{(1.3 \times 10^{-4})}{(8.7 \times 10^{-4})(8.7 \times 10^{-4})} = 1.7 \times 10^2$$

## The Reaction Quotient

The **reaction quotient,** *Q*, is a measure of how different the current state of a reaction is from equilibrium at a given temperature.

The reaction quotient expression is the same as the equilibrium expression. It is calculated using actual concentrations and partial pressures for a reaction that isn't at equilibrium.

$$a\,\text{A} + b\,\text{B} \rightleftharpoons c\,\text{C} + d\,\text{D}$$

$$Q_C = \frac{[\text{C}]^c[\text{D}]^d}{[\text{A}]^a[\text{B}]^b}$$

or

$$Q_p = \frac{P_C{}^c P_D{}^d}{P_A{}^a P_B{}^b}$$

The reaction quotient is compared with *K* to determine whether a reaction will shift toward products or reactants to reach equilibrium:

- When $Q > K$, the reaction shifts toward reactants. The rate of the reverse reaction is greater than the rate of the forward reaction until equilibrium is established.
- When $Q < K$, the reaction shifts toward products. The rate of the forward reaction is greater than the rate of the reverse reaction until equilibrium is established.
- When $Q = K$, the reaction is at equilibrium. At equilibrium, the rates of the forward and reverse reactions are equal.

This relationship is illustrated for a hypothetical reaction for which $K = 0.75$. The use of particulate representations for the determination of $Q$ and $K$ is likely to appear on the AP Chemistry exam.

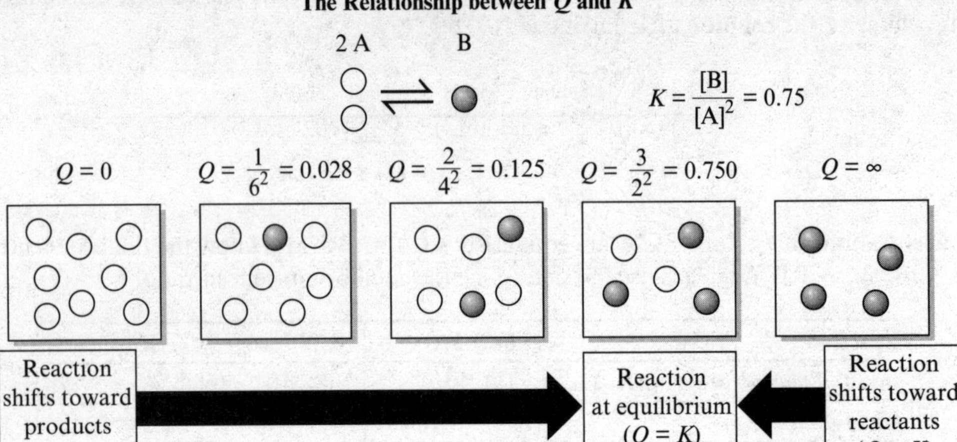


## Practice

The equilibrium constant, $K_p$, for the reaction of CO(*g*) with $Cl_2(g)$ at 1000. K is $2.9 \times 10^{-2}$.

$$CO(g) + Cl_2(g) \rightleftharpoons COCl_2(g)$$

The partial pressures of the gases are as follows:

$$P_{CO} = 0.40 \text{ atm}$$
$$P_{Cl_2} = 0.60 \text{ atm}$$
$$P_{COCl_2} = 0.020 \text{ atm}$$

Will the reaction shift to produce reactants or products?

The equilibrium quotient, $Q$, must be calculated and compared with $K_p$.

$$Q = \frac{P_{COCl_2}}{P_{CO}P_{Cl_2}} = \frac{(0.020)}{(0.40)(0.60)} = 8.3 \times 10^{-2}$$

Because $Q > K$, the reaction shifts toward reactants until equilibrium is established.

## Practice

For the reaction between reactants A and B to form C, $K_{eq} = 0.50$.

$$2\text{ A} + \text{B} \rightleftharpoons \text{C}$$

Examine the particle diagram showing the current state of the reaction below. Choose the option that best describes how the reaction will proceed.

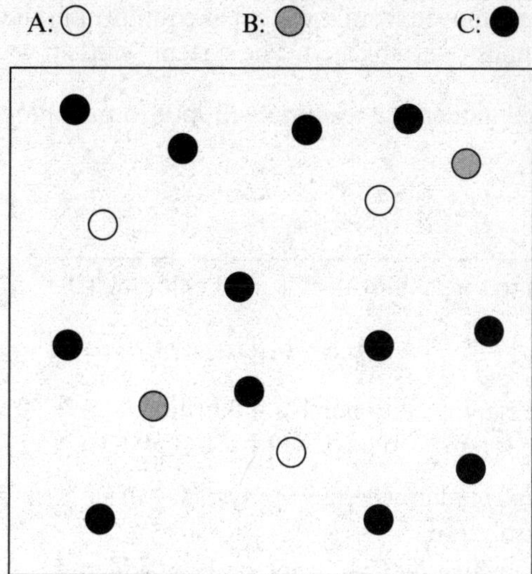


**A.** The reaction is at equilibrium.
**B.** The reaction will proceed in the direction of products until equilibrium is established.
**C.** The reaction will proceed in the direction of reactants until equilibrium is established.

---

The correct answer is choice C.

In the picture, there are 3 circles representing reactant A, 2 circles representing reactant B, and 12 circles representing reactant C. From this information, we can calculate $Q$.

$$Q = \frac{[C]}{[A]^2[B]} = \frac{12}{3^2 \times 2} = \frac{2}{3} = 0.67$$

Since $Q > K$, the reaction must proceed toward reactants to establish equilibrium.

# Le Chatelier's Principle

When a change is made to a system at equilibrium, the system will respond in a way that re-establishes the equilibrium. This is **Le Chatelier's principle,** and it is used to make qualitative predictions about how a system will react to stress.

Changes in conditions that shift a system from equilibrium may include

- addition or removal of a reactant or product, or
- a change in pressure if gases are produced or consumed

A change in temperature causes the equilibrium constant itself to change, but Le Chatelier's principle can still be used to qualitatively predict the effect of a change in temperature on a process that is exothermic or endothermic.

## Concentration Changes

When a reactant or product is added to a system at equilibrium, that reactant or product will be consumed until a new equilibrium state is reached. The reaction "shifts away" from the added species.

For example, if reactant A is added to the following reaction at equilibrium, C will be produced and A and B will be consumed until equilibrium is reached.

$$A + B \rightleftharpoons C$$

Likewise, if a reactant or product is removed from a system at equilibrium, the system will produce more of that reactant or product until equilibrium is re-established. The system "shifts toward" the removed species.

If C is removed from the reaction container, the reaction will shift toward products.

## Practice

Hypochlorite, $ClO^-$, decomposes to form chloride, $Cl^-$, and chlorate, $ClO_3^-$, in an aqueous solution.

$$3\ ClO^-(aq) \rightleftharpoons 2\ Cl^-(aq) + ClO_3^-(aq)$$

A 1.0-L solution of hypochlorite is allowed to reach equilibrium at 25 °C. The concentrations of the three species in a reaction vessel are determined to be $[ClO^-] = 1.0 \times 10^{-9}\ M$, $[Cl^-] = 2.0\ M$, and $[ClO_3^-] = 2.5\ M$.

**a.** Without doing any calculations, predict whether the system will shift toward reactants or products when 0.5 mole of NaCl is added to the solution.
**b.** Calculate $K_c$ for the reaction in which NaCl has not yet been added.
**c.** Calculate $Q_c$ for the reaction in which NaCl has been added.
**d.** Explain whether the calculated values of $K_c$ and $Q_c$ support the prediction made in part (a).

**(a)** When NaCl, a water-soluble salt, is added to the solution, it dissociates into $Na^+$ and $Cl^-$. While the $Na^+$ ions do not have any effect on the equilibrium, the $Cl^-$ ions do. According to Le Chatelier's principle, addition of one of the products ($Cl^-$) to the reaction vessel will cause the reaction to shift toward reactants in order to re-establish equilibrium.

**(b)** The value of $K_c$ is

$$K_c = \frac{[Cl^-]^2[ClO_3^-]}{[ClO^-]^3} = \frac{(2.0)^2(2.5)}{(1.0\times10^{-9})^3} = 1.0\times10^{28}$$

**(c)** The added 0.5 mole of $Cl^-$ results in a new $Cl^-$ concentration of 2.5 *M*. The value of $Q_c$ is

$$Q_c = \frac{[Cl^-]^2[ClO_3^-]}{[ClO^-]^3} = \frac{(2.5)^2(2.5)}{(1.0\times10^{-9})^3} = 1.6\times10^{28}$$

**(d)** Because $Q_c > K_c$, the reaction will shift toward reactants as predicted.

# Pressure Changes

If a container containing a reaction that produces or consumes a gas is compressed, the system will respond by shifting in the direction that reduces the number of moles of gas present.

For example, assume that A, B, and C are all gaseous species. An increase in the pressure in the reaction vessel that decreases the volume of the container will cause the reaction to shift toward products, and the total number of moles of gas present will decrease.

$$A(g) + B(g) \rightleftharpoons C(g)$$

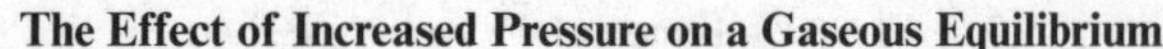

**The Effect of Increased Pressure on a Gaseous Equilibrium**

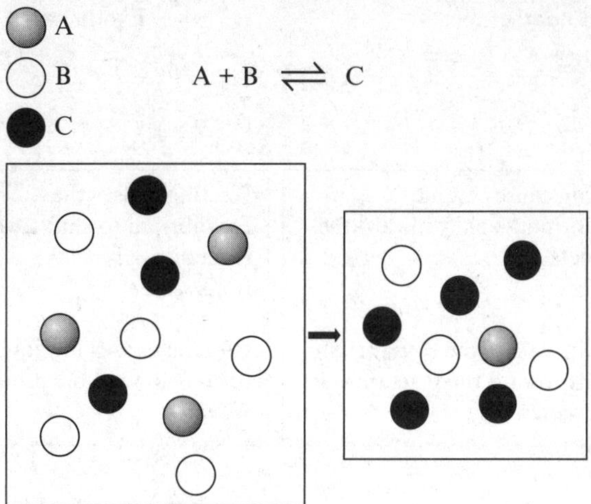


## Practice

For the reactions below, predict the direction that the reaction will shift as a result of an increase in pressure.

**a.** $2\ Cl_2(g) + O_2(g) \rightleftharpoons 2\ Cl_2O(g)$
**b.** $H_2(g) + I_2(g) \rightleftharpoons 2\ HI(g)$
**c.** $CH_4(g) + H_2O(g) \rightleftharpoons CO(g) + 3\ H_2(g)$

An increase in pressure causes a reaction to shift toward the side with fewer gas molecules. Due to an increase in pressure, reaction (a) will shift toward products, reaction (b) will undergo no change, and reaction (c) will shift toward reactants.

**TIP: It should be noted that addition of an inert or nonreactive gas to a system at constant volume will not shift the equilibrium since it has no effect on the concentrations or partial pressures of the other gases present.**

# Temperature Changes

An equilibrium constant for a reaction is specific for a particular temperature. A change in the temperature of a system (unlike a change in concentration or pressure) changes the value of the equilibrium constant.

An *endothermic* reaction consumes heat. Therefore, heat is treated like a reactant. So, raising the temperature is like adding a reactant: It shifts the reaction toward products. This causes an increase in the value of the equilibrium constant.

On the other hand, an *exothermic* reaction produces heat. Therefore, heat is treated like a product. Raising the temperature is like adding a product: It shifts the reaction toward reactants. This causes a decrease in the value of the equilibrium constant.

**NOTE: It is very important to remember that ONLY temperature changes affect the value of *K*.**

**The Effect of Temperature on Equilibrium**

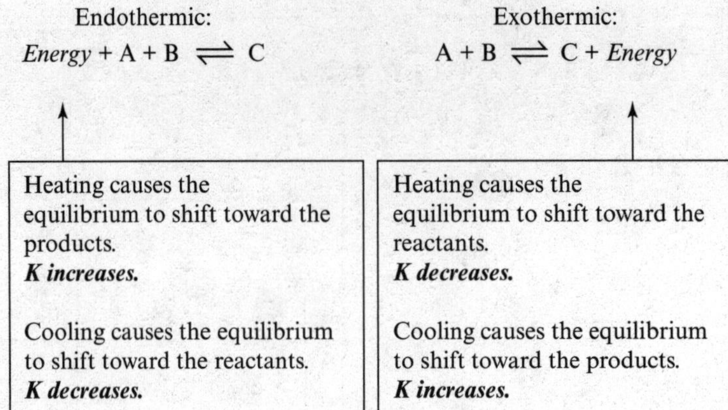


## Practice

Use the bond energy data to determine whether $K$ will increase or decrease for the reaction of $F_2$ with $O_2$ as temperature increases.

$$F_2(g) + \frac{1}{2}O_2(g) \rightleftharpoons OF_2(g)$$

| Bond | Bond Energy (kJ·mol$^{-1}$) |
|---|---|
| F–F | 159 |
| O=O | 498 |
| O–F | 190 |

First, draw the Lewis structures of each of the reactants and products.

$:\ddot{F}-\ddot{F}:$ $\quad \ddot{O}=\ddot{O}$ $\quad :\ddot{F}-\ddot{O}:-F$

Next, use the bond energies to determine whether the reaction is endothermic or exothermic, as described in Chapter 6, "Thermodynamics."

$$\Delta H_{rxn} = \Sigma \text{ bond energies of reactants} - \Sigma \text{ bond energies of products}$$

$$\Delta H_{rxn} = \left[1\text{ BE}_{F-F} + \frac{1}{2}\text{BE}_{O=O}\right] - [2\text{ BE}_{O-F}]$$

$$\Delta H_{rxn} = \left[1(159\text{ kJ}\cdot\text{mol}^{-1}) + \frac{1}{2}(498\text{ kJ}\cdot\text{mol}^{-1})\right] - \left[2(190\text{ kJ}\cdot\text{mol}^{-1})\right] = 28\text{ kJ}\cdot\text{mol}^{-1}$$

The positive enthalpy of reaction indicates an endothermic reaction, so as temperature increases, the reaction will shift toward the products. The equilibrium constant, $K$, increases.

# The Solubility Product

The **solubility** of a compound is the amount of the compound that dissociates in a given volume of solvent.

Ionic compounds exhibit a range of solubilities. Some compounds, like NaCl, are very soluble. The solubility of NaCl in water at 25 °C is 359 grams (6.14 moles) per liter. Other compounds, such as AgCl, are described as "insoluble." In actuality, AgCl has a slight solubility of 1.9 mg ($1.3 \times 10^{-5}$ moles) per liter of water at 25 °C.

The solubility product, $K_{sp}$, is the equilibrium constant that describes the dissociation of an ionic compound, such as $Mg(OH)_2$, to form a saturated solution at a particular temperature. Magnesium hydroxide, $Mg(OH)_2$, has a $K_{sp}$ value of $8.9 \times 10^{-12}$ at 25 °C. The equation for the dissolution of $Mg(OH)_2$ is shown.

$$Mg(OH)_2(s) \rightleftharpoons Mg^{2+}(aq) + 2\ OH^-(aq) \qquad K_{sp} = [Mg^{2+}][OH^-]^2$$

Notice that $Mg(OH)_2$ does not appear in the solubility product expression since it is a solid.

The number of moles of a substance that will dissolve in 1 liter of water is the **molar solubility.**

The molar solubility of $Mg(OH)_2$ in moles per liter of water at 25 °C can be determined from the $K_{sp}$ using a RICE table. In the RICE table below, the $Mg(OH)_2$ column is crossed out because it is a solid that does not contribute to the mass action expression.

Notice that $x$ is equal to the molar solubility of $Mg(OH)_2$.

| **Reaction** | **$Mg(OH)_2$** | **$Mg^{2+}$** | **$2\ OH^-$** |
|---|---|---|---|
| **Initial Concentrations ($M$)** | | 0 | 0 |
| **Change** | | $+x$ | $+2x$ |
| **Equilibrium Concentrations ($M$)** | | $x$ | $2x$ |

The $K_{sp}$ can be used to determine $x$.

$$K_{sp} = [Mg^{2+}][OH^-]^2$$

$$K_{sp} = x(2x)^2$$

$$4x^3 = 8.9 \times 10^{-12}$$

$$x = \sqrt[3]{\frac{8.9 \times 10^{-12}}{4}} = 1.3 \times 10^{-4}\ M$$

The molar solubility of $Mg(OH)_2$ at 25 °C is $1.3 \times 10^{-4}\ M$.

## Practice

Calculate the $K_{sp}$ for $Co(OH)_3$, which has a molar solubility of $5.52 \times 10^{-5}\ M$ at 25 °C.

The dissolution of $Co(OH)_3$ occurs according to the following equation.

$$Co(OH)_3(s) \rightleftharpoons Co^{3+}(aq) + 3\ OH^-(aq) \qquad K_{sp} = [Co^{3+}][OH^-]^3$$

The solubility is equal to the maximum number of moles of $Co(OH)_3$ that will dissolve in pure water at this temperature, so it is also equal to the equilibrium concentration of $Co^{3+}$ for a saturated solution.

| **Reaction** | **$Co(OH)_3$** | **$Co^{3+}$** | **$3\ OH^-$** |
|---|---|---|---|
| **Initial Concentrations ($M$)** | | 0 | 0 |
| **Change** | | $+5.52 \times 10^{-5}$ | $+(3 \times 5.52 \times 10^{-5})$ |
| **Equilibrium Concentrations ($M$)** | | $5.52 \times 10^{-5}$ | $1.65 \times 10^{-4}$ |

$$K_{sp} = [Co^{3+}][OH^-]^3 = (5.52 \times 10^{-5})(1.65 \times 10^{-4})^3 = 2.51 \times 10^{-16}$$

## The Common Ion Effect

The **common ion effect** refers to the decreased dissociation of an ionic compound in solution due to the presence of one of the ions produced.

For example, consider the dissociation of solid $CaF_2$ in a solution of 1.00 *M* $Ca(NO_3)_2$. Since nitrate salts are soluble (they have very large $K_{sp}$ values), the solution is already 1.00 *M* in $Ca^{2+}$. Calcium fluoride, on the other hand, has a very small $K_{sp}$ of $4.0 \times 10^{-11}$ at 25 °C.

$$CaF_2(s) \rightleftharpoons Ca^{2+}(aq) + 2\,F^-(aq) \qquad K_{sp} = 4.0 \times 10^{-11}$$

Because $Ca^{2+}$ is already present, the equilibrium shifts to the left, and less $CaF_2$ dissolves in the solution than would dissolve in pure water.

The number of moles of $CaF_2$ that could be dissolved in 1.00 L of 1.00 *M* $Ca(NO_3)_2$ solution can be determined using a RICE table.

| **Reaction** | $CaF_2$ | $Ca^{2+}$ | $2\,F^-$ |
|---|---|---|---|
| **Initial Concentrations (*M*)** | | 1.00 | 0 |
| **Change** | | $+x$ | $+2x$ |
| **Equilibrium Concentrations (*M*)** | | $1.00 + x$ | $2x$ |

Again, the solubility product expression is used, but the algebra required to solve for $x$ is complicated beyond the scope of the AP Chemistry exam.

$$K_{sp} = [Ca^{2+}][F^-]^2$$

$$K_{sp} = (1.00 + x)(2x)^2 = 4.0 \times 10^{-11}$$

Rather than solving the resulting polynomial expression for $x$, the simplifying assumption can be made that $x$, the number of moles of $CaF_2$ that dissolve in a 1.00 *M* solution of $Ca^{2+}$ solution, is negligibly small. In that case, it can be omitted from the $(1.00 + x)$ term.

$$4.0 \times 10^{-11} = (1.00 + \cancel{x})(2x)^2 = 4x^2$$

$$x = \sqrt{\frac{4.0 \times 10^{-11}}{4}} = 3.2 \times 10^{-6}\ M$$

The solubility of $CaF_2$ in an aqueous solution of 1.00 *M* $Ca(NO_3)_2$ at 25 °C is $3.2 \times 10^{-6}$ *M*.

### Practice

Zinc hydroxide, $Zn(OH)_2$, has a $K_{sp}$ of $4.5 \times 10^{-17}$ at 25 °C.

**a.** Qualitatively compare the solubility of $Zn(OH)_2$ in pure water with the solubility of $Zn(OH)_2$ in a 0.250 *M* solution of $Zn(NO_3)_2$.
**b.** Calculate the solubility of $Zn(OH)_2$ in pure water.
**c.** Calculate the solubility of $Zn(OH)_2$ in 0.250 *M* $Zn(NO_3)_2$ solution.

**(a)** The dissolution of $Zn(OH)_2$ occurs according to the reaction shown.

$$Zn(OH)_2(s) \rightleftharpoons Zn^{2+}(aq) + 2\,OH^-(aq)$$

Zinc hydroxide will be less soluble in a solution containing $Zn(NO_3)_2$ because, according to Le Chatelier's principle, the $Zn^{2+}$ ion present will shift the dissolution equilibrium to the left, in the direction of the reactant.

**(b)** The solubility of $Zn(OH)_2$ in pure water is calculated as before.

| Reaction | $Zn(OH)_2$ | $Zn^{2+}$ | $2\ OH^-$ |
|---|---|---|---|
| Initial Concentrations ($M$) | | 0 | 0 |
| Change | | $+x$ | $+2x$ |
| Equilibrium Concentrations ($M$) | | $x$ | $2x$ |

$$K_{sp} = [Zn^{2+}][OH^-]^2$$

$$K_{sp} = (x)(2x)^2 = 4.5 \times 10^{-17}$$

$$4.5\times10^{-17} = (x)(2x)^2 = 4x^3$$

$$\sqrt[3]{\frac{4.5\times10^{-17}}{4}} = x = 2.2\times10^{-6}\ M$$

**(c)** Because nitrate salts are highly soluble in water, the concentration of $Zn^{2+}$ is equal to the concentration of $Zn(NO_3)_2$, 0.250 $M$. The calculation of the solubility of $Zn(OH)_2$ in a 0.250 $M$ $Zn^{2+}$ solution takes the common ion effect into account.

| Reaction | $Zn(OH)_2$ | $Zn^{2+}$ | $2\ OH^-$ |
|---|---|---|---|
| Initial Concentrations ($M$) | | 0.250 | 0 |
| Change | | $+x$ | $+2x$ |
| Equilibrium Concentrations ($M$) | | $0.250 + x$ | $2x$ |

The value of $x$ is negligible compared with the initial $Zn^{2+}$ concentration.

$$K_{sp} = [Zn^{2+}][OH^-]^2$$

$$K_{sp} = 4.5\times10^{-17} = (0.250 + \cancel{x})(2x)^2$$

$$x = \sqrt{\frac{4.5\times10^{-17}}{(0.250)(4)}} = 6.7\times10^{-9}\ M$$

# Precipitation Reactions

The $K_{sp}$ for an ionic compound can be used to determine whether or not a precipitation reaction will occur when aqueous solutions of the ions are combined.

This solubility product can be compared with the reaction quotient, $Q$, to determine whether a precipitation reaction will occur under a specified set of conditions.

The $K_{sp}$ of $BaSO_4$, for example, is very small at 25 °C. Barium sulfate is minimally soluble in water. A student adds 25.0 mL of a $1.0 \times 10^{-4}$ $M$ solution of $Ba(NO_3)_2$ to a 25.0-mL solution of a $1.0 \times 10^{-4}$ $M$ solution of $Na_2SO_4$. Is a precipitation reaction expected to occur?

$$BaSO_4(s) \rightleftharpoons Ba^{2+}(aq) + SO_4^{\ 2-}(aq) \qquad K_{sp} = 1.1\times10^{-10}$$

First, find the number of moles of $Ba^{2+}$ and $SO_4^{\ 2-}$.

$$(0.0250\ \cancel{L})\left(\frac{1.0\times10^{-4}\ \text{mol}}{1\ \cancel{L}}\right) = 2.5\times10^{-6}\ \text{mol Ba}^{2+} = 2.5\times10^{-6}\ \text{mol SO}_4^{\ 2-}$$

Next, find the concentrations of $Ba^{2+}$ and $SO_4^{2-}$ in the combined solution (which has a total of 50.0 mL).

$$[Ba^{2+}] = [SO_4^{2-}] = \frac{2.5 \times 10^{-6}\ \text{mol}}{0.050\ \text{L}} = 5.0 \times 10^{-5}\ M$$

Finally, calculate $Q$.

$$Q = [Ba^{2+}][SO_4^{2-}] = (5.0 \times 10^{-5})^2 = 2.5 \times 10^{-9}$$

Because $Q$ is greater than $K_{sp}$, the dissociation reaction shifts toward the solid reactants. In other words, a precipitation reaction can be predicted to occur.

## Practice

Predict whether a precipitation reaction will occur when 125 mL of a 0.0420 $M$ solution of KI are added to 125 mL of 0.0210 $M$ $Pb(NO_3)_2$. The $K_{sp}$ for $PbI_2$ is $1.4 \times 10^{-8}$ at 25 °C.

The dissolution reaction for $PbI_2$ is

$$PbI_2(s) \rightleftharpoons Pb^{2+}(aq) + 2\ I^-(aq)$$

The reaction quotient must be determined using the initial concentrations of $Pb^{2+}$ and $I^-$ that result upon the addition of the two solutions.

First, we find the number of moles of $Pb^{2+}$ and calculate its concentration in the 250-mL combined solution.

$$(0.125\ \text{L})\left(\frac{0.0210\ \text{mol}\ Pb^{2+}}{1\ \text{L}}\right) = 0.002625\ \text{mol}\ Pb^{2+}$$

$$[Pb^{2+}] = \frac{0.002625\ \text{mol}}{0.250\ \text{L}} = 1.05 \times 10^{-2}\ M$$

Next, we find the number of moles of $I^-$ and calculate its concentration in the 250-mL combined solution.

$$(0.125\ \text{L})\left(\frac{0.0420\ \text{mol}\ I^-}{1\ \text{L}}\right) = 0.00525\ \text{mol}\ I^-$$

$$[I^-] = \frac{0.00525\ \text{mol}}{0.250\ \text{L}} = 2.10 \times 10^{-2}\ M$$

Finally, $Q$ is calculated and compared with $K_{sp}$.

$$Q = [Pb^{2+}][I^-]^2 = (1.05 \times 10^{-2})(2.10 \times 10^{-2})^2 = 4.63 \times 10^{-6}$$

Because $Q$ is greater than $K_{sp}$, the equilibrium favors the solid $PbI_2$ on the reactant side of the equation, and a precipitation reaction occurs.

# Review Questions

## Multiple Choice

**1.** Given the data in the table below, determine the equilibrium constant for the reaction of $Br_2(g)$ with $NO(g)$ to form $NOBr(g)$.

$$Br_2(g) + 2\ NO(g) \rightleftharpoons 2\ NOBr(g)$$

| Reaction | $K$ |
| --- | --- |
| $N_2(g) + Br_2(g) + O_2(g) \rightleftharpoons 2\ NOBr(g)$ | $2.0 \times 10^{-27}$ |
| $\frac{1}{2}N_2(g) + \frac{1}{2}O_2(g) \rightleftharpoons NO(g)$ | $1.0 \times 10^{-15}$ |

A. $2.0 \times 10^{-57}$
B. $5.0 \times 10^{-2}$
C. $5.0 \times 10^{2}$
D. $2.0 \times 10^{3}$

**2.** Which of the following statements is FALSE?

A. A reaction with a large $K_{eq}$ value always proceeds to completion.
B. A reaction with a very small $K_{eq}$ has reactant equilibrium quantities that are much larger than product equilibrium quantities.
C. $K_{eq}$ gives no information about the overall rate of the reaction.
D. A value of $K_{eq}$ near 1 indicates that significant quantities of reactants and products are present.

*Questions 3–4 refer to the reaction of $CO(g)$ with $H_2O(g)$.*

A 1.0-mole sample of $CO(g)$ and a 2.0-mole sample of $H_2O(g)$ are combined in a rigid, evacuated 1-L vessel. They react to form $CO_2(g)$ and $H_2(g)$. At equilibrium, 0.67 mole of $CO_2(g)$ is present in the container.

**3.** What is the equilibrium constant, $K_c$, for the reaction?

A. 0.5
B. 1.0
C. 2.0
D. 4.0

**4.** The reaction is set up such that the concentrations of the species are as follows.

| Species | Concentration |
| --- | --- |
| [CO] | 2.0 *M* |
| $[H_2O]$ | 1.5 *M* |
| $[CO_2]$ | 3.0 *M* |
| $[H_2]$ | 2.0 *M* |

Which of the following best describes the results of the reaction?

A. The reaction will shift toward products to reach equilibrium because $Q > K$.
B. The reaction will shift toward reactants to reach equilibrium because $Q > K$.
C. The reaction will shift toward products to reach equilibrium because $Q < K$.
D. The reaction will shift toward reactants to reach equilibrium because $Q < K$.

*Questions 5–7 refer to the decomposition of BrCl(g).*

$$2\ BrCl(g) \rightleftharpoons Br_2(g) + Cl_2(g) \qquad K_c = 36 \text{ at a certain temperature}$$

**5.** A closed, rigid 1.0-L container is charged with 2.00 moles of BrCl(*g*). What is the equilibrium concentration of BrCl(*g*) in the container?

**A.** 0.15 *M*
**B.** 0.92 *M*
**C.** 1.1 *M*
**D.** 1.8 *M*

**6.** Which of the following statements is TRUE of the reaction as it progresses from pure reactants toward equilibrium?

**A.** The rate of the forward reaction decreases until it is equal to the rate of the reverse reaction.
**B.** The rate of the forward reaction increases until it is equal to the rate of the reverse reaction.
**C.** The rate of the forward reaction and the rate of the reverse reaction are identical throughout the experiment.
**D.** The rate of the forward reaction decreases until equilibrium is reached and the reaction stops.

**7.** If additional $Cl_2(g)$ is added after equilibrium has been established, which graph best shows the effect on the concentration of $Cl_2(g)$?

**A.**

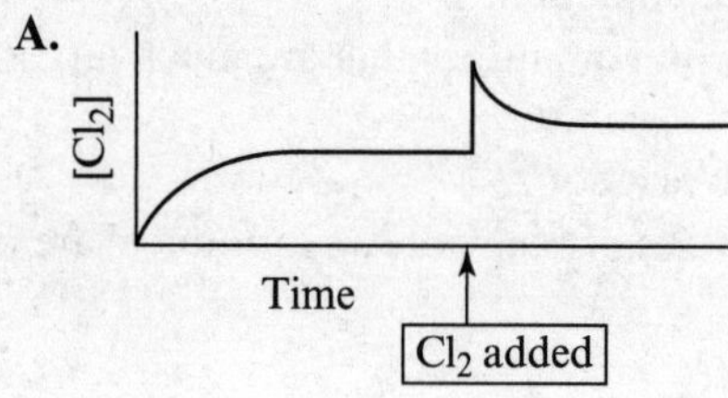


**B.**

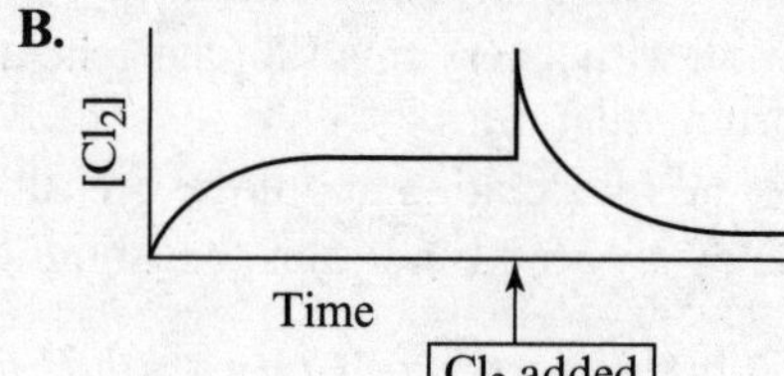


**C.**

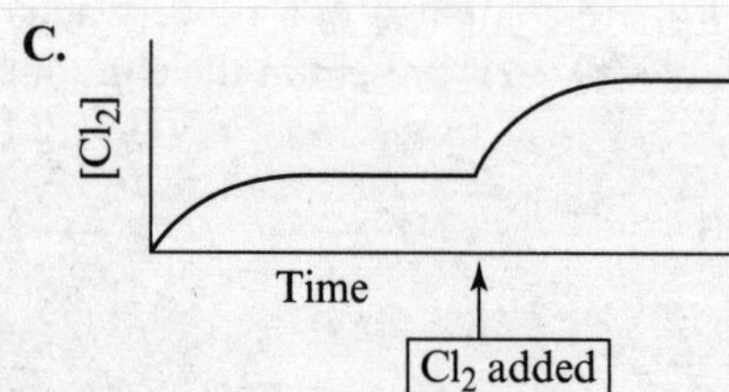


**D.**

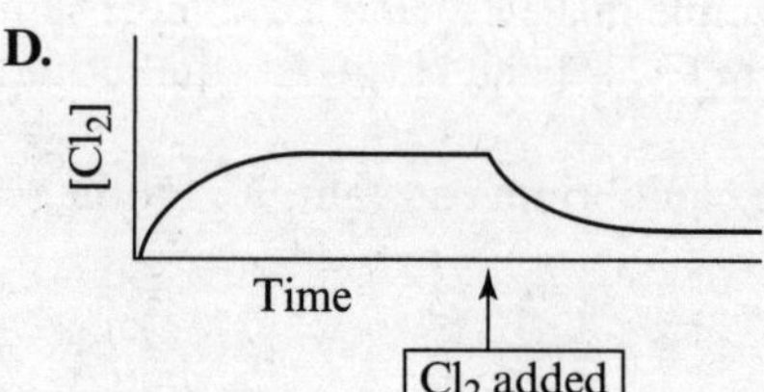


**8.** Hemoglobin, Hb, is an iron-containing metalloprotein that transports oxygen in the blood of vertebrates. Carbon monoxide poisoning occurs when CO coordinates with the iron in hemoglobin.

$$Hb(O_2)_4(aq) + 4\ CO(g) \rightleftharpoons Hb(CO)_4(aq) + 4\ O_2(g)$$

Which strategy would be effective for reversing the effects of carbon monoxide poisoning?

**A.** Lowering the total pressure of the system
**B.** Raising the total pressure of the system
**C.** Exposing the system to a low partial pressure of $O_2(g)$
**D.** Exposing the system to a high partial pressure of $O_2(g)$

**9.** The reaction of $Cl_2$ with $H_2O$ forms HCl and $O_2$. The value of $K_c$ is $1.9 \times 10^{-15}$ at 298 K.

$$2\ Cl_2(g) + 2\ H_2O(g) \rightleftharpoons 4\ HCl(g) + O_2(g) \qquad \Delta H° = 115\ kJ{\cdot}mol^{-1}$$

What effect will an increase in temperature have on the equilibrium constant?

**A.** $K_c$ will decrease because there are more product molecules than reactant molecules.
**B.** $K_c$ will increase because the reaction is endothermic.
**C.** $K_c$ will be unaffected because it is constant.
**D.** The effect on $K_c$ depends on the change in enthalpy at the new temperature.

**10.** Each solid sphere below represents a molecule of A, and each hollow sphere represents a molecule of B. The volume of the container is 1.0 L. Which drawing best illustrates an equilibrium mixture of A and B if A decomposes to form B according to the reaction below?

$$A \rightleftharpoons 2\ B \qquad K_c = 6$$

A = ●
B = ○

**A.**

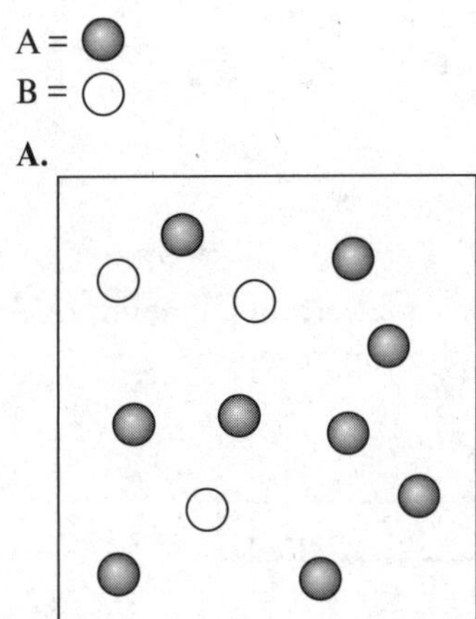

**B.**

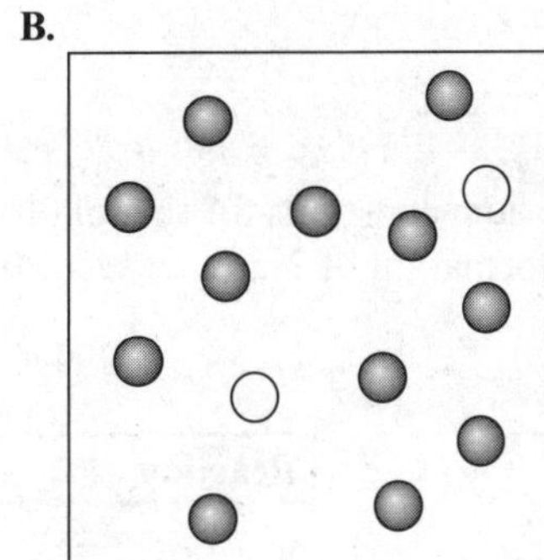

**C.**

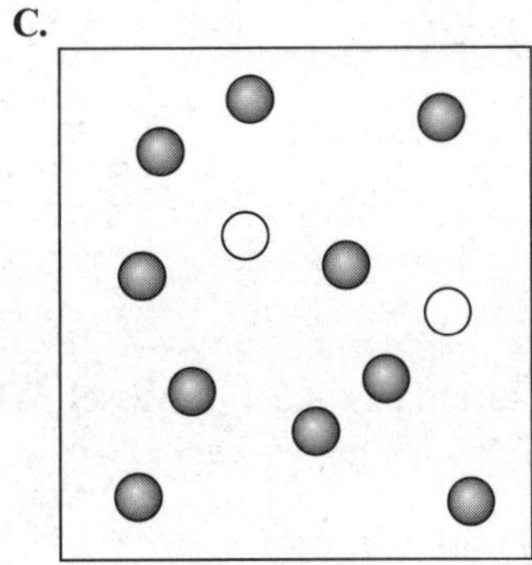

**D.**

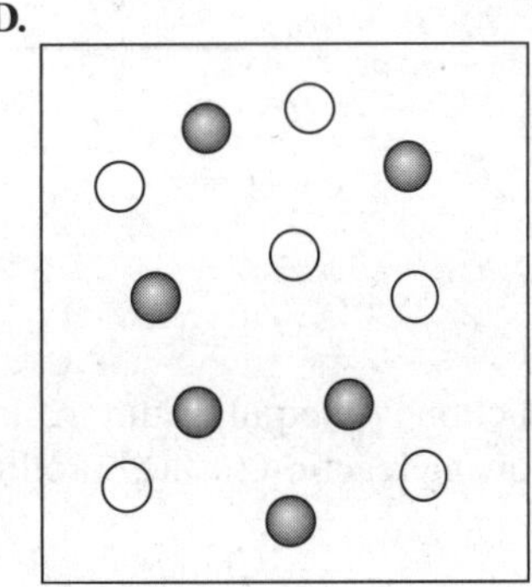

**11.** Which of the following is the correct expression for the $K_{sp}$ of magnesium hydroxide?

**A.** $\dfrac{[Mg^{2+}][OH^-]^2}{[Mg(OH)_2]}$
**B.** $[Mg^{2+}][OH^-]$
**C.** $[Mg^{2+}][OH^-]^2$
**D.** $\dfrac{1}{[Mg^{2+}][OH^-]^2}$

*Questions 12–13 refer to the following reaction between iron(III) and thiocyanate ions.*

$$Fe^{3+}(aq) + SCN^{-}(aq) \rightleftharpoons FeSCN^{2+}(aq)$$

**12.** For a reaction between $Fe^{3+}$ and $SCN^{-}$, the initial concentration of $Fe^{3+}$ is 0.4 *M* and the initial concentration of $SCN^{-}$ is 0.6 *M*. When the reaction reaches equilibrium, the concentration of $FeSCN^{2+}$ is found to be 0.2 *M*. What are the equilibrium concentrations of $Fe^{3+}$ and $SCN^{-}$?

| | $[Fe^{3+}]$ | $[SCN^{-}]$ |
|---|---|---|
| **A.** | 0.2 *M* | 0.4 *M* |
| **B.** | 0.6 *M* | 0.8 *M* |
| **C.** | 0.4 *M* | 0.2 *M* |
| **D.** | 0.8 *M* | 0.6 *M* |

**13.** What is $K_c$ for the reaction between $Fe^{3+}$ and $SCN^{-}$?

**A.** 0.4
**B.** 0.5
**C.** 2.0
**D.** 2.5

**14.** Given the equilibrium constants for the following reactions at a certain temperature, find the equilibrium constant for the formation of $Na_2O_2(s)$ according to the following reaction.

$$2\ NaO(g) \rightleftharpoons Na_2O_2(s)$$

| Reaction | *K* |
|---|---|
| $Na_2O_2(s) \rightleftharpoons 2\ Na(l) + O_2(g)$ | $5.0 \times 10^{-29}$ |
| $NaO(g) \rightleftharpoons Na(l) + \frac{1}{2}O_2(g)$ | $2.0 \times 10^{-5}$ |

**A.** $2.0 \times 10^{-34}$
**B.** $4.0 \times 10^{19}$
**C.** $8.0 \times 10^{18}$
**D.** $8.0 \times 10^{-20}$

**15.** The following reaction is at equilibrium in an aqueous solution in a reaction vessel. Which of the following will shift the following reaction toward products?

$$Cl_2(aq) + 2\ Br^{-}(aq) \rightleftharpoons 2\ Cl^{-}(aq) + Br_2(aq)$$

**A.** Reducing the volume of the reaction by half
**B.** Doubling the volume of the reaction
**C.** Addition of NaBr(*s*) to the reaction
**D.** Addition of NaCl(*s*) to the reaction

## Long Free Response

**1.** A 1.00-gram sample of solid $CO_2$ is placed in a rigid, evacuated 1.00-L vessel that contains a negligible volume of a solid metal catalyst. The reaction is heated to 3000. K, and the reaction shown below occurs.

$$2\ CO_2(g) + \text{heat} \rightleftharpoons 2\ CO(g) + O_2(g)$$

a. What is the pressure inside the container at 3000. K before any reaction has occurred?
b. When the system is at equilibrium, the total pressure in the container is 6.15 atm. Calculate the partial pressures of $CO_2(g)$, $CO(g)$, and $O_2(g)$ in the container at 3000. K.
c. Write the equilibrium constant expression, $K_p$, for the reaction.
d. Calculate the value of the equilibrium constant, $K_p$, for the reaction at 3000. K.
e. If the temperature of the reaction is decreased to 2000. K, will the value of $K_p$ increase, decrease, or remain constant? Explain your response.
f. In a separate experiment, the original partial pressures at 3000. K are shown in the table below. Will the amount of $CO_2$ in the container increase, decrease, or remain constant? Justify your response with a calculation.

| Species | Partial Pressure (atm) |
|---|---|
| $CO_2$ | 5.0 |
| CO | 0.20 |
| $O_2$ | 0.20 |

## Short Free Response

**2.** A saturated solution of $CaF_2$ has an $F^-$ concentration of $4.31 \times 10^{-4}\ M$ at 25 °C.

a. Write the expression for the solubility product, $K_{sp}$, for $CaF_2$.
b. Determine the $K_{sp}$ for $CaF_2$ at 25 °C.
c. How many moles of $CaF_2$ will dissolve in a 1.00-L solution that is 0.250 $M$ in NaF?
d. At 25 °C, the pH of a saturated aqueous solution of $CaF_2$ is 7.10 and the pH of a saturated aqueous solution of $MgF_2$ is 7.85. A solution of 1.0 $M$ NaF is slowly added to an aqueous solution that is 0.100 $M$ $Mg(NO_3)_2$ and 0.100 $M$ $Ca(NO_3)_2$. Write the net ionic equation for the precipitation reaction that occurs first and explain your answer. (This is a challenging question that requires knowledge of Chapter 8.)

# Answers and Explanations

## Multiple Choice

**1. D.** When the second equation is reversed and multiplied by 2, the sum of the two equations gives the overall transformation.

$$\left[NO(g) \rightleftharpoons \frac{1}{2}N_2(g) + \frac{1}{2}O_2(g)\right] \times 2 = 2\ NO(g) \rightleftharpoons N_2(g) + O_2(g)$$

The equilibrium constant for the reaction must reflect the change to that equation. First, the equation is reversed.

$$(1.0 \times 10^{-15})^{-1} = 1.0 \times 10^{15}$$

Second, the equation is multiplied by 2, so the equilibrium constant must be squared.

$$(1.0 \times 10^{15})^2 = 1.0 \times 10^{30}$$

The equations can be added together to result in the overall equation, so their equilibrium constants are multiplied.

$$\cancel{N_2(g)} + Br_2(g) + \cancel{O_2(g)} \rightleftharpoons 2\ NOBr(g)$$
$$2\ NO(g) \rightleftharpoons \cancel{N_2(g)} + \cancel{O_2(g)}$$
$$\overline{Br_2(g) + 2\ NO(g) \rightleftharpoons 2\ NOBr(g)}$$

$(1.0 \times 10^{30})(2.0 \times 10^{-27}) = 2.0 \times 10^3$, choice D

2. **A.** A reaction with a large equilibrium constant has a much larger proportion of products at equilibrium than reactants; however, not all reactions with very large equilibrium constants necessarily reach completion, so choice A is the false statement.
3. **B.** The balanced chemical equation is

$$CO(g) + H_2O(g) \rightleftharpoons CO_2(g) + H_2(g)$$

The equilibrium concentrations can be determined using a RICE table.

| **Reaction** | **CO** | **$H_2O$** | **$CO_2$** | **$H_2$** |
|---|---|---|---|---|
| **Initial Concentrations (*M*)** | 1.0 | 2.0 | 0 | 0 |
| **Change** | –0.67 | –0.67 | +0.67 | +0.67 |
| **Equilibrium Concentrations (*M*)** | 0.33 | 1.33 | 0.67 | 0.67 |

The math is simplified by recognizing that 0.33 is approximately $\frac{1}{3}$.

$$K_c = \frac{[CO_2][H_2]}{[CO][H_2O]} = \frac{\left(\frac{2}{3}\right)^2}{\left(\frac{1}{3}\right)\left(\frac{4}{3}\right)} = 1.0, \text{ choice B}$$

4. **B.** A comparison of *K* with *Q* gives the direction in which the reaction will proceed to establish equilibrium.

$$Q = \frac{[CO_2][H_2]}{[CO][H_2O]} = \frac{(3.0)(2.0)}{(2.0)(1.5)} = 2.0$$

Since *Q* is greater than *K*, the reaction shifts to the left toward reactants, choice B. Reactants are formed faster than products are consumed until equilibrium is reached.

5. **A.** The equilibrium concentrations are determined using a RICE table.

| **Reaction** | **2 BrCl** | **$Br_2$** | **$Cl_2$** |
|---|---|---|---|
| **Initial Concentrations (*M*)** | 2.00 | 0 | 0 |
| **Change** | $-2x$ | $+x$ | $+x$ |
| **Equilibrium Concentrations (*M*)** | $2.00 - 2x$ | $x$ | $x$ |

$$K_c = \frac{[Br_2][Cl_2]}{[BrCl]^2}$$

$$\sqrt{\frac{x^2}{(2-2x)^2}} = \sqrt{36}$$

$$\frac{x}{2-2x} = 6$$

$$x = \frac{12}{13}$$

This is equal to approximately 0.92 $M$. The equilibrium concentration of BrCl is

$$2.00 - 2\left(\frac{12}{13}\right) = 0.15M\text{, choice A}$$

**6.** **A.** When the reaction begins, the rate of the reverse reaction is negligible compared with the rate of the forward reaction since there is little product to react. As the reaction proceeds, the rate of the forward reaction decreases and the rate of the reverse reaction increases until both rates are equal and there is no net change in the forward or reverse direction, choice A (although reaction in both the forward and reverse directions is still continuously occurring).

**7.** **A.** As depicted in the graph in choice A, when chlorine gas is added, the reaction consumes some of the added chlorine as it shifts back to the reactants. The final chlorine concentration, however, is still greater than it was after the initial equilibrium is established.

**8.** **D.** Increasing the partial pressure of $O_2$, choice D, will have the effect of shifting the equilibrium toward $Hb(O_2)_4$, releasing CO.

**9.** **B.** The reaction is endothermic, so an increase in temperature will shift the reaction toward products, increasing the equilibrium constant, choice B.

**10.** **D.** The equilibrium expression for the hypothetical reaction is

$$K = \frac{[B]^2}{[A]}$$

The figure that depicts equilibrium is the one in which [A] = 6 $M$ and [B] = 6 $M$, choice D.

$$6 = \frac{6^2}{6}$$

**11.** **C.** The dissociation equation for $Mg(OH)_2$ is

$$Mg(OH)_2(s) \rightleftharpoons Mg^{2+}(aq) + 2\,OH^-(aq)$$

The expression for $K_{sp}$ for $Mg(OH)_2$ is

$$K_{sp} = [Mg^{2+}][OH^-]^2\text{, choice C}$$

**12.** **A.** The equilibrium concentrations are determined using a RICE table. Since the equilibrium concentration of $FeSCN^{2+}$ is 0.2 $M$, this is equal to the change in concentrations.

| **Reaction** | **$Fe^{3+}$** | **$SCN^-$** | **$FeSCN^{2+}$** |
|---|---|---|---|
| **Initial Concentrations ($M$)** | 0.4 | 0.6 | 0 |
| **Change** | –0.2 | –0.2 | +0.2 |
| **Equilibrium Concentrations ($M$)** | 0.2 | 0.4 | 0.2 |

The equilibrium concentration of $Fe^{3+}$ is 0.2 $M$, and the equilibrium concentration of $SCN^-$ is 0.4 $M$, choice A.

**13.** **D.** The $K_c$ for this reaction can be determined from the equilibrium concentrations.

$$K_c = \frac{\cancel{0.2}}{(\cancel{0.2})(0.4)} = \frac{1}{\left(\frac{2}{5}\right)} = \frac{5}{2} = 2.5\text{, choice D}$$

**14.** C. The first equation must be reversed, and its equilibrium constant is the reciprocal of the original $K$.

$$2\ \mathrm{Na}(l) + \mathrm{O_2}(g) \rightleftharpoons \mathrm{N_2O_2}(s) \qquad K = \frac{1}{5.0\times10^{-29}} = 2.0\times10^{28}$$

The second equation must be multiplied by 2. Its equilibrium constant is squared as a result.

$$\mathrm{NaO}(g) \rightleftharpoons \mathrm{Na}(l) + \frac{1}{2}\mathrm{O_2}(g) \qquad K = (2.0\times10^{-5})^2 = 4.0\times10^{-10}$$

The sum of the reactions is equal to the overall reaction.

$$2\ \cancel{\mathrm{Na}(l)} + \cancel{\mathrm{O_2}(g)} \rightleftharpoons \mathrm{N_2O_2}(s)$$
$$2\ \mathrm{NaO}(g) \rightleftharpoons 2\ \cancel{\mathrm{Na}(l)} + \cancel{\mathrm{O_2}(g)}$$
$$\overline{2\ \mathrm{NaO}(g) \rightleftharpoons \mathrm{N_2O_2}(s)}$$

$$K = (2.0 \times 10^{28})(4.0 \times 10^{-10}) = 8.0 \times 10^{18}\text{, choice C}$$

**15.** C. The addition of NaBr, a soluble salt, to the aqueous solution released $Br^-$ ions, choice C. According to Le Chatelier's principle, addition of a reactant shifts the equilibrium toward products.

## Long Free Response

**1.** **a.** The pressure inside the container is determined using the ideal gas law.

$$1.00\ \cancel{\mathrm{g\ CO_2}} \times \frac{1\ \mathrm{mol\ CO_2}}{44.01\ \cancel{\mathrm{g\ CO_2}}} = 0.0227\ \mathrm{mol\ CO_2}$$

$$P = \frac{nRT}{V} = \frac{(0.0227\ \cancel{\mathrm{mol}})(0.08206\ \cancel{\mathrm{L}}\cdot\mathrm{atm}\cdot\cancel{\mathrm{mol^{-1}}}\cdot\cancel{\mathrm{K^{-1}}})(3000\ \cancel{\mathrm{K}})}{1\ \cancel{\mathrm{L}}} = 5.59\ \mathrm{atm}$$

**b.** The equilibrium partial pressures are related to one another by the stoichiometric coefficients in the balanced chemical equation. The total pressure in the container is the sum of the partial pressures.

$$P_{\text{total}} = P_{\mathrm{CO_2}} + P_{\mathrm{CO}} + P_{\mathrm{O_2}}$$

| Reaction | $2\ CO_2$ | $2\ CO$ | $O_2$ |
|---|---|---|---|
| **Initial Partial Pressure (atm)** | 5.59 | 0 | 0 |
| **Change** | $-2x$ | $+2x$ | $+x$ |
| **Equilibrium Partial Pressure (atm)** | $5.59 - 2x$ | $2x$ | $x$ |

The value of $x$ can be determined from the total pressure. Then the partial pressures can be calculated.

$$P_{\text{total}} = (5.59 - 2x) + 2x + x = 6.15\ \mathrm{atm}$$

$$x = 0.56\ \mathrm{atm}$$
$$P_{\mathrm{CO_2}} = 4.47\ \mathrm{atm}$$
$$P_{\mathrm{CO}} = 1.12\ \mathrm{atm}$$
$$P_{\mathrm{O_2}} = 0.56\ \mathrm{atm}$$

**c.** The equilibrium constant expression is

$$K_p = \frac{P_{CO}^2 P_{O_2}}{P_{CO_2}^2}$$

**d.** The value of the equilibrium constant is determined from the partial pressures of the species present at equilibrium.

$$K_p = \frac{(1.12)^2(0.56)}{(4.47)^2} = 0.035$$

**e.** A decrease in temperature for an endothermic reaction decreases the equilibrium constant. Because heat can be thought of as a reactant in an endothermic process, removing heat will shift the equilibrium toward reactants according to Le Chatelier's principle. This shift toward reactants decreases the equilibrium constant.

**f.** In order to determine whether the reaction shifts toward reactants or products, it is necessary to evaluate $Q$.

$$Q = \frac{(0.20)^2(0.20)}{(5.0)^2} = 3.2\times10^{-4}$$

Because $Q < K$, the reaction will shift to the right. This means that the amount of $CO_2$ in the container will decrease.

## Short Free Response

**2. a.** The expression for $K_{sp}$ is based on the dissociation reaction of $CaF_2$ in an aqueous solution.

$$CaF_2(s) \rightleftharpoons Ca^{2+}(aq) + 2\,F^-(aq)$$

$$K_{sp} = [Ca^{2+}][F^-]^2$$

**b.** The concentration of $F^-$ is twice the concentration of $Ca^{2+}$.

$$[F^-] = 4.31\times10^{-4}\ M$$

$$[Ca^{2+}] = \frac{4.31\times10^{-4}\ M}{2} = 2.16\times10^{-4}\ M$$

$$K_{sp} = (2.16\times10^{-4})(4.31\times10^{-4})^2 = 4.00\times10^{-11}$$

**c.** The number of moles of $CaF_2$ that could be dissolved in 1.00 L of 0.250 $M$ NaF solution can be determined using a RICE table.

| **Reaction** | $CaF_2$ | $Ca^{2+}$ | $2\,F^-$ |
|---|---|---|---|
| **Initial Concentrations ($M$)** | | 0 | 0.250 |
| **Change** | | $+x$ | $+2x$ |
| **Equilibrium Concentrations ($M$)** | | $x$ | $0.250 + 2x$ |

$$K_{sp} = [Ca^{2+}][F^-]^2$$

$$K_{sp} = (x)(0.250 + 2x)^2$$

Rather than solving the resulting polynomial expression for $x$, the simplifying assumption can be made that $x$, the number of moles of $CaF_2$ that dissolve in 1.0 L of a 0.250 $M$ solution of $F^-$, is negligibly small.

$$4.00 \times 10^{-11} = (x)(0.250 + \cancel{2x})^2 = 0.0625x$$
$$x = 6.40 \times 10^{-10}\ M$$

A total of $6.40 \times 10^{-10}$ moles of $CaF_2$ will dissolve in 1.0 L of a 0.250 $M$ solution of NaF at 25 °C.

**d.** (Please notice that this question, while included in the AP Chemistry learning objectives for the equilibrium unit, requires knowledge of Chapter 8, "Acids and Bases." It may be helpful to return to this question after reviewing Chapter 8.)

The fluoride ion, $F^-$, is the conjugate base of a weak acid, HF, so a solution of $F^-$ will have a pH that is greater than 7 due to the reaction

$$F^-(aq) + H_2O(l) \rightleftharpoons HF(aq) + OH^-(aq)$$

Because the saturated $MgF_2$ solution has a higher pH than the saturated $CaF_2$ solution, more $F^-$ must be present in the $MgF_2$ solution. The solubility of $MgF_2$ is greater than the solubility of $CaF_2$, so $CaF_2$ precipitates first.

The net ionic equation for the precipitation reaction that occurs is just the reverse of the dissociation reaction.

$$Ca^{2+} + 2\,F^- \rightarrow CaF_2$$

State symbols are not required for credit, but if they are included, the reaction becomes

$$Ca^{2+}(aq) + 2\,F^-(aq) \rightarrow CaF_2(s)$$

# Chapter 8

# Acids and Bases

**Reactions between acids and bases can be understood in terms of chemical equilibrium.**

Chemists employ several definitions of acids and bases. The most important for AP Chemistry is the Brønsted-Lowry definition, but familiarity with the older Arrhenius acid-base theory facilitates understanding of Brønsted-Lowry acids and bases.

Acid-base dissociation reactions in water are equilibrium reactions. Strong acids and bases have large equilibrium constants for dissociation in water. The ionization reactions of weak acids and bases have relatively small equilibrium constants.

An **acid-base reaction** in an aqueous solution is a reaction in which an acid reacts with a base to form a solution of salt and water.

$$NaOH(aq) + HCl(aq) \rightarrow NaCl(aq) + H_2O(l)$$

## Definitions of Acids and Bases

An **Arrhenius acid** is a species that releases a proton. The word "proton" is synonymous with "hydrogen ion," or "$H^+$." The ionization of an acid in water produces a **hydrogen ion,** $H^+$, and the anion of the acid.

$$HCl(aq) \rightarrow H^+(aq) + Cl^-(aq)$$

**Arrhenius bases** are species that release a hydroxide ion, $OH^-$, upon dissociation. Strong soluble bases such as NaOH, KOH, and $Ba(OH)_2$ are Arrhenius bases.

$$NaOH(aq) \rightarrow Na^+ (aq) + OH^-(aq)$$

A **Brønsted-Lowry acid** is a proton donor, exactly like an Arrhenius acid. A **Brønsted-Lowry base** is a proton acceptor.

The anion that is produced upon the release of the proton from an acid is the **conjugate base** of the acid. It is called the conjugate base because it could accept a proton under appropriate conditions.

In the ionization reaction of HCl shown below, $Cl^-$ is the conjugate base of HCl and $H_3O^+$ is the conjugate acid of $H_2O$. A conjugate acid/base pair differs by one $H^+$.

**Conjugate Acid/Conjugate Base Relationships**

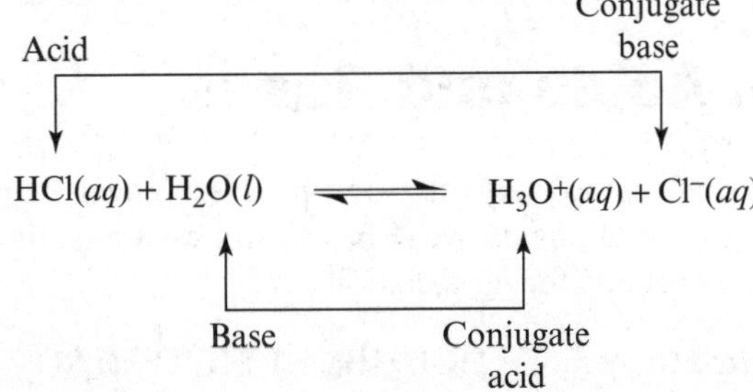

A Brønsted-Lowry base such as ammonia, $NH_3$, reacts with water by accepting a proton. The resulting ammonium ion, $NH_4^+$, is called the **conjugate acid** of the base, because it could donate a proton under appropriate conditions. The conjugate base of water, $OH^-$, is produced in the reaction.

**Conjugate Acid/Conjugate Base Relationships**

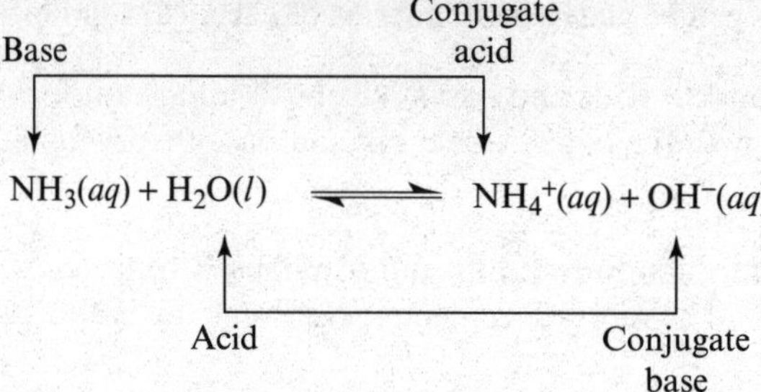


Notice that water is behaving as a base in the reaction with HCl and as an acid in the reaction with $NH_3$. Water is an amphoteric species. An **amphoteric** species can act as both an acid and a base. It is also an **amphiprotic** species, one that can either donate or accept a proton. (Since all amphiprotic species are amphoteric, but not all amphoteric species are amphiprotic, it is best to use "amphoteric" in free-response answers.)

## Practice

Which of the following species are amphoteric? Write two equations for each amphoteric species. One equation should show the species reacting with water as a Brønsted-Lowry acid, and the other equation should show the species reacting with water as a Brønsted-Lowry base.

$$HCO_3^-,\ SO_4^{2-},\ HPO_4^{2-}$$

$HCO_3^-$ and $HPO_4^{2-}$ are both amphoteric species. $SO_4^{2-}$ can accept a proton, but it does not have a proton to donate, so it is not amphoteric.

The reactions of $HCO_3^-$ and $HPO_4^{2-}$, in which the species are behaving as acids, are

$$HCO_3^- + H_2O \rightarrow CO_3^{2-} + H_3O^+$$
$$HPO_4^{2-} + H_2O \rightarrow PO_4^{3-} + H_3O^+$$

The reactions of $HCO_3^-$ and $HPO_4^{2-}$, in which the species are behaving as bases, are

$$HCO_3^- + H_2O \rightarrow H_2CO_3 + OH^-$$
$$HPO_4^{2-} + H_2O \rightarrow H_2PO_4^- + OH^-$$

# Strong and Weak Acids and Bases

**Strong acids** and **strong bases** dissociate completely into ions in water; they are strong electrolytes with large equilibrium constants for dissociation. Weak acids, weak bases, and weakly-soluble hydroxides are weak electrolytes with small equilibrium constants for dissociation.

Six strong acids should be committed to memory: HCl, HBr, HI, $HClO_4$, $H_2SO_4$, and $HNO_3$. In addition, hydronium is always considered to be a strong acid.

Other strong, oxygen-containing ternary acids (made up of three elements) can be recognized by the fact that they are composed of at least two more oxygen atoms than hydrogen atoms. Both $HClO_3$ and $HBrO_4$ are strong acids. They each have at least two more oxygen atoms than hydrogen atoms.

Strong bases dissociate completely in water. They are the group 1 hydroxides, LiOH, NaOH, KOH, RbOH, and CsOH. Two hydroxides from group 2, $Ba(OH)_2$ and $Sr(OH)_2$, also dissociate completely and are considered to be strong bases.

| Strong Acids to Memorize: | Strong Bases to Memorize: |
|---|---|
| HCl | LiOH |
| HBr | NaOH |
| HI | KOH |
| $HClO_4$ | RbOH |
| $H_2SO_4$ | CsOH |
| $HNO_3$ | $Ba(OH)_2$ |
| | $Sr(OH)_2$ |

A **weak acid** is any acid that partially ionizes in an aqueous solution. Some weak acids are HF, $H_2CO_3$, $H_3PO_4$, and $HNO_2$, among many others. In addition, all **carboxylic acids,** which are organic acids containing the functional group $-CO_2H$, are weak acids. An example of an organic acid is acetic acid, $CH_3CO_2H$. When it loses its acidic proton, $H^+$, it is converted to its conjugate base, an acetate ion.

**A Weak Acid and Its Conjugate Base**

Weak acid

Conjugate base

Acidic hydrogen

Acetic acid

Acetate anion

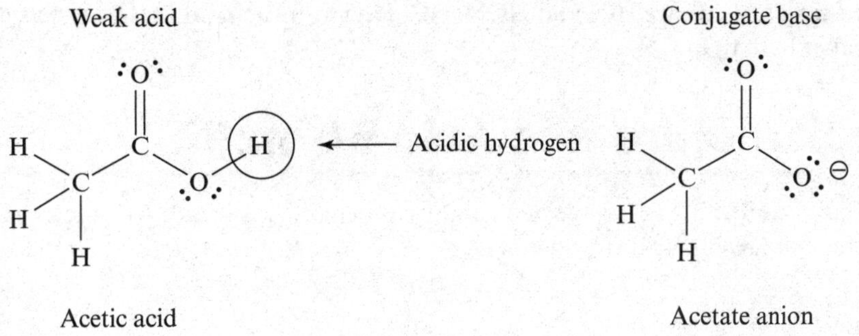

Weak bases are either weakly-soluble hydroxides or amines. Weakly-soluble hydroxides include compounds such as $Mg(OH)_2$ or $Al(OH)_3$. Ammonia, $NH_3$, and the amines are common weak bases discussed in AP Chemistry. The amines have the general formula $NR_3$, where R is either hydrogen or a hydrocarbon group. Examples of amines are $CH_3NH_2$, $(CH_3)_2NH$, and $(CH_3)_3N$.

The stronger an acid, the weaker its conjugate base. The stronger a base, the weaker its conjugate acid.

- HCl is a very strong acid, and its conjugate base, $Cl^-$, is a very weak base.
- Acetic acid, $CH_3CO_2H$, is a weaker acid than HCl. Its conjugate base, acetate, $CH_3CO_2^-$, is a stronger base than $Cl^-$.
- Ammonia, $NH_3$, is a stronger base than acetate, $CH_3CO_2^-$. The conjugate acid of ammonia, $NH_4^+$, is a weaker acid than acetic acid, $CH_3CO_2H$, the conjugate acid of acetate.

Notice that these generalizations refer to relative strength. Acetate, $CH_3CO_2^-$, is not a strong base. It is merely a stronger base than $Cl^-$.

## Practice

Determine the weaker acid in each of the following pairs:

**a.** $H_2O$ or $NH_4^+$
**b.** $(CH_3)_2NH$ or $(CH_3)_2NH_2^+$

To determine the weaker acid, look for the stronger conjugate base in each pair.

**(a)** In this pair, the conjugate bases are $OH^-$ and $NH_3$. The hydroxide ion is a much stronger base than ammonia, a weak base. Water, the conjugate acid of hydroxide, is a weaker acid than $NH_4^+$, the conjugate acid of ammonia.

**(b)** In this pair, the conjugate bases are $(CH_3)_2N^-$ and $(CH_3)_2NH$, respectively. The stronger conjugate base is $(CH_3)_2N^-$. When a weakly basic amine such as $(CH_3)_2NH$ loses a proton, the resulting anionic species is strongly basic. The conjugate base of the second species is $(CH_3)_2NH$. It is a weak base, so its conjugate acid is the stronger acid. The first species, $(CH_3)_2NH$, is the weaker acid.

## Practice

Determine the weaker base in each of the following pairs:

**a.** $H_2O$ or $HSO_3^-$
**b.** $Cl^-$ or $CN^-$

To determine the weaker base, look for the stronger conjugate acid in each pair.

**(a)** In this pair, the conjugate acids are $H_3O^+$ and $H_2SO_3$. Hydronium, $H_3O^+$, is a strong acid, and $H_2SO_3$ is a weak acid. Therefore, water is a weaker base than $HSO_3^-$.

**(b)** In this pair, the conjugate acids are HCl and HCN. HCl is a strong acid and HCN is a weak acid. Therefore, chloride, $Cl^-$, is a weaker base than $CN^-$.

# Molecular Structure and Acid Strength

Several structural factors contribute to a molecule's ability to behave as an acid. Any molecular feature that stabilizes the conjugate base results in a stronger acid.

## Resonance

A molecule's ability to donate a proton is influenced by resonance stabilization of its conjugate base.

For example, when a proton is removed from ethanol, resonance stabilization of the conjugate base is not possible. When a proton is removed from acetic acid, however, the resulting conjugate base is stabilized by resonance. Because the deprotonated form of acetic acid is more stable, the loss of its proton is more favorable; acetic acid is a stronger acid than ethanol.

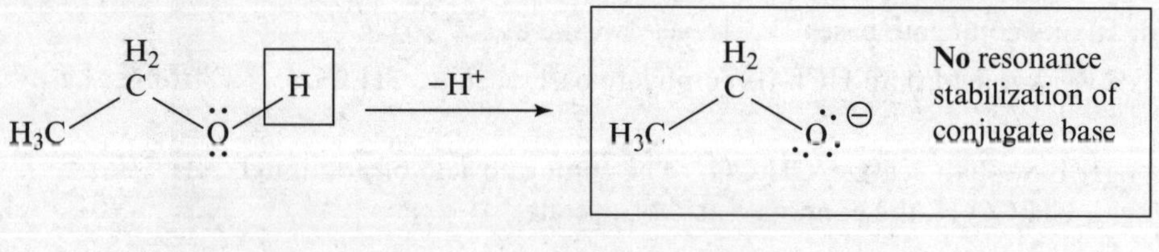

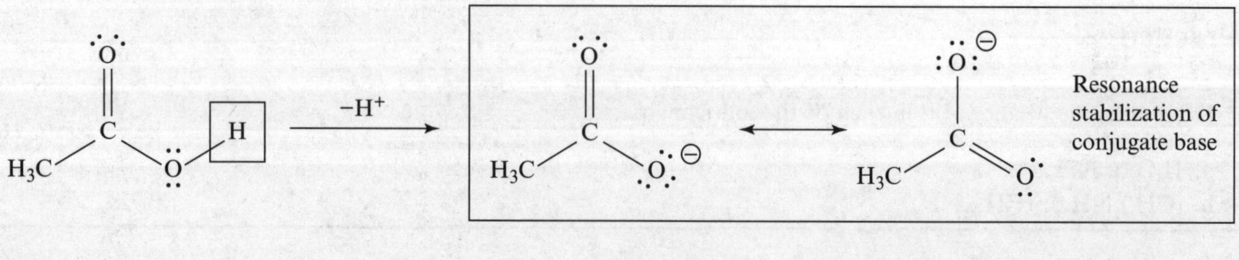

$H_3C$–$CH_2$–O–H $\xrightarrow{-H^+}$ $H_3C$–$CH_2$–$O^-$

No resonance stabilization of conjugate base

$H_3C$–C(=O)–O–H $\xrightarrow{-H^+}$ $H_3C$–C(=O)–$O^-$ $\longleftrightarrow$ $H_3C$–C($O^-$)=O

Resonance stabilization of conjugate base

## Electronegativity

The more electronegative an atom, the less it tends to share its electrons. Anions of more electronegative atoms tend to be weaker bases that have stronger conjugate acids.

Consider HF, $H_2O$, $NH_3$, and $CH_4$. The conjugate bases of these acids are $F^-$, $OH^-$, $NH_2^-$, and $CH_3^-$, respectively. The order of electronegativity is F > O > N > C.

Fluorine is the most electronegative of the atoms, so $F^-$ is the weakest conjugate base, followed by $OH^-$ and $NH_2^-$. Carbon is the least electronegative, so the conjugate base $CH_3^-$ is the strongest conjugate base.

The order of the acid strength is HF > $H_2O$ > $NH_3$ > $CH_4$.

## Inductive Effects

Conjugate bases can also be influenced by **inductive effects,** which are effects that occur in one part of the molecule due to the presence of electronegative elements in another part of the molecule.

Consider $H_2FC–CO_2H$, $H_2ClC–CO_2H$, and $H_2BrC–CO_2H$, as shown in the figure below.

The electronegativity ranking for the atoms bound to carbon is F > Cl > Br. The conjugate base $H_2FC–CO_2^-$ is the most stabilized by the inductive effect due to the presence of F, so it is the weakest of the conjugate bases. The conjugate base $H_2BrC–CO_2^-$ is the least stabilized by the inductive effect due to the presence of Br, so it is the strongest of the conjugate bases.

Increasing electronegativity of circled atom

Increasing stabilization of conjugate base

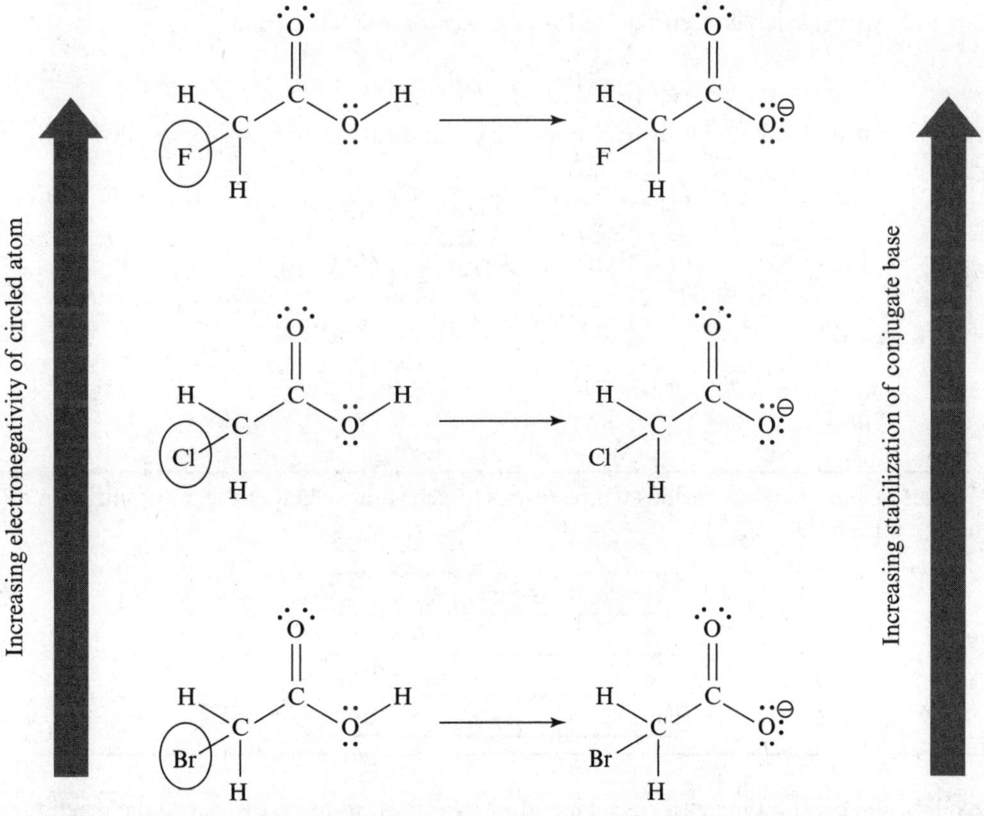

The order of acidity of the acids is $H_2FC–CO_2H$ > $H_2ClC–CO_2H$ > $H_2BrC–CO_2H$.

### Practice

Place the following acids in order of increasing acid strength: H–O–Cl, H–O–F, H–O–Br.

The order of acid strength is H–O–Br < H–O–Cl < H–O–F.

Increasing electronegativity of the halogen bound to oxygen stabilizes the conjugate base of the acid when the proton is lost. The order of increasing electronegativity of the halogens is Br < Cl < F.

## The Autoionization of Water

As discussed earlier, water is amphoteric; it can behave as either an acid or a base. Pure water undergoes **autoionization** according to the reaction illustrated below. ***Note:*** Both $H_3O^+$ and $H^+$ are found on the AP Chemistry curriculum. Be prepared to use either. It is also important to understand that free $H^+$ does not exist in reality. In the presence of water, $H^+$ always exists as $H_3O^+$.

$$2\ H_2O(l) \rightleftharpoons H_3O^+(aq) + OH^-(aq) \qquad \text{or} \qquad H_2O(l) \rightleftharpoons H^+(aq) + OH^-(aq)$$

$$K_w = [H_3O^+][OH^-] = [H^+][OH^-] = 1.0 \times 10^{-14}$$

The equilibrium constant for this reaction, $K_w$, is $1.0 \times 10^{-14}$ at 25 °C. This extremely small value means that neutral water is ionized only to a very small degree.

The hydronium and hydroxide ion concentrations for pure, *neutral* water are equal.

$$[H_3O^+] = [OH^-]$$

The concentrations of hydronium and hydroxide ions at 25 °C are $1.0 \times 10^{-7}\ M$, as determined by $K_w$:

$$K_w = \left[H_3O^+\right]\left[OH^-\right] = \left[H_3O^+\right]^2 = 1.0 \times 10^{-14}$$

$$\sqrt{\left[H_3O^+\right]^2} = \sqrt{1.0 \times 10^{-14}}$$

$$\left[H_3O^+\right] = \left[OH^-\right] = 1.0 \times 10^{-7}\ M$$

### Practice

Use the $K_w$ values of pure water at various temperatures to determine whether the autoionization of water is endothermic or exothermic.

| Temperature (°C) | $K_w$ |
|---|---|
| 25 | $1.0 \times 10^{-14}$ |
| 30 | $1.5 \times 10^{-14}$ |
| 50 | $5.5 \times 10^{-14}$ |

Because the equilibrium constant increases as temperature increases, addition of heat to the reaction causes the equilibrium to shift toward the products. The autoionization of water is endothermic.

## The pH and pOH Scales

The **pH scale** is a commonly used measure of the acidity of a solution. The **pH** is defined as

$$pH = -\log[H^+] = -\log[H_3O^+]$$

The pH scale ranges from below 0 for strongly acidic solutions to over 14 for strongly basic solutions. At 25 °C, a solution with a pH that is less than 7 is acidic, and a solution with a pH that is greater than 7 is basic.

As discussed previously, for pure, neutral water at 25 °C,

$$[H_3O^+] = 1.0 \times 10^{-7}\ M$$

The pH of pure, neutral water at 25 °C is

$$pH = -\log[H_3O^+] = -\log(1.0 \times 10^{-7}) = 7.00$$

Because pH is based on a logarithm scale, a solution with a pH of 6 is ten times more acidic than pure water at room temperature, which has a pH of 7. A solution with a pH of 5 has 100 times the hydronium ion concentration of pure water ($1.0 \times 10^{-5}\ M$), and a solution with a pH of 4 has 1,000 times the hydronium ion concentration ($1.0 \times 10^{-4}\ M$) of pure water.

The **pOH** of a solution is defined as

$$pOH = -\log[OH^-]$$

Because pure water at room temperature has a hydroxide ion concentration of $1.0 \times 10^{-7}\ M$, the pOH of water at 25 °C is

$$pOH = -\log[1.0 \times 10^{-7}] = 7.00$$

Remember that for water at 25 °C,

$$[H_3O^+][OH^-] = 1.0 \times 10^{-14}$$

Therefore,

$$pH + pOH = 14.00$$

**TIP: When taking the logarithm of a value, the number of significant figures in the value becomes the number of digits to the right of the decimal. For example, log $2.734 \times 10^5$ = 5.4368. The value 2.734 has four significant figures, and the answer, 5.4368, is rounded to four digits to the right of the decimal.**

### Practice

What is the pH of a solution that has a hydronium ion concentration of 0.0200 $M$ at 25 °C?

The pH of a solution is the negative logarithm of the hydronium ion concentration.

$$pH = -\log[H_3O^+] = -\log 0.0200 = 1.699$$

### Practice

The pH of a solution is 3.20 at 25 °C. What is the hydroxide ion concentration at this temperature?

The pH is first converted to pOH.

$$pOH = 14.00 - 3.20 = 10.80$$

The pOH is then used to determine the $[OH^-]$.

$$10.80 = -\log[OH^-]$$

$$[OH^-] = 10^{-10.80} = 1.6 \times 10^{-11}\ M$$

## Strong Acids and pH

The six strong acids, HCl, HBr, HI, $HClO_4$, $H_2SO_4$, and $HNO_3$, have very large equilibrium constants for the dissociation of a proton. (Recall that **dissociation** means that the proton is released completely from the acid to form a charged species, $H_3O^+$ or $H^+$, in an aqueous solution.)

Because the equilibrium constants for dissociation of the protons of strong acids are so large, the pH of any solution of a strong acid is calculated by assuming that the dissociation is complete, and the concentration of the acid is equal to the concentration of hydronium.

For a sample of nitric acid in water, for instance, we assume that all of the $HNO_3$ has been dissociated to form $H_3O^+$.

$$[HNO_3] = [H_3O^+]$$

### Practice

What is the pH of a 0.00010 *M* aqueous solution of HBr at 25 °C? (Notice that you should be able to do this without a calculator.)

Hydrobromic acid is one of the strong acids. It has a very large equilibrium constant for dissociation, so it essentially dissociates completely in an aqueous solution.

$$HBr(aq) + H_2O(l) \rightarrow Br^-(aq) + H_3O^+(aq)$$

The concentration of $H_3O^+$ after the dissociation occurs is equal to the initial concentration of HBr, $1.0 \times 10^{-4}$ *M*.

$$pH = -\log(1.0 \times 10^{-4}) = 4.00$$

## Strong Bases and pOH

The strong bases that should be memorized are the hydroxides of the alkali metals: LiOH, NaOH, KOH, RbOH, and CsOH. In addition to the group 1 hydroxides, hydroxides of the alkaline earth metals, $Sr(OH)_2$ and $Ba(OH)_2$, are considered to be strong.

These species have very large equilibrium constants for their dissociation reactions. In this case, dissociation refers to the release of the hydroxide ion from the metal cation in solution.

$$LiOH(aq) \rightarrow Li^+(aq) + OH^-(aq)$$

The pOH of any strongly basic solution is calculated using the hydroxide ion concentration.

### Practice

What is the pH of a $1.00 \times 10^{-4}$ *M* aqueous solution of $Ba(OH)_2$ at 25 °C?

At low concentrations, $Ba(OH)_2$, a strong base, dissociates completely into $Ba^{2+}$ and $OH^-$.

$$Ba(OH)_2(aq) \rightarrow Ba^{2+}(aq) + 2\ OH^-(aq)$$

It is apparent from the reaction stoichiometry that 2 hydroxide ions are released for every $Ba(OH)_2$ molecule dissolved, so the concentration of $OH^-$ is $2.0 \times 10^{-4}$ *M*.

The pOH can be determined from the $OH^-$ concentration.

$$pOH = -\log(2.0 \times 10^{-4}) = 3.70$$

Finally, in order to find the pH, the pOH is subtracted from 14.

$$pOH + pH = 14.00$$
$$pH = 14.00 - pOH$$
$$pH = 14.00 - 3.70 = 10.30$$

## Weak Acids, $K_a$, and Percent Ionization

**Weak acids** are acids that do not dissociate completely in an aqueous solution.

The equilibrium constant for the dissociation of a weak acid such as acetic acid is known as the **acid dissociation constant,** $K_a$. The reaction of the weak acid, HA, with water to form the conjugate base, $A^-$, and hydronium is illustrated by the equation below. Recall that pure liquids ($H_2O(l)$ in this case) are excluded from the equilibrium mass action expression.

$$HA(aq) + H_2O(l) \rightleftharpoons H_3O^+(aq) + A^-(aq)$$
$$K_a = \frac{[H_3O^+][A^-]}{[HA]}$$

It is also acceptable in AP Chemistry to write $H^+$ in the equation for the dissociation of an acid.

$$HA(aq) \rightleftharpoons H^+(aq) + A^-(aq)$$
$$K_a = \frac{[H^+][A^-]}{[HA]}$$

The $K_a$ for acetic acid is $1.8 \times 10^{-5}$, which is quite small. This means that of the acetic acid dissolved in water, only a few of the molecules are dissociated into ions.

It is often useful to calculate percent ionization and pH for solutions of weak acids given the initial concentration of acid. The **percent ionization** is the percentage of a species in solution that is dissociated.

For example, suppose a question asks for the percent ionization of a 0.100 *M* solution of acetic acid, $CH_3CO_2H$, in addition to the pH of the solution, given the acid dissociation constant, $K_a = 1.8 \times 10^{-5}$. The first step is to write the equation for the dissociation of acetic acid in water.

$$CH_3CO_2H(aq) + H_2O(l) \rightleftharpoons H_3O^+(aq) + CH_3CO_2^-(aq)$$

The next step is to set up a RICE table to determine the equilibrium concentrations of the species in solution.

| **Reaction** | $CH_3CO_2H$ | $H_3O^+$ | $CH_3CO_2^-$ |
|---|---|---|---|
| **Initial Concentrations (*M*)** | 0.100 | 0 | 0 |
| **Change** | $-x$ | $+x$ | $+x$ |
| **Equilibrium Concentrations (*M*)** | $0.100 - x$ | $x$ | $x$ |

The equilibrium constant, $1.8 \times 10^{-5}$, is equal to the mass action expression for the dissociation of acetic acid.

$$K_a = \frac{[H_3O^+][CH_3CO_2^-]}{[CH_3CO_2H]}$$

$$1.8 \times 10^{-5} = \frac{x^2}{0.100 - x}$$

Instead of using the quadratic formula to solve for $x$, it is acceptable to use the shortcut that was discussed in Chapter 7, "Chemical Equilibrium." An equilibrium constant of $1.8 \times 10^{-5}$ indicates that very little of the acid is dissociated. So little is dissociated that $x$ may be neglected in the denominator since subtraction of $x$ from 0.100 does not significantly change the value of the denominator. Students are expected to be able to provide this justification for simplifying the calculation.

$$1.8 \times 10^{-5} = \frac{x^2}{0.100 - \not{x}}$$

$$(1.8 \times 10^{-5})(0.100) = x^2$$

$$\sqrt{(1.8 \times 10^{-5})(0.100)} = \sqrt{x^2}$$

$$x = 1.3 \times 10^{-3}\ M = [H_3O^+] = [CH_3CO_2^-]$$

The pH of the solution is determined using the $H_3O^+$ concentration. Even though only two significant figures can be reported for the concentration of $H_3O^+$, the entire calculator result for $x$ is used to calculate the pH.

$$pH = -\log[H_3O^+] = -\log(1.34164 \times 10^{-3}) = 2.87$$

The percent ionization is the percentage of ionized acetate ion in the total acetic acid/acetate concentration. Again, the unrounded value of the concentration should be used.

$$\%\text{ ionization} = \frac{[CH_3CO_2^-]}{[CH_3CO_2^-] + [CH_3CO_2H]} \times 100\% = \frac{1.3 \times 10^{-3}\ \not{M}}{0.100\ \not{M}} \times 100\% = 1.3\%\text{ ionized}$$

## Practice

A sample of 0.0100 $M$ hypochlorous acid, HClO, is 0.17 % ionized at 25 °C. What is the acid dissociation constant for HClO?

The dissociation of HClO is shown by the following equation.

$$HClO(aq) + H_2O(l) \rightleftharpoons ClO^-(aq) + H_3O^+(aq)$$

The concentration of $ClO^-$ is 0.17% of the initial concentration of HClO. This is equal to the hydronium ion concentration.

$$[ClO^-] = (0.0017)(0.0100\ M) = 1.7 \times 10^{-5}\ M = [H_3O^+]$$

The HClO concentration at equilibrium is the initial concentration minus the portion that ionized.

$$[HClO] = 0.0100\ M - 1.7 \times 10^{-5}\ M \approx 0.0100\ M$$

$$K_a = \frac{[H_3O^+][ClO^-]}{[HClO]} = \frac{(1.7 \times 10^{-5})^2}{0.0100} = 2.9 \times 10^{-8}$$

# Weak Bases and $K_b$

**Weak bases** are those that only weakly ionize in an aqueous solution. They ionize according to the reaction illustrated by the equation shown.

$$B(aq) + H_2O(l) \rightleftharpoons BH^+(aq) + OH^-(aq)$$

The equilibrium constant for base dissociation according to this equation is known as $K_b$.

$$K_b = \frac{[BH^+][OH^-]}{[B]}$$

Weak bases include **amines,** compounds that contain a nitrogen atom bound to carbon or hydrogen. Ammonia, for instance, is a weak base. It is the simplest amine. Other examples of weakly basic amines include methylamine, $CH_3NH_2$, and pyridine, $C_5H_5N$. The structures of these bases are shown in the figure below.

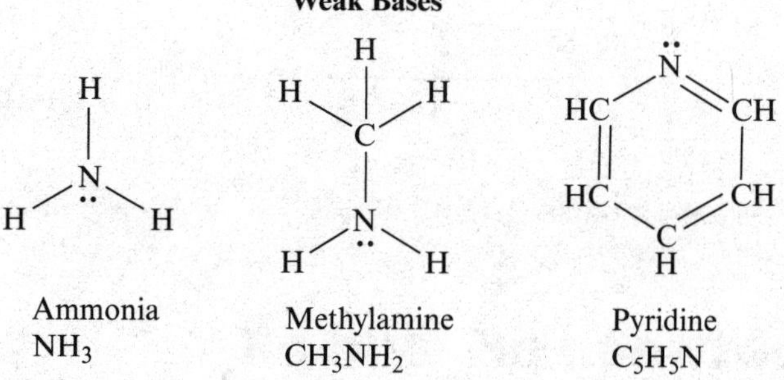


When one of these bases accepts a proton, the result is a positive formal charge on the nitrogen atom.

**Conjugate Acids of Weak Bases**

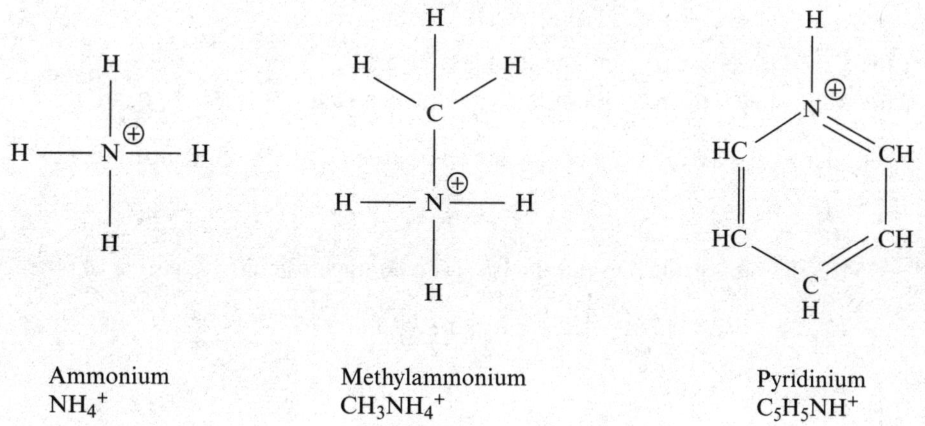

Ammonium
$NH_4^+$

Methylammonium
$CH_3NH_4^+$

Pyridinium
$C_5H_5NH^+$

The pH and percent ionization of a weak base solution are determined in a manner similar to those of a weak acid solution.

## Practice

The $K_b$ of aniline, $C_6H_5NH_2$, is $3.8 \times 10^{-10}$. What is the pH of a 0.150 *M* solution of aniline?

Aniline is a weak base. The dissociation reaction for aniline in an aqueous solution is represented by:

$$C_6H_5NH_2(aq) + H_2O(l) \rightleftharpoons C_6H_5NH_3^+(aq) + OH^-(aq)$$

A RICE table is used to determine the concentrations.

| Reaction | $C_6H_5NH_2$ | $C_6H_5NH_3^+$ | $OH^-$ |
|---|---|---|---|
| Initial Concentrations (*M*) | 0.150 | 0 | 0 |
| Change | $-x$ | $+x$ | $+x$ |
| Equilibrium Concentrations (*M*) | $0.150 - x$ | $x$ | $x$ |

The mass action expression for the $K_b$ is

$$K_b = \frac{[C_6H_5NH_3^+][OH^-]}{[C_6H_5NH_2]}$$

Substitution of the equilibrium concentrations into the expression for $K_b$ and the assumption that $x$ is much smaller than the initial concentration of aniline allow for the calculation of the hydroxide ion concentration.

$$3.8 \times 10^{-10} = \frac{x^2}{0.150 - \cancel{x}}$$

$$x = \sqrt{(3.8 \times 10^{-10})(0.150)} = 7.6 \times 10^{-6}\ M = [OH^-]$$

$$pOH = -\log 7.6 \times 10^{-6} = 5.12$$

$$pH = 14.00 - 5.12 = 8.88$$

## Relationship between p$K_a$ and p$K_b$

A convenient expression for the acidity or basicity of a substance is given by its **p$K_a$** or its **p$K_b$**.

$$pK_a = -\log K_a$$

$$pK_b = -\log K_b$$

- The higher the $K_a$, the lower the p$K_a$ and the stronger the acid.
- The higher the $K_b$, the lower the p$K_b$ and the stronger the base.

The relationship between $K_w$, $K_a$, and $K_b$ for a conjugate acid-base pair at a given temperature is

$$K_w = K_a \cdot K_b$$

Since $K_w = 10^{-14}$ at 25 °C, the sum of the p$K_a$ and the p$K_b$ for a conjugate acid-base pair is 14.00 at 25 °C.

$$pK_w = pK_a + pK_b = 14.00$$

### Practice

Based on the provided data, is benzoic acid or formic acid a stronger acid? Determine the $K_a$ of each acid.

| Acid | p$K_a$ at 25 °C |
|---|---|
| Benzoic acid | 4.20 |
| Formic acid | 3.75 |

The lower p$K_a$ of formic acid indicates that it is a stronger acid.

The $K_a$ of formic acid is

$$K_a = 10^{-pK_a} = 10^{-3.75} = 1.8 \times 10^{-4}$$

The $K_a$ of benzoic acid is

$$K_a = 10^{-pK_a} = 10^{-4.20} = 6.3 \times 10^{-5}$$

The larger $K_a$ and lower p$K_a$ of formic acid confirm that it ionizes to a greater extent (and produces more hydronium) than benzoic acid.

### Practice

The $K_a$ of hydrofluoric acid, HF, is $7.2 \times 10^{-4}$ at 25 °C. What is the p$K_b$ of the fluoride ion?

The $K_b$ for $F^-$, the conjugate base of HF, is determined using the $K_a$ and $K_w$ at 25 °C.

$$K_w = K_a \cdot K_b$$

$$K_b = \frac{K_w}{K_a} = \frac{1.00 \times 10^{-14}}{7.2 \times 10^{-4}} = 1.4 \times 10^{-11}$$

$$pK_b = -\log\ K_b = -\log\ 1.4 \times 10^{-11} = 10.86$$

# Acidic and Basic Salts

Many salts, when added to neutral water, have no effect on the pH of the water. If sodium chloride is added to a sample of pure water at 25 °C, the pH of the water does not change. When a sample of sodium acetate is added to pure water, however, the pH increases. If ammonium chloride is added to water, the pH decreases. Sodium acetate, therefore, is referred to as a basic salt and ammonium chloride is an acidic salt.

## Cation of Strong Base/Anion of Strong Acid

A salt that is composed of the cation of a strong base and the anion of a strong acid is a **neutral salt.**

Sodium chloride, the neutral salt mentioned above, is composed of sodium cations and chloride anions. A sodium ion is the cation of a strong base, NaOH, and a chloride anion is the anion of a strong acid, HCl. When added to pure water, sodium chloride does not cause a change in the pH. For this reason, sodium chloride is referred to as a neutral salt. Other examples of neutral salts include $KNO_3$, LiBr, and $BaCl_2$.

## Cation of Strong Base/Anion of Weak Acid

A salt that is composed of the cation of a strong base and the anion of a weak acid, such as sodium acetate, $NaCH_3CO_2$, is a **basic salt.** Sodium acetate dissociates completely in water, and the acetate anions that are released increase the pH of an aqueous solution by extracting protons from the water molecules. This results in an increase in the hydroxide ion concentration.

$$CH_3CO_2^-(aq) + H_2O(l) \rightleftharpoons CH_3CO_2H(aq) + OH^-(aq)$$

Other examples of basic salts include KCN, LiF, and $Cs_2CO_3$.

## Cation of Weak Base/Anion of Strong Acid

A salt that is composed of the cation of a weak base and the anion of a strong acid, such as ammonium chloride, $NH_4Cl$, is an **acidic salt.** Ammonium chloride is soluble in water, and it dissociates completely. The ammonium ion releases a proton into the solution and decreases the pH.

$$NH_4^+(aq) + H_2O(l) \rightleftharpoons NH_3(aq) + H_3O^+(aq)$$

### Practice

Predict whether solutions of the following salts will be acidic, basic, or neutral when added to pure water. Support your predictions with balanced equations for hydrolysis to produce the ions that contribute to an acidic or basic pH.

**a.** $LiNO_2$
**b.** $NH_4Br$
**c.** $NaClO_4$

**(a)** $Li^+$ is the cation of a strong base. $NO_2^-$ is the conjugate base of a weak acid. $LiNO_2$ is a basic salt.

$$NO_2^-(aq) + H_2O(l) \rightleftharpoons HNO_2(aq) + OH^-(aq)$$

**(b)** $NH_4^+$ is the conjugate acid of a weak base. $Br^-$ is the anion of a strong acid. $NH_4Br$ is an acidic salt.

$$NH_4^+(aq) + H_2O(l) \rightleftharpoons NH_3(aq) + H_3O^+(aq)$$

**(c)** $Na^+$ is the cation of a strong base. $ClO_4^-$ is the anion of a strong acid. $NaClO_4$ is a neutral salt.

### Practice

The $K_b$ for ammonia is $1.8 \times 10^{-5}$. Find the pH of a 0.0500 $M$ solution of ammonium chloride.

Ammonium chloride is an acidic salt, since ammonium is the conjugate acid of a weak base and chloride is the conjugate base of a strong acid.

To find the pH, it is necessary to work with $K_a$ since the ammonium ion is the conjugate acid of ammonia.

$$K_w = K_a \cdot K_b$$

$$K_a = \frac{K_w}{K_b} = \frac{1.0\times10^{-14}}{1.8\times10^{-5}} = 5.6\times10^{-10}$$

The dissociation reaction for ammonium is

$$NH_4^+(aq) + H_2O(l) \rightleftharpoons NH_3(aq) + H_3O^+(aq)$$

A RICE table can be used to find the concentration of $H_3O^+$.

| Reaction | $NH_4^+$ | $NH_3$ | $H_3O^+$ |
|---|---|---|---|
| Initial Concentrations ($M$) | 0.0500 | 0 | 0 |
| Change | $-x$ | $+x$ | $+x$ |
| Equilibrium Concentrations ($M$) | $0.0500 - x$ | $x$ | $x$ |

Because the acid dissociation constant is small, the $K_a$ expression can be simplified and $x$ can be neglected from the $(0.0500 - x)$ term.

$$K_a = \frac{[NH_3][H_3O^+]}{[NH_4^+]}$$

$$5.6\times10^{-10} = \frac{x^2}{0.0500 - \cancel{x}}$$

$$\sqrt{(5.6\times10^{-10})(0.0500)} = \sqrt{x^2}$$

$$x = 5.3\times10^{-6}\ M = [H_3O^+]$$

$$pH = -\log[H_3O^+] = -\log 5.3\times10^{-6} = 5.28$$

## Practice

The $K_a$ for formic acid, $H_2CO_2$, is $1.77 \times 10^{-4}$. What concentration of sodium formate, $NaHCO_2$, would have a pH of 8.00 at 25 °C?

Sodium formate is a basic salt since the sodium ion is the cation of a strong base and formate is the anion of a weak acid. The species of interest is the formate ion, the conjugate base of formic acid. Therefore, $K_a$ must be converted to $K_b$.

$$K_w = K_a \cdot K_b$$

$$K_b = \frac{K_w}{K_a} = \frac{1.00\times10^{-14}}{1.77\times10^{-4}} = 5.65\times10^{-11}$$

The hydrolysis reaction of the formate ion is

$$HCO_2^-(aq) + H_2O(l) \rightleftharpoons H_2CO_2(aq) + OH^-(aq)$$

This time the pH is given, so the final concentration of the hydroxide ion can be calculated.

$$pOH = 14.00 - 8.00 = 6.00$$

$$[OH^-] = 10^{-6.00} = 1.0 \times 10^{-6}\ M$$

Let the initial concentration of the formate ion be $x$.

| Reaction | $HCO_2^-$ | $H_2CO_2$ | $OH^-$ |
|---|---|---|---|
| Initial Concentrations ($M$) | $x$ | 0 | 0 |
| Change | $-1.0 \times 10^{-6}$ | $+1.0 \times 10^{-6}$ | $+1.0 \times 10^{-6}$ |
| Equilibrium Concentrations ($M$) | $x - 1.0 \times 10^{-6}$ | $1.0 \times 10^{-6}$ | $1.0 \times 10^{-6}$ |

$$K_b = \frac{[OH^-][H_2CO_2]}{[HCO_2^-]} = \frac{(1.0\times10^{-6})^2}{x - 1.0\times10^{-6}} = 5.65\times10^{-11}$$

$$x = 1.8\times10^{-2}\ M$$

The initial concentration of $NaHCO_2$ is $1.8 \times 10^{-2}\ M$.

## Cation of Weak Base/Anion of Weak Acid

Sometimes a salt is composed of the conjugate base of a weak acid and the conjugate acid of a weak base. An example of this kind of salt is ammonium acetate, $NH_4CH_3CO_2$. The pH of a solution of this type of salt is determined by the ionization constants of the parent acid and base.

The parent base for ammonium, $NH_4^+$, is ammonia, $NH_3$. The acid dissociation constant for the hydrolysis of the ammonium ion is determined from the $K_b$ for ammonia.

$$NH_4^+(aq) + H_2O(l) \rightleftharpoons NH_3(aq) + H_3O^+(aq) \qquad K_b\text{ (ammonia)} = 1.8\times10^{-5}$$

$$K_a = \frac{K_w}{K_b} = \frac{1.0\times10^{-14}}{1.8\times10^{-5}} = 5.6\times10^{-10}$$

The parent acid for the acetate ion is acetic acid. The base dissociation constant for the hydrolysis of the acetate ion is determined by the $K_a$ for acetic acid.

$$CH_3CO_2^-(aq) + H_2O(l) \rightleftharpoons CH_3CO_2H(aq) + OH(aq) \qquad K_a\text{ (acetic acid)} = 1.8\times10^{-5}$$

$$K_b = \frac{K_w}{K_a} = \frac{1.0\times10^{-14}}{1.8\times10^{-5}} = 5.6\times10^{-10}$$

Because the hydrolysis of ammonium produces the same concentration of hydronium as the concentration of hydroxide produced by the hydrolysis of acetate, as shown by the fact that the dissociation constants are equal, an ammonium acetate salt solution is neutral.

If, however, the $K_a$ for the conjugate acid of the weak base is greater than the $K_b$ for the conjugate base of the weak acid, $H_3O^+$ will be produced to a greater extent than $OH^-$, and the solution will be acidic. To see why this is the case, consider pyridinium formate, $C_5H_5NHCHO_2$. Pyridinium formate is composed of a pyridinium cation, $C_5H_5NH^+$, and a formate anion, $CHO_2^-$. The weak base and weak acid are $C_5H_5N$ ($K_b = 1.8 \times 10^{-9}$) and $CHO_2H$ ($K_a = 1.8 \times 10^{-4}$), respectively. The hydrolysis reactions for the two ions are shown below.

$$C_5H_5NH^+(aq) + H_2O(l) \rightleftharpoons C_5H_5N(aq) + H_3O^+(aq)$$

$$K_a = \frac{K_w}{K_b} = \frac{1.0\times10^{-14}}{1.8\times10^{-9}} = 5.6\times10^{-6}$$

$$CHO_2^-(aq) + H_2O(l) \rightleftharpoons CHO_2H(aq) + OH^-(aq)$$

$$K_b = \frac{K_w}{K_a} = \frac{1.0\times10^{-14}}{1.8\times10^{-4}} = 5.6\times10^{-11}$$

Because the hydrolysis of pyridinium to form hydronium occurs to a much greater extent than the formation of hydroxide by formate ($10^{-6}$ is much greater than $10^{-11}$), a solution of the pyridinium formate is acidic.

### Practice

The $K_b$ for ammonia, $NH_3$, is $1.8 \times 10^{-5}$. Given the following acid dissociation constants, predict whether the following salts will form acidic or basic solutions when added to pure water.

**a.** $NH_4IO_3$
**b.** $NH_4CN$

| Acid | $K_a$ |
|---|---|
| $HIO_3$ | $1.6 \times 10^{-1}$ |
| HCN | $6.2 \times 10^{-10}$ |

Since the conjugate acid of ammonia is the cation for each of these species, the $K_a$ for ammonium is needed for comparison.

$$NH_4^+(aq) + H_2O(l) \rightleftharpoons NH_3(aq) + H_3O^+(aq)$$

$$K_a = \frac{K_w}{K_b} = \frac{1.0 \times 10^{-14}}{1.8 \times 10^{-5}} = 5.6 \times 10^{-10}$$

**(a)** The $K_b$ for $IO_3^-$ is needed.

$$IO_3^-(aq) + H_2O(l) \rightleftharpoons HIO_3(aq) + OH^-(aq)$$

$$K_b = \frac{K_w}{K_a} = \frac{1.0 \times 10^{-14}}{1.6 \times 10^{-1}} = 6.25 \times 10^{-14}$$

Because the base dissociation constant for $IO_3^-$ is less than the acid dissociation constant for $NH_4^+$, in an $NH_4IO_3$ solution, the amount of hydronium produced by the hydrolysis of ammonium is greater than the amount of hydroxide produced by the hydrolysis of iodate. Therefore, ammonium iodate is an acidic salt.

**(b)** The $K_b$ for $CN^-$ is needed.

$$CN^-(aq) + H_2O(l) \rightleftharpoons HCN(aq) + OH^-(aq)$$

$$K_b = \frac{K_w}{K_a} = \frac{1.0 \times 10^{-14}}{6.2 \times 10^{-10}} = 1.6 \times 10^{-5}$$

The $K_b$ for $CN^-$ is greater than the $K_a$ for $NH_4^+$, so $NH_4CN$ is a basic salt.

# Buffer Solutions

A **buffer** is an aqueous solution that resists a change in pH upon the addition of small quantities of acids or bases.

A buffer is either composed of

- a **weak acid and its conjugate base** in similar concentrations (HA and $A^-$), or
- a **weak base and its conjugate acid** in similar concentrations (B and $BH^+$)

Consider the buffer solution composed of formic acid, $H_2CO_2$, and its conjugate base, sodium formate, $NaHCO_2$. Formic acid is a weak acid with a $K_a$ of $1.77 \times 10^{-4}$. Sodium formate dissociates completely in an aqueous solution to form $HCO_2^-$.

The illustration below represents an equimolar mixture of formic acid and formate ion. The two situations illustrated show why the buffer system resists a pH change. *Sodium ions and water molecules are omitted from the diagrams for clarity.*

In the first situation, acid is added to the solution in the form of $H^+$ ions (two $H^+$ ions are added in this representation). Instead of lowering the pH by increasing the hydronium concentration, the protons react with free formate ions to produce formic acid molecules. This results in a negligible pH change.

In the second situation, base is added to the solution in the form of $OH^-$ ions (two in this representation). Protons are removed from two of the formic acid molecules to form water, and again, the pH changes very little.

**Resistance of a Buffer to a Change in pH**

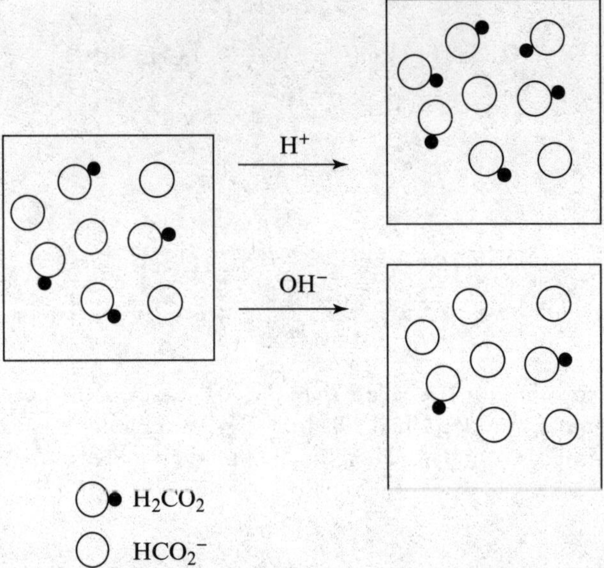


## Practice

Which of the following reagents *could* produce a buffer? Explain your reasoning.

**a.** $NaNO_3$ and $HNO_3$
**b.** $NaNO_2$ and $HNO_3$
**c.** $NaNO_2$ and $HNO_2$
**d.** $NaNO_3$ and $HNO_2$

**(a)** A buffer could not be formed. $NaNO_3$ is the salt of a strong acid and a strong base. $HNO_3$ is a strong acid. These reagents could not be combined to produce a buffer.

**(b)** A buffer could be formed. $NaNO_2$ is the salt of a weak acid and $HNO_3$ is a strong acid. If $HNO_3$ is added to the $NO_2^-$ from the salt, some $HNO_2$ will be formed. The amount of $HNO_3$ added could be adjusted so that a 1:1 molar ratio of $[HNO_2]$:$[NO_2^-]$ would result. This resulting weak acid/conjugate base pair is a buffer.

**(c)** A buffer could be formed. $NaNO_2$ is the salt of a weak acid, and $HNO_2$ is its conjugate acid. This pair forms a buffer when combined in approximately equimolar quantities.

**(d)** A buffer could not be formed. $NaNO_3$ is a neutral salt. It will have no effect on a solution of $HNO_2$, which is a weak acid. No buffer formation is possible.

# The Henderson-Hasselbalch Equation

The **Henderson-Hasselbalch equation** is a derivation of the acid dissociation equilibrium expression that relates the pH of a buffer solution composed of a weak acid, HA, and the anion of the weak acid, $A^-$, to the $pK_a$ of the weak acid. This equation is included on the *AP Chemistry Equations and Constants* sheet. Students will not be asked to derive the Henderson-Hasselbalch equation, but they are expected to know how to use it.

$$pH = pK_a + \log \frac{[A^-]}{[HA]}$$

A weak acid/conjugate base system produces the most effective buffer when [HA] = [A⁻]. Under these conditions, the pH of the solution is equal to the p$K_a$ of the weak acid since log 1 = 0.

$$\text{pH} = \text{p}K_a\text{: [HA]} = [\text{A}^-]$$

When the pH of the buffer solution is less than the p$K_a$ of the weak acid form of the buffer, the concentration of protonated weak acid is greater than the concentration of the anion. (The log of a number less than 1 is a negative number.)

$$\text{pH} < \text{p}K_a\text{: [HA]} > [\text{A}^-]$$

When the pH of the buffer solution is greater than the p$K_a$, the concentration of the anion is greater than the concentration of the weak acid. (The log of a number greater than 1 is a positive number.)

$$\text{pH} > \text{p}K_a\text{: [HA]} < [\text{A}^-]$$

The Henderson-Hasselbalch equation can also be modified to describe a buffer composed of a weak base, B, and its conjugate acid, $BH^+$. Notice the similarities and differences between the acid and base forms of the equation. Because the pOH version of the Henderson-Hasselbalch equation is *not* included on the *AP Chemistry Equations and Constants* sheet, this convenient equation must be memorized in order to be used.

$$\text{pOH} = \text{p}K_b + \log\frac{[\text{BH}^+]}{[\text{B}]}$$

A weak base/conjugate acid system produces the most effective buffer when [B] = [$BH^+$]. At this point the pOH of the solution is equal to the p$K_b$ of the weak base. Again, the pOH is dependent on the relative concentrations of base and conjugate acid.

$$\text{pOH} = \text{p}K_b\text{: [B]} = [\text{BH}^+]$$

$$\text{pOH} < \text{p}K_b\text{: [B]} > [\text{BH}^+]$$

$$\text{pOH} > \text{p}K_b\text{: [B]} < [\text{BH}^+]$$

## Practice

Use the $K_a$ values of the following substances to determine which can be added to one-half of an equivalent of strong base to produce a buffer with a pH of 7.

| Acid | $K_a$ at 25 °C |
|---|---|
| HF | $7.2 \times 10^{-4}$ |
| $HN_3$ | $1.9 \times 10^{-5}$ |
| $H_2S$ | $1.0 \times 10^{-7}$ |

These acids can be partially reacted with a strong base such as NaOH to produce buffers with pH ranges near the p$K_a$ values of the acids.

The p$K_a$ values of HF, $HN_3$, and $H_2S$ are 3.14, 4.72, and 7.00, respectively. The p$K_a$ of $H_2S$ is equal to the pH that is needed, so $H_2S$ is the correct choice.

By adding one-half of an equivalent of a strong base to $H_2S$ such that $[H_2S] = [HS^-]$, the pH of the solution will be equal to the p$K_a$.

$$H_2S(aq) + OH^-(aq) \rightarrow HS^-(aq) + H_2O(l)$$

$$pH = 7.00 + \log \frac{[HS^-]}{[H_2S]} = 7.00 + \log 1 = 7.00$$

## Practice

The p$K_a$ of benzoic acid, $C_6H_5CO_2H$, is 4.19 at 25 °C.

**a.** What is the pH of a 0.500 *M* solution of benzoic acid?
**b.** When 0.030 mole of NaOH is added to 100.0 mL of a 0.500 *M* solution of benzoic acid, what is the resulting pH? Assume the volume of the solution does not change when the NaOH is added.

**(a)** The initial pH is determined by the weakly-dissociating $C_6H_5CO_2H$.

$$C_6H_5CO_2H(aq) + H_2O(l) \rightleftharpoons C_6H_5CO_2^-(aq) + H_3O^+(aq)$$

The $K_a$ of benzoic acid can be calculated.

$$K_a = 10^{-pK_a} = 10^{-4.19} = 6.5 \times 10^{-5}$$

Next, a RICE table can be used to determine the concentration of hydronium.

| Reaction | $C_6H_5CO_2H$ | $C_6H_5CO_2^-$ | $H_3O^+$ |
|---|---|---|---|
| Initial Concentrations (*M*) | 0.500 | 0 | 0 |
| Change | $-x$ | $+x$ | $+x$ |
| Equilibrium Concentrations (*M*) | $0.500 - x$ | $x$ | $x$ |

Because $K_a$ is small, $x$ can be neglected in the denominator of the $K_a$ expression.

$$K_a = \frac{[C_6H_5CO_2^-][H_3O^+]}{[C_6H_5CO_2H]}$$

$$6.5 \times 10^{-5} = \frac{x^2}{0.500 - \cancel{x}}$$

$$\sqrt{(6.5 \times 10^{-5})(0.500)} = \sqrt{x^2}$$

$$x = 5.7 \times 10^{-3}\ M = [H_3O^+]$$

Finally, the pH can be determined from the hydronium concentration.

$$pH = -\log [H_3O^+] = -\log 5.7 \times 10^{-3} = 2.24$$

**(b)** When NaOH is added, some of the benzoic acid is converted to benzoate ion.

$$C_6H_5CO_2H(aq) + OH^-(aq) \rightarrow C_6H_5CO_2^-(aq) + H_2O(l)$$

The initial number of moles of benzoic acid is

$$0.1000\ \cancel{L} \times \frac{0.500 \text{ mol } C_6H_5CO_2H}{1\ \cancel{L}} = 0.0500 \text{ mol } C_6H_5CO_2H$$

After the reaction, 0.030 mole of benzoic acid has been converted to benzoate. The number of moles of benzoic acid remaining is

$$0.0500\text{ mol} - 0.030\text{ mol} = 0.020\text{ mol } C_6H_5CO_2H$$

The pH can be determined using the Henderson-Hasselbalch equation.

$$\text{pH} = \text{p}K_a + \log\frac{[C_6H_5CO_2^-]}{[C_6H_5CO_2H]} = 4.19 + \log\frac{\left(\frac{0.030\text{ mol}}{0.100\text{ L}}\right)}{\left(\frac{0.020\text{ mol}}{0.100\text{ L}}\right)} = 4.37$$

## Buffer Capacity

**Buffer capacity** refers to the amount of strong acid or strong base that can be added to a buffer system before the pH of the buffer system begins to change dramatically. The absolute concentrations of acid and conjugate base or base and conjugate acid determine the buffer capacity of the solution. The higher the absolute concentration of the buffer, the more acid or base can be added before the pH begins to change.

Dilution of a buffer system *does not affect the pH of the solution.* To see why this is the case, let's look at the pH values of two buffer systems. Buffer A contains 0.40 mole of HA and 0.60 mole of $A^-$ in *1.00 L* of solution, and buffer B contains 0.40 mole of HA and 0.60 mole of $A^-$ in *0.100 L* of solution. The pH values of the two solutions are calculated as follows:

Buffer A:

$$\text{pH} = \text{p}K_a + \log\frac{[A^-]}{[HA]} = \text{p}K_a + \log\frac{\left(\frac{0.60\text{ mol}}{\cancel{1.00\text{ L}}}\right)}{\left(\frac{0.40\text{ mol}}{\cancel{1.00\text{ L}}}\right)} = \text{p}K_a + \log\frac{0.60}{0.40}$$

Buffer B:

$$\text{pH} = \text{p}K_a + \log\frac{[A^-]}{[HA]} = \text{p}K_a + \log\frac{\left(\frac{0.60\text{ mol}}{\cancel{0.100\text{ L}}}\right)}{\left(\frac{0.40\text{ mol}}{\cancel{0.100\text{ L}}}\right)} = \text{p}K_a + \log\frac{0.60}{0.40}$$

Despite the fact that the pH values of the two buffers are identical, buffer B is ten times as concentrated as buffer A, so it has a much larger buffer capacity. It is capable of reacting with a larger quantity of added acid or base before the pH begins to change appreciably.

## Practice

Which of the following weak acid/conjugate base ($HA/A^-$) representations has the highest buffer capacity for addition of strong acids or strong bases?

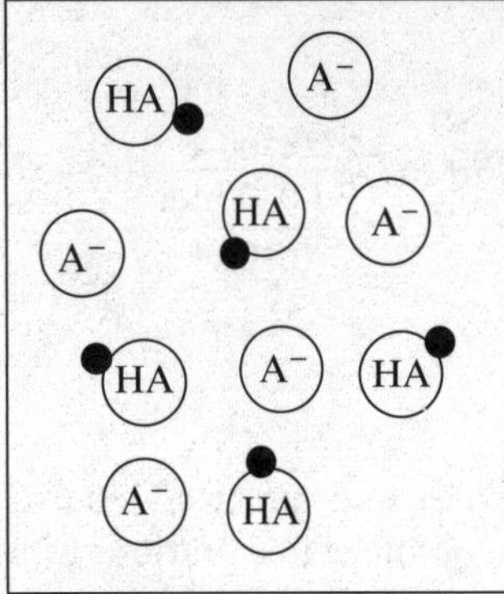


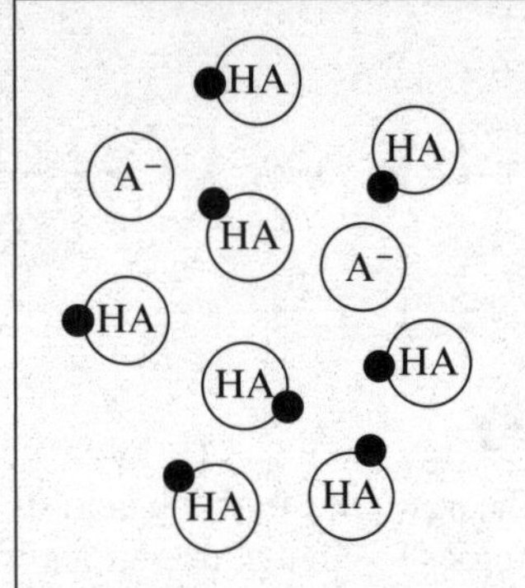


Both diagrams show ten particles. The diagram on the left has an equal ratio of HA and $A^-$, so it has a greater ability to neutralize both $H_3O^+$ and $OH^-$. The diagram on the right has a smaller capability to neutralize added $H_3O^+$ since it has a larger number of HA molecules and relatively few $A^-$.

# Acid-Base Titrations

**Titration** is a technique by which an unknown concentration of an **analyte** is determined by reaction with an accurately known amount of a standard, or **titrant.**

When exactly enough titrant has been added to react with all the analyte molecules, the **equivalence point** of the titration has been reached.

The **endpoint** is some indication that the reaction is complete. The endpoint is often a color change of an **indicator** such as **phenolphthalein** that occurs near the equivalence point of the reaction of a strong acid with a strong base.

**Titration curves** are plots of pH versus the amount of titrant added to an acid or base solution.

## Strong Acid/Strong Base

When a strong acid is titrated with a strong base, the products are water and a neutral salt.

When a strong acid solution of unknown concentration is titrated with a known concentration of a strong base, the volume of base required to reach the equivalence point of the titration is determined by the volume of base required to reach the endpoint of the titration.

The net ionic equation for the reaction is

$$H_3O^+(aq) + OH^-(aq) \rightarrow 2\ H_2O(l)$$

The equilibrium constant for this process is very large, $K_{eq} \approx 1 \times 10^3$, so the equilibrium greatly favors the products.

Consider the titration curve below, in which 50.0 mL of an unknown concentration of HCl is titrated with a 0.100 *M* solution of NaOH. When 20.0 mL of the NaOH have been added, the pH rises sharply and passes through an inflection point at a pH of 7. At this point, the number of moles of $OH^-$ added are equal to the number of moles of $H_3O^+$ present initially. This is the equivalence point of the titration. The amount of $OH^-$ added to reach the equivalence point is sometimes referred to as an **equivalent.**

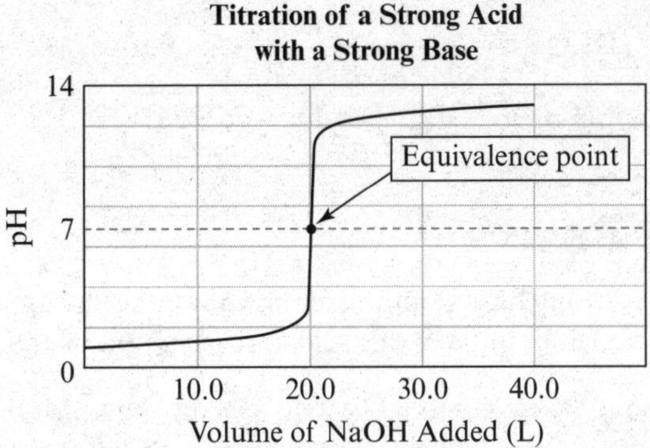


Notice that as base is added, the pH gradually increases until just before the equivalence point. Within 10% of the volume of base required to reach the equivalence point, the pH begins its sharp rise before leveling off when the hydroxide concentration becomes larger than the hydronium concentration.

## Practice

A student titrates a 25.0-mL aqueous sample of $HNO_3$ with 0.200 *M* standardized KOH. The titration requires 30.1 mL of base to reach the equivalence point.

**a.** How many moles of $HNO_3$ have reacted when the equivalence point has been reached?
**b.** What is the initial concentration of $HNO_3$?
**c.** What is the pH of the solution after 15.0 mL of KOH have been added?

**(a)** The number of moles of base added at the equivalence point is equal to the number of moles of $HNO_3$ that were initially present. The initial $[HNO_3]$ can be calculated.

$$HNO_3(aq) + KOH(aq) \rightarrow KNO_3(aq) + H_2O(l)$$

$$0.0301\ \cancel{\text{L KOH}} \times \frac{0.200\ \cancel{\text{mol KOH}}}{1\ \cancel{\text{L KOH}}} \times \frac{1\ \text{mol HNO}_3}{1\ \cancel{\text{mol KOH}}} = 0.00602\ \text{mol HNO}_3$$

**(b)** The initial concentration of $HNO_3$ can be calculated from the number of moles of $HNO_3$ reacted and the volume.

$$[HNO_3]_{initial} = \frac{0.00602\ \text{mol HNO}_3}{0.0250\ \text{L}} = 0.241\ M$$

**(c)** The number of moles of $HNO_3$ remaining after the addition of the base determines the pH. The number of moles of base added is calculated next.

$$0.0150\ \cancel{\text{L KOH}} \times \frac{0.200\ \text{mol KOH}}{1\ \cancel{\text{L KOH}}} = 0.00300\ \text{mol KOH}$$

The number of moles of KOH added is equal to the number of moles of $HNO_3$ that have reacted. Since 0.00602 mole of $HNO_3$ was initially present, the number of moles that remain can be calculated.

$$0.00602\ \text{mol} - 0.00300\ \text{mol} = 0.00302\ \text{mol H}_3\text{O}^+$$

The concentration of hydronium is equal to the concentration of $HNO_3$ remaining. The hydronium concentration can then be used to determine the pH.

$$[H_3O^+] = \frac{0.00302 \text{ mol}}{0.0250 \text{ L} + 0.0150 \text{ L}} = 0.0755\ M$$

$$\text{pH} = -\log [H_3O^+] = -\log 0.755 = 1.122$$

## Weak Acid/Strong Base

The titration of a weak acid with a strong base is more complex than the strong acid/strong base titration since there is a large **buffering region** in which both the weak acid and the anion of the weak acid are present.

The titration curve for the titration of a weak acid, HA, with a strong base, MOH, looks like the one shown below. Several key facts can be gleaned from this plot without any calculations, so it is important to be able to recognize this type of curve and predict which species dominate at each point along the titration curve.

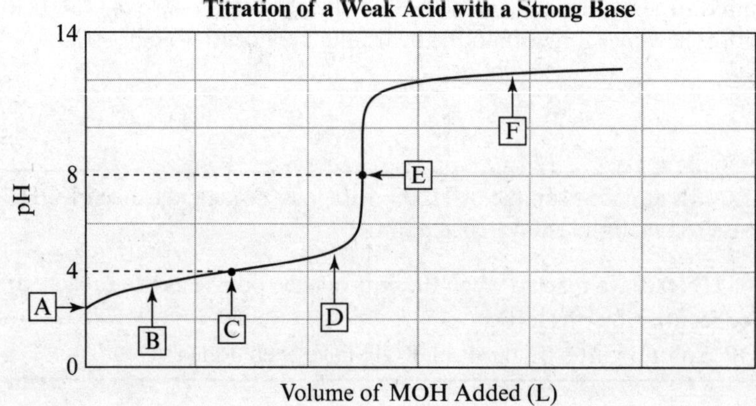


- At point **A,** the pH depends only on the concentration of the weak acid and the acid dissociation constant, $K_a$.

$$HA + H_2O \rightleftharpoons A^- + H_3O^+$$

- Within regions **B** and **D,** the solution is buffered and the pH is calculated using the Henderson-Hasselbalch equation. The base that has been added has converted some of the HA to $A^-$.

$$OH^- + HA \rightarrow H_2O + A^-$$

The relative concentrations of HA and $A^-$, along with the $pK_a$, determine the pH of the solution.

$$\text{pH} = pK_a + \log \frac{[A^-]}{[HA]}$$

- At point **C,** the half-equivalence point, exactly enough base has been added to produce a buffer solution with 1:1 ratio of [HA]:[$A^-$]. At this point in the titration, the pH of the solution is equal to the $pK_a$ of the acid.
- At point **E,** the equivalence point, the number of moles of $OH^-$ that have been added is equal to the number of moles of HA that were initially present. The pH at the equivalence point is determined by the basic salt, $A^-$, which is the dominant species.

$$A^- + H_2O \rightleftharpoons HA + OH^-$$

- Within region **F,** the $OH^-$ in excess of the HA initially present determines the pOH and thus the pH of the solution.

Consider, for example, the titration of 100.0 mL of a 0.250 $M$ solution of nitrous acid, $HNO_2$, at 25 °C with a strong base. Nitrous acid has a $K_a$ of $4.0 \times 10^{-4}$.

Before the titration begins, the dominant species in the solution is $HNO_2$. A very small proportion of the weak acid is dissociated, so the minor components of the system are $H_3O^+$ and $NO_2^-$. The $H_3O^+$ concentration determines the pH of the solution.

$$HNO_2(aq) + H_2O(l) \rightleftharpoons H_3O^+(aq) + NO_2^-(aq) \qquad K_a = 4.0 \times 10^{-4}$$

| Reaction | $HNO_2$ | $H_3O^+$ | $NO_2^-$ |
|---|---|---|---|
| Initial Concentrations ($M$) | 0.250 | 0 | 0 |
| Change | $-x$ | $+x$ | $+x$ |
| Equilibrium Concentrations ($M$) | $0.250 - x$ | $x$ | $x$ |

$$K_a = \frac{[H_3O^+][NO_2^-]}{[HNO_2]}$$

$$4.0 \times 10^{-4} = \frac{x^2}{0.250 - x}$$

The amount of acid that dissociates is small compared to the initial amount, so $x$ in the denominator may be neglected.

$$4.0 \times 10^{-4} = \frac{x^2}{0.250 - \cancel{x}}$$

$$\sqrt{(4.0 \times 10^{-4})(0.250)} = \sqrt{x^2}$$

$$x = 0.010\ M$$

$$[H_3O^+] = 0.010\ M$$

The pH of the solution at point **A** is

$$-\log(0.010) = 2.00$$

Now suppose the solution is titrated with 0.250 $M$ NaOH. How can the pH be determined at the equivalence point? It is first necessary to know how much NaOH must be added. This can be determined stoichiometrically. The volume of NaOH that must be added to reach the equivalence point is equal to the number of moles of $HNO_2$ initially present.

$$HNO_2(aq) + OH^-(aq) \rightarrow NO_2^-(aq) + H_2O(l)$$

$$\text{volume NaOH added} = 0.1000\ \text{L}\ \cancel{HNO_2} \times \frac{0.250\ \cancel{\text{mol}\ HNO_2}}{1\ \cancel{\text{L}\ HNO_2}} \times \frac{1\ \cancel{\text{mol NaOH}}}{1\ \cancel{\text{mol}\ HNO_2}} \times \frac{1\ \text{L NaOH}}{0.250\ \cancel{\text{mol NaOH}}}$$

$$= 0.100\ \text{L NaOH}$$

Since all of the $OH^-$ and $HNO_2$ have reacted at the equivalence point, the major species in the solution is $NO_2^-$. The number of moles of $NO_2^-$ is equal to the number of moles of $HNO_2$ that were present initially.

$$\text{moles of } NO_2^- = 0.100\ \text{L}\ OH^- \times \frac{0.250\ \text{mol}\ OH^-}{1\ \text{L}\ OH^-} \times \frac{1\ \text{mol}\ NO_2^-}{1\ \text{mol}\ OH^-} = 0.0250\ \text{mol}\ NO_2^-$$

The concentration of $NO_2^-$ is determined from the number of moles and the total volume of the solution after the addition of 100. mL of NaOH to the original 100.0 mL of $HNO_2$ solution.

$$[NO_2^-] = \frac{0.0250 \text{ mol } NO_2^-}{0.100 \text{ L} + 0.100 \text{ L}} = 0.125\ M\ NO_2^-$$

The next step in finding the pH is to evaluate $NO_2^-$ as a basic species. The $K_b$ for $NO_2^-$ is $K_w$ at 25 °C ($1.0 \times 10^{-14}$) divided by the $K_a$ of $HNO_2$ ($4.0 \times 10^{-4}$).

$$NO_2^-(aq) + H_2O(l) \rightleftharpoons HNO_2(aq) + OH^-(aq)$$

$$K_w = K_a \cdot K_b = 1.0 \times 10^{-14}$$

$$K_b = \frac{1.0 \times 10^{-14}}{4.0 \times 10^{-4}} = 2.5 \times 10^{-11}$$

The $K_b$ can be used to find the $OH^-$ concentration.

| Reaction | $NO_2^-$ | $HNO_2$ | $OH^-$ |
|---|---|---|---|
| Initial Concentrations ($M$) | 0.125 | 0 | 0 |
| Change | $-x$ | $+x$ | $+x$ |
| Equilibrium Concentrations ($M$) | $0.125 - x$ | $x$ | $x$ |

$$K_b = \frac{[HNO_2][OH^-]}{[NO_2^-]}$$

$$2.5 \times 10^{-11} = \frac{x^2}{0.125 - \cancel{x}}$$

The amount of hydroxide is small compared to the amount of $NO_2^-$, so $x$ in the denominator may be neglected.

$$\sqrt{(2.5 \times 10^{-11})(0.125)} = \sqrt{x^2}$$

$$x = 1.8 \times 10^{-6}\ M = [OH^-]$$

The pOH of the solution at the equivalence point is determined using the unrounded value, and the pOH can be used to determine the pH.

$$pOH = -\log(1.76776695 \times 10^{-6}) = 5.75$$

$$14.00 = pH + pOH$$

$$pH = 14.00 - 5.75 = 8.25$$

*It is very important to note that at the equivalence point of a titration of a weak acid with a strong base, the pH is not 7.* The solution has a pH higher than 7 because a basic salt is the major species in solution.

To determine the pH at any point within the buffering region (any point in region **B** or **D** of the titration curve), the Henderson-Hasselbalch equation may be used. The concentrations of $HNO_2$ and $NO_2$ are determined stoichiometrically.

$$HNO_2(aq) + OH^-(aq) \rightarrow NO_2^-(aq) + H_2O(l)$$

A buffer is present after 40.00 mL of the 0.250 $M$ $OH^-$ solution have been added to 100.0 mL of 0.250 $M$ $HNO_2$ solution. Suppose the pH of this buffer is needed. Because the equivalence point has not yet been reached, $OH^-$ is the limiting reagent.

The *numbers of moles* of reactants and products can be tabulated, as shown below (although this should not be confused with a RICE table, which is used to determine equilibrium *concentrations*).

The number of moles of $HNO_2$ decreases by the amount that reacted with $OH^-$ to form $NO_2^-$. The $OH^-$ is the limiting reagent, so it reacts completely. The number of moles of $NO_2^-$ increases by the same amount of $OH^-$ that was added. The final concentrations are determined by dividing by the total volume, 140. mL.

| | $HNO_2$ | $OH^-$ | $NO_2^-$ |
|---|---|---|---|
| **Initial Moles** | $0.100\ \cancel{L} \times \frac{0.250\text{ mol}}{1\ \cancel{L}} = 0.0250\text{ mol}$ | $0.0400\ \cancel{L} \times \frac{0.250\text{ mol}}{1\ \cancel{L}} = 0.0100\text{ mol}$ | $\approx 0$ |
| **Change in Moles** | –0.0100 | –0.0100 | +0.0100 |
| **Final Number of Moles** | 0.0150 | 0 | 0.0100 |
| **Final Concentration (*M*)** | $\frac{0.0150\text{ mol}}{0.140\text{ L}} = 0.107\ M$ | 0 | $\frac{0.0100\text{ mol}}{0.140\text{ L}} = 0.0714\ M$ |

The pH of the solution can now be determined using the Henderson-Hasselbalch equation.

$$\text{pH} = \text{p}K_a + \log\frac{[NO_2^-]}{[HNO_2]} = (-\log 4.0\times10^{-4}) + \log\frac{0.0714}{0.107} = 3.22$$

## Practice

A 100.-mL solution of formic acid, $H_2CO_2$, is titrated with a 1.00 *M* solution of NaOH. A volume of 38.0 mL of the NaOH solution is required to reach the equivalence point. The $K_a$ of formic acid is $1.77 \times 10^{-4}$.

**a.** What is the pH at the equivalence point?
**b.** What is the pH of the solution after 12.0 mL of NaOH have been added?

**(a)** The number of moles of NaOH added is equal to the number of moles of formate ion that are present at the end of the titration.

$$H_2CO_2(aq) + OH^-(aq) \rightarrow HCO_2^-(aq) + H_2O(l)$$

$$0.0380\ \cancel{\text{L NaOH}} \times \frac{1.00\ \cancel{\text{mol NaOH}}}{1\ \cancel{\text{L NaOH}}} \times \frac{1\text{ mol }HCO_2^-}{1\ \cancel{\text{mol NaOH}}} = 0.0380\text{ mol }HCO_2^-$$

$$[HCO_2^-] = \frac{0.0380\text{ mol }HCO_2^-}{0.100\text{ L} + 0.0380\text{ L}} = 0.275\ M$$

The pH at the equivalence point depends on the behavior of $HCO_2^-$ as a weak base.

$$HCO_2^-(aq) + H_2O(l) \rightleftharpoons H_2CO_2(aq) + OH^-(aq)$$

The $K_a$ of formic acid must be converted to the $K_b$ of formate.

$$K_w = K_a \cdot K_b = 1.0\times10^{-14}$$

$$K_b = \frac{1.00\times10^{-14}}{1.77\times10^{-4}} = 5.65\times10^{-11}$$

| Reaction | $HCO_2^-$ | $H_2CO_2$ | $OH^-$ |
|---|---|---|---|
| Initial Concentrations ($M$) | 0.275 | 0 | 0 |
| Change | $-x$ | $+x$ | $+x$ |
| Equilibrium Concentrations ($M$) | $0.275 - x$ | $x$ | $x$ |

The very small $K_b$ allows for the assumption that $x$ is negligible compared with the concentration of $HCO_2^-$.

$$K_b = \frac{[H_2CO_2][OH^-]}{[HCO_2^-]}$$

$$5.65 \times 10^{-11} = \frac{x^2}{0.275 - \cancel{x}}$$

$$x = \sqrt{(5.65 \times 10^{-11})(0.275)} = 3.94 \times 10^{-6} = [OH^-]$$

$$\text{pOH} = -\log 3.94 \times 10^{-6} = 5.404$$

$$\text{pH} = 14.000 - 5.404 = 8.596$$

**(b)** The number of moles of $OH^-$ added is equal to the number of moles of formic acid converted to formate.

$$0.0120 \ \cancel{\text{L NaOH}} \times \frac{1.00 \ \cancel{\text{mol NaOH}}}{1 \ \cancel{\text{L NaOH}}} \times \frac{1 \text{ mol } HCO_2^-}{1 \ \cancel{\text{mol NaOH}}} = 0.0120 \text{ mol } HCO_2^-$$

The number of moles of formic acid remaining is the difference between the initial number and the number of moles converted to formate.

$$0.0380 \text{ mol} - 0.0120 \text{ mol} = 0.0260 \text{ mol } HCO_2H$$

Because the total volume of the solution (100. mL + 12.0 mL) cancels out of the equation, the numbers of moles of formic acid and formate can be used directly in the Henderson-Hasselbalch equation.

$$\text{pH} = \text{p}K_a + \log \frac{[HCO_2^-]}{[HCO_2H]} = (-\log 1.77 \times 10^{-4}) + \log \frac{0.0120}{0.0260} = 3.416$$

## Weak Base/Strong Acid

When a weak base, B, is titrated with a strong acid, $H_3O^+$, the reaction produces $BH^+$ and $H_2O$.

The curve starts at a high pH value. As acid, HX, is added and converts the base, B, to its conjugate acid form, $BH^+$, the pH has a value above 7 in the buffering region. At the half-equivalence point, the pH is equal to the p$K_b$, and at the equivalence point, the pH depends on the concentration of the acidic species, $BH^+$.

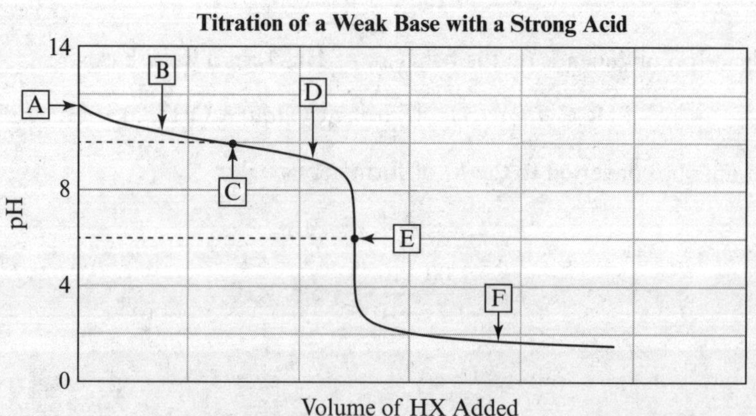

- At point **A,** the pOH depends only on the concentration of the weak base and the base dissociation constant, $K_b$.

$$B + H_2O \rightleftharpoons BH^+ + OH^-$$

- Within regions **B** and **D,** the solution is buffered and the pOH is calculated using the appropriate form of the Henderson-Hasselbalch equation. The acid that has been added has converted some of the B to $BH^+$. The relative concentrations of B and $BH^+$, along with the $pK_b$, determine the pOH of the solution.

$$pOH = pK_b + \log \frac{[BH^+]}{[B]}$$

- At point **C,** the half-equivalence point, exactly enough acid has been added to produce a 1:1 ratio of [B]:[ $BH^+$]. At this point in the titration, the pOH of the solution is equal to the $pK_b$ of the weak base.
- At point **E,** the equivalence point, the number of moles of $H^+$ that have been added is equal to the number of moles of B that were initially present. The pH at the equivalence point is determined by the acidic species, $BH^+$, which is the dominant species.

$$BH^+ + H_2O \rightleftharpoons B + H_3O^+$$

- Within region **F,** the excess $H_3O^+$ determines the pH of the solution.

## Polyprotic Acid/Strong Base

A **polyprotic acid** is one that has more than one proton to donate in an aqueous solution. Examples of polyprotic acids include phosphoric acid, $H_3PO_4$, carbonic acid, $H_2CO_3$, and sulfuric acid, $H_2SO_4$.

The acid dissociation constant for the loss of the first proton is called $K_{a_1}$. The second proton dissociates with a dissociation constant of $K_{a_2}$, and so on. The dissociation reactions for phosphoric acid are shown below.

$$H_3PO_4 + H_2O \rightleftharpoons H_2PO_4^- + H_3O^+ \qquad K_{a_1} = 7.5 \times 10^{-3}$$
$$H_2PO_4^- + H_2O \rightleftharpoons HPO_4^{2-} + H_3O^+ \qquad K_{a_2} = 6.2 \times 10^{-8}$$
$$HPO_4^{2-} + H_2O \rightleftharpoons PO_4^{3-} + H_3O^+ \qquad K_{a_3} = 4.8 \times 10^{-13}$$

As one equivalent of base is added to a polyprotic acid, one proton from each molecule of acid is removed. Once all the molecules of acid have lost a single proton, a second equivalent of base will remove a second proton from the acid molecules, and so on. The inflection points on the titration curve for the reaction of a strong base with a weak polyprotic acid represent the equivalence points, just as they do in the titration curves for the reaction of a strong base with a monoprotic weak acid. The number of acidic protons in a molecule is equal to the number of inflections in the titration curve. The curve shown below is for a diprotic acid.

**Titration of a Diprotic Acid with a Strong Base**

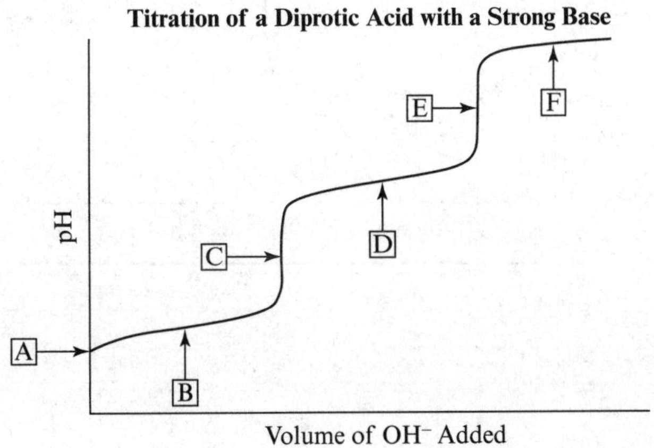

- At point **A,** the pH depends only on the concentration of the weak acid and the first acid dissociation constant, $K_{a_1}$.

  The pH is determined by the ionization of the weak acid.

$$H_2A + H_2O \rightleftharpoons HA^- + H_3O^+$$

- Between points **A** and **C,** the base that has been added has converted some of the $H_2A$ to $HA^-$. The relative concentrations of $H_2A$ and $HA^-$ determine the pH of the solution.
- At point **B,** the half-equivalence point, exactly enough base has been added to produce a 1:1 ratio of $[H_2A]$:$[HA^-]$. At this point in the titration, the pH of the solution is equal to the $pK_{a_1}$ of the weak acid.
- At point **C,** the equivalence point, the number of moles of $OH^-$ that have been added is equal to the number of moles of $H_2A$ that were initially present. The pH at the equivalence point is determined by the species, $HA^-$, which is the dominant species.

$$HA^- + H_2O \rightleftharpoons A^{2-} + H_3O^+$$

- At point **D,** the second half-equivalence point, exactly enough base has been added to produce a 1:1 ratio of $[HA^-]$:$[A^{2-}]$. At this point in the titration, the pH of the solution is equal to the $pK_{a_2}$ of the weak acid.
- Between points **C** and **E,** the base that has been added has converted some of the $HA^-$ to $A^{2-}$. The relative concentrations of $HA^-$ and $A^{2-}$ determine the pH of the solution.
- At point **E,** the second equivalence point, the number of moles of $OH^-$ that have been added is equal to twice the number of moles of $H_2A$ that were initially present. The pH at the equivalence point is determined by the species, $A^{2-}$, which is the dominant species.

$$A^{2-} + H_2O \rightleftharpoons OH^- + HA^-$$

- Within region **F,** the excess $OH^-$ determines the pOH of the solution.

## Practice

The curve below is observed when oxalic acid, $C_2O_4H_2$, is titrated with sodium hydroxide.

**a.** Write the equations for the acid dissociation reactions and the equilibrium expressions for $K_{a_1}$ and $K_{a_2}$.
**b.** Use the curve to determine the approximate numerical values of $K_{a_1}$ and $K_{a_2}$. Show how you arrived at your answers.

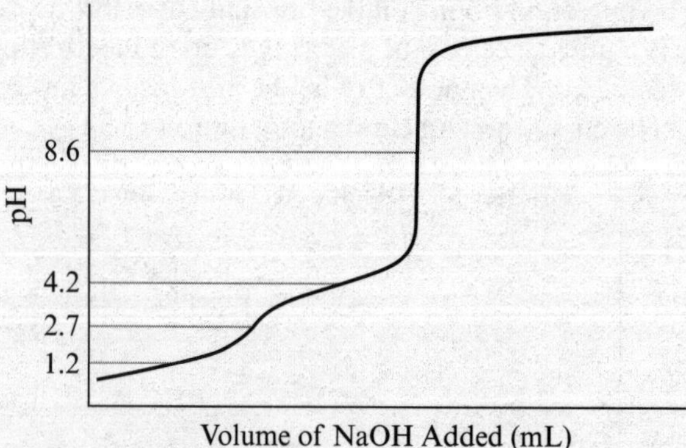

**(a)** The equations and equilibrium expressions for the hydrolysis of oxalic acid are

$$C_2O_4H_2(aq) + H_2O(l) \rightleftharpoons C_2O_4H^-(aq) + H_3O^+(aq) \qquad K_{a_1} = \frac{[C_2O_4H^-][H_3O^+]}{[C_2O_4H_2]}$$

$$C_2O_4H^-(aq) + H_2O(l) \rightleftharpoons C_2O_4^{2-}(aq) + H_3O^+(aq) \qquad K_{a_2} = \frac{[C_2O_4^{2-}][H_3O^+]}{[C_2O_4H^-]}$$

**(b)** The acid dissociation constants are most easily determined by the pH at the half-equivalence points of the titration, where pH = p$K_a$.

At pH = 1.2, $[C_2O_4H_2] = [C_2O_4H^-]$, so $pK_{a_1} = 1.2$.

$$pK_{a_1} = -\log K_{a_1}$$

$$K_{a_1} = 10^{-pK_{a1}} = 10^{-1.2} = 6\times10^{-2}$$

At pH = 4.2, $[C_2O_4H^-] = [C_2O_4^{2-}]$, so $pK_{a_2} = 4.2$.

$$pK_{a_2} = -\log K_{a_2}$$

$$K_{a_2} = 10^{-pK_{a2}} = 10^{-4.2} = 6\times10^{-5}$$

## Acid-Base Indicators

**Indicators** are usually weak acids that exhibit a color change when they are converted from protonated (HIn) to deprotonated form ($In^-$). An example of an indicator is bromocresol green, shown below. In the protonated form, HIn, bromocresol green is yellow. In the deprotonated form, $In^-$, it is blue. The $K_a$ for the deprotonation of bromocresol green at 25 °C is $1.3 \times 10^{-5}$, so the p$K_a$ is 4.89.

$$HIn + H_2O \rightleftharpoons In^- + H_3O^+ \qquad K_a = \frac{[In^-][H_3O^+]}{[HIn]}$$

**Bromocresol Green, an Acid-Base Indicator**

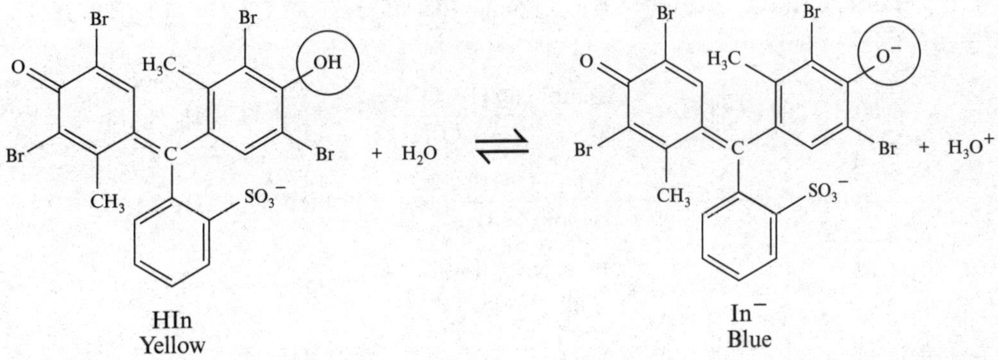

HIn
Yellow

$In^-$
Blue

The useful range for an indicator is generally within one pH unit of the p$K_a$ of the indicator. For bromocresol green, the equivalence point of a reaction that occurs at a pH of 4.89 ± 1 can be detected effectively.

When hydroxide is added to an acidic solution of HIn, the $H_3O^+$ is neutralized and the equilibrium shifts to the right. When approximately 10% of the indicator molecules have been converted to $In^-$, the beginning of the change from a yellow solution to a green solution is detectable. As additional base is added, more HIn is converted to the $In^-$ form.

If, on the other hand, hydronium is added to a basic solution of $In^-$, the equilibrium shifts to the left and the indicator is protonated. When approximately 10% of the $In^-$ molecules have been converted to HIn, the change from a blue solution to a green solution is detectable.

A wide range of indicators are available. The most appropriate indicator for a particular reaction is one with a $pK_a$ that is closest to the pH at the equivalence point of the reaction under investigation.

## Practice

A researcher has access to the indicators tabulated below.

| Indicator (In) | $pK_a$ | Color of HIn | Color of $In^-$ |
|---|---|---|---|
| Methyl orange | 3.4 | Red | Yellow |
| Methyl red | 5.0 | Red | Yellow |
| Phenol red | 7.9 | Yellow | Red |
| Alizarin yellow | 11.2 | Yellow | Violet |

**a.** What is the pH at the equivalence point of the titration of 1.00 L of 0.500 *M* chloroacetic acid, $ClCH_2CO_2H$, with 0.500 *M* hydroxide? The $K_a$ of chloroacetic acid is $1.4 \times 10^{-3}$.

**b.** What is an appropriate indicator for this titration?

**c.** What is the color of the solution containing the chosen indicator when the pH is equal to the $pK_a$ of chloroacetic acid?

**(a)** At the equivalence point of the titration, the concentration of chloroacetate ion, $ClCH_2CO_2^-$, determines the pH of the solution. The anion behaves as a weak base, so the $K_b$ is needed.

$$ClCH_2CO_2^-(aq) + H_2O(l) \rightleftharpoons ClCH_2CO_2H(aq) + OH^-(aq)$$

$$K_w = K_a \cdot K_b = 1.0 \times 10^{-14}$$

$$K_b = \frac{1.0 \times 10^{-14}}{1.4 \times 10^{-3}} = 7.1 \times 10^{-12}$$

In order to reach the equivalence point, 1.00 L of 0.500 *M* $OH^-$ solution must be added. At the equivalence point, the number of moles of $ClCH_2CO_2^-$ is equal to the initial number of moles of $ClCH_2CO_2H$. The molarity is determined using this information.

$$1.00\ \cancel{\text{L ClCH}_2\text{CO}_2\text{H}} \times \frac{0.500\ \text{mol ClCH}_2\text{CO}_2\text{H}}{1.00\ \cancel{\text{L ClCH}_2\text{CO}_2\text{H}}} = 0.500\ \text{mol ClCH}_2\text{CO}_2\text{H}$$

$$0.500\ \cancel{\text{mol ClCH}_2\text{CO}_2\text{H}} \times \frac{1\ \text{mol ClCH}_2\text{CO}_2^-}{1\ \cancel{\text{mol ClCH}_2\text{CO}_2\text{H}}} = 0.500\ \text{mol ClCH}_2\text{CO}_2^-$$

$$\frac{0.500\ \text{mol ClCH}_2\text{CO}_2^-}{1.00\ \text{L} + 1.00\ \text{L}} = 0.250\ M\ \text{ClCH}_2\text{CO}_2^-$$

Next, the concentration of $OH^-$ must be determined.

| Reaction | $ClCH_2CO_2^-$ | $ClCH_2CO_2H$ | $OH^-$ |
|---|---|---|---|
| Initial Concentrations (*M*) | 0.250 | 0 | 0 |
| Change | $-x$ | $+x$ | $+x$ |
| Equilibrium Concentrations (*M*) | $0.250 - x$ | $x$ | $x$ |

The value of $K_b$ is small, so $x$ is assumed to be negligible compared to the chloroacetate ion concentration.

$$K_b = \frac{[ClCH_2CO_2H][OH^-]}{[ClCH_2CO_2^-]}$$

$$7.142857 \times 10^{-12} = \frac{x^2}{0.250 - \cancel{x}}$$

$$x = \sqrt{(7.142857 \times 10^{-12})(0.250)} = 1.3363 \times 10^{-6} = [OH^-]$$

$$pOH = -\log 1.3363 \times 10^{-6} = 5.87 \quad (5.874094...)$$

$$pH = 14.00 - 5.874094 = 8.12$$

**(b)** The pH at the equivalence point of the titration lies within the color-change range of methyl red. It shows a color change at pH = 5.0 ± 1, so phenol red is the best choice.

**(c)** The p$K_a$ of chloroacetic acid is

$$pK_a = -\log 1.4 \times 10^{-3} = 2.85$$

At a pH of 2.85, methyl red will be in its protonated form, HIn, which will result in a red solution.

# Review Questions

## Multiple Choice

*Questions 1–3 refer to the following reaction.*

$$H_2S(aq) + H_2O(l) \rightleftharpoons HS^-(aq) + H_3O^+(aq) \qquad K_a = 9 \times 10^{-8}$$

**1.** What is the approximate pH of a 1.00 *M* solution of $H_2S$?

**A.** 1.5
**B.** 3.5
**C.** 5.5
**D.** 7.5

**2.** Which species are acting as Brønsted-Lowry acids in the reaction?

**A.** $H_2S$ and $H_2O$
**B.** $H_2O$ and $HS^-$
**C.** $H_2O$ and $H_3O^+$
**D.** $H_2S$ and $H_3O^+$

**3.** Which species in the reaction has the largest $K_b$?

**A.** $H_2S$
**B.** $H_2O$
**C.** $HS^-$
**D.** $H_3O^+$

**4.** Which of the following statements is false?

**A.** Strong acids and strong bases are approximately 100% ionized in dilute aqueous solutions.
**B.** A conjugate acid is formed when a proton is added to a base.
**C.** Water is considered to be amphiprotic, but not amphoteric.
**D.** All strong bases appear to have the same strength in an aqueous solution.

**5.** Based on its structural features, which of the following acids is expected to be the strongest?

A.

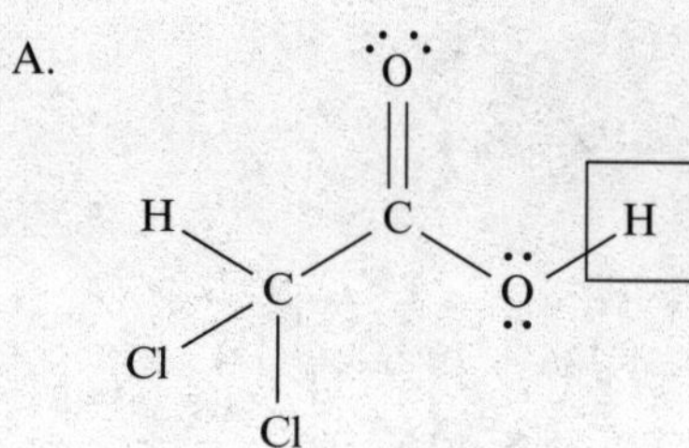


B.

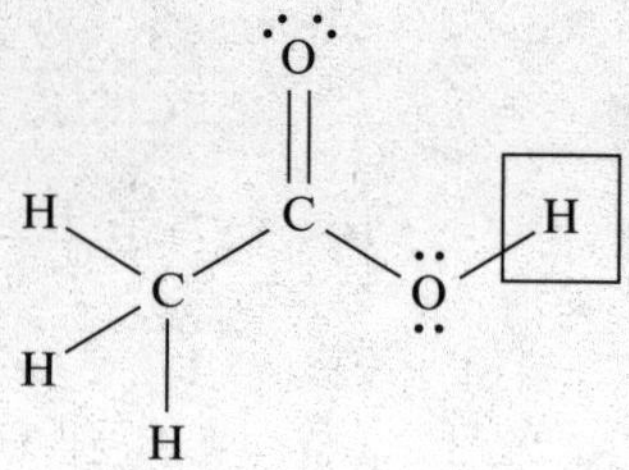

C.

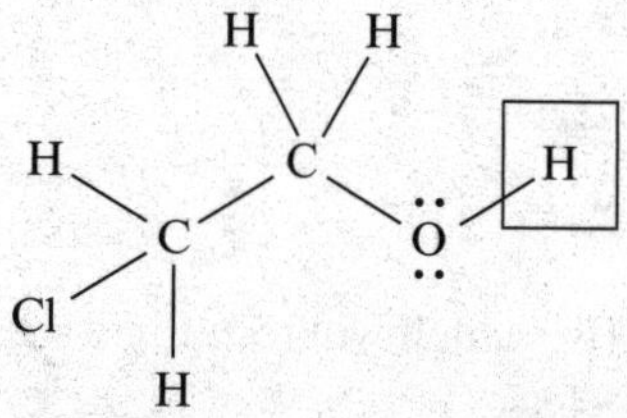

D.

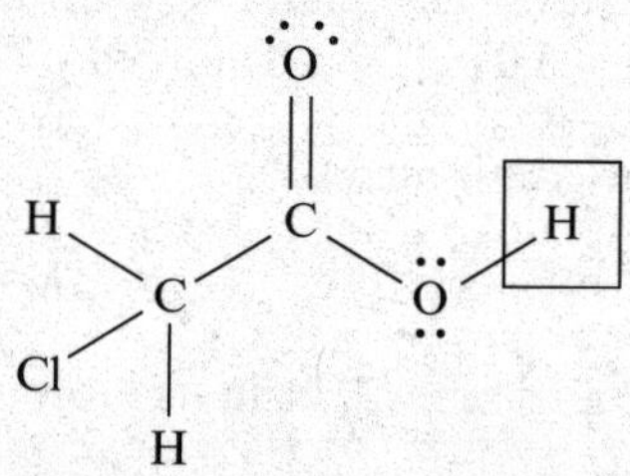

**6.** Which of the following solutions has the highest pH at 25 °C?

**A.** 1.00 *M* NaF
**B.** 1.00 *M* NaBr
**C.** 1.00 *M* $CH_3OH$
**D.** 1.00 *M* $CH_3CO_2H$

**7.** What is the pH of a $5.00 \times 10^{-2}$ *M* aqueous solution of $Ba(OH)_2$?

**A.** 9
**B.** 12
**C.** 13
**D.** 14

**8.** An acid, HA, has a $pK_a$ of 4.00. What is the approximate percent ionization of a 0.010 *M* aqueous HA solution?

**A.** 0.10%
**B.** 1.0%
**C.** 10%
**D.** 100%

**9.** Which of the following salts is expected to have the highest pH in a 1.00 *M* aqueous solution?

**A.** $NaClO_4$
**B.** KBr
**C.** $MgI_2$
**D.** $Ca(CN)_2$

**10.** Given the following acid dissociation constants, select the correct statement.

| Acid | $K_{a_1}$ | $K_{a_2}$ | $K_{a_3}$ |
|---|---|---|---|
| $H_3PO_4$ | $7.5 \times 10^{-3}$ | $6.2 \times 10^{-8}$ | $4.8 \times 10^{-13}$ |
| $H_3C_6H_5O_7$ | $8.4 \times 10^{-4}$ | $1.8 \times 10^{-5}$ | $4.0 \times 10^{-6}$ |

**A.** $H_3PO_4$ is a weaker acid than $H_3C_6H_5O_7$.
**B.** $HPO_4^{2-}$ is a weaker base than $C_6H_5O_7^{3-}$.
**C.** $H_2PO_4^-$ is a stronger acid than $H_2C_6H_5O_7^-$.
**D.** $PO_4^{3-}$ is a stronger base than $C_6H_5O_7^{3-}$.

*Questions 11–12 refer to the amino acid glycine.*

Glycine, $H_2NCH_2CO_2H$, is the simplest of the twenty amino acids that are commonly found in proteins. It contains a weakly basic amine group, $-NH_2$, and a weakly acidic carboxylic acid group, $-CO_2H$. In very low pH solutions, the amine group is in conjugate acid form and the carboxylic acid group is neutral. In very high pH solutions, the carboxylic acid group is in conjugate base form and the amine group is neutral. The neutral carboxylic acid group is more strongly acidic than the conjugate acid form of the amine group. A 20.0-mL aqueous solution of 0.100 *M* glycine is titrated with 0.100 *M* NaOH at 25 °C. The following titration curve results.

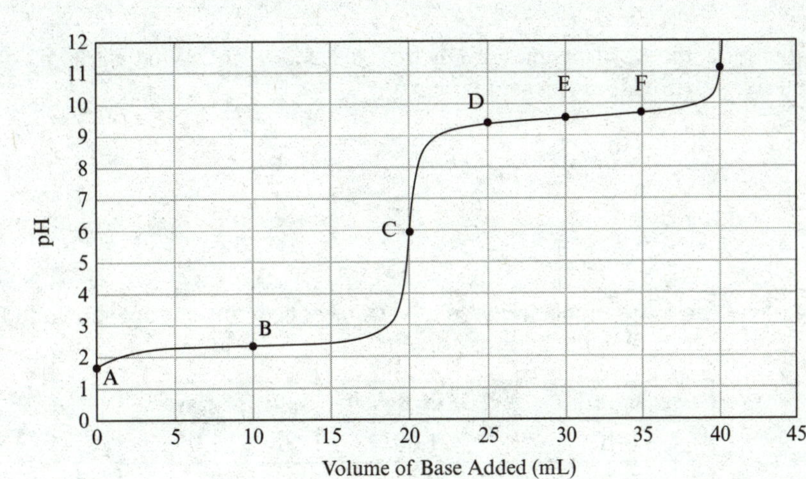


**11.** Which of the structures below correctly shows the form of glycine at point A in the titration?

**A.** $H_3N^+{-}CH_2{-}C(OH){=}O^+H$

**B.** $H_3N^+{-}CH_2{-}C(=O){-}OH$

**C.** $H_3N^+{-}CH_2{-}C(=O){-}O^-$

**D.** $H_2N{-}CH_2{-}C(=O){-}O^-$

**12.** Estimate the p$K_a$ for the following reaction of glycine.

$$^{+}H_3NCH_2CO_2^-(aq) + H_2O(l) \rightleftharpoons H_2NCH_2CO_2^-(aq) + H_3O^+(aq)$$

A. 2.3
B. 6.0
C. 9.6
D. 11.0

**13.** Use the data in the table to determine which scenario would lead to a buffer with a pH of 6.

| Acid | $K_a$ |
|---|---|
| HF | $1 \times 10^{-3}$ |
| HOCl | $1 \times 10^{-8}$ |

A. 1.0 mole of NaOCl is added to 0.01 mole of HOCl in enough water to make 1.0 L of solution.
B. 0.01 mole of NaF is added to 1.0 mole of HF in enough water to make 1.0 L of solution.
C. 1.0 mole of NaF is added to 0.01 mole of HF in enough water to make 1.0 L of solution.
D. 0.01 mole of NaOCl is added to 1.0 mole of HOCl in enough water to make 1.0 L of solution.

**14.** At a certain temperature, the $K_{sp}$ of magnesium hydroxide is $4.0 \times 10^{-12}$. What is the pH of a saturated solution of magnesium hydroxide at this temperature?

A. 8.6
B. 10.3
C. 11.0
D. 12.0

**15.** The pOH values of two basic solutions prepared at various concentrations are shown in the table below. Which of the statements regarding the bases is correct?

| Concentration | pOH of Base 1 | pOH of Base 2 |
|---|---|---|
| 0.10 | 1.00 | 2.17 |
| 0.50 | 0.30 | 1.83 |
| 1.0 | 0.00 | 1.68 |

A. The conjugate acid of base 1 has a higher p$K_a$ than the conjugate acid of base 2.
B. Base 1 has a higher p$K_b$ than base 2.
C. The concentration of $H^+$ is greater in a 1.0 $M$ solution of base 1 than in a 1.0 $M$ solution of base 2.
D. If equal volumes of 0.5 $M$ solutions of each base are titrated with strong acid, base 1 will require a greater volume of acid to reach the equivalence point than base 2.

## Long Free Response

**1.** The graph below shows the results of the titration of 50.0 mL of a 0.500 *M* solution of a weak acid, HA, with 0.500 *M* NaOH.

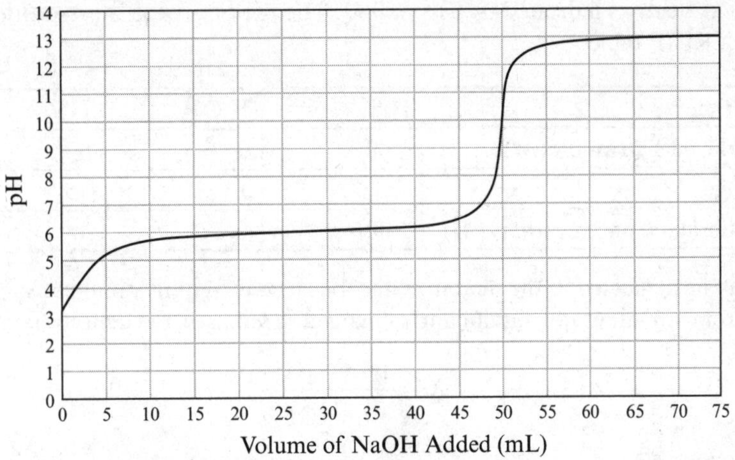


**a.** Describe three features of the graph that identify HA as a weak acid.

**b.** **i.** Describe three different ways that the acid dissociation constant, $K_a$, for HA may be determined.
**ii.** Choose one of the methods that you described and use it to calculate the $K_a$ for HA.

**c.** On the graph above, sketch the curve that would result if 50.0 mL of 0.500 *M* $HNO_3$ was titrated instead of HA.

**d.** How will the pH at the equivalence point differ if 1.00 *M* NaOH is used instead of 0.500 *M*?

**e.** What is the pH of the solution after 10.0 mL of 0.500 *M* NaOH have been added to the HA solution?

**f.** Would the pH at the endpoint of the titration be higher or lower than the pH at the equivalence point if an indicator with a $K_a$ of $1 \times 10^{-6}$ had been used instead of a pH meter? Justify your response with a calculation.

## Short Free Response

**2.** At 25 °C, the pH of 25.0 mL of a 0.00300 *M* aqueous solution of aniline, $C_6H_5NH_2$, is 8.174.

**a.** Write the balanced net ionic equation for the base dissociation reaction of aniline with water.

**b.** Determine the $pK_b$ of aniline.

**c.** What is the pH of a 25.0-mL solution of 0.00300 *M* aniline to which 25.0 mL of 0.00100 *M* HCl have been added?

**d.** When a few drops of additional HCl are added to the solution in part c, the pH does not change detectably. Write a net ionic equation that shows why this is true.

# Answers and Explanations

## Multiple Choice

1. **B.** $H_2S$ is a very weak acid, as indicated by its small $K_a$. The pH of a weak acid solution is formally determined using a RICE table.

| Reaction | $H_2S$ | $H_3O^+$ | $HS^-$ |
|---|---|---|---|
| Initial Concentrations ($M$) | 1.00 | 0 | 0 |
| Change | $-x$ | $+x$ | $+x$ |
| Equilibrium Concentrations ($M$) | $1.00 - x$ | $x$ | $x$ |

The value of $x$ may be neglected in the denominator. Because the equilibrium constant is so small, subtraction of $x$ from 1.00 does not significantly change the value of the denominator.

$$K_a = \frac{[H_3O^+][HS^-]}{[H_2S]}$$

$$9 \times 10^{-8} = \frac{x^2}{1.00 - \cancel{x}}$$

The pH of the solution is estimated to be between 3 and 4.

$$\sqrt{(9 \times 10^{-8})} = \sqrt{x^2}$$

$$x = 3 \times 10^{-4}\ M = [H_3O^+]$$

$$\text{pH} = -\log(3 \times 10^{-4}) \approx 3\text{–}4$$

Choice B, 3.5, falls in this range.

2. **D.** Brønsted-Lowry acids are proton donors in solution. The species that behave as Brønsted-Lowry acids are $H_2S$ in the reactants and $H_3O^+$ in the products, choice D.

3. **C.** Brønsted-Lowry bases are proton acceptors in solution. The species that behave as bases in the reaction are $HS^-$ in the products and $H_2O$ in the reactants. Because the acid dissociation constant is small, the equilibrium lies to the left. Because the equilibrium favors the side of the reaction with the weaker acid and base, $H_2O$ is a weaker base than $HS^-$. The $K_b$ of $HS^-$, choice C, is larger than the $K_b$ of $H_2O$.

4. **C.** Water is both amphoteric, which means it can behave as both an acid and a base, and amphiprotic, which means it can behave as both a proton donor and a proton acceptor. Therefore, the statement in choice C is false.

5. **A.** The conjugate base of the carboxylic acid with the largest number of inductively electron-withdrawing chlorine atoms, choice A, is the most stable of the structures drawn, so its conjugate acid is the strongest.

6. **A.** The $F^-$ ion is the conjugate base of a weak acid, which means it behaves as a base in an aqueous solution. The more basic a solution, the higher its pH. Therefore, 1.00 $M$ NaF, choice A, has the highest pH at 25 °C.

7. **C.** Barium hydroxide is a strong, soluble base. For every mole of $Ba(OH)_2$ dissolved, two moles of $OH^-$ are released in the solution. The concentration of $OH^-$ is

$$\frac{5.00 \times 10^{-2}\ \text{mol}\ \cancel{Ba(OH)_2}}{1\ \text{L}} \times \frac{2\ \text{mol OH}^-}{1\ \text{mol}\ \cancel{Ba(OH)_2}} = 0.100\ M\ \text{OH}^- = 1.00 \times 10^{-1}\ M\ \text{OH}^-$$

$$\text{pOH} = -\log[\text{OH}^-] = -\log 1.00 \times 10^{-1} = 1$$

$$\text{pH} = 14 - \text{pOH} = 14 - 1 = 13, \text{ choice C}$$

**8.** **C.** The $K_a$ for the weak acid is $10^{-4}$.

$$HA(aq) + H_2O(l) \rightleftharpoons A^-(aq) + H_3O^+(aq)$$

The concentration of ionized HA is determined using a RICE table.

| Reaction | HA | $H_3O^+$ | $A^-$ |
|---|---|---|---|
| Initial Concentrations (*M*) | 0.0100 | 0 | 0 |
| Change | $-x$ | $+x$ | $+x$ |
| Equilibrium Concentrations (*M*) | $0.0100 - x$ | $x$ | $x$ |

$$K_a = \frac{[H_3O^+][A^-]}{[HA]}$$

$$1\times10^{-4} = \frac{x^2}{0.0100 - \cancel{x}}$$

$$\sqrt{(1\times10^{-4})(1\times10^{-2})} = \sqrt{x^2}$$

$$x = 1\times10^{-3}\ M = [A^-]$$

$$\%\ \text{ionization} = \frac{1\times10^{-3}}{1\times10^{-2}}\times100\% = 10\%, \text{ choice C}$$

**9.** **D.** The solution with the highest pH is the one in which the anion is the conjugate base of the weakest acid. The weakest conjugate acid is HCN, so the $CN^-$ released by $Ca(CN)_2$, choice D, produces the most basic solution of the choices given.

**10.** **D.** Given a list of $K_a$ values, the conjugate acid strength of the bases must be compared. The conjugate acid of $PO_4^{3-}$ is $HPO_4^{2-}$, with a $K_a$ of $4.8\times10^{-13}$. The conjugate acid of $C_6H_5O_7^{3-}$ is $HC_6H_5O_7^{2-}$, with a $K_a$ of $4.0\times10^{-6}$. Since $HPO_4^{2-}$ is a weaker acid than $HC_6H_5O_7^{2-}$, $PO_4^{3-}$ is a stronger base than $C_6H_5O_7^{3-}$. choice D.

**11.** **B.** At the lowest pH in the titration, the carboxylic acid group has the form $-CO_2H$, and the amine group has the form $-NH_3^+$, where the nitrogen atom has a formal charge of +1. This is depicted in choice B.

**12.** **C.** The form of glycine that is present at the end of the titration is the one in which both the carboxylic acid and the amine are deprotonated, $H_2NCH_2CO_2^-$. This form is the conjugate base of the zwitterionic species, $^+H_3NCH_2CO_2^-$, for which the p$K_a$, ~9.6 (choice C), can be determined halfway between the first and second equivalence points.

**13.** **D.** The Henderson-Hasselbalch equation can be used to determine which ratio of conjugate acid and conjugate base is appropriate. For the $[OCl^-]$:[HOCl] ratio of 0.01:1.0, choice D, the pH is equal to 6.

$$pH = pK_a + \log\frac{[A^-]}{[HA]} = 8 + \log\left(\frac{1.0\times10^{-2}}{1.0}\right) = 8 - 2 = 6$$

**14.** **B.** The dissociation equation for $Mg(OH)_2$ is

$$Mg(OH)_2(s) \rightleftharpoons Mg^{2+}(aq) + 2\ OH^-(aq)$$

$$K_{sp} = [Mg^{2+}][OH^-]^2 = x(2x)^2 = 4x^3$$

$$4x^3 = 4.0\times10^{-12}$$

$$x = 1.0\times10^{-4}$$

$$[OH^-] = 2.0\times10^{-4}$$

$$pOH \approx 3-4$$

$$pH \approx 10-11$$

Choice B, 10.3, falls in this range.

**15.** A. Base 1 is a strong base (pOH = 0, pH = 14 for a 1.0 *M* solution) and base 2 is a weaker base. The stronger the conjugate base, the weaker its conjugate acid, so the conjugate acid of base 1 has a lower $K_a$ and a higher p$K_a$ than the conjugate acid of base 2.

## Long Free Response

**1. a.** The three features of the graph that indicate that HA is a weak acid rather than a strong acid are 1) the initial pH is greater than 0.30, which is the expected pH for a 0.500 *M* solution of a strong acid; 2) there is an initial rise in the pH when base is first added, and then the pH levels off in the buffering region before it undergoes another sharp rise just prior to the equivalence point; and 3) the pH at the equivalence point is greater than 7.

**b.** **i.** The acid dissociation constant may be determined by 1) the pH and [HA] of the solution before any base has been added (the $K_a$ is equal to $\frac{(10^{-pH})^2}{[HA]}$). 2) At the half-equivalence point of the titration, the $K_a$ is equal to $10^{-pH}$. 3) The $K_b$ at the equivalence point can be determined. (It is equal to $\frac{\left(10^{-(14-pH)}\right)^2}{[A^-]}$.) The $K_a$ can then be determined by $\frac{K_w}{K_b}$.

**ii.** The simplest method is the second one, in which pH = p$K_a$ at the half-equivalence point. It can be seen from the plot that the pH at the half-equivalence point, when 25.0 mL of base have been added, is 6.

$$K_a = 10^{-6}$$

**c.** If $HNO_3$, a strong acid, was titrated, the initial pH would be around 0.3, the curve would not rise initially and then enter a buffer region, and the pH at the equivalence point would be 7.

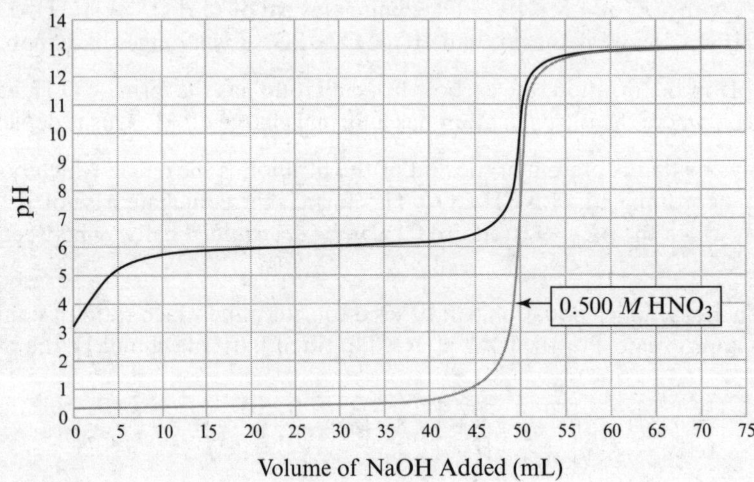


**d.** If 1.00 *M* NaOH were used instead of 0.500 *M*, a smaller volume of NaOH would be required to reach the equivalence point. The concentration of $A^-$ would be greater since the total volume would be less, resulting in a slightly higher $[OH^-]$, a lower pOH, and a higher pH.

**e.** The number of moles of HA present before the titration begins is

$$0.0500 \cancel{\text{L HA}} \times \frac{0.500 \text{ mol HA}}{1 \cancel{\text{L HA}}} = 0.0250 \text{ mol HA}$$

After 10.0 mL of 0.500 *M* NaOH have been added to the solution, the number of moles of HA that have been converted to $A^-$ is

$$0.0100 \cancel{\text{L NaOH}} \times \frac{0.500 \cancel{\text{mol NaOH}}}{1 \cancel{\text{L NaOH}}} \times \frac{1 \text{ mol A}^-}{1 \cancel{\text{mol NaOH}}} = 0.00500 \text{ mol A}^-$$

The number of moles of HA remaining are

$$0.0250 \text{ mol} - 0.00500 \text{ mol} = 0.0200 \text{ mol HA}$$

The Henderson-Hasselbalch equation, given on the *AP Chemistry Equations and Constants* sheet, is used to determine the pH of the solution at this point. Notice that the total volume, 60.0 mL, cancels, so number of moles can be used directly instead of concentrations in this calculation.

$$\text{pH} = \text{p}K_a + \log\frac{[\text{A}^-]}{[\text{HA}]} = 6.0 + \log\left(\frac{0.00500}{0.0200}\right) = 5.4$$

**f.** The pH at the endpoint of the titration is determined by the concentration and $K_b$ of $A^-$. The concentration is

$$\frac{0.0250 \text{ mol A}^-}{0.100 \text{ L}} = 0.250\ M\ \text{A}^-$$

Note that the volume used is the total volume after the addition of 50.0 mL of NaOH to the original 50.0 mL of HA.

The $K_b$ is determined using the $K_a$ of HA and $K_w$.

$$K_b = \frac{K_w}{K_a} = \frac{1\times10^{-14}}{1\times10^{-6}} = 1\times10^{-8}$$

The base dissociation reaction of $A^-$ is as follows.

$$\text{A}^-(aq) + \text{H}_2\text{O}(l) \rightleftharpoons \text{HA}(aq) + \text{OH}^-(aq)$$

$$K_b = \frac{[\text{HA}][\text{OH}^-]}{[\text{A}^-]}$$

A RICE table is used to determine the equilibrium concentrations.

| Reaction | $A^-$ | HA | $OH^-$ |
|---|---|---|---|
| **Initial Concentrations (*M*)** | 0.250 | 0 | 0 |
| **Change** | $-x$ | $+x$ | $+x$ |
| **Equilibrium Concentrations (*M*)** | $0.250 - x$ | $x$ | $x$ |

Substitution of the equilibrium concentrations into the expression for $K_b$ and the assumption that $x$ is much smaller than the initial concentration of $A^-$ allow for the calculation of the hydroxide ion concentration.

$$1.0\times10^{-8} = \frac{x^2}{0.250 - \cancel{x}}$$

$$x = \sqrt{(1.0\times10^{-8})(0.250)} = 5.0\times10^{-5}\ M = [\text{OH}^-]$$

$$\text{pOH} = 5.0\times10^{-5} = 4.30$$

$$\text{pH} = 14.00 - 4.30 = 9.70$$

If an indicator with a p$K_a$ of 6 had been used, the pH at the endpoint of the titration would be about 5–7, which is lower than the pH at the equivalence point.

# Short Free Response

**2.** **a.** The net ionic equation for the dissociation reaction of aniline with water is shown below.

$$C_6H_5NH_2 + H_2O \rightleftharpoons C_6H_5NH_3^+ + OH^-$$

State symbols are not required for credit, but if they are included, the reaction becomes

$$C_6H_5NH_2(aq) + H_2O(l) \rightleftharpoons C_6H_5NH_3^+(aq) + OH^-(aq)$$

**b.** The $pK_b$ can be determined from the $K_b$. The expression for the $K_b$ is

$$K_b = \frac{[C_6H_5NH_3^+][OH^-]}{[C_6H_5NH_2]}$$

The value of the pH can be used to determine the pOH of the solution.

$$pOH = 14.000 - pH = 5.826$$

The $[OH^-]$ concentration is $10^{-5.826}$ *M*, or $1.49 \times 10^{-6}$ *M*. This is equal to $[C_6H_5NH_3^+]$. The aniline concentration is the difference between the initial concentration and the concentration of $[C_6H_5NH_3^+]$.

$$[C_6H_5NH_2] = 0.00300\ M - 1.49 \times 10^{-6}\ M \approx 0.00300\ M$$

$$K_b = \frac{(1.49 \times 10^{-6})^2}{0.00300} = 7.43 \times 10^{-10}$$

$$pK_b = -\log(7.43 \times 10^{-10}) = 9.13$$

**c.** The result of the addition is a buffer solution. The number of moles of $C_6H_5NH_2$ and $C_6H_5NH_3^+$ can be substituted into the base version of the Henderson-Hasselbalch equation.

$$C_6H_5NH_2(aq) + H^+(aq) \rightarrow C_6H_5NH_3^+(aq)$$

$$0.0250\ \cancel{L\ HCl} \times \frac{0.00100\ \cancel{mol\ HCl}}{1\ \cancel{L\ HCl}} \times \frac{1\ mol\ C_6H_5NH_3^+}{1\ \cancel{mol\ HCl}} = 2.50 \times 10^{-5}\ mol\ C_6H_5NH_3^+$$

$$0.0250\ L\ C_6H_5NH_2 \times \frac{0.00300\ mol\ C_6H_5NH_2}{1\ L\ C_6H_5NH_2} = 7.50 \times 10^{-5}\ mol\ C_6H_5NH_2 \text{ initially present}$$

$$7.50 \times 10^{-5}\ mol\ C_6H_5NH_2 - 2.50 \times 10^{-5}\ mol\ C_6H_5NH_3^+ = 5.00 \times 10^{-5}\ mol\ C_6H_5NH_2 \text{ present at equilibrium}$$

$$pOH = pK_b + \log\frac{[C_6H_5NH_3^+]}{[C_6H_5NH_2]} = 9.13 + \log\left(\frac{2.50 \times 10^{-5}}{5.00 \times 10^{-5}}\right) = 8.83$$

$$pH = 14.00 - 8.83 = 5.17$$

**d.** The solution is a buffer since it has a base, $C_6H_5NH_2$, and its conjugate acid, $C_6H_5NH_3^+$. The reaction between HCl and aniline allows the solution to resist a change in pH as HCl is added.

$$C_6H_5NH_2 + H^+ \rightleftharpoons C_6H_5NH_3^+$$

State symbols are not required for credit, but if they are included, the reaction becomes

$$C_6H_5NH_2(aq) + H^+(aq) \rightleftharpoons C_6H_5NH_3^+(aq)$$

# Chapter 9

# Applications of Thermodynamics

**The relationship between enthalpy, entropy, and Gibbs free energy can be used to predict the thermodynamic favorability of a process.**

A reaction or process that is thermodynamically favorable will proceed in the forward direction. Once a thermodynamically favored process begins, no intervention is necessary for it to continue.

Consider a balloon filled with air. If the balloon bursts, it releases its gaseous contents to the surroundings. The release of gas is thermodynamically favored. The reverse of this process is thermodynamically disfavored. The gas molecules will never re-assemble into a balloon-shaped volume.

Enthalpy changes alone do not predict whether a process will be thermodynamically favored or disfavored. In the balloon-bursting example, little if any heat is exchanged between the system and the surroundings, yet the process is thermodynamically favored.

Examples of both exothermic processes, like the combustion of hydrocarbons, and endothermic processes, like the melting of ice on a warm countertop, are thermodynamically favored.

If thermodynamic favorability is to be predicted, both enthalpy and **entropy,** the range of positional states experienced by the particles in a system or the dispersal of energy in a system, have to be taken into account.

Although "spontaneous" is a description of thermodynamic favorability that is no longer used on the AP Chemistry exam, a very large number of chemistry (and physics) resources use the word to describe a thermodynamically favored process. Spontaneity should never be confused with speed or "suddenness." Some spontaneous processes, like the rusting of a solid sheet of iron in the presence of oxygen and moisture, are very slow.

## Entropy

**Entropy,** $S$, is analogous to disorder or randomness at a molecular level. It is generally reported with units of J/(mol·K). (Be aware that this differs from the units of $\Delta H$, which are usually kJ/mol.)

Any change that increases the range of motion of a collection of molecules increases their entropy. An increase in the entropy of a system is denoted by the expression $\Delta S_{sys} > 0$.

Changes that increase the entropy of a system include phase transitions that allow molecules to move more freely—like melting and boiling, increases in temperature or volume, and mixing. Also, the system entropy increases for reactions in which gaseous products outnumber gaseous reactants.

It is important to remember that while the entropy of the system increases when solids or liquids are dissolved in liquids, when gases are dissolved in liquids, entropy of the system decreases.

| **Processes in Which Entropy Increases ($\Delta S_{sys} > 0$)** | **Processes in Which Entropy Decreases ($\Delta S_{sys} < 0$)** |
|---|---|
| Melting | Freezing |
| Boiling | Condensation |
| Sublimation | Deposition |
| Dissolving a solid or liquid in a solvent | Dissolving a gas in a solvent |
| Overall generation of moles of gas by a reaction | Overall consumption of moles of gas by a reaction |

## Practice

For each of the following processes, determine if the process likely involves an increase in entropy of the system or a decrease.

**a.** The freezing of ethanol, $C_2H_6O$
**b.** Dissolving a crystal of sodium chloride in water
**c.** Solid calcium carbonate decomposes to form solid calcium oxide and gaseous carbon dioxide
**d.** A sample of helium at constant temperature is compressed from a volume of 3 liters to a volume of 2 liters
**e.** A sample of helium in a rigid 3-liter container is heated from 25 °C to 35 °C

**(a)** The freezing of ethanol involves a decrease in system entropy, $\Delta S_{sys} < 0$. When ethanol freezes, the molecules' positions become fixed, and this decreases their dispersal.

$$C_2H_6O(l) \rightarrow C_2H_6O(s)$$

**(b)** System entropy increases, $\Delta S_{sys} > 0$, upon the dissolution of a crystal of sodium chloride in water. The sodium and chlorine ions are more dispersed and free to move in an aqueous solution than they are in fixed crystal lattice points.

$$NaCl(s) \rightarrow Na^+(aq) + Cl^-(aq)$$

**(c)** When a solid molecule of calcium carbonate decomposes, system entropy increases, $\Delta S_{sys} > 0$. Reactions in which the number of moles of gaseous products is greater than the number of moles of gaseous reactants generally involve an increase in system entropy.

$$CaCO_3(s) \rightarrow CaO(s) + CO_2(g)$$

**(d)** When a container holding a fixed number of helium atoms is decreased in volume at a constant temperature, the positional possibilities available to each atom decreases. This causes the entropy of the system to decrease, $\Delta S_{sys} < 0$.

**(e)** At constant volume, an increase in the temperature of a sample causes the kinetic energy of the molecules to increase. An increase in the motion of the particles increases the entropy of the system, $\Delta S_{sys} > 0$.

# The Second Law of Thermodynamics

The **second law of thermodynamics** states that the entropy of the universe is always increasing. The quantity of matter and energy in the universe is fixed (first law), but the entropy is not (second law).

Stated mathematically, the second law of thermodynamics says that the change in the combined entropy of the system plus the surroundings is always a positive quantity in a thermodynamically favored process.

$$\Delta S_{univ} > 0$$

The thermodynamic favorability of a process is a function of the effect of the process on the entropy of the universe.

The entropy of the universe is made up of the entropy of a system plus the entropy of its surroundings.

$$\Delta S_{univ} = \Delta S_{sys} + \Delta S_{surr}$$

At constant temperature and pressure, the entropy change of the surroundings is related to the enthalpy change of a process.

$$\Delta S_{surr} = \frac{\Delta H_{surr}}{T}$$

The enthalpy change of the surroundings is determined by the enthalpy change of the system. The amount of heat transferred between system and surroundings is equal in magnitude and opposite in sign, so the change in entropy of the surroundings is opposite in sign to the heat transferred by the system.

$$\Delta H_{surr} = -\Delta H_{sys}$$

$$\Delta S_{surr} = \frac{-\Delta H_{sys}}{T}$$

The entropy change of a system can be calculated from tabulated values of the standard entropies of the products and reactants.

$$\Delta S^{\circ}_{sys} = \Sigma S^{\circ}_{products} - \Sigma S^{\circ}_{reactants}$$

Consider the highly exothermic reaction of solid sodium metal with chlorine gas.

$$\text{Na}(s) + \frac{1}{2}\text{Cl}_2(g) \rightarrow \text{NaCl}(s)$$

| **Substance** | **$\Delta H^{\circ}_f$ (kJ/mol)** | **$S^{\circ}$ (J/(mol·K))** |
|---|---|---|
| NaCl(*s*) | –411.2 | 72.1 |
| Na(*s*) | 0 | 51.3 |
| $Cl_2$(*g*) | 0 | 165 |

In terms of the entropy of the system, this reaction is disfavored since the entropy decreases. This can be predicted qualitatively since the number of moles of gaseous reactants is greater than the number of moles of gaseous products. This prediction can also be quantified.

$$\Delta S^{\circ}_{sys} = \Sigma S^{\circ}_{products} - \Sigma S^{\circ}_{reactants}$$

$$\Delta S^{\circ}_{sys} = 72.1 \text{ J/(mol}\cdot\text{K)} - \left[1(51.3 \text{ J/(mol}\cdot\text{K)}) + \frac{1}{2}(165 \text{ J/(mol}\cdot\text{K)})\right]$$

$$\Delta S^{\circ}_{sys} = -62 \text{ J/(mol}\cdot\text{K)}$$

Even though the $\Delta S_{sys}$ is negative, the reaction is thermodynamically favorable under appropriate conditions. It generates a large amount of heat, and the heat that is released is transferred to the surroundings, which ultimately leads to an increase in the entropy of the universe. The enthalpy change for the reaction can be used to determine the entropy change for the surroundings at 298 K.

$$\Delta H^{\circ}_{rxn} = \Sigma n\Delta H^{\circ}_{f(products)} - \Sigma n\Delta H^{\circ}_{f(reactants)}$$

$$\Delta H^{\circ}_{sys} = -411.2 \text{ kJ/mol} - \left[1(0 \text{ kJ/mol}) + \frac{1}{2}(0 \text{ kJ/mol})\right]$$

$$\Delta H^{\circ}_{sys} = -411.2 \text{ kJ/mol}$$

$$\Delta H_{surr} = -\Delta H_{sys} = 411.2 \text{ kJ/mol}$$

$$\Delta S_{surr} = \frac{\Delta H_{surr}}{T} = \frac{411.2 \text{ kJ/mol}}{298 \text{ K}} = 1.38 \text{ kJ/(mol}\cdot\text{K)}$$

The larger entropy increase of the surroundings compensates for the entropy decrease of the system; the change in the entropy of the universe increases, and the second law of thermodynamics is obeyed.

$$\Delta S_{univ} = \Delta S_{sys} + \Delta S_{surr}$$

$$\Delta S_{univ} = \left[(-62 \cancel{J}/(mol \cdot K)) \times \frac{1\ kJ}{1000\ \cancel{J}}\right] + 1.38\ kJ/(mol \cdot K) = 1.32\ kJ/(mol \cdot K)$$

## Practice

The following reaction is thermodynamically favored.

$$NH_3(g) + HCl(g) \rightarrow NH_4Cl(s)$$

Given the information tabulated below,

**a.** calculate the standard enthalpy of the reaction.
**b.** calculate the entropy change of the system.
**c.** explain qualitatively why this reaction occurs as written.

| Substance | $\Delta H^\circ_f$ (kJ/mol) | $S^\circ$ (J/(mol·K)) |
|---|---|---|
| $NH_4Cl(s)$ | –314.4 | 94.6 |
| $NH_3(g)$ | –45.9 | 192.8 |
| $HCl(g)$ | –92.3 | 186.9 |

**(a)** The standard enthalpy of the reaction is determined from the standard enthalpies of formation.

$$\Delta H^\circ_{rxn} = \Sigma n\Delta H^\circ_{f(products)} - \Sigma n\Delta H^\circ_{f(reactants)}$$

$$\Delta H^\circ_{sys} = -314.4\ kJ/mol - [-45.9\ kJ/mol + (-92.3\ kJ/mol)]$$

$$\Delta H^\circ_{sys} = -176.2\ kJ/mol$$

**(b)** The entropy change associated with the reaction is determined from the enthalpies of the product and reactants.

$$\Delta S^\circ_{sys} = \Sigma S^\circ_{(products)} - \Sigma S^\circ_{(reactants)}$$

$$\Delta S^\circ_{sys} = 94.6\ J/(mol \cdot K) - [192.8\ J/(mol \cdot K) + 186.9\ J/(mol \cdot K)]$$

$$\Delta S^\circ_{sys} = -285.1\ J/(mol \cdot K)$$

**(c)** Despite the entropy decrease of the system, the reaction is thermodynamically favored. The reaction must be sufficiently exothermic to increase the entropy of the surroundings enough to offset the entropy decrease of the system, since the change in entropy of the universe is always positive.

# The Third Law of Thermodynamics

The **third law of thermodynamics** states that at 0 K, when a substance is a perfect, motionless crystal, the entropy of the substance is zero. As the temperature is raised, the vibrational motion increases the dispersion of the particles making up the crystal. Since this variation of entropy with temperature is understood, the absolute entropies of substances can be reported. The absolute entropy of a substance at a temperature is the entropy change as the substance goes from 0 K to the new temperature.

The values of **standard absolute entropies, $S°$,** or **standard molar entropies,** are reported in tables of thermodynamic values. The tabulated $S°$ value is the entropy of one mole of a substance in its standard state at 1 atm pressure and 298 K.

Unlike the enthalpies of formation of pure elements in their thermodynamic standard states, the absolute entropies of elements and molecules are nonzero. This is because these elements are not perfectly ordered at temperatures above 0 K.

Absolute entropy increases from solid to liquid to gas. Also, it increases with increasing molecular complexity; the larger the number of electrons in a molecule, the higher the absolute entropy of the molecule, but the effect of complexity is not as pronounced as the effect of the phase of the molecule.

### Practice

Rank the compounds in order of increasing absolute entropy at 298 K: $H_2(g)$, $Cl_2(g)$, $H_2O(l)$, $HCl(g)$.

Gaseous compounds tend to have higher absolute entropies than liquids, so the entropy of $H_2O(l)$ is predicted to be the lowest. The three gases must be ranked in terms of molecular complexity. Hydrogen gas, $H_2$, is the least complex. It has only two electrons, so it has a lower entropy than $Cl_2$ or HCl. Chlorine gas, $Cl_2$, has the greatest complexity and the largest number of electrons, so it has the highest absolute entropy. The correct ranking of the molecules in terms of increasing absolute entropy is $H_2O(l) < H_2(g) < HCl(g) < Cl_2(g)$.

The actual absolute entropy values are shown for comparison.

| Substance | $S°$ (J/(mol·K)) |
|---|---|
| $H_2O(l)$ | 70 |
| $H_2(g)$ | 131 |
| $HCl(g)$ | 187 |
| $Cl_2(g)$ | 223 |

## Gibbs Free Energy

The change in **Gibbs free energy, $\Delta G$,** associated with a process describes the thermodynamic favorability of the process at constant temperature and pressure. Gibbs free energy provides a value for the maximum amount of useful work that can be obtained from a system. The Gibbs free energy of a system, at constant temperature and pressure, is given by

$$\Delta G = \Delta H - T\Delta S$$

The standard molar Gibbs free energy, $\Delta G°$, for thermodynamic standard conditions (1 atm, 1 $M$ solutions, and 298 K) is given by

$$\Delta G° = \Delta H° - T\Delta S°$$

The products of a thermodynamically favored, or spontaneous, reaction have final Gibbs energies that are lower than those of the reactants. In other words, $\Delta G$ is negative for a thermodynamically favored process. For a process that is thermodynamically disfavored, $\Delta G$ is positive, and for a process in which reactants and products are at equilibrium, $\Delta G = 0$.

| | |
|---|---|
| $\Delta G < 0$ | Thermodynamically favored |
| $\Delta G > 0$ | Thermodynamically disfavored |
| $\Delta G = 0$ | Reaction at equilibrium |

Standard molar Gibbs free energy values are experimentally determined and are tabulated in the scientific literature along with $\Delta H°$ and $S°$ values. Like $\Delta H°$ values, $\Delta G°$ values are zero for elements in their standard states.

The standard molar Gibbs free energy change for a reaction can be determined in the same ways that the enthalpy and entropy changes can be calculated.

$$\Delta G^\circ_{\text{rxn}} = \sum n\Delta G^\circ_{\text{f(products)}} - \sum n\Delta G^\circ_{\text{f(reactants)}}$$

For example, a prediction can be made regarding whether the decomposition of pure hydrogen peroxide is a spontaneous process at 1 atm and 298 K using tabulated values of $\Delta G°_f$.

$$2\ H_2O_2(l) \rightarrow 2\ H_2O(l) + O_2(g)$$

| Substance | $\Delta G°_f$ (kJ/mol) |
|---|---|
| $H_2O_2(l)$ | –120.4 |
| $H_2O(l)$ | –237.1 |
| $O_2(g)$ | 0 |

The reaction is carried out under standard conditions, so these free energies of formation may be used to determine the overall free energy change for this reaction.

$$\Delta G^\circ_{\text{rxn}} = \sum n\Delta G^\circ_{\text{f(products)}} - \sum n\Delta G^\circ_{\text{f(reactants)}}$$

$$\Delta G^\circ_{\text{rxn}} = [(2(-237.1\text{ kJ/mol})) + (1(0\text{ kJ/mol}))] - (2(-120.4\text{ kJ/mol}))$$

$$\Delta G^\circ_{\text{rxn}} = -233.4\text{ kJ/mol}$$

The negative value of the change in free energy for the reaction indicates that this is a thermodynamically favored process under standard conditions.

A second method for determining the Gibbs free energy change for a reaction is to use a procedure similar to the one used to find $\Delta H°$ using Hess's law. The steps for which the free energy changes are known must be manipulated so that they add up to the overall process.

When graphite is added to solid copper(II) oxide, the reduction of CuO(*s*) to solid copper becomes thermodynamically favored under standard conditions.

$$2\ CuO(s) + C(s) \rightarrow 2\ Cu(s) + CO_2(g)$$

The reactions that make up this transformation are:

$$CuO(s) \rightarrow Cu(s) + \frac{1}{2}\ O_2(g) \qquad \Delta G^\circ_{298\text{ K}} = 129.7\text{ kJ/mol}$$

$$C(s) + O_2(g) \rightarrow CO_2(g) \qquad \Delta G^\circ_{298\text{ K}} = -394.4\text{ kJ/mol}$$

When the first equation is multiplied by 2, it can be added to the second equation to result in the overall transformation. The $\Delta G°$ for the first reaction is also multiplied by 2. The Gibbs free energy changes can be added to give the overall $\Delta G°$ for the reaction. (This is valid only at 298 K.)

$$2\ CuO(s) \rightarrow 2\ Cu(s) + \cancel{O_2(g)} \qquad \Delta G^\circ_{298\text{ K}} = 2(129.7\text{ kJ/mol})$$

$$C(s) + \cancel{O_2(g)} \rightarrow CO_2(g) \qquad \Delta G^\circ_{298\text{ K}} = -394.4\text{ kJ/mol}$$

$$2\ CuO(s) + C(s) \rightarrow 2\ Cu(s) + CO_2(g) \qquad \Delta G^\circ_{298\text{ K}} = -135.0\text{ kJ/mol}$$

This example serves to illustrate another important point. The decomposition of copper(II) oxide to solid copper and oxygen gas under standard conditions is thermodynamically disfavored ($\Delta G°_{298\ K} = 129.7$ kJ/mol). The **coupling** of this reaction to one in which the oxygen produced forms bonds with carbon results in a sufficient decrease in Gibbs free energy to make the transformation of copper(II) oxide to solid copper possible.

## Practice

Use the standard molar free energies of formation below to determine whether the reaction between solid $B_2O_3$ and hydrogen gas to form gaseous $B_2H_6$ and oxygen is spontaneous under standard thermodynamic conditions.

| Substance | $\Delta G°_f$ (kJ/mol) |
|---|---|
| $B_2O_3(s)$ | –1194.3 |
| $B_2H_6(g)$ | 87.6 |

The first step is to write and balance the chemical equation.

$$2\ B_2O_3(s) + 6\ H_2(g) \rightarrow 2\ B_2H_6(g) + 3\ O_2(g)$$

The standard free energies of formation can be used to determine the overall change in Gibbs free energy for the process under standard thermodynamic conditions.

$$\Delta G^{\circ}_{\text{rxn}} = \Sigma n\Delta G^{\circ}_{\text{f(products)}} - \Sigma n\Delta G^{\circ}_{\text{f(reactants)}}$$

$$\Delta G^{\circ}_{\text{rxn}} = [(2(87.6\text{ kJ/mol})) + (3(0\text{ kJ/mol}))] - [(2(-1194.3\text{ kJ/mol})) + (6(0\text{ kJ/mol}))]$$

$$\Delta G^{\circ}_{\text{rxn}} = 2564\text{ kJ/mol}$$

The positive change in free energy for the reaction indicates that this is a thermodynamically disfavored process under standard conditions.

## Practice

Calculate $\Delta G°$ for the formation of gaseous $NO_2$.

$$NO(g) + O(g) \rightarrow NO_2(g)$$

Use the following information.

$$2\ O_3(g) \rightarrow 3\ O_2(g) \qquad \Delta G^{\circ}_{298\text{ K}} = -326.4\text{ kJ/mol}$$

$$O_2(g) \rightarrow 2\ O(g) \qquad \Delta G^{\circ}_{298\text{ K}} = 463.4\text{ kJ/mol}$$

$$NO(g) + O_3(g) \rightarrow NO_2(g) + O_2(g) \qquad \Delta G^{\circ}_{298\text{ K}} = -199.5\text{ kJ/mol}$$

The first and second equations, along with their associated $\Delta G°$ values, are algebraically manipulated in the same manner as for $\Delta H°$. The first and second equations, along with their associated $\Delta G°$ values, are multiplied by $-\frac{1}{2}$. Addition of the three equations and their $\Delta G°$ values gives the overall change in Gibbs free energy for the transformation at 298 K.

$$\frac{3}{2}\cancel{O_2}(g) \rightarrow \cancel{O_3(g)} \qquad \Delta G^\circ_{298\text{ K}} = \left(-\frac{1}{2}\right)(-326.4\text{ kJ/mol})$$

$$O(g) \rightarrow \frac{1}{2}\cancel{O_2}(g) \qquad \Delta G^\circ_{298\text{ K}} = \left(-\frac{1}{2}\right)(463.4\text{ kJ/mol})$$

$$NO(g) + \cancel{O_3(g)} \rightarrow NO_2(g) + \cancel{O_2(g)} \qquad \Delta G^\circ_{298\text{ K}} = -199.5\text{ kJ/mol}$$

$$NO(g) + O(g) \rightarrow NO_2(g) \qquad \Delta G^\circ_{298\text{ K}} = -268.0\text{ kJ/mol}$$

## The Effect of $\Delta H$, $\Delta S$, and $T$ on $\Delta G$

Standard Gibbs free energy may be determined using the relationship between $\Delta S^\circ$ and $\Delta H^\circ$ at a particular temperature, $T$.

$$\Delta G^\circ = \Delta H^\circ - T\Delta S^\circ$$

The reaction between hydrogen and oxygen gases to form water vapor is thermodynamically favorable at room temperature. (The activation energy for the reaction is high, so a spark must be applied to initiate the reaction. Once the reaction has begun, however, it proceeds with no intervention.)

$$H_2(g) + \frac{1}{2}O_2(g) \rightarrow H_2O(g)$$

The change in Gibbs free energy for this process can be found by determining the enthalpy and entropy changes at this temperature.

| Substance | $\Delta H^\circ_f$ (kJ/mol) | $S^\circ$ (J/(mol·K)) |
|---|---|---|
| $H_2O(g)$ | –237.1 | 188.8 |
| $H_2(g)$ | 0 | 130.7 |
| $O_2(g)$ | 0 | 205.2 |

The enthalpy change for the process is the standard enthalpy of formation of one mole of $H_2O$.

$$\Delta H^\circ = -237.1\text{ kJ/mol}$$

The entropy change for this reaction is the difference between the standard entropies of products and reactants.

$$\Delta S^\circ_{rxn} = \Sigma S^\circ_{(products)} - \Sigma S^\circ_{(reactants)}$$

$$\Delta S^\circ_{rxn} = 188.8\text{ J/(mol}\cdot\text{K)} - \left[1(130.7\text{ J/(mol}\cdot\text{K))} + \left(\frac{1}{2}(205.2\text{ J/(mol}\cdot\text{K))}\right)\right]$$

$$\Delta S^\circ_{rxn} = 44.5\text{ J/mol K}$$

The standard Gibbs free energy change is determined using these values at 298 K. Notice that a conversion factor must be used, since $\Delta H^\circ$ and $\Delta S^\circ$ are reported in different units.

$$\Delta G^\circ = \Delta H^\circ - T\Delta S^\circ$$

$$\Delta G^\circ = -237.1\text{ kJ/mol} - \left[(298\ \cancel{K})(-44.5\ \cancel{J}/(\text{mol}\cdot\cancel{K}))\left(\frac{1\text{ kJ}}{1000\ \cancel{J}}\right)\right]$$

$$\Delta G^\circ = -223.8\text{ kJ/mol}$$

This equation is useful for determining the thermodynamic favorability of a process at temperatures other than the thermodynamic standard temperature, 298 K. The thermodynamic favorability of a process at one temperature can differ greatly from the favorability of the process at a different temperature.

Consider the boiling of liquid water. At atmospheric pressure, this process is thermodynamically favored ($\Delta G < 0$) above 100 °C, but it is thermodynamically disfavored ($\Delta G > 0$) at temperatures below 100 °C.

Since standard enthalpies of formation are relatively constant across a wide range of temperatures, the tabulated enthalpies of formation and standard entropies can be used to predict whether or not a process is favorable at a given temperature.

The Gibbs free energy for boiling water at 80 °C (353 K) and at 120 °C (393 K) can be calculated as examples. The pressure in both situations is 1 atm.

$$H_2O(l) \rightarrow H_2O(g)$$

| Substance | $\Delta H°_f$ (kJ/mol) | $S°$ (J/mol·K)) |
|---|---|---|
| $H_2O(l)$ | –285.8 | 70.0 |
| $H_2O(g)$ | –241.8 | 188.8 |

The values of $\Delta S°$ and $\Delta H°_f$ are determined for the process.

$$\Delta H^\circ_{\text{rxn}} = \Sigma n\Delta H^\circ_{\text{f(products)}} - \Sigma n\Delta H^\circ_{\text{f(reactants)}}$$
$$\Delta H^\circ_{\text{rxn}} = [1(-241.8 \text{ kJ/mol})] - [1(-285.8 \text{ kJ/mol})]$$
$$\Delta H^\circ_{\text{rxn}} = 44.0 \text{ kJ/mol}$$

$$\Delta S^\circ_{\text{rxn}} = \Sigma S^\circ_{\text{(products)}} - \Sigma S^\circ_{\text{(reactants)}}$$
$$\Delta S^\circ_{\text{rxn}} = [1(188.8 \text{ J/(mol}\cdot\text{K))}] - [1(70.0 \text{ J/(mol}\cdot\text{K))}]$$
$$\Delta S^\circ_{\text{rxn}} = 118.8 \text{ J/(mol}\cdot\text{K)}$$

These values are used to determine the Gibbs free energy change at the two different temperatures.

At 80 °C (353 K):

$$\Delta G = \Delta H - T\Delta S = 44.0 \text{ kJ/mol} - \left[(353 \cancel{\text{K}})(118.8 \cancel{\text{J}}/(\text{mol}\cdot\cancel{\text{K}}))\left(\frac{1 \text{ kJ}}{1000 \cancel{\text{J}}}\right)\right]$$
$$\Delta G = 2.1 \text{ kJ/mol}$$

The positive change in Gibbs free energy indicates that boiling water is thermodynamically disfavored at 80 °C.

At 120 °C (393 K):

$$\Delta G = \Delta H - T\Delta S = 44.0 \text{ kJ/mol} - \left[(393 \cancel{\text{K}})(118.8 \cancel{\text{J}}/(\text{mol}\cdot\cancel{\text{K}}))\left(\frac{1 \text{ kJ}}{1000 \cancel{\text{J}}}\right)\right]$$
$$\Delta G = -2.7 \text{ kJ/mol}$$

The negative value for the change in Gibbs free energy indicates that water boiling is a thermodynamically favored process at 120 °C.

At equilibrium, the change in Gibbs free energy for a process is zero. This fact can be used to determine the temperature at which phase changes occur.

At the boiling point of water at 1 atm (the **normal boiling point**), liquid water is in equilibrium with water vapor. The values of $\Delta H°$ and $S°$ have already been determined (44.0 kJ/mol and 118.8 J/(mol·K), respectively). The boiling point of water, $T$, can be determined by setting the change in Gibbs free energy to zero.

$$\Delta G = 0 = \Delta H - T\Delta S = 44.0 \text{ kJ/mol} - \left[ T(118.8 \cancel{\text{J}}/(\text{mol}\cdot\cancel{\text{K}}))\left(\frac{1 \text{ kJ}}{1000 \cancel{\text{J}}}\right)\right]$$

$$T = 370 \text{ K}$$

This is very close to the true normal boiling point of water, 373 K. The difference between the two values can be accounted for by the slight differences in $\Delta H°$ and $S°$ at temperatures other than 298 K.

In general, the signs of $\Delta H$ and $\Delta S$ can be indicators of whether the change in Gibbs free energy of a process will be greater than or less than zero, as summarized below. The $T\Delta S$ term in the Gibbs free energy equation relates the temperature to the spontaneity of a process.

| $\Delta H$ | $\Delta S$ | $\Delta G$ |
|---|---|---|
| Negative (–) | Positive (+) | Negative at all temperatures. *Always* thermodynamically favorable. |
| Negative (–) | Negative (–) | Negative at low temperatures. Positive at high temperatures. Only thermodynamically favorable at *low temperatures*. |
| Positive (+) | Positive (+) | Negative at high temperatures. Positive at low temperatures. Only thermodynamically favorable at *high temperatures*. |
| Positive (+) | Negative (–) | Positive at all temperatures. *Never* thermodynamically favorable. |

## Practice

The following reaction is endothermic. Is it thermodynamically favorable at all temperatures, at high temperatures only, at low temperatures only, or at no temperature?

$$2\ CuCl_2(s) \rightarrow 2\ CuCl(s) + Cl_2(g)$$

The enthalpy change for an endothermic process is positive.

This reaction also involves an increase in entropy, since a gas is formed from a solid. This means that $\Delta S$ is positive and that $T\Delta S > 0$. The higher the temperature, the larger the positive value that is subtracted from the enthalpy term, and the more negative the overall Gibbs free energy change becomes. The reaction will be thermodynamically favorable only at high temperatures.

## Practice

Use the data provided to calculate the minimum temperature at which the decomposition of copper(II) chloride is thermodynamically favorable according to the reaction in the previous practice problem.

$$2\ CuCl_2(s) \rightarrow 2\ CuCl(s) + Cl_2(g)$$

Assume that the values of $\Delta H°$ and $S°$ are independent of temperature.

| Substance | $\Delta H°_f$ (kJ/mol) | $S°$ (J/(mol·K)) |
|---|---|---|
| $CuCl_2(s)$ | –220.1 | 108.1 |
| $CuCl(s)$ | –137.2 | 86.2 |
| $Cl_2(g)$ | 0 | 223.1 |

The enthalpy and entropy changes for the reaction are calculated first.

$$\Delta H^\circ_{rxn} = \Sigma n\Delta H^\circ_{f(products)} - \Sigma n\Delta H^\circ_{f(reactants)}$$
$$\Delta H^\circ_{rxn} = [2(-137.2\ \text{kJ/mol}) + 1(0\ \text{kJ/mol})] - [2(-220.1\ \text{kJ/mol})]$$
$$\Delta H^\circ_{rxn} = 165.8\ \text{kJ/mol}$$

$$\Delta S^\circ_{rxn} = \Sigma S^\circ_{(products)} - \Sigma S^\circ_{(reactants)}$$
$$\Delta S^\circ_{rxn} = [2(86.2\ \text{J/(mol}\cdot\text{K))} + 1(223.1\ \text{J/(mol}\cdot\text{K))}] - [2(108.1\ \text{J/(mol}\cdot\text{K))}]$$
$$\Delta S^\circ_{rxn} = 179.3\ \text{J/(mol}\cdot\text{K)}$$

Since both $\Delta H^\circ_{rxn}$ and $\Delta S^\circ_{rxn}$ are positive, this process is only thermodynamically favorable at high temperatures. The minimum temperature at which this reaction is favorable is determined by setting $\Delta G^\circ$ to zero.

$$\Delta G^\circ = 0 = \Delta H^\circ - T\Delta S^\circ = 165.8\ \text{kJ/mol} - \left[T(179.3\ \text{J/(mol}\cdot\text{K))}\left(\frac{1\ \text{kJ}}{1000\ \text{J}}\right)\right]$$
$$T = 924.7\ \text{K}$$

# Kinetic Control

As discussed in Chapter 5, "Kinetics," the activation energy of a reaction determines its rate at a particular temperature. Some thermodynamically favorable reactions do not occur at a measurable rate due to an insurmountable activation energy for a particular temperature. Factors such as activation energy and presence or absence of a catalyst have no effect on the Gibbs free energy change of a process.

The conversion of carbon in the form of diamond to carbon in the form of graphite is thermodynamically favored. The Gibbs free energy of graphite is slightly lower than that of diamond, but diamond does not generally convert to graphite under familiar conditions. The reason for this is the very high activation energy for the reaction. This is an example of **kinetic control.**

**Kinetic Control**

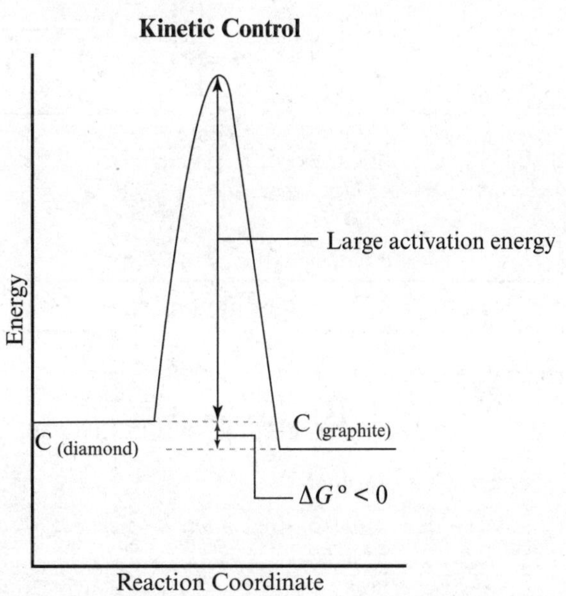

# Free Energy and Equilibrium

The equilibrium constant is an indication of the thermodynamic favorability of a process at a particular temperature. If a reaction has an equilibrium constant larger than 1, products are favored over reactants. The reaction will occur in the forward direction with thermodynamic favorability. In other words, when $K > 1$, $\Delta G° < 0$.

On the other hand, if a reaction has an equilibrium constant less than 1, $K < 1$, then $\Delta G° > 0$ for the process.

For a reaction that is at equilibrium, $K = 1$ and $\Delta G° = 0$.

The relationship between $\Delta G°$ and $K$ is given on the *AP Chemistry Equations and Constants* sheet.

$$\Delta G° = -RT \ln K$$

Note that the units of $R$ must be compatible with the units of $\Delta G°$ units. Therefore, the form of $R$ used in the above equation is

$$R = 8.314 \text{ J/mol K}$$

When the reaction quotient, $Q$, is less than $K$, $\Delta G < 0$; the conversion of reactants to products is thermodynamically favored. When $Q$ is greater than $K$, $\Delta G > 0$; the conversion of reactants to products is thermodynamically disfavored.

Below, the free energy of a reaction is plotted against the reaction coordinate with pure reactants on the left and pure products on the right. At equilibrium, $Q = K$ and $\Delta G = 0$.

**The Relationship between Free Energy and Equilibrium**

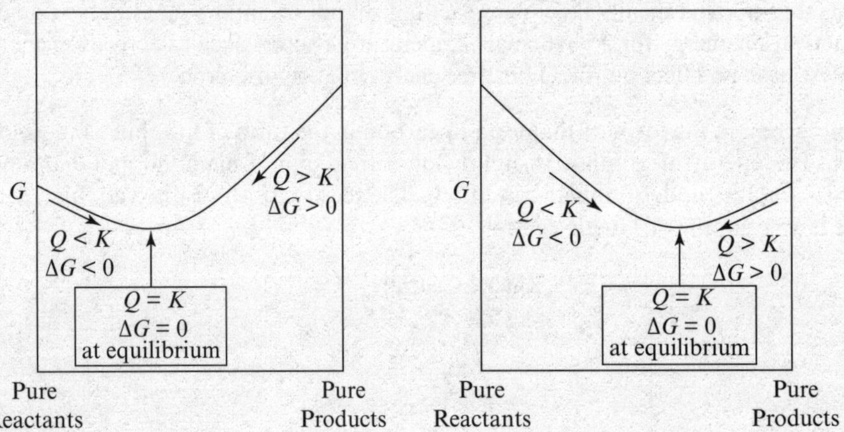


## Practice

Determine the change in Gibbs free energy, $\Delta G°$, for the autoionization of water at 298 K.

The autoionization of water occurs with a $K_w$ of $1.00 \times 10^{-14}$ at 298 K.

$$\Delta G° = -\left(8.314 \text{ J/(mol}\cdot\cancel{\text{K}})\right)\left(298\ \cancel{\text{K}}\right) \ln 1.00 \times 10^{-14}$$

$$\Delta G° = 7.99 \times 10^{4} \text{ J/mol}$$

## Practice

For the formation of gaseous ammonia by the reaction of nitrogen and hydrogen gases, $\Delta H°$ is –92.6 kJ/mol and $\Delta S°$ is –199 J/(mol·K). If the partial pressures of $NH_3$, $H_2$, and $N_2$ are 1.5 atm, 0.080 atm, and 0.060 atm, respectively, will the reaction shift toward reactants or products in order to reach equilibrium at 298 K?

The balanced equation is

$$N_2(g) + 3\,H_2(g) \rightleftharpoons 2\,NH_3(g)$$

The value of $\Delta G°$ must first be determined from $\Delta H°$ and $\Delta S°$.

$$\Delta G° = \Delta H° - T\Delta S°$$

$$\Delta G° = \left(-92.6\ \cancel{\text{kJ}}/\text{mol} \times \frac{1000\ \text{J}}{1\ \cancel{\text{kJ}}}\right) - \left((298\ \cancel{\text{K}})(-199\ \text{J/(mol}\cdot\cancel{\text{K}}))\right) = -3.33\times10^4\ \text{J/mol}$$

The value of $\Delta G°$ is used to calculate $K$.

$$\Delta G° = -RT\ln K$$

$$-3.33\times10^4\ \text{J/mol} = -(8.314\ \text{J/(mol·K)})(298\ \text{K})\ \ln K$$

$$\ln K = \frac{-3.33\times10^4\ \cancel{\text{J/mol}}}{-(8.314\ \cancel{\text{J/(mol·K)}})(298\ \cancel{\text{K}})} = 13.4$$

$$K = e^{13.4} = 7\times10^5$$

In order to determine whether the reaction proceeds toward reactants or products under the conditions given, $K$ must be compared with $Q$. The value of $Q$ is determined using the partial pressures.

$$Q = \frac{P^2_{NH_3}}{P_{N_2}P^3_{H_2}} = \frac{1.5^2}{(0.060)(0.080)^3} = 7.3\times10^4$$

Since $Q$ is less than $K$, the reaction will proceed toward the product.

# Electrochemistry

**Electrochemistry** is the study of the electron movement and chemical change. In other words, it is the study of oxidation-reduction (redox) processes, which may be thermodynamically favorable or unfavorable.

An **electrochemical cell** is an apparatus in which the oxidation and reduction half reactions of a redox process are separated in space. An electrochemical cell in which the redox process is thermodynamically favorable is a **galvanic cell,** also called a voltaic cell. A cell in which the redox process is unfavorable and must be supplied with energy from an external power source, such as a battery, is an **electrolytic cell.**

In electrochemical cells, there are two electrodes where the half reactions take place. The **cathode** is the electrode where reduction occurs, and the **anode** is the electrode where oxidation occurs. A useful mnemonic for this relationship is "red cat": **Red**uction always occurs at the **cat**hode.

## Standard Cell Potential

A charge difference between two electrodes is the driving force for the flow of electrons between the electrodes. This charge difference is called **potential difference,** and it is measured in **volts, V,** equal to joules of work per coulomb of charge transferred. Because potential difference gives rise to the force that results in the motion of electrons, it is also referred to as **electromotive force, emf.**

The **standard cell potential, $E°_{cell}$,** or **standard emf** for a redox process measures the favorability of that process when reactants in solutions are at concentrations of 1 $M$ or gaseous reactants are at 1 atm of pressure. The more positive the standard cell potential, the more favorable the process. An unfavorable process has a negative cell potential.

$$E°_{cell} = E°_{reduction} - E°_{oxidation}$$

or

$$E°_{cell} = E°_{cathode} - E°_{anode}$$

$E°_{cathode}$ and $E°_{anode}$ represent the standard reduction potentials for the electrodes.

Two standard reduction potentials are tabulated below.

| Half Reaction | Standard Reduction Potential, $E°$ (V) |
|---|---|
| $Al^{3+} + 3\,e^- \rightarrow Al$ | −1.66 |
| $Ca^{2+} + 2\,e^- \rightarrow Ca$ | −2.76 |

These values can be used to determine the standard cell potential for the electrochemical cell in which $Al^{3+}$ is reduced to Al and Ca is oxidized to $Ca^{2+}$. (The oxidation reaction of calcium is just the reverse of the reduction, $Ca \rightarrow Ca^{2+} + 2\,e^-$.)

$$2\,Al^{3+} + 3\,Ca \rightarrow 2\,Al + 3\,Ca^{2+}$$

$$E°_{cell} = E°_{reduction} - E°_{oxidation} = -1.66 - (-2.76) = +1.10\text{ V}$$

The positive cell potential indicates a thermodynamically favored process. This procedure will always work for finding the correct cell potential, but there are other procedures for finding cell potentials that involve changing the sign of one of the cell half reaction reduction potentials. *If you have been presented with one of these other methods by your instructor and are comfortable with it, use it to avoid confusion.*

**Standard reduction potentials** for half reactions, such as those given for $Ca^{2+}$ and $Al^{3+}$ above, are measured against a **standard hydrogen electrode,** which is an inert platinum electrode immersed in a 1 *M* solution of $H_3O^+$ under 1 atm partial pressure of hydrogen gas. The reaction that is defined by chemists to have a cell potential of zero is the reduction of $H^+$.

$$2\,H^+ + 2\,e^- \rightarrow H_2 \qquad E°_{cell} = 0.00\text{ V}$$

## Practice

Based on their standard reduction potentials, which of the following species have a greater tendency to be reduced than $H^+$?

| Half Reaction | Standard Reduction Potential, $E°$ (V) |
|---|---|
| $O_2 + 4\,H^+ + 4\,e^- \rightarrow 2\,H_2O$ | 1.23 |
| $Ag^+ + e^- \rightarrow Ag$ | 0.80 |
| $Co^{2+} + 2\,e^- \rightarrow Co$ | −0.28 |

As indicated by their positive standard reduction potentials, both $O_2$ and $Ag^+$ have a greater tendency to be reduced than $H^+$. Cobalt ion, $Co^{2+}$, has a negative reduction potential, so it has a lesser tendency to undergo reduction than $H^+$.

# Galvanic/Voltaic Cells

A **galvanic cell,** also known as a **voltaic cell,** is one in which a redox process is thermodynamically favorable. A galvanic cell has a positive cell potential, $E°_{cell}$. Galvanic cells produce electrical energy by causing electrical current to move through the wire connecting the anode and the cathode. Batteries are galvanic cells.

*The change in Gibbs free energy, $\Delta G°_{rxn}$, is less than zero for a redox process with a positive cell potential.*

Shown below is a depiction of a galvanic cell. In this cell, a strip of copper immersed in a solution of 1 *M* copper(II) sulfate is connected by a wire to a strip of zinc in a solution of 1 *M* zinc sulfate.

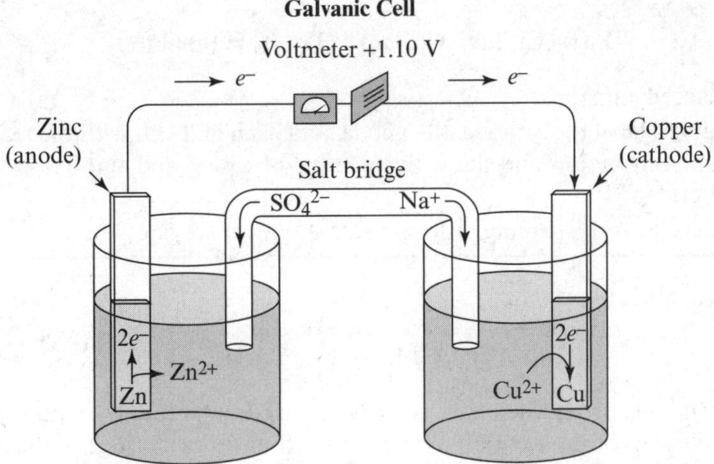


The standard reduction potentials for the $Cu^{2+}$ and $Zn^{2+}$ ions are given below. Because the cell is galvanic, $E°_{cell}$ is positive. In order for this to be true, copper, which has the higher standard reduction potential, must be reduced at the cathode, and zinc, which has the lower reduction potential, must be oxidized at the anode.

| Half Reaction | Standard Reduction Potential, $E°$ (V) |
|---|---|
| $Cu^{2+} + 2e^- \rightarrow Cu$ | 0.34 |
| $Zn^{2+} + 2e^- \rightarrow Zn$ | –0.76 |

$$E°_{cell} = E°_{cathode} - E°_{anode} = 0.34\text{ V} - (-0.76\text{ V}) = 1.10\text{ V}$$

Overall equation: $Cu^{2+} + Zn \rightarrow Cu + Zn^{2+}$

Electrons flow from the zinc electrode, where oxidation is occurring, to the copper electrode, where reduction is occurring. *Electrons always flow from the anode to the cathode, regardless of whether the cell is galvanic or electrolytic.*

As electrons are gained by the $Cu^{2+}$ in solution at the cathode, the $Cu^{2+}$ ions leave the solution and become solid copper, which is deposited on the Cu electrode. This adds copper mass to the cathode. As electrons leave the zinc electrode, solid zinc atoms leave the electrode as they become $Zn^{2+}$ ions in solution. The zinc anode loses mass.

When both the reactant and the product of a half reaction remain in solution, or when a reactant or product is a gas, an inert (unreactive) electrode such as platinum or graphite is used. Inert electrodes do not participate in the redox reaction; they just provide the surface where half reactions occur.

In order to maintain charge balance, a **salt bridge** connects the two half-cells. The salt bridge consists of a soluble salt, such as $Na_2SO_4$, embedded in a gel. The ions of the salt can move into the solutions, but the solutions themselves are prevented from mixing. Because $Cu^{2+}$ ions are leaving the solution at the cathode, the positive charges are replaced by cations ($Na^+$) from the salt bridge. Because $Zn^{2+}$ ions are entering the solution at the anode, the positive charges are balanced by anions ($SO_4^{2-}$) from the salt bridge.

A picture like the galvanic cell image above is *not* referred to as a "cell diagram." A **cell diagram** is a shorthand method for describing an electrochemical cell in which the anode is written first, the cathode is written last, the cathode and the anode are separated by a double vertical line, ||, and phase boundaries are denoted by single vertical lines, |. The diagram for the galvanic cell pictured above is shown below.

$$Zn(s)\ |\ Zn^{2+}(1\ M)\ ||\ Cu^{2+}(1\ M)\ |\ Cu(s)$$

## Practice

Consider the following cell diagram:

$$Pt(s) \mid Cr^{2+}(aq), Cr^{3+}(aq) \parallel H^{+}(aq) \mid H_2(g) \mid Pt(s)$$

**a.** Write the overall balanced equation.
**b.** Sketch the cell. Be sure to label the cathode, the anode, and each half-cell with the reaction taking place. Show the direction of electron flow and show the direction of cation and anion flow to and from the solution and the salt bridge.
**c.** Finally, is either of the electrodes gaining or losing mass? Why or why not?

**(a)** The overall balanced equation is

$$2\,Cr^{2+} + 2\,H^{+} \rightarrow 2\,Cr^{3+} + H_2$$

**(b)** The electron flow is from the anode, where $Cr^{2+}$ is being oxidized to $Cr^{3+}$, to the cathode, where $H^{+}$ is being reduced to $H_2$, which bubbles out of the solution. Anions flow from the salt bridge to the anode to offset the positive charges that result from the oxidation of $Cr^{2+}$. Cations flow from the salt bridge to the cathode to offset the loss of $H^{+}$ from the solution. (*Electrons should never be shown flowing to or from the salt bridge.*)

**Sketch of a Galvanic Cell**

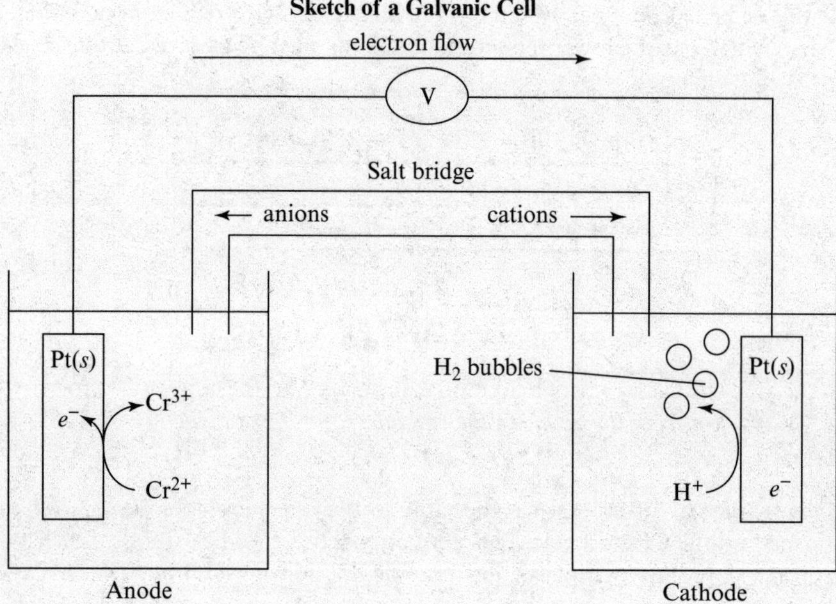


**(c)** Neither of the platinum electrodes is gaining mass. The $Cr^{2+}$, when oxidized to $Cr^{3+}$, remains in solution. The $H^{+}$, when reduced, forms a gas that bubbles out of the solution.

# Electrolytic Cells

**Electrolytic cells** are electrochemical cells that involve a thermodynamically disfavored redox process. Electrolytic cells require an external source of energy, such as a battery, in order to force the electrolysis reaction to proceed. An electrolytic cell has a negative cell potential and a change in Gibbs free energy, $\Delta G^{\circ}_{rxn}$, greater than zero.

The electrolysis of molten sodium chloride, which occurs industrially at high temperature in a Downs cell, is depicted in the figure below. The cell potential is negative, and the species with the more negative standard reduction potential, $Na^{+}$, is being reduced to molten sodium at the cathode. Chloride ions, $Cl^{-}$, are being oxidized to chlorine gas at the anode. Electrons flow from the anode to the cathode, similar to a galvanic cell, but this movement requires a driving force.

The battery or electrical energy source forces the electrons to move. The two half reactions do not require separation in an electrolytic cell, so there is no need for a salt bridge.

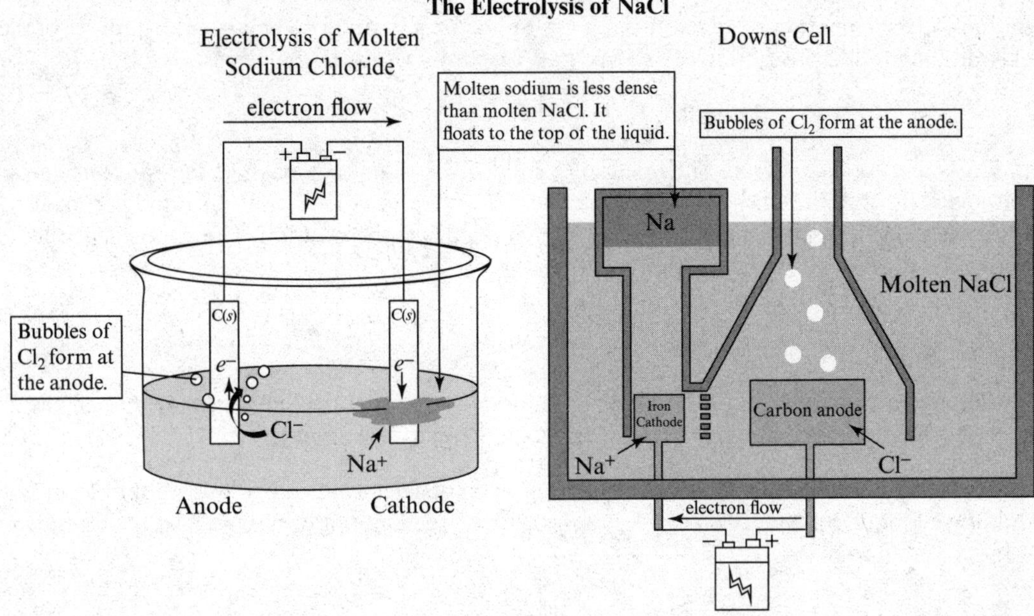


We can predict the products of electrolysis of molten salts. The cation is reduced at the cathode and the anion is oxidized at the anode. In the example above, $Na^+$ is reduced to Na and $Cl^-$ is oxidized to $Cl_2$.

Prediction of the products of the electrolysis of aqueous solutions of electrolytes is more complicated because the electrolysis of water competes with the reaction of the electrolyte.

At the cathode, the reduction of water is more easily accomplished than the reduction of the cations of active metals. The **active metals** are all of the group 1 alkali metals in addition to the group 2 alkaline earth metals Ca, Sr, and Ba. Formation of $H_2$ and $OH^-$ occurs.

$$2\,H_2O + 2\,e^- \rightarrow H_2 + 2\,OH^- \quad \text{(occurs preferentially at the cathode)}$$

$$Na^+ + e^- \rightarrow Na \quad \text{(does not occur in aqueous solution)}$$

At the anode, water is oxidized to form $O_2$. Oxidation of polyatomic ions such as $SO_4^{2-}$ and $NO_3^-$ does not occur. Notice that the sulfur and nitrogen atoms in these anions are already highly oxidized. Sulfur has an oxidation number of +6 and nitrogen has an oxidation number of +5.

$$2\,H_2O \rightarrow O_2 + 4\,H^+ + 4\,e^- \quad \text{(occurs preferentially at the anode)}$$

## Practice

An electrolytic cell is assembled in which an aqueous solution of silver nitrate undergoes electrolysis.

**a.** Write and label the half reaction occurring at the cathode.
**b.** Write and label the half reaction occurring at the anode.
**c.** Write the overall balanced equation.
**d.** Are either of the electrodes gaining or losing mass? Why or why not?

**(a)** The reduction of $Ag^+$ ions occurs at the cathode. Since silver is not an active metal, the reduction of water is not a competing process.

$$\text{Cathode: } Ag^+ + e^- \rightarrow Ag$$

**(b)** The oxidation of water occurs at the anode. Nitrate, a polyatomic anion, is not oxidized.

$$\text{Anode: } 2\ H_2O \rightarrow O_2 + 4\ H^+ + 4\ e^-$$

**(c)** The equation for the reduction of silver must be multiplied by 4 in order to balance the number of electrons in the reactants and the products. Addition of the two half reactions results in the overall equation.

$$\text{Overall: } 4\ Ag^+ + 2\ H_2O \rightarrow 4\ Ag + O_2 + 4\ H^+$$

**(d)** The cathode, where $Ag^+$ is undergoing reduction to solid silver, is gaining mass; as the $Ag^+$ ions gain electrons, the neutral atoms plate onto the surface of the electrode. The oxidation of water at the anode produces oxygen gas, which bubbles out of the solution, and $H^+$ ions, which remain in the solution. The anode is neither gaining nor losing mass.

## Faraday's Law

**Faraday's law of electrolysis** states that the number of moles of a product that can be formed in an electrolytic reaction has a stoichiometric relationship with the number of electrons supplied.

The number of electrons supplied in an electrolysis reaction can be determined by the amount of current that flows over a known amount of time. **Current, *I*,** is a measure of the amount of charge, *q*, transferred per unit time, *t*. It is measured in **amperes, A.**

$$I = \frac{q}{t}$$

One ampere is equal to one **coulomb, C,** the SI unit of charge, per second.

$$1 \text{ ampere} = \frac{1 \text{ coulomb}}{1 \text{ second}}$$

The **Faraday constant, *F*,** is the amount of charge supplied by one mole of electrons. The equation for current and the value of Faraday's constant are both included on the *AP Chemistry Equations and Constants* sheet.

$$F = \frac{96{,}485 \text{ coulombs}}{1 \text{ mol } e^-}$$

The answer to any quantitative question regarding the amount of product formed, reactants consumed, current, time, or charge can be determined using the equation for current and the balanced equation for the redox process under analysis.

For example, suppose the electrolysis of an aqueous solution of silver nitrate results in the deposit of a mass of 10.8 grams of silver at the cathode. What current is required in order for this process to be complete in exactly 5.00 hours?

The balanced equation for the reduction of silver cations is

$$Ag^+ + e^- \rightarrow Ag$$

One mole of electrons is required to reduce one mole of silver cations. The charge, *q*, in coulombs is calculated first.

$$10.8 \cancel{\text{g Ag}} \times \frac{1 \cancel{\text{mol Ag}}}{107.87 \cancel{\text{g Ag}}} \times \frac{1 \cancel{\text{mol } e^-}}{1 \cancel{\text{mol Ag}}} \times \frac{96{,}485 \text{ C}}{1 \cancel{\text{mol } e^-}} = 9{,}660 \text{ C}$$

Since the current is expressed as coulombs per second, the time must be converted to seconds.

$$5.00\ \cancel{h} \times \frac{3600\ \text{s}}{1\ \cancel{h}} = 1.80 \times 10^4\ \text{s}$$

The current is determined using the appropriate equation.

$$I = \frac{q}{t} = \frac{9660\ \text{C}}{1.80 \times 10^4\ \text{s}} = 0.537\ \text{A}$$

## Practice

An electrolytic cell is constructed in which electrodes are placed into an aqueous solution of $CuSO_4$. A current of 3.00 A is applied. How long must the electrolysis proceed in order to plate 5.00 grams of copper at the cathode?

The balanced half reaction is needed to give the number of moles of electrons required for the reduction of one mole of copper atoms.

$$Cu^{2+} + 2\,e^- \rightarrow Cu$$

In order to solve for time, begin with the number of grams of copper needed. Convert the grams of copper to moles, convert moles of copper to moles of electrons, and then use Faraday's constant to find the charge transfer required, $q$.

$$5.00\ \cancel{\text{g Cu}} \times \frac{1\ \cancel{\text{mol Cu}}}{63.55\ \cancel{\text{g Cu}}} \times \frac{2\ \cancel{\text{mol } e^-}}{1\ \cancel{\text{mol Cu}}} \times \frac{96{,}485\ \text{C}}{1\ \cancel{\text{mol } e^-}} = 1.52 \times 10^4\ \text{C}$$

$$(1.518253 \times 10^4)$$

The equation for current is rearranged to solve for time.

$$t = \frac{q}{I} = \frac{1.518253 \times 10^4\ \cancel{\text{C}}}{3.00\ \cancel{\text{C}} \cdot \text{s}^{-1}} = 5.06 \times 10^3\ \text{s}$$

# Cell Potential and Gibbs Free Energy

The thermodynamic favorability of an oxidation-reduction reaction is a function of the standard cell potential, $E°$. The relationship between the change in Gibbs free energy, $\Delta G°$, and the standard cell potential is given on the *AP Chemistry Equations and Constants* sheet.

$$\Delta G° = -nFE°$$

In this relationship, $n$ is the number of moles of electrons transferred and $F$ is Faraday's constant, also given on the *AP Chemistry Equations and Constants* sheet.

A positive standard cell potential results in a negative change in Gibbs free energy. This corresponds to a thermodynamically favored process. A negative standard cell potential leads to a positive change in Gibbs free energy, indicating a thermodynamically unfavorable process.

For example, the standard cell potential can be used to calculate the change in Gibbs free energy for the reaction of cobalt with iron(II) ions.

$$Co(s) + Fe^{2+}(aq) \rightarrow Co^{2+}(aq) + Fe(s)$$

| Half Reaction | Standard Reduction Potential, $E°$ (V) |
|---|---|
| $Co^{2+} + 2\,e^- \rightarrow Co$ | –0.28 |
| $Fe^{2+} + 2\,e^- \rightarrow Fe$ | –0.44 |

$$E°_{cell} = E°_{reduction} - E°_{oxidation} = -0.44\text{ V} - (-0.28\text{ V}) = -0.16\text{ V}$$

The standard cell potential is then used to determine the Gibbs free energy change for this two-electron transfer. The conversion factor between volts and joules per coulomb is also given on the *AP Chemistry Equations and Constants* sheet.

$$\Delta G° = -nFE° = -(2)\left(\frac{96{,}485\ \cancel{C}}{1\text{ mol }e^-}\right)(-0.16\ \cancel{V})\left(\frac{1\text{ J}\cdot\cancel{C^{-1}}}{1\ \cancel{V}}\right) = 31{,}000\text{ J}\cdot\text{mol}^{-1}$$

The negative cell potential and the positive $\Delta G°$ indicate the reduction of $Fe^{2+}(aq)$ by $Co(s)$ is a thermodynamically unfavorable process.

## Practice

Calculate the change in Gibbs free energy for the following unbalanced reaction in acidic solution. Is this reaction thermodynamically favorable or unfavorable?

$$Ce^{4+} + I^- \rightarrow I_2 + Ce^{3+}$$

| Half Reaction | Standard Reduction Potential, $E°$ (V) |
|---|---|
| $Ce^{4+} + e^- \rightarrow Ce^{3+}$ | 1.61 |
| $I_2 + 2\,e^- \rightarrow 2\,I^-$ | 0.53 |

For a review of balancing oxidation-reduction equations, see Chapter 4, "Chemical Reactions." First, divide the reaction into oxidation and reduction half reactions and then balance the half reactions. Reduction of $Ce^{4+}$ is occurring at the cathode, and oxidation of $I_2$ is occurring at the anode.

<u>Oxidation:</u> $2\,I^- \rightarrow I_2 + 2\,e^-$

<u>Reduction:</u> $Ce^{4+} + e^- \rightarrow Ce^{3+}$

Reduction of $Ce^{4+}$ is occurring at the cathode, and oxidation of $I_2$ is occurring at the anode. In order to balance the overall equation, the reduction of $Ce^{4+}$ is multiplied by 2.

$$2\,I^- \rightarrow I_2 + \cancel{2e^-}$$

$$2\,Ce^{4+} + \cancel{2e^-} \rightarrow 2\,Ce^{3+}$$

Overall, this reaction involves a two-electron transfer, so $n = 2$.

$$2\,Ce^{4+} + 2\,I^- \rightarrow 2\,Ce^{3+} + I_2$$

The standard cell potential is calculated from the half reaction reduction potentials.

$$E°_{cell} = E°_{reduction} - E°_{oxidation} = 1.61\text{ V} - (0.53\text{ V}) = 1.08\text{ V}$$

This value is used to calculate the change in Gibbs free energy.

$$\Delta G^\circ = -nFE^\circ = -(2)\left(\frac{96{,}485\ \cancel{C}}{1\text{ mol } e^-}\right)(1.08\ \cancel{V})\left(\frac{1\text{ J}\cdot\cancel{C^{-1}}}{1\ \cancel{V}}\right) = -2.10\times10^5\text{ J/mol}$$

The negative $\Delta G^\circ$ indicates a thermodynamically favorable reaction.

## Cell Potential under Nonstandard Conditions

The nonstandard cell potential, $E$, or the free energy change, $\Delta G$, for an electrochemical cell under nonstandard conditions must sometimes be predicted. (Remember, under standard thermodynamic conditions, the concentrations of solutions are exactly 1 $M$, pressures are 1 atm, and temperature is 298 K.)

The relationship between the nonstandard Gibbs free energy change and the nonstandard cell potential is similar to the relationship at standard conditions.

$$\Delta G = -nFE$$

It is important to realize that as a reaction proceeds in a cell, $E$ and $\Delta G$ both approach zero. The concentrations of all reactants and products approach those that satisfy the expression for $K$. For a battery (or galvanic cell), once the reaction has reached equilibrium, the battery is said to be "dead."

AP Chemistry students should be able to make qualitative predictions regarding cell potentials under nonstandard conditions. For example, consider the following redox reaction under standard cell conditions at 25 °C.

$$3\ Sn^{4+}(aq) + 2\ Cr(s) \rightarrow 3\ Sn^{2+}(aq) + 2\ Cr^{3+}(aq)$$

| Half Reaction | Standard Reduction Potential, $E^\circ$ (V) |
|---|---|
| $Sn^{4+} + 2\ e^- \rightarrow Sn^{2+}$ | 0.15 |
| $Cr^{3+} + 3\ e^- \rightarrow Cr$ | –0.74 |

The standard reduction potential is calculated as usual. For a thermodynamically favorable process, $Sn^{4+}$, having a greater $E^\circ$ value, is reduced at the cathode, while Cr is oxidized at the anode.

$$E^\circ_{cell} = E^\circ_{reduction} - E^\circ_{oxidation} = 0.15\text{ V} - (-0.74\text{ V}) = 0.89\text{ V}$$

This value of $E^\circ$ is valid when $[Sn^{4+}] = [Sn^{2+}] = [Cr^{3+}] = 1.00\ M$. Consider the case where $[Sn^{2+}]$ is greater than 1.00 $M$. In terms of the nonstandard reduction potential, $E$ is now *less positive* than $E^\circ$.

A more serviceable method for determining the nonstandard cell potential is to use the Nernst equation.

$$E = E^\circ - \frac{RT}{nF}\ln Q \qquad \text{(only valid at 298 K)}$$

Here, $E$ and $E^\circ$ are the nonstandard and standard cell potentials, respectively; $n$ is the number of electrons transferred in the balanced redox equation; and $Q$ is the reaction quotient. The values of $R$ and $F$, the gas constant and Faraday constant, are located on the *AP Chemistry Equations and Constants* sheet. The equation is derived from the relationship between the change in Gibbs free energy and $Q$.

For example, suppose a galvanic cell is composed of the following two half reactions.

| Half Reaction | Standard Reduction Potential, $E^\circ$ (V) |
|---|---|
| $Pb^{2+}(aq) + 2\ e^- \rightarrow Pb(s)$ | –0.13 |
| $Cr^{3+}(aq) + e^- \rightarrow Cr^{2+}(aq)$ | –0.41 |

The cell potential when $[Pb^{2+}] = 1.00 \times 10^{-2}$ *M*, $[Cr^{3+}] = 3.00 \times 10^{-4}$ *M*, and $[Cr^{2+}] = 3.00 \times 10^{-4}$ *M* at 25 °C can be calculated using the Nernst equation.

The first step is to determine the standard cell potential. For a thermodynamically favorable process, $Pb^{2+}$ must be reduced, and $Cr^{3+}$ must be oxidized.

$$E^\circ_{cell} = E^\circ_{reduction} - E^\circ_{oxidation} = -0.13\text{ V} - (-0.41\text{ V}) = 0.28\text{ V}$$

The balanced equation is

$$Pb^{2+}(aq) + 2\,Cr^{2+}(aq) \rightarrow Pb(s) + 2\,Cr^{3+}(aq)$$

There are a total of two electrons transferred in the balanced process ($n = 2$). Notice that 1 volt is equal to 1 J/coulomb. This information is located on the *AP Chemistry Equations and Constants* sheet.

$$E = E^\circ - \frac{RT}{nF}\ln\frac{[Cr^{3+}]^2}{[Pb^{2+}][Cr^{2+}]^2}$$

$$= 0.28\text{ J/C} - \left[\frac{(8.314\text{ J/}(\cancel{\text{mol}}\cdot\cancel{\text{K}}))(298\ \cancel{\text{K}})}{2(96{,}485\text{ C/}\cancel{\text{mol}})}\ln\frac{[3.00\times10^{-4}]^2}{[1.00\times10^{-2}][3.00\times10^{-4}]^2}\right] = 0.22\text{ V}$$

The nonstandard cell potential is 0.22 V.

# Coupled Reactions

Thermodynamically unfavorable processes occur often in biological systems and in technology. In order for a thermodynamically unfavorable process to proceed, energy must be supplied. Consider again the thermodynamically disfavored reaction in which copper is produced from copper(II) oxide.

$$CuO(s) \rightarrow Cu(s) + \frac{1}{2}O_2(g) \qquad \Delta G^\circ = 129.7\text{ kJ/mol}$$

When this thermodynamically unfavorable reaction is coupled with a favorable process, the thermodynamically disfavored reaction becomes favorable.

$$CuO(s) \rightarrow Cu(s) + \frac{1}{2}O_2(g) \qquad \Delta G^\circ = 129.7\text{ kJ/mol}$$
$$C(s) + O_2(g) \rightarrow CO_2(g) \qquad \Delta G^\circ = -394.4\text{ kJ/mol}$$

The overall change in Gibbs free energy for the reaction of copper(II) oxide with carbon is –67.5 kJ/mol, in contrast to 129.7 kJ/mol for the uncoupled process.

Technology can be applied to thermodynamically unfavorable systems to make them favorable. Electrolysis reactions have negative standard cell potentials and positive Gibbs free energy changes. They do not proceed in the absence of an energy source because they are thermodynamically unfavorable. A battery is a device that provides the driving force for such a disfavored process.

An example is the refinement of copper metal by electrolysis. An impure copper anode and a pure copper cathode are immersed in an aqueous solution of copper(II) sulfate. Oxidation of copper at the anode produces copper ions in solution. The copper ions are reduced to pure copper at the cathode by coupling the reaction with an external power source provided by a thermodynamically favored reaction.

Coupled reactions are very important for the spontaneity of biological reactions. Many reactions in biology are thermodynamically disfavored and rely on the sharing of intermediates with thermodynamically favored processes. **Adenosine triphosphate, ATP,** is a common molecule used by living systems to provide energy for thermodynamically disfavored processes. ATP is hydrolyzed to ADP, adenosine diphosphate, and inorganic phosphate, $P_i$. This is a thermodynamically favorable process with a $\Delta G°$ of –30.5 kJ/mol.

**The Hydrolysis of ATP**

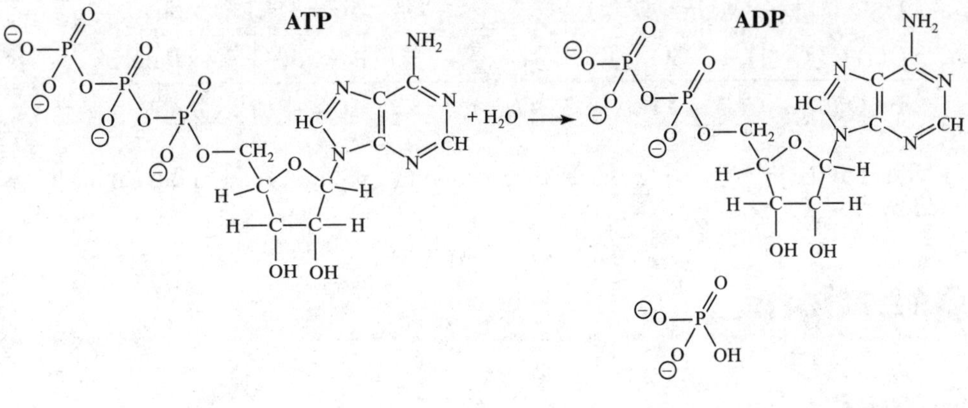

ATP + $H_2O$ → ADP

One example of reaction coupling occurs in the first step of glycolysis in the cell. The uncoupled reaction is thermodynamically unfavorable.

$$\text{Glucose} + P_i \rightarrow \text{glucose-6-phosphate} + H_2O \qquad \Delta G° = 13.8 \text{ kJ/mol}$$

By coupling this reaction with the hydrolysis of ATP, the cell drives the synthesis of glucose-6-phosphate.

$$\text{glucose} + \cancel{P_i} \rightarrow \text{glucose-6-phosphate} + \cancel{H_2O} \qquad \Delta G° = 13.8 \text{ kJ/mol}$$

$$\text{ATP} + \cancel{H_2O} \rightarrow \text{ADP} + \cancel{P_i} \qquad \Delta G° = -30.5 \text{ kJ/mol}$$

$$\text{glucose} + \text{ATP} \rightarrow \text{glucose-6-phosphate} + \text{ADP} \qquad \Delta G° = -16.7 \text{ kJ/mol}$$

Another example of reaction coupling is found in **photosynthesis,** the thermodynamically unfavorable synthesis of glucose, $C_6H_{12}O_6$, from carbon dioxide and water, which occurs in green plants.

$$6\ CO_2(g) + 6\ H_2O(l) \xrightarrow{\text{light}} C_6H_{12}O_6(s) + 6\ O_2(g)$$

This reaction has a positive change in Gibbs free energy, $\Delta G°$ = +2872 kJ/mol. The plant cells absorb visible photons of light, which excite the electrons of pigment molecules to higher-energy states. The energy is transferred through a complex system of receptor molecules to ultimately drive the synthesis of ATP. The cell can then use the ATP to provide the energy for the formation of glucose in photosynthesis.

## Practice

The conversion of iron ore to pure iron is thermodynamically unfavorable at temperatures below about 1250 °C. The change in Gibbs free energy at 1225 °C is +824.1 kJ/mol.

$$2\ Fe_2O_3(s) \rightarrow 4\ Fe(s) + 3\ O_2(g) \qquad \Delta G = 824.1 \text{ kJ/mol}$$

The oxygen that is produced reacts with coke, a carbon-containing substance similar to charcoal.

$$O_2(g) + C(s) \rightarrow CO_2(g) \qquad \Delta G = -397.8 \text{ kJ/mol}$$

**a.** What is the overall reaction and change in Gibbs free energy for the conversion of iron ore to pure iron in the presence of coke?

**b.** Is the overall process thermodynamically favorable or unfavorable?

**(a)** The second equation and its associated $\Delta G$ value must be multiplied by a factor of 3 in order for carbon to react with all the $O_2$ produced in the first equation. The two reactions and their $\Delta G$ values can be added to obtain the overall reaction and the change in Gibbs free energy.

$$2\ Fe_2O_3(s) \rightarrow 4\ Fe(s) + \cancel{3\ O_2(g)} \qquad \Delta G = 824.1\ \text{kJ/mol}$$

$$3\ C(s) + \cancel{3\ O_2(g)} \rightarrow 3\ CO_2(g) \qquad \Delta G = 3(-397.8\ \text{kJ/mol})$$

$$2\ Fe_2O_3(s) + 3\ C(s) \rightarrow 4\ Fe(s) + 3\ CO_2(g) \qquad \Delta G = -369.3\ \text{kJ/mol}$$

**(b)** When carbon is added to the iron(II) oxide, the decomposition to solid iron is thermodynamically favorable at 1225 °C, as indicated by $\Delta G < 0$.

# Review Questions

## Multiple Choice

**1.** Which of the following processes involves a decrease in entropy?

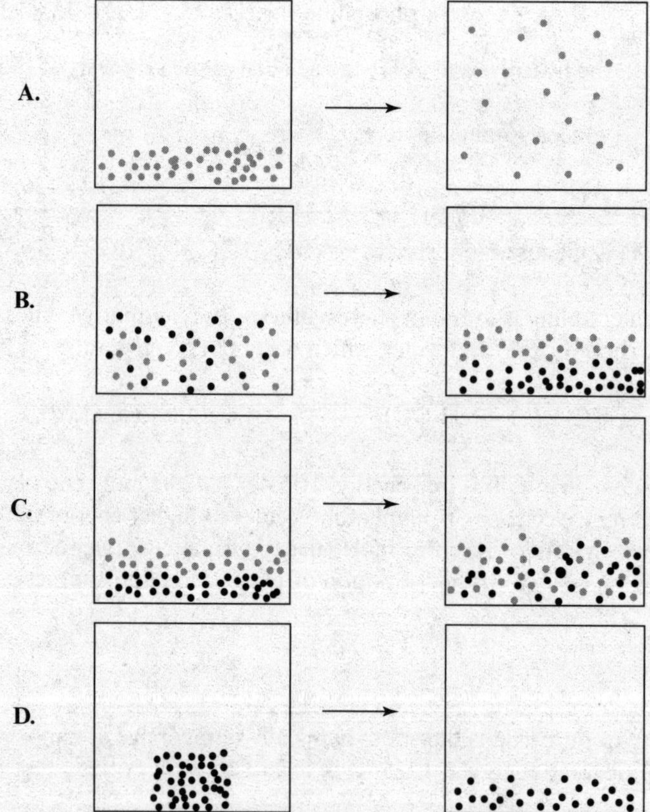

*Refer to the following reaction at 25 °C for questions 2–3.*

$$2\ N_2(g) + O_2(g) + 163.2\ \text{kJ} \rightarrow 2\ N_2O(g)$$

**2.** Which of the following is true of the entropy change for system and surroundings for the reaction between $N_2$ and $O_2$?

A. $\Delta S_{sys} > 0, \Delta S_{surr} < 0$
B. $\Delta S_{sys} < 0, \Delta S_{surr} > 0$
C. $\Delta S_{sys} > 0, \Delta S_{surr} > 0$
D. $\Delta S_{sys} < 0, \Delta S_{surr} < 0$

**3.** Which of the following statements is true regarding the thermodynamic favorability of the reaction between $N_2$ and $O_2$?

A. The reaction is thermodynamically favorable at no temperature.
B. The reaction is thermodynamically favorable at all temperatures.
C. The reaction is thermodynamically favorable at high temperatures.
D. The reaction is thermodynamically favorable at low temperatures.

**4.** Which choice correctly ranks the substances in order of increasing standard molar entropy?

A. $Ne(g) < He(g) < CH_3CH_2OH(g) < SO_2(g)$
B. $CH_3CH_2OH(g) < SO_2(g) < Ne(g) < He(g)$
C. $SO_2(g) < Ne(g) < CH_3CH_2OH(g) < He(g)$
D. $He(g) < Ne(g) < SO_2(g) < CH_3CH_2OH(g)$

**5.** Examine the following reactions.

$$2\ Fe + \frac{3}{2}\ O_2 \rightarrow Fe_2O_3 \qquad \Delta G° = -742\ \text{kJ/mol}$$

$$CO + \frac{1}{2}\ O_2 \rightarrow CO_2 \qquad \Delta G° = -257\ \text{kJ/mol}$$

Which choice is correct in terms of the value of $\Delta G°$ and the thermodynamic favorability for the reaction of $Fe_2O_3$ with CO to form Fe and $CO_2$?

$$Fe_2O_3 + 3\ CO \rightarrow 2\ Fe + 3\ CO_2$$

A. $\Delta G° = -29$ kJ/mol; the reaction is thermodynamically unfavorable
B. $\Delta G° = -29$ kJ/mol; the reaction is thermodynamically favorable
C. $\Delta G° = 485$ kJ/mol; the reaction is thermodynamically unfavorable
D. $\Delta G° = 485$ kJ/mol; the reaction is thermodynamically favorable

**6.** Given the following information for the boiling of an element, X, estimate the boiling point of the element.

$$X(l) \rightarrow X(g)$$
$$\Delta H° = 40\ \text{kJ/mol}$$
$$\Delta S° = 20\ \text{J/(mol·K)}$$

A. 1727 °C
B. 2000 °C
C. 2 °C
D. –271 °C

**7.** The reaction of $SO_2$ with $O_2$ to form $SO_3$ has a $\Delta G°$ of –71 kJ/mol at 25 °C.

$$SO_2(g) + \frac{1}{2}O_2(g) \rightarrow SO_3(g)$$

Which of the following is true of the equilibrium constant for this reaction?

**A.** $K > 1$
**B.** $K < 1$
**C.** $K = 0$
**D.** No conclusion can be drawn about the equilibrium constant based on the information given.

**8.** The rusting of a sheet of iron in air at room temperature is unobservable in a short period of time. Which choice best explains this observation?

**A.** Moisture must be present to catalyze the reaction.
**B.** The reaction is exothermic and thermodynamically unfavorable under these conditions.
**C.** The reaction has a high activation energy and is slow under these conditions.
**D.** The partial pressure of oxygen in air is such that $Q > K$ for the reaction.

**9.** The sodium-potassium pump is a trans-membrane protein that transports potassium into a cell and sodium out of a cell. This process is thermodynamically unfavorable. Hydrolysis of ATP to form ADP by the protein is necessary for potassium transport to occur. Based on this information, what is the most likely explanation for the ability of the protein to function?

**A.** The hydrolysis of ATP to form ADP is thermodynamically unfavorable, and the overall Gibbs free energy change for the process is positive.
**B.** The change in Gibbs free energy for the conversion of ADP to ATP is zero.
**C.** The hydrolysis of ATP to ADP by the trans-membrane protein has a negative change in Gibbs free energy, making the overall process thermodynamically favorable.
**D.** The higher sodium concentration outside the cell is favored by the resulting increase in the entropy of the system and decrease in the enthalpy of the surroundings.

**10.** An electrochemical cell was constructed with a tin electrode and tin(II) nitrate at the cathode and a silver electrode and silver nitrate at the anode. Which statement regarding this system is true?

| Half Reaction | $E°$ (V) |
|---|---|
| $Sn^{2+} + 2e^- \rightarrow Sn$ | –0.83 |
| $Ag^+ + e^- \rightarrow Ag$ | 1.07 |

**A.** The cell is galvanic and $\Delta G° > 0$.
**B.** The cell is galvanic and $\Delta G° < 0$.
**C.** The cell is electrolytic and $\Delta G° > 0$.
**D.** The cell is electrolytic and $\Delta G° < 0$.

**11.** A student places solutions of copper(II) sulfate, magnesium sulfate, and lead(II) nitrate into several small test tubes and adds samples of solid metals, as shown below.

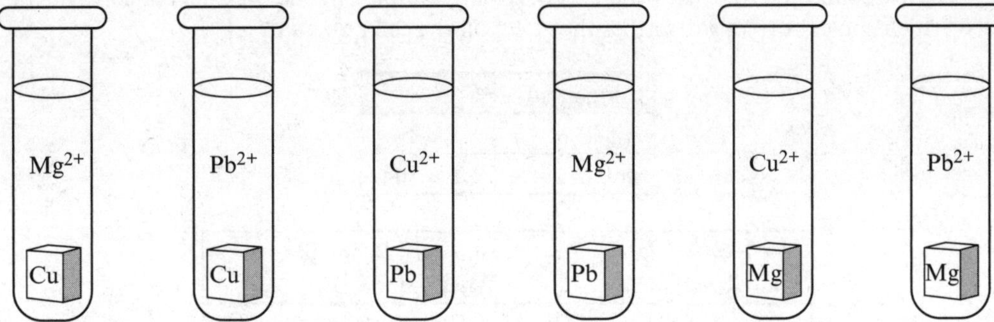


The results of the experiment are tabulated.

| | $Cu^{2+}(aq)$ | $Mg^{2+}(aq)$ | $Pb^{2+}(aq)$ |
|---|---|---|---|
| **Cu(*s*)** | | No reaction | No reaction |
| **Mg(*s*)** | Reaction occurs | | Reaction occurs |
| **Pb(*s*)** | Reaction occurs | No reaction | |

Based on this data, which of the following rankings of the half reactions is the correct order of standard reduction potentials, from most negative to most positive?

| | Lowest (Most Negative) Reduction Potential | | Highest (Most Positive) Reduction Potential |
|---|---|---|---|
| **A.** | $Cu^{2+} + 2\,e^- \rightarrow Cu$ | $Pb^{2+} + 2\,e^- \rightarrow Pb$ | $Mg^{2+} + 2\,e^- \rightarrow Mg$ |
| **B.** | $Mg^{2+} + 2\,e^- \rightarrow Mg$ | $Pb^{2+} + 2\,e^- \rightarrow Pb$ | $Cu^{2+} + 2\,e^- \rightarrow Cu$ |
| **C.** | $Pb^{2+} + 2\,e^- \rightarrow Pb$ | $Cu^{2+} + 2\,e^- \rightarrow Cu$ | $Mg^{2+} + 2\,e^- \rightarrow Mg$ |
| **D.** | $Mg^{2+} + 2\,e^- \rightarrow Mg$ | $Cu^{2+} + 2\,e^- \rightarrow Cu$ | $Pb^{2+} + 2\,e^- \rightarrow Pb$ |

**12.** A 2.0-A current is passed through a solution of $M(NO_3)_2$ for exactly 2 hours and 42 minutes. If 6.4 grams of metal are plated onto the cathode, what is the identity of the metal cation, $M^{2+}$?

**A.** $Fe^{2+}$
**B.** $Cu^{2+}$
**C.** $Cd^{2+}$
**D.** $Sn^{3+}$

**13.** The dissolution of sodium nitrate is endothermic, yet sodium nitrate is highly soluble in water. Which of the following statements provides the best explanation for this observation?

**A.** The dissolution has a sufficiently positive $\Delta S°$ to result in thermodynamic favorability.
**B.** The dissolution has a sufficiently negative $\Delta S°$ to result in thermodynamic favorability.
**C.** The activation energy of the dissolution is sufficiently high to result in an overall decrease in $\Delta G°$.
**D.** The activation energy of the dissolution is sufficiently low to result in an overall increase in $\Delta G°$.

**14.** Two reactions have enthalpy changes that are almost identical, yet one of the reactions has an equilibrium constant that is much larger than the other. Which of the following must be true?

**A.** The reaction with the larger equilibrium constant has the more positive entropy change.
**B.** The reaction with the larger equilibrium constant has the less positive entropy change.
**C.** The reaction with the larger equilibrium constant has the larger activation energy.
**D.** The reaction with the larger equilibrium constant has the smaller activation energy.

**15.** A galvanic cell is constructed in which a silver electrode is immersed in 50.0 mL of an aqueous 1.0 *M* $AgNO_3$ solution and a zinc electrode is immersed in 50.0 mL of an aqueous 1.0 *M* $Zn(NO_3)_2$ solution. The two half-cells are connected by a salt bridge. A 0.010-mg sample of solid NaCl is added to only one of the half-cells. Which choice correctly describes the effect on the cell potential?

| Compound | $K_{sp}$ |
|---|---|
| AgCl | $1.3 \times 10^{-10}$ |
| $ZnCl_2$ | $1.3 \times 10^{2}$ |

| Half Reaction | Standard Reduction Potential, $E°$ (V) |
|---|---|
| $Zn^{2+}(aq) + 2\,e^- \rightarrow Zn(s)$ | –0.76 |
| $Ag^{+}(aq) + e^- \rightarrow Ag(s)$ | 0.80 |

**A.** The cell potential is decreased because the reaction quotient, *Q*, is increased by the addition of NaCl to the zinc half-cell.

**B.** The cell potential is increased because the reaction quotient is decreased by the addition of NaCl to the zinc half-cell.

**C.** The cell potential is decreased because the reaction quotient, *Q*, is increased by the addition of NaCl to the silver half-cell.

**D.** The cell potential is increased because the reaction quotient is decreased by the addition of NaCl to the silver half-cell.

## Long Free Response

**1.** An electrochemical cell is constructed according to the cell diagram: Ni | $Ni^{2+}$ (1.00 *M*) || $Ag^+$ (1.00 *M*) | Ag.

| Half Reaction | Standard Reduction Potential, $E°$ (V) |
|---|---|
| $Ag^+ + e^- \rightarrow Ag$ | 0.80 |
| $Ni^{2+} + 2\,e^- \rightarrow Ni$ | –0.25 |

**a.** Write the balanced net ionic equation for the reaction.

**b.** **i.** Determine the cell potential, $E°_{cell}$.
**ii.** Is the cell a voltaic/galvanic cell or an electrochemical cell? Explain how you know.

**c.** Which electrode is losing mass? Explain what is happening to this mass.

**d.** Determine the value of $\Delta G°$ for the reaction that is occurring in the cell.

**e.** **i.** Determine the value of *K* for the reaction at 25 °C.
**ii.** If the cell is constructed such that the concentration of $Ag^+$ is $8.22 \times 10^{-1}$ *M*, and the concentration of $Ni^{2+}$ is $7.69 \times 10^{-3}$ *M*, is $E_{cell}$ greater or less than $E°_{cell}$? Justify your answer.

## Short Free Response

2. When white phosphorus under an atmosphere of oxygen at 25 °C is touched with a hot wire, the following reaction occurs with evolution of heat and bright white light.

$$P_4(s) + 5\ O_2(g) \rightarrow 2\ P_2O_5(s)$$

a. Predict the sign of $\Delta S°$ for the system and discuss how you arrived at your prediction.
b. What is the sign of $\Delta G°$ and how would $\Delta G$ change as the temperature of this reaction is increased from 25 °C? Justify your response.
c. Is the potential energy of the reactants higher or lower than that of the products? Explain your answer.
d. When the reactants are placed together at room temperature, no observable reaction takes place until the hot wire touches the phosphorus. Explain why this is true.

# Answers and Explanations

## Multiple Choice

1. **B.** The diagram in choice B shows a system that is increasing its organization. There is a greater dispersal of matter in the initial diagram than there is in the final diagram, so the entropy change is negative.

2. **D.** Because the number of moles of gaseous products (2) is fewer than the number of moles of gaseous reactants (3), the entropy of the system is decreasing, $\Delta S_{sys} < 0$. The reaction is endothermic, as evidenced from the inclusion of heat as a reactant, so $\Delta H_{sys} > 0$. The entropy change of the surroundings is opposite in sign from the enthalpy change of the system, $\Delta S_{surr} = \frac{-\Delta H_{sys}}{T}$, so the entropy change of the surroundings in this case is negative. Therefore, choice D is true: $\Delta S_{sys} < 0$, $\Delta S_{surr} < 0$.

3. **A.** Thermodynamic favorability should be evaluated in terms of change in Gibbs free energy.

$$\Delta G = \Delta H - T\Delta S$$

Because the entropy change, $\Delta S$, for the system is negative and the enthalpy change, $\Delta H$, is positive, there are no temperatures at which the change in Gibbs energy is negative. The reaction is thermodynamically favored at no temperature, choice A.

4. **D.** From He to Ne, there is an increase in the number of electrons and an increase in absolute entropy. From Ne to $SO_2$ to $CH_3CH_2OH$, there is an increase in molecular complexity, which also leads to an increase in absolute entropy. The substances in order of increasing standard molar entropy are correctly shown in choice D:

$$He(g) < Ne(g) < SO_2(g) < CH_3CH_2OH(g)$$

5. **B.** For the reactions to add up to the overall transformation, the reaction of $Fe_2O_3$ must be reversed and its $\Delta G°$ multiplied by –1. The reaction of CO with $O_2$ must be multiplied by 3, and the $\Delta G°$ for that reaction must be multiplied by 3. The reactions and $\Delta G°$ values are then added to obtain the overall reaction and $\Delta G°$. A negative $\Delta G°$ indicates a thermodynamically favored process.

$$Fe_2O_3 \rightarrow 2\ Fe + \frac{3}{2}\ O_2 \qquad \Delta G° = 742\ \text{kJ/mol}$$

$$3\left(CO + \frac{1}{2}O_2 \rightarrow CO_2\right) \qquad \Delta G° = -3(257\ \text{kJ/mol})$$

$$Fe_2O_3 + 3\ CO \rightarrow 2\ Fe + 3\ CO_2 \qquad \Delta G° = -29\ \text{kJ/mol}$$

Choice B is correct: $\Delta G° = -29$ kJ/mol; the reaction is thermodynamically favorable.

**6.** **A.** Because the liquid and vapor phases are in equilibrium at the boiling point, $\Delta G = 0$. We can solve for the temperature. Remember to convert the units of $\Delta H°$ from kJ to J.

$$\Delta G° = \Delta H° - T\Delta S°$$
$$0 = \Delta H° - T\Delta S°$$
$$0 = 40{,}000 \text{ J/mol} - T(20 \text{ (J/(mol·K))}$$
$$T = 2000 \text{ K} - 273 = 1727 \text{ °C, choice A}$$

**7.** **A.** The negative value of $\Delta G°$ is associated with a $K$ value that is greater than 1, according to the equation

$$\Delta G° = -RT\ln K$$

When $\Delta G° < 0$, $K > 1$, choice A.

**8.** **C.** The rusting of iron proceeds according to the equation:

$$4 \text{ Fe}(s) + 3 \text{ O}_2(g) \rightarrow 2 \text{ Fe}_2\text{O}_3(s)$$

This reaction is exothermic and proceeds toward products under ordinary conditions ($\Delta G < 0$), but the activation energy is relatively high and the reaction is slow, choice C.

**9.** **C.** Reactions that are thermodynamically unfavorable, like the transport of potassium into the cell against its concentration gradient, can be driven by coupling to another source of energy. Since the reaction is coupled to the hydrolysis of ATP, the hydrolysis reaction must have sufficiently negative Gibbs free energy to overcome the positive free energy of the transport, choice C.

**10.** **C.** Because the tin electrode is the cathode and the silver electrode is the anode, the cell potential can be calculated using the standard reduction potentials.

$$E°_{\text{cell}} = E°_{\text{cathode}} - E°_{\text{anode}} = -0.83 \text{ V} - (1.07 \text{ V}) = -1.9 \text{ V}$$

The standard cell potential is negative, so the reaction is thermodynamically unfavorable. This is an electrolytic cell with $\Delta G° > 0$, choice C.

**11.** **B.** The solids in this question have oxidation states of zero, and the ions have oxidation states of +2. If a reaction occurs, the solid metal is oxidized and the ion is reduced in a single replacement reaction. $Cu^{2+}$ has the highest reduction potential because it is reduced by both Mg and Pb. $Mg^{2+}$ has the lowest reduction potential; it is reduced by neither Cu nor Pb. The correct order is shown in choice B.

**12.** **B.** The number of electrons transferred is determined using the equation for current. The equation relating current, charge, and time $\left(I = \frac{q}{t}\right)$ is found on the *AP Chemistry Equations and Constants* sheet. The number of electrons transferred is used to calculate the number of moles of metal, M. The molar mass, ***M***, of the metal can be obtained by dividing the mass by the number of moles of metal. The molar mass, 64 g/mol, most closely matches that of copper, so the metal cation is $Cu^{2+}$, choice B.

$$q = t \times I = (162 \cancel{\text{min}}) \frac{60 \cancel{\text{s}}}{1 \cancel{\text{min}}} \times \frac{2.0 \cancel{\text{C}}}{1 \cancel{\text{s}}} \times \frac{1 \text{ mol } e^-}{96{,}485 \cancel{\text{C}}} \times \frac{1 \text{ mol M}}{2 \text{ mol } e^-} = 0.101 \text{ mol } M$$

$$\boldsymbol{M} = \frac{6.4 \text{ g}}{0.101 \text{ mol}} = 64 \text{ g/mol}$$

Without a calculator, the calculation can be simplified. Using scientific notation is one way to deal with large numbers.

$$\frac{160 \times 60 \times 2}{96{,}000 \times 2} = \frac{9.6 \times 10^3}{9.6 \times 10^4} \approx 0.1$$

**13.** **A.** For sodium nitrate to be soluble, the change in Gibbs free energy for the dissolution process must be negative. Because $\Delta H° > 0$, $\Delta S°$ must be greater than zero, choice A, to result in a negative $\Delta G°$, according to the equation $\Delta G° = \Delta H° - T\Delta S°$.

**14.** **A.** The reaction with the larger equilibrium constant has a more negative change in Gibbs free energy, $\Delta G°$. Since both reactions have similar values of $\Delta H°$, the reaction with the more positive entropy change, choice A, will have a smaller $\Delta G°$ according to $\Delta G° = \Delta H° - T\Delta S°$.

**15.** **C.** The standard cell potential is positive for a galvanic cell, so $Ag^+$ is reduced at the cathode and Zn is oxidized at the anode.

$$E°_{cell} = E°_{reduction} - E°_{oxidation} = 0.80\text{ V} - (-0.76\text{ V}) > 0$$

The overall balanced equation is

$$2\,Ag^+ + Zn \rightarrow 2\,Ag + Zn^{2+}$$

Addition of NaCl to the $Ag^+$ half-cell will cause AgCl to precipitate, decreasing the concentration of $Ag^+$ and increasing the reaction quotient, $Q$.

The Nernst equation can be used to perform a quick logic check.

$$E = E° - \frac{RT}{nF}\ln\frac{[Zn^{2+}]}{[Ag^+]^2}$$

A decrease in $[Ag^+]$ will increase the ln $Q$ term and decrease $E$. Choice C is correct.

## Long Free Response

**1.** **a.** In order to balance the net ionic equation, the two half reactions must be balanced. In cell notation, the anode is on the left (oxidation) and the cathode is on the right (reduction).

$$Ni \rightarrow Ni^{2+} + 2\,e^-$$
$$2\,Ag^+ + 2\,e^- \rightarrow 2\,Ag$$

The balanced net ionic equation is

$$Ni + 2\,Ag^+ \rightarrow Ni^{2+} + 2\,Ag$$

**b.** **i.** The cell potential may be calculated as follows.

$$E°_{cell} = E°_{cathode} - E°_{anode} = 0.80\text{ V} - (-0.25\text{ V}) = 1.05\text{ V}$$

**ii.** This is a voltaic/galvanic cell; this is clear from the positive cell potential.

**c.** Mass is being lost from the anode, the electrode where oxidation of solid nickel to form $Ni^{2+}$ is occurring. As the electrons leave the nickel and it is converted to cations, it becomes soluble in an aqueous solution. The $Ni^{2+}$ ions leave the solid anode and enter the solution.

**d.** The relationship between the Gibbs free energy and the cell potential is given on the *AP Chemistry Equations and Constants* sheet, as is the Faraday constant, $F$, and the definition of volts, which is joules/coulomb. For this two-electron transfer process, $n = 2$.

$$\begin{aligned}\Delta G° &= -nFE° \\ &= -2(96{,}485\ \text{C̸/mol})(1.05\ \text{J/C̸}) \\ &= -202{,}618.5\ \text{J/mol} \\ &= -203\ \text{kJ/mol}\end{aligned}$$

**e.** **i.** The equilibrium constant can be determined using the value of $\Delta G°$.

$$\Delta G° = -RT \ln K$$

$$\ln K = -\frac{\Delta G°}{RT} = -\frac{-202{,}618.5\ \cancel{\text{J/mol}}}{(8.314\ \cancel{\text{J/mol}}\cdot\cancel{\text{K}})(298\ \cancel{\text{K}})} = 81.78107...$$

$$K = e^{81.78107} = 3.29\times10^{35}$$

**ii.** One method of determining whether the cell potential will be higher or lower at this set of nonstandard conditions is to calculate $Q$ and compare it with $K$.

$$Q = \frac{[\text{Ni}^{2+}]}{[\text{Ag}^{+}]^2} = \frac{7.69\times10^{-3}}{(8.22\times10^{-1})^2} = 0.0187$$

Because $Q < K$, the value of $E_{cell}$ is greater than $E°_{cell}$.

Alternatively, we can compare the value of $E_{cell}$ using the Nernst equation.

$$E = E° - \frac{RT}{nF}\ln\frac{[\text{Ni}^{2+}]}{[\text{Ag}^{+}]^2}$$

Because $\ln(0.0187) < 0$, the term subtracted from $E°_{cell}$ is negative. This results in an $E_{cell}$ that is greater than $E°$.

## Short Free Response

**2. a.** The standard entropy change is likely to be negative because the number of moles of gaseous reactants is greater than the number of moles of gaseous products. This corresponds to a decrease in dispersion of matter.

**b.** The reaction is thermodynamically favorable since it proceeds in the direction that is written. For a spontaneous process, the standard change in Gibbs free energy, $\Delta G°$, is negative. The reaction is also exothermic since it gives off heat. Since $\Delta H < 0$, $\Delta S < 0$, and $\Delta G = \Delta H - T\Delta S$, the reaction becomes less spontaneous ($\Delta G$ becomes more positive) as the temperature increases.

**c.** The reaction is exothermic, which means that less energy is consumed when bonds in the reactants are broken than is released in the process of new bond formation. The potential energy of reactants is greater than that of products.

**d.** The reactants do not have sufficient energy to overcome the activation barrier until their energy is increased by contact with the hot wire. Once the reaction is initiated, it proceeds spontaneously. In other words, the reaction is under kinetic control.

Chapter 10

# Laboratory Review

Science is not simply a body of facts; it is a system of making sense of the world around us through investigation and observation. Approximately 25% of a student's time in AP Chemistry is devoted to hands-on laboratory work, and a number of questions on the exam are intended to assess student understanding of experimental chemistry. A selection of AP Chemistry laboratory techniques are briefly summarized here.

**NOTE: The review of laboratory experiments in this book is not meant to replace detailed laboratory procedures and safety concerns. Under NO circumstances are the scenarios for the labs to be used as directions for performing the laboratory work.**

## Measurement

Any measurement has an inherent uncertainty. For example, when reading the volume of a liquid in a graduated cylinder, the bottom of the meniscus can be read to the nearest marking, and then one more digit is estimated to the researcher's best approximation. No more than one digit should be estimated.

**Estimation of a Volume**

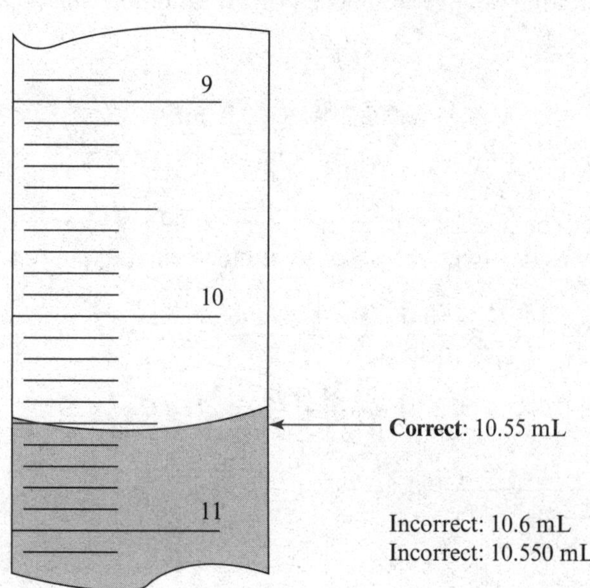


The number of significant figures for a reported measurement communicates the level of precision with which the chemist made a measurement. In the picture above, the measurement 10.55 mL has four significant figures. The higher the number of significant figures, the more precise the measurement. It is important to be familiar with the conventions for working with significant figures, as discussed on page 4 of the Introduction.

## Determining the Composition of an Alloy

Spectrophotometry may be used to determine the amount of copper in a sample of brass. Brass is an alloy of copper and zinc.

A sample of brass of unknown composition is converted to a solution of $Cu^{2+}$ ions in a precise volume of water.

The mass of a brass sample is determined using a balance, and then the brass is treated with a minimum volume of concentrated nitric acid. This completely dissolves the brass and converts the copper metal to blue $Cu^{2+}$ ions in solution. Zinc ions are also released, but they are colorless.

$$Cu(s) + 4\ HNO_3(aq) \rightarrow Cu(NO_3)_2(aq) + 2\ NO_2(g) + 2\ H_2O(l)$$

The resulting solution is diluted to a precise volume. The absorbance of the sample is measured and compared to a standard curve.

The standard curve is generated using various solutions of known $Cu^{2+}$ concentration, so Beer's law may be used to determine the concentration of the $Cu^{2+}$ in the sample. From the concentration and the volume of the dilution, the amount of copper in the brass may be determined.

## Practice

A 2.28-g sample of brass is dissolved in approximately 20 mL of concentrated excess $HNO_3$. The resulting blue solution is diluted to 100.0 mL in a volumetric flask. A 1.00-mL volumetric pipet is used to transfer a sample of the $Cu^{2+}$ solution to another volumetric flask, and the sample is again diluted to 100.0 mL. Comparison of the absorbance of the resulting solution with a standard curve indicates a $Cu^{2+}$ concentration of 0.00269 *M*. What is the mass percentage of copper in the original sample of brass?

The final measured concentration of $Cu^{2+}$ was the result of dilution of 1.00 mL of the original solution. The concentration of the original solution can be determined using the dilution equation.

$$M_1V_1 = M_2V_2$$

$$M_1 = \frac{M_2V_2}{V_1} = \frac{(0.00269\ M)(100.0\ \cancel{mL})}{1.00\ \cancel{mL}} = 0.269\ M$$

The number of moles of $Cu^{2+}$ can be calculated from the volume and the concentration of the original solution. The number of moles is then converted to grams using the atomic weight of copper.

$$100.0\ \cancel{mL} \times \frac{1\ \cancel{L}}{1000\ \cancel{mL}} \times \frac{0.269\ mol}{1\ \cancel{L}} = 0.0269\ mol\ Cu^{2+}$$

$$0.0269\ \cancel{mol\ Cu^{2+}} \times \frac{63.55\ g\ Cu}{1\ \cancel{mol\ Cu^{2+}}} = 1.71\ g\ Cu$$

$$\frac{1.71\ g\ Cu}{2.28\ g\ brass} \times 100\% = 75.0\%$$

# Gravimetric Analysis

**Gravimetric analysis** is a technique in which an analyte is converted to an insoluble solid, filtered from solution, and dried. The mass of the solid is measured and the original concentration of the analyte in solution is deduced.

The concentration of calcium ions present in a sample of water can be determined gravimetrically by precipitation, filtration, and drying. An excess of sodium carbonate is added to the unknown $Ca^{2+}$ solution. The product is solid calcium carbonate, which has a $K_{sp}$ of $8.7 \times 10^{-9}$. The net ionic equation is

$$Ca^{2+}(aq) + CO_3^{2-}(aq) \rightarrow CaCO_3(s)$$

The solid is filtered through a vacuum filtration apparatus, as shown on page 9. A piece of pre-weighed filter paper is used to collect the solid.

Once the solvent has been removed, the solid is washed with a small amount of deionized water. The filter paper with the wet solid is placed on a watch glass in a drying oven and allowed to dry completely. When the solid is dry, the mass of the filter paper with the solid is determined and the mass of the empty filter paper is subtracted to give the mass of the solid.

It is important to note that if the sample is not rinsed with deionized water before drying, the mass will be too high due to the presence of other ions. If the sample is rinsed with too much deionized water, however, the mass will be too low due to some dissolving of the slightly soluble calcium carbonate.

In addition, it is important to break up the pieces of solid calcium carbonate with a metal spatula between periods of drying in order to release water that is trapped in the interior of the solid.

### Practice

A 1.00-L sample of mineral water is boiled down until about 20 mL remain. Approximately 40 mL of 0.5 *M* $Na_2CO_3$ are added to the solution and a precipitate forms. The solid is filtered and dried, and the following data is collected. What is the concentration of $Ca^{2+}$ in the original sample? Report your answer in mol/L.

| | |
|---|---|
| Mass of filter paper | 0.238 g |
| Mass of filter paper + solid after first drying | 2.196 g |
| Mass of filter paper + solid after second drying | 2.048 g |
| Mass of filter paper + solid after third drying | 2.042 g |
| Mass of filter paper + solid after fourth drying | 2.041 g |

The mass does not change very much between the third and fourth drying periods, so the final mass of the solid plus the filter paper is 2.041 g. The mass of the sodium carbonate solid is

$$2.041 \text{ g} - 0.238 \text{ g} = 1.803 \text{ g}$$

This mass can be divided by the molar mass of $CaCO_3$ to find the number of moles of $Ca^{2+}$ in the 20-mL sample.

$$1.803 \text{ g } \cancel{CaCO_3} \times \frac{1 \text{ } \cancel{\text{mol } CaCO_3}}{100.09 \text{ } \cancel{\text{g } CaCO_3}} \times \frac{1 \text{ mol } Ca^{2+}}{1 \text{ } \cancel{\text{mol } CaCO_3}} = 0.01801 \text{ mol } Ca^{2+}$$

The molarity of $Ca^{2+}$ in the 1.00-L sample can be determined by recognizing that the total number of moles of $Ca^{2+}$ were present in the 1.00-L sample.

$$\frac{0.01801 \text{ mol } Ca^{2+}}{1.00 \text{ L}} = 0.0180 \; M \; Ca^{2+}$$

## Ionic, Covalent, and Metallic Bonds

Various unknown solids are classified by bonding type: ionic, polar covalent, nonpolar covalent, or metallic. Unknown substances may include the following:

| Ionic | Covalent (polar) | Covalent (nonpolar) | Metallic |
|---|---|---|---|
| $NH_4Cl$ | $C_6H_5CO_2H$ (benzoic acid) | Wax | Aluminum |
| MgO | $C_{12}H_{22}O_{11}$ (sucrose) | Iodine ($I_2$) | Magnesium |

Methods for determining the identity of unknown solids by bonding type include assessment of water solubility, pH, melting point, and electrical conductivity.

Solubility in water can provide information about bonding types present in the compound. All ammonium salts are soluble, so $NH_4Cl$ will dissolve. MgO, $C_6H_5CO_2H$, and $C_{12}H_{22}O_{11}$ are also soluble. Wax, iodine, aluminum, and magnesium are insoluble. Remember that a compound is not necessarily soluble in water merely because it is ionic. Some ionic compounds have low $K_{sp}$ values.

- The pH of the resulting aqueous solution may be higher or lower than 7. The pH values of solutions of $NH_4^+$ and benzoic acid are slightly lower than 7. The pH value of MgO is higher than 7 due to the formation of a hydroxide in an aqueous solution: $MgO + H_2O \rightarrow Mg(OH)_2$. The sucrose solution is neutral.
- The conductivity of the aqueous solution indicates the presence of ions. $NH_4Cl$ and MgO ($Mg(OH)_2$) are highly conductive strong electrolytes in an aqueous solution. Benzoic acid, a weak acid, is weakly conductive, and sucrose is a nonelectrolyte.

Melting point determination can distinguish between molecular solids and ionic solids. The coulombic attraction between oppositely charged ions is much stronger than the intermolecular forces that hold the molecules of a molecular solid together. The melting point of benzoic acid, a molecular solid, is much lower than an ionic compound. A simple test for a low melting point is to place a small amount of the solid on the clean lid of an aluminum can. The lid is placed on a hot plate and heated. If the solid melts, the melting point is reasonably low, and the solid is not ionic.

Metallic solids can be distinguished from nonpolar covalent solids by their electrical conductivity. If a sample is conductive in the solid state, the electron sea model of bonding applies. A solid that is not conductive and not soluble in water is likely to be a nonpolar covalent compound.

A final test for a nonpolar covalent compound is dissolution of the compound in a nonpolar solvent such as hexane. Iodine and wax are soluble in hexane, while metals such as aluminum and magnesium are not.

### Practice

A solid green crystalline sample appears to be insoluble in deionized water. The solution exhibits slight electrical conductivity, and the pH of the water is greater than 7. When a few drops of aqueous HCl are added to the solution, more of the solid dissolves. The sample is insoluble in hexane, chloroform, and ethanol. When the solid is heated to the maximum temperature that can be achieved with a hot plate, it does not melt. Is the solid ionic, polar covalent, nonpolar covalent, or metallic?

The insolubility of the compound in water does not rule out any of the four possible bonding types.

The high pH of the solution rules out weak acids but not weak bases. The sample is unlikely to be metallic since it is a green crystal. The slight electrical conductivity of the solution confirms that the sample is either an ionic compound with a low $K_{sp}$ or a weakly basic polar covalent compound.

The high melting point and insolubility in solvents such as hexane, chloroform, and ethanol suggest that this solid is an insoluble ionic compound.

## Decomposition Stoichiometry

In this experiment, the composition of a mixture of $NaHCO_3$ and $Na_2CO_3$ is determined. One of the compounds, $NaHCO_3$, undergoes decomposition upon heating to temperatures that can be achieved in a classroom laboratory, but the sample is not heated to a temperature high enough to decompose $Na_2CO_3$.

$$2\ NaHCO_3(s) \xrightarrow{\Delta} Na_2CO_3(s) + H_2O(g) + CO_2(g)$$

By comparing the mass of the mixture with the mass of the decomposed sample, the number of moles of $CO_2(g)$ lost can be determined. This information may be used to find the mass of the $NaHCO_3(s)$ in the original sample.

The technique for heating the mixture involves placing it in a pre-weighed crucible, covering it with the lid askew so that gases do not build up and blow off the lid, and heating the crucible in a ceramic triangle over a Bunsen burner. After the sample is removed from the heat, it is allowed to cool completely before the mass is determined.

**Heating a Laboratory Crucible over a Bunsen Burner**

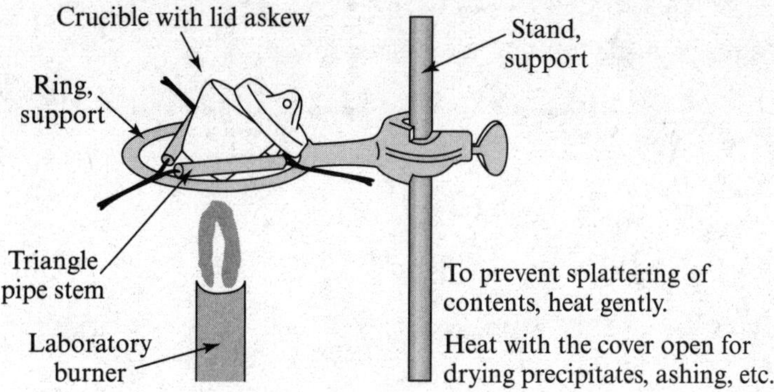


## Practice

A mixture of solid $Na_2CO_3$ and $NaHCO_3$ has a mass of 1.073 g. The mixture is placed in a crucible with a mass of 23.940 g. After heating to constant mass and cooling, the mass of the crucible plus the sample is 24.728 g.

**a.** What is the percentage of $NaHCO_3$ in the original mixture?

**b.** Would the calculated percentage be too high or too low if an additional impurity with a very high melting point was present in the mixture?

**(a)** The mass of the crucible plus the original sample is

$$23.940\text{ g} + 1.073\text{ g} = 25.013\text{ g}$$

The mass of the sample that was lost is

$$25.013\text{ g} - 24.728\text{ g} = 0.285\text{ g}$$

The mass of the sample that was lost is due to a 1:1 ratio of $CO_2$ (MW = 44.01 g/mol) and $H_2O$ (MW = 18.02 g/mol). Let $x$ be equal to the number of moles of $CO_2$ lost and the number of moles of $H_2O$ lost.

$$\text{mass lost} = x(44.01\text{ g/mol}) + x(18.02\text{ g/mol}) = 0.285\text{ g}$$
$$x(44.01 + 18.02) = 0.285$$
$$x = 0.00459\text{ mol H}_2\text{O} = 0.00459\text{ mol CO}_2$$

The balanced chemical equation shows that the stoichiometric ratio of $NaHCO_3$ to $CO_2$ is 2:1. Therefore,

$$0.00459\ \cancel{\text{mol CO}_2} \times \frac{2\ \cancel{\text{mol NaHCO}_3}}{1\ \cancel{\text{mol CO}_2}} \times \frac{84.01\text{ g NaHCO}_3}{1\ \cancel{\text{mol NaHCO}_3}} = 0.771\text{ g NaHCO}_3$$

The mass percentage of $NaHCO_3$ in the original sample is

$$\frac{0.771\text{ g NaHCO}_3}{1.073\text{ g sample}} \times 100\% = 71.9\%\text{ NaHCO}_3$$

**(b)** A second impurity would not affect the calculated mass of $NaHCO_3$. The calculated mass percentage, therefore, would also be unaffected.

# Liquid-Liquid Extraction

Liquid-liquid extraction is the technique of partitioning a solute between two immiscible liquid phases due to selective solubility in only one of the phases. In this experiment, students separate aspirin from acetaminophen in a separatory funnel.

The analgesics are first dissolved in ethyl acetate, an organic solvent.

**The Structures of Aspirin, Acetaminophen, and Ethyl Acetate**

Aspirin
$C_9H_7CO_4H$

Acetaminophen
$C_8H_9NO_2$

Ethyl acetate solvent

The solution is added to a separatory funnel, and then an aqueous solution of $NaHCO_3$ is added. Because water does not mix with ethyl acetate and because water is more dense than ethyl acetate, water forms the bottom layer in the separatory funnel. Neither aspirin nor acetaminophen is soluble in water.

The separatory funnel is next shaken gently (with frequent venting to prevent pressure buildup), exposing the solutes in the ethyl acetate layer to the aqueous $NaHCO_3$. Because aspirin is a weak acid, it is deprotonated by $HCO_3^-$, forming the sodium salt. The acetaminophen does not react and remains dissolved in the ethyl acetate layer.

The sodium salt of aspirin is no longer soluble in ethyl acetate. As a result of further shaking and venting, the aspirin anion, which is water-soluble, is partitioned into the water layer, or aqueous phase. The resulting $H_2CO_3$ decomposes into water and carbon dioxide.

$$C_9H_7CO_4H + HCO_3^- \rightarrow C_9H_7CO_4^- + H_2CO_3 \rightarrow C_9H_7CO_4^- + H_2O + CO_2$$

**Use of a Separatory Funnel to Separate Aspirin from Acetaminophen**

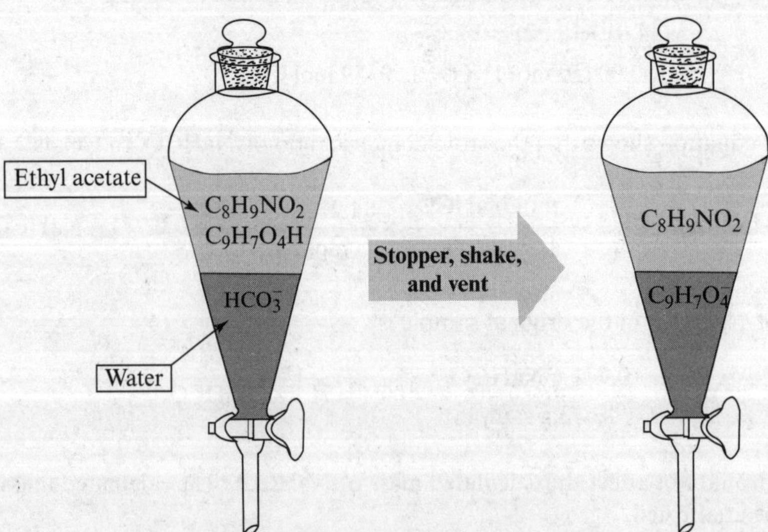

The aspirin sodium salt solution is drained out of the separatory funnel, separating it from the ethyl acetate layer. Aqueous HCl is then added to the drained aqueous layer. This protonates the aspirin so that it is, once again, insoluble in water. It forms a solid precipitate that can be filtered.

The ethyl acetate layer still contains some dissolved $H_2O$ in addition to acetaminophen. The ethyl acetate can be dried by the addition of a small amount of insoluble $MgSO_4$. Magnesium sulfate is insoluble in ethyl acetate, but it attracts the water molecules present in the ethyl acetate. The hydrated solid $MgSO_4$ can then be filtered off, taking the water with it. The ethyl acetate can then be evaporated, leaving pure acetaminophen.

## Practice

A solution of dichloromethane, $CH_2Cl_2$, in a separatory funnel contains biphenyl, $C_{12}H_{10}$, a neutral compound, and 3-nitroaniline, $C_6H_4NO_2NH_2$, a weak base. Describe how the biphenyl might be separated from the 3-nitroaniline.

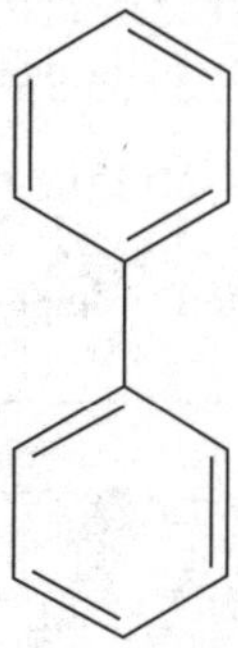

$NH_2$

$NO_2$

Biphenyl

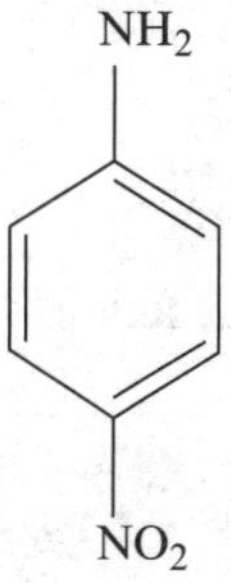

Nitroaniline

An aqueous solution of HCl can be added to the separatory funnel. The contents of the funnel are shaken and vented to expose the solutes in the $CH_2Cl_2$ layer to the HCl in the aqueous layer. The 3-nitroaniline will be protonated $-NH_2$ nitrogen, resulting in a water-soluble cation, $-NH_3^+$, while biphenyl will remain in the $CH_2Cl_2$ layer. The two layers are then separated, and the $CH_2Cl_2$ layer is dried. The aqueous layer can be neutralized with aqueous $NaHCO_3$, resulting in a neutral 3-nitroaniline precipitate, which can be removed from the water by filtration.

# Measurement of the Rate of a Chemical Reaction

This experiment examines the effect of the surface area of a solid reactant and concentration of an aqueous reactant on the rate of a gas-producing reaction. The reaction under investigation is between solid marble chips, composed of $CaCO_3$, and aqueous HCl.

$$CaCO_3(s) + 2\,HCl(aq) \rightarrow CaCl_2(aq) + H_2O(l) + CO_2(g)$$

The volume of gas collected can be measured as a function of time using either a syringe in a rubber stopper or a eudiometer (see below). Alternatively, the reaction flask can be placed on a balance, and the mass, which decreases as $CO_2$ is released, can be measured at regular time intervals.

**Two Pieces of Apparatus for Measuring the Rate of Gas Formation**

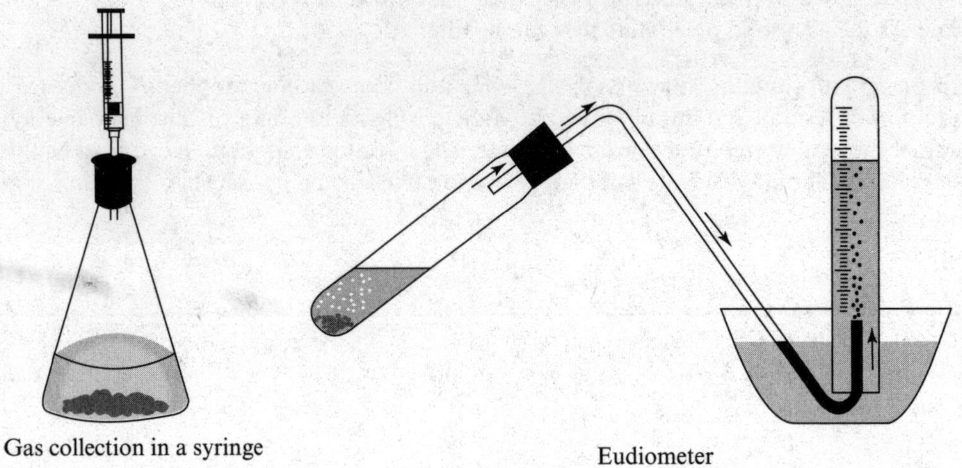


The volume of gas collected is measured at regular, timed intervals, and the data is graphed as shown below.

**The Rate of Calcium Carbonate Decomposition**

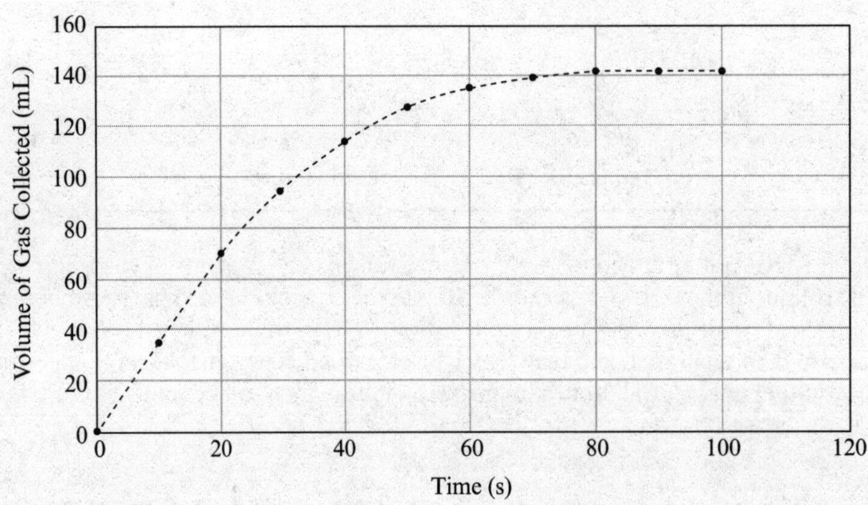


## Practice

A student places 1-$cm^3$ marble chips in an Erlenmeyer flask, adds excess 1.0 *M* HCl, stoppers the flask with a gas collection syringe, and records the volume of gas collected at regular intervals. What are two ways in which the student might cause the rate of the reaction to increase?

Two simple ways to increase the rate of the reaction are to increase the surface area of the marble chips by breaking or grinding or to increase the concentration of the HCl used. Both of these actions increase the frequency of sufficiently energetic collisions that have the necessary orientations of reactants for the reaction to occur.

# Review Questions

**1.** What volume should be reported for the buret reading shown below?

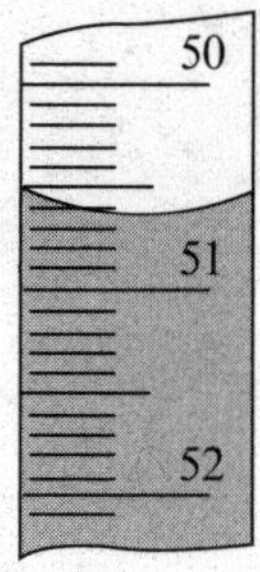


A. 50.65 mL
B. 50.6 mL
C. 50.650 mL
D. 51.3 mL

*Refer to the following experimental description to answer questions 2–5.*

Ostalloy is an alloy composed of various combinations of indium, tin, and bismuth, lead, or cadmium.

| Alloy | % Indium by Mass | % Tin by Mass | % Bismuth by Mass | % Lead by Mass | % Cadmium by Mass |
|---|---|---|---|---|---|
| Ostalloy No. 117 | 19.1 | 8.3 | 44.7 | 22.6 | 5.3 |
| Ostalloy No. 136 | 21.0 | 12.0 | 49.0 | 18.0 | 0 |
| Ostalloy No. 178 | 29.6 | 16.3 | 51.4 | 0 | 0 |
| Ostalloy No. 200 | 44.0 | 42.0 | 0 | 0 | 14.0 |

Ostalloy may be dissolved in acid with a reagent that forms a yellow complex with $Sn^{2+}$ ions. The absorbance of this complex was found to have a maximum at 428 nm. A Beer's law calibration curve was constructed by diluting this solution to various concentrations of $Sn^{2+}$ complex. Each sample's absorbance was measured in a 1.00-cm cuvette in a spectrophotometer. The results are plotted below.

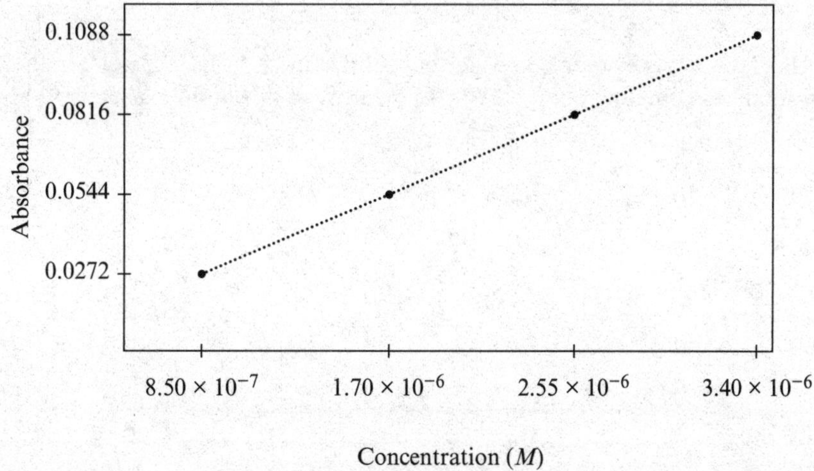


A 0.500-gram sample of an unknown Ostalloy was dissolved in approximately 5 mL of the acid solution. The solution was diluted to 500.0 mL in a volumetric flask. A 1.00-mL sample was transferred with a pipette to a second 500.0-mL volumetric flask and diluted to the mark.

**2.** The absorbance of the diluted solution was found to be 0.0640 in a 1.00-cm cuvette. What is the approximate concentration of the solution?

A. $1.80 \times 10^{-6}$ *M*
B. $2.00 \times 10^{-6}$ *M*
C. $2.50 \times 10^{-6}$ *M*
D. $3.30 \times 10^{-6}$ *M*

**3.** What is the approximate molar absorptivity, $\varepsilon$, of the yellow $Sn^{2+}$ complex at 428 nm?

A. $3 \times 10^{-7}$ L/(cm·mol)
B. $3 \times 10^{-5}$ L/(cm·mol)
C. $3 \times 10^{4}$ L/(cm·mol)
D. $3 \times 10^{5}$ L/(cm·mol)

**4.** How many moles of tin were present in the original sample?

A. $1.00 \times 10^{-3}$ mol
B. $5.00 \times 10^{-4}$ mol
C. $1.00 \times 10^{-4}$ mol
D. $2.00 \times 10^{-6}$ mol

**5.** Which alloy was the unknown sample composed of?

A. Ostalloy No. 117
B. Ostalloy No. 136
C. Ostalloy No. 178
D. Ostalloy No. 200

**6.** A student is performing a gravimetric analysis on a solution of $Ca^{2+}$. The student causes the calcium to precipitate by adding $Na_2CO_3$. The student filters the solid using a Büchner funnel and a piece of filter paper. Which of the following is LEAST likely to lead to an error in the mass of $CaCO_3$ determined?

A. Neglecting to wash the sample with a small amount of deionized, neutral $H_2O$
B. Failing to add an excess of $Na_2CO_3$
C. Breaking up the wet solid with a spatula before allowing it to dry for an additional length of time
D. Washing the sample with large amounts of deionized, neutral $H_2O$

**7.** A lustrous, brittle gray solid is insoluble in water but soluble in $CCl_4$. It does not conduct electricity in its solid form. The solid has a melting point of 114 °C. What kind of bonding is likely to be present in the substance?

A. Metallic
B. Ionic
C. Polar covalent
D. Nonpolar covalent

**8.** A sample containing $NaHCO_3(s)$ is heated to constant mass in a crucible. It decomposes according to the following reaction:

$$2\ NaHCO_3(s) \xrightarrow{\Delta} Na_2CO_3(s) + H_2O(g) + CO_2(g)$$

The mass data before and after heating is shown in the table below. How many moles of $NaHCO_3$ were initially present?

| Mass of crucible | 5.80 g |
|---|---|
| Mass of crucible + $NaHCO_3$ before heating | 9.16 g |
| Mass of crucible + $Na_2CO_3$ after heating | 7.92 g |

**A.** 0.0150 mol
**B.** 0.0200 mol
**C.** 0.0300 mol
**D.** 0.0400 mol

**9.** The rate of a reaction with $CaCO_3$ and HCl was monitored by collection of $CO_2$ gas.

In trial I, a 1.5-gram chip of $CaCO_3$ was added to 50.0 mL of 3.0 *M* HCl.

In trial II, 1.5 grams of $CaCO_3$ powder were added to 50.0 mL of 3.0 *M* HCl.

Which graph correctly shows the expected change in volume of $CO_2$ with time?

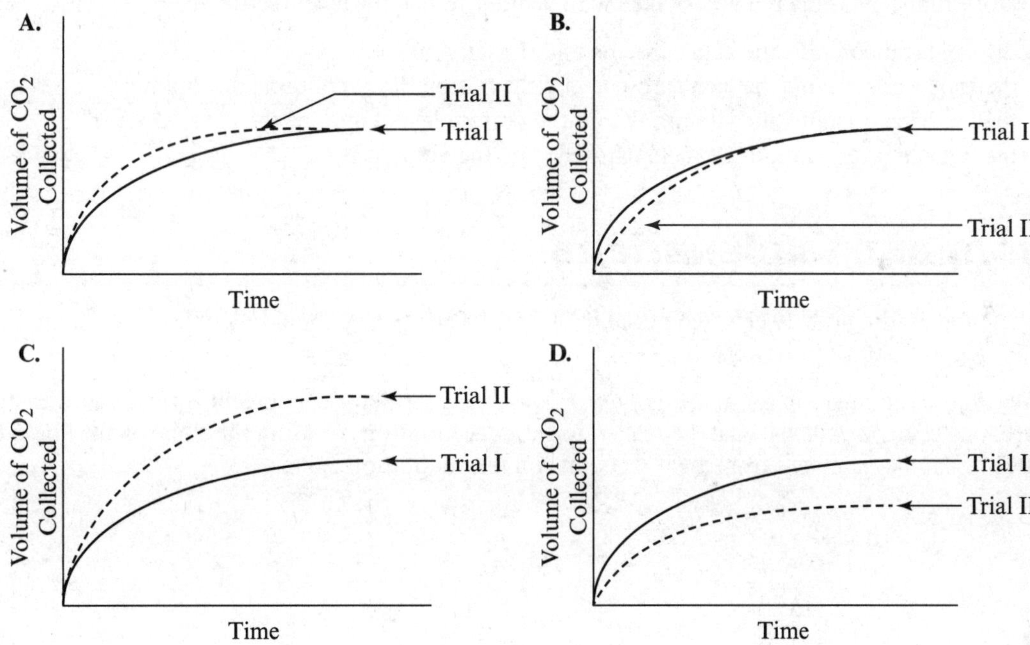

**10.** Beta carotene and vitamin $B_5$ are both compounds that are found in sweet potatoes. The structures of the two compounds are shown below.

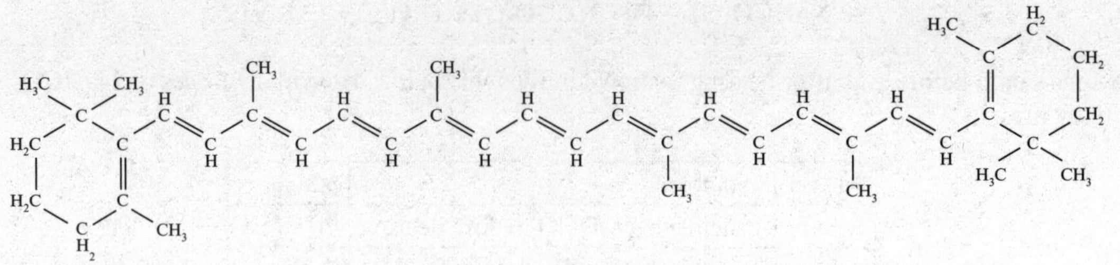

**Beta Carotene**

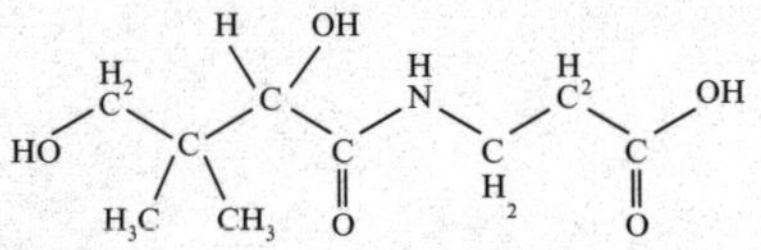

**Vitamin $B_5$**

A food scientist mixes a 1-g portion of finely ground sweet potato in a dish with a 10-mL portion of ethanol. The solids are filtered, and the ethanol filtrate is added to a separatory funnel. A 100-mL portion of isooctane, $(CH_3)_3CCH_2CH(CH_3)_2$ (density = 0.692 g/mL), and a 100-mL portion of water are added. The separatory funnel is stoppered and shaken with venting. When the layers settle,

**A.** the top layer contains the beta carotene and the vitamin $B_5$
**B.** the top layer contains the beta carotene and the bottom layer contains the vitamin $B_5$
**C.** the top layer contains the vitamin $B_5$ and the bottom layer contains the beta carotene
**D.** the bottom layer contains the beta carotene and the vitamin $B_5$

# Answers and Explanations

**1. A.** The buret reading is between 50.6 mL and 50.7 mL. The reading should include one estimated digit, and 50.65 is a good estimate, choice A.

**2. B.** We can use the equation for a line, $y = mx + b$, and assume that the intercept is zero. The measured absorbance value, $A = 0.0640$, can be related to the concentration, $C$, using the slope of the line. The slope of the line can be estimated from any two points on the calibration curve.

$$A = \left(\frac{\Delta y}{\Delta x}\right)C$$

$$C = A\left(\frac{\Delta x}{\Delta y}\right)$$

$$C = 0.0640\left(\frac{\left(3.40\times10^{-6} - 0.850\times10^{-6}\right)M}{0.1088 - 0.0272}\right) = 2.00\times10^{-6}\,M, \text{ choice B}$$

The answer can be estimated without a calculator as follows.

$$C \approx \cancel{0.0640}^{\,0.032}\left(\frac{\left(2.6\times10^{-6}\right)M}{\cancel{0.08}^{\,0.04}}\right) = 3.2\left(\frac{\left(2.6\times10^{-6}\right)M}{4}\right) \approx 2\times10^{-6}\,M$$

**3. C.** The molar absorptivity, $\varepsilon$, can be estimated using Beer's law, which is located on the *AP Chemistry Equations and Constants* sheet. Any of the points on the plot may be chosen for the estimation. The point with the absorbance of 0.1088 is chosen here as an example.

$$A = \varepsilon bc$$

$$\varepsilon = \frac{A}{bc} = \frac{0.1088}{(1.00 \text{ cm})(3.40 \times 10^{-6} \text{ mol/L})} = 3.20 \times 10^{4} \text{ L/(cm} \cdot \text{mol)}$$

This answer is closest to $3 \times 10^4$ L/(cm·mol), choice C. The answer can be estimated without a calculator as follows.

$$\varepsilon \approx \frac{1 \times 10^{-1}}{3 \times 10^{-6} \text{ ((cm} \cdot \text{mol)/L)}} = \frac{1}{3} \times 10^{[-1-(-6)]} \text{ L/(cm} \cdot \text{mol)} \approx 3 \times 10^{5} \text{ L/(cm} \cdot \text{mol)}$$

**4. B.** The concentration of the solution that was determined in the spectrophotometer was solved previously in question 2.

$$[Sn^{2+}] = 2.00 \times 10^{-6}\ M$$

This sample had been diluted to 500.0 mL from 1.00 mL. We can use the dilution equation to find the molarity of the 1.00-mL sample that was transferred.

$$M_1V_1 = M_2V_2$$

$$(2.00 \times 10^{-6}\ M)(500.0 \ \cancel{\text{mL}}) = M_2(1.00 \ \cancel{\text{mL}})$$

$$M_2 = 1.00 \times 10^{-3}\ M$$

This sample contained the original quantity of dissolved tin and had been diluted to 500.0 mL.

$$0.5000 \ \cancel{\text{L}} \times \frac{1.00 \times 10^{-3} \text{ mol}}{1 \ \cancel{\text{L}}} = 5.00 \times 10^{-4} \text{ mol, choice B}$$

**5. B.** The grade of Ostalloy may be determined by the mass percent of tin. We know that the original unknown sample had a mass of 0.500 gram. The mass percent may be determined by converting the number of moles of tin to grams of tin.

$$5.00 \times 10^{-4} \text{ mol} \times \frac{118.71 \text{ g}}{1 \text{ mol}} = 0.0594 \text{ g}$$

Without a calculator, the mass can be estimated.

$$(5 \times 10^{-4}) \times (1.2 \times 10^{2}) = 6 \times 10^{-2} = 0.06 \text{ g}$$

Division of the estimated mass of tin by the mass of the sample gives the mass percent.

$$\frac{0.06 \text{ g tin}}{0.500 \text{ g sample}} \times \frac{2}{2} = \frac{0.12}{1} = \frac{12}{100} = 12\%$$

The alloy is Ostalloy No. 136, choice B, which is composed of 12.0% tin by mass.

**6. C.** Breaking up the wet solid to allow the interior to be exposed to the atmosphere will decrease the likelihood that trapped water will cause the measured mass to be artificially high. Therefore, choice C is the least likely to introduce an error in the mass calculation. Neglecting to wash the sample (choice A), failing to add an excess of $Na_2CO_3$ (choice B), and washing the sample with too much $H_2O$ (choice D) are all likely to lead to substantial errors in the mass calculation.

**7. D.** Despite the compound's luster, its inability to conduct electricity in the solid state rules out the possibility of a metal. The low melting point rules out the possibility of ionic bonding. The solubility of the compound in $CCl_4$, a nonpolar solvent, provides evidence that the bond is nonpolar covalent, choice D.

**8. D.** The mass that is driven off may be found by subtracting the mass of the crucible and sample after heating from the mass of the crucible and sample before heating. (Notice that we do not need the mass of the crucible alone.)

$$9.16\text{ g} - 7.92\text{ g} = 1.24\text{ g}$$

The mass that is driven off of the sample when it is heated is a 1:1 molar ratio of $H_2O$ and $CO_2$. This means that the number of grams driven off is equal to the number of moles times the sum of the molar masses of $H_2O$ and $CO_2$. Let $x$ be the number of moles.

$$\text{mass lost} = x(44.01\text{ g}\cdot\text{mol}^{-1}) + x(18.02\text{ g}\cdot\text{mol}^{-1}) = 1.24\text{ g}$$

$$x(44.01 + 18.02) = 1.24$$

$$x = \frac{1.24}{62.03} = 0.0200\text{ mol H}_2\text{O} = 0.0200\text{ mol CO}_2$$

From the stoichiometry of the reaction, we see that for every mole of $H_2O$ or $CO_2$ lost, two moles of $NaHCO_3$ have reacted, so there must have been 0.0400 mol of $NaHCO_3$ present before heating, choice D.

**9. A.** The reaction in trial II should have proceeded at a faster rate than in trial I because the powdered form of $CaCO_3$ has a higher surface area than a single chip of $CaCO_3$. The frequency of successful collisions between HCl molecules and the higher surface area reactant is greater because more $CaCO_3$ formula units are exposed to the solution. For this reason, the reaction reaches completion at a higher rate, indicated by the graph in choice A.

**10. B.** When the layers settle, the top layer contains the beta carotene and the bottom layer contains the vitamin $B_5$, choice B. Because isooctane is less dense than water, it forms the top layer. Isooctane is a hydrocarbon, so beta carotene is likely to be soluble in isooctane, while vitamin $B_5$, which has polar- and hydrogen-bonding functional groups, is not.

## PERIODIC TABLE OF THE ELEMENTS

| 1 | 2 | 3 | 4 | 5 | 6 | 7 | 8 | 9 | 10 | 11 | 12 | 13 | 14 | 15 | 16 | 17 | 18 |
|---|---|---|---|---|---|---|---|---|---|---|---|---|---|---|---|---|---|
| 1<br>**H**<br>1.008 | | | | | | | | | | | | | | | | | 2<br>**He**<br>4.00 |
| 3<br>**Li**<br>6.94 | 4<br>**Be**<br>9.01 | | | | | | | | | | | 5<br>**B**<br>10.81 | 6<br>**C**<br>12.01 | 7<br>**N**<br>14.01 | 8<br>**O**<br>16.00 | 9<br>**F**<br>19.00 | 10<br>**Ne**<br>20.18 |
| 11<br>**Na**<br>22.99 | 12<br>**Mg**<br>24.30 | | | | | | | | | | | 13<br>**Al**<br>26.98 | 14<br>**Si**<br>28.09 | 15<br>**P**<br>30.97 | 16<br>**S**<br>32.06 | 17<br>**Cl**<br>35.45 | 18<br>**Ar**<br>39.95 |
| 19<br>**K**<br>39.10 | 20<br>**Ca**<br>40.08 | 21<br>**Sc**<br>44.96 | 22<br>**Ti**<br>47.90 | 23<br>**V**<br>50.94 | 24<br>**Cr**<br>52.00 | 25<br>**Mn**<br>54.94 | 26<br>**Fe**<br>55.85 | 27<br>**Co**<br>58.93 | 28<br>**Ni**<br>58.69 | 29<br>**Cu**<br>63.55 | 30<br>**Zn**<br>65.39 | 31<br>**Ga**<br>69.72 | 32<br>**Ge**<br>72.59 | 33<br>**As**<br>74.92 | 34<br>**Se**<br>78.96 | 35<br>**Br**<br>79.90 | 36<br>**Kr**<br>83.80 |
| 37<br>**Rb**<br>85.47 | 38<br>**Sr**<br>87.62 | 39<br>**Y**<br>88.91 | 40<br>**Zr**<br>91.22 | 41<br>**Nb**<br>92.91 | 42<br>**Mo**<br>95.94 | 43<br>**Tc**<br>(98) | 44<br>**Ru**<br>101.1 | 45<br>**Rh**<br>102.91 | 46<br>**Pd**<br>106.42 | 47<br>**Ag**<br>107.87 | 48<br>**Cd**<br>112.41 | 49<br>**In**<br>114.82 | 50<br>**Sn**<br>118.71 | 51<br>**Sb**<br>121.75 | 52<br>**Te**<br>127.60 | 53<br>**I**<br>126.91 | 54<br>**Xe**<br>131.29 |
| 55<br>**Cs**<br>132.91 | 56<br>**Ba**<br>137.33 | 57<br>***La**<br>138.91 | 72<br>**Hf**<br>178.49 | 73<br>**Ta**<br>180.95 | 74<br>**W**<br>183.85 | 75<br>**Re**<br>186.21 | 76<br>**Os**<br>190.2 | 77<br>**Ir**<br>192.2 | 78<br>**Pt**<br>195.08 | 79<br>**Au**<br>196.97 | 80<br>**Hg**<br>200.59 | 81<br>**Tl**<br>204.38 | 82<br>**Pb**<br>207.2 | 83<br>**Bi**<br>208.98 | 84<br>**Po**<br>(209) | 85<br>**At**<br>(210) | 86<br>**Rn**<br>(222) |
| 87<br>**Fr**<br>(223) | 88<br>**Ra**<br>226.02 | 89<br>**†Ac**<br>227.03 | 104<br>**Rf**<br>(261) | 105<br>**Db**<br>(262) | 106<br>**Sg**<br>(266) | 107<br>**Bh**<br>(264) | 108<br>**Hs**<br>(277) | 109<br>**Mt**<br>(268) | 110<br>**Ds**<br>(271) | 111<br>**Rg**<br>(272) | | | | | | | |

| | | | | | | | | | | | | | | |
|---|---|---|---|---|---|---|---|---|---|---|---|---|---|---|
| *Lanthanide Series | 58<br>**Ce**<br>140.12 | 59<br>**Pr**<br>140.91 | 60<br>**Nd**<br>144.24 | 61<br>**Pm**<br>(145) | 62<br>**Sm**<br>150.4 | 63<br>**Eu**<br>151.97 | 64<br>**Gd**<br>157.25 | 65<br>**Tb**<br>158.93 | 66<br>**Dy**<br>162.50 | 67<br>**Ho**<br>164.93 | 68<br>**Er**<br>167.26 | 69<br>**Tm**<br>168.93 | 70<br>**Yb**<br>173.04 | 71<br>**Lu**<br>174.97 |
| †Actinide Series | 90<br>**Tn**<br>232.04 | 91<br>**Pa**<br>231.04 | 92<br>**U**<br>238.03 | 93<br>**Np**<br>(237) | 94<br>**Pu**<br>(244) | 95<br>**Am**<br>(243) | 96<br>**Cm**<br>(247) | 97<br>**Bk**<br>(247) | 98<br>**Cf**<br>(251) | 99<br>**Es**<br>(252) | 100<br>**Fm**<br>(257) | 101<br>**Md**<br>(258) | 102<br>**No**<br>(259) | 103<br>**Lr**<br>(262) |

# AP Chemistry Equations and Constants

Throughout the exam the following symbols have the definitions specified unless otherwise noted.

| | | | | | |
|---|---|---|---|---|---|
| L, mL | = | liter(s), milliliter(s) | mm Hg | = | millimeters of mercury |
| g | = | gram(s) | J, kJ | = | joule(s), kilojoule(s) |
| nm | = | nanometer(s) | V | = | volt(s) |
| atm | = | atmosphere(s) | mol | = | mole(s) |

## Atomic Structure

$E = hv$

$c = \lambda v$

$E$ = energy

$v$ = frequency

$\lambda$ = wavelength

Planck's constant, $h = 6.626 \times 10^{-34}$ J s

Speed of light, $c = 2.998 \times 10^{8}$ m s$^{-1}$

Avogadro's number $= 6.022 \times 10^{23}$ mol$^{-1}$

Electron charge, $e = -1.602 \times 10^{-19}$ coulomb

## Equilibrium

$$K_c = \frac{[C]^c[D]^d}{[A]^a[B]^b}, \text{ where } a\,A + b\,B \rightleftharpoons c\,C + d\,D$$

$$K_p = \frac{(P_C)^c(P_D)^d}{(P_A)^a(P_B)^b}$$

$$K_a = \frac{[H^+][A^-]}{HA}$$

$$K_b = \frac{[OH^-][HB^+]}{[B]}$$

$$K_w = [H^+][OH^-] = 1.0 \times 10^{-14} \text{ at } 25\ °C$$
$$= K_a \times K_b$$

$pH = -\log[H^+]$, $pOH = -\log[OH^-]$

$14 = pH + pOH$

$$pII - pK_a + \log\frac{[A^-]}{[HA]}$$

$pK_a = -\log K_a$, $pK_b = -\log K_b$

**Equilibrium Constants:**

$K_c$ molar concentrations

$K_p$ gas pressures

$K_a$ weak acid

$K_b$ weak base

$K_w$ water

## Kinetics

$[A]_t - [A]_0 = -kt$

$\ln[A]_t - \ln[A]_0 = -kt$

$$\frac{1}{[A]_t} - \frac{1}{[A]_0} = kt$$

$$t_{1/2} = \frac{0.693}{k}$$

$k$ = rate constant

$t$ = time

$t_{1/2}$ = half-life

| Gases, Liquids, and Solutions | |
|---|---|
| $PV = nRT$<br>$P_A = P_{total} \times X_A$, where $X_A = \frac{\text{moles A}}{\text{total moles}}$<br>$P_{total} = P_A + P_B + P_C + \ldots$<br>$n = \frac{m}{M}$<br>$K = {}^\circ C + 273$<br>$D = \frac{m}{v}$<br>$KE_{molecule} = \frac{1}{2}mv^2$<br>Molarity, $M$ = moles of solute per liter of solution<br>$A = \varepsilon bc$ | $P$ = pressure<br>$V$ = volume<br>$T$ = temperature<br>$n$ = number of moles<br>$m$ = mass<br>$M$ = molar mass<br>$D$ = density<br>$KE$ = kinetic energy<br>$v$ = velocity<br>$A$ = absorbance<br>$\varepsilon$ = molar absorptivity<br>$b$ = path length<br>$c$ = concentration<br><br>Gas constant, $R = 8.134$ J mol$^{-1}$ K$^{-1}$<br>$= 0.08206$ L atm mol$^{-1}$ K$^{-1}$<br>$= 62.36$ L torr mol$^{-1}$ K$^{-1}$<br><br>1 atm = 760 mm Hg = 760 torr<br>STP = 273.15 K and 1.0 atm<br>Ideal gas at STP = 22.4 L · mol$^{-1}$ |
| **Thermodynamics/Electrochemistry** | |
| $q = mc\Delta T$<br>$\Delta S^\circ = \Sigma S^\circ$ products $- \Sigma S^\circ$ reactants<br>$\Delta H^\circ = \Sigma H^\circ_f$ products $- \Sigma H^\circ_f$ reactants<br>$\Delta G^\circ = \Sigma G^\circ_f$ products $- \Sigma G^\circ_f$ reactants<br>$\Delta G^\circ = \Delta H^\circ - T\Delta S^\circ$<br>$= -RT \ln K$<br>$= -nFE^\circ$<br>$I = \frac{q}{t}$<br>$E_{cell} = E^\circ_{cell} - \frac{RT}{nF}\ln Q$ | $q$ = heat<br>$m$ = mass<br>$c$ = specific heat capacity<br>$T$ = temperature<br>$S^\circ$ = standard entropy<br>$H^\circ$ = standard enthalpy<br>$G^\circ$ = standard Gibbs free energy<br>$n$ = number of moles<br>$E^\circ$ = standard reduction potential<br>$I$ = current (amperes)<br>$q$ = charge (coulombs)<br>$t$ = time (seconds)<br>Q = reaction quotient<br><br>Faraday's constant, $F$ = 96,485 coulombs per mole of electrons<br><br>1 volt = $\frac{1 \text{ joule}}{1 \text{ coulomb}}$ |

# Practice Exam 1 Answer Sheet for Section I

| | | |
|---|---|---|
| 1 Ⓐ Ⓑ Ⓒ Ⓓ | 21 Ⓐ Ⓑ Ⓒ Ⓓ | 41 Ⓐ Ⓑ Ⓒ Ⓓ |
| 2 Ⓐ Ⓑ Ⓒ Ⓓ | 22 Ⓐ Ⓑ Ⓒ Ⓓ | 42 Ⓐ Ⓑ Ⓒ Ⓓ |
| 3 Ⓐ Ⓑ Ⓒ Ⓓ | 23 Ⓐ Ⓑ Ⓒ Ⓓ | 43 Ⓐ Ⓑ Ⓒ Ⓓ |
| 4 Ⓐ Ⓑ Ⓒ Ⓓ | 24 Ⓐ Ⓑ Ⓒ Ⓓ | 44 Ⓐ Ⓑ Ⓒ Ⓓ |
| 5 Ⓐ Ⓑ Ⓒ Ⓓ | 25 Ⓐ Ⓑ Ⓒ Ⓓ | 45 Ⓐ Ⓑ Ⓒ Ⓓ |
| 6 Ⓐ Ⓑ Ⓒ Ⓓ | 26 Ⓐ Ⓑ Ⓒ Ⓓ | 46 Ⓐ Ⓑ Ⓒ Ⓓ |
| 7 Ⓐ Ⓑ Ⓒ Ⓓ | 27 Ⓐ Ⓑ Ⓒ Ⓓ | 47 Ⓐ Ⓑ Ⓒ Ⓓ |
| 8 Ⓐ Ⓑ Ⓒ Ⓓ | 28 Ⓐ Ⓑ Ⓒ Ⓓ | 48 Ⓐ Ⓑ Ⓒ Ⓓ |
| 9 Ⓐ Ⓑ Ⓒ Ⓓ | 29 Ⓐ Ⓑ Ⓒ Ⓓ | 49 Ⓐ Ⓑ Ⓒ Ⓓ |
| 10 Ⓐ Ⓑ Ⓒ Ⓓ | 30 Ⓐ Ⓑ Ⓒ Ⓓ | 50 Ⓐ Ⓑ Ⓒ Ⓓ |
| 11 Ⓐ Ⓑ Ⓒ Ⓓ | 31 Ⓐ Ⓑ Ⓒ Ⓓ | 51 Ⓐ Ⓑ Ⓒ Ⓓ |
| 12 Ⓐ Ⓑ Ⓒ Ⓓ | 32 Ⓐ Ⓑ Ⓒ Ⓓ | 52 Ⓐ Ⓑ Ⓒ Ⓓ |
| 13 Ⓐ Ⓑ Ⓒ Ⓓ | 33 Ⓐ Ⓑ Ⓒ Ⓓ | 53 Ⓐ Ⓑ Ⓒ Ⓓ |
| 14 Ⓐ Ⓑ Ⓒ Ⓓ | 34 Ⓐ Ⓑ Ⓒ Ⓓ | 54 Ⓐ Ⓑ Ⓒ Ⓓ |
| 15 Ⓐ Ⓑ Ⓒ Ⓓ | 35 Ⓐ Ⓑ Ⓒ Ⓓ | 55 Ⓐ Ⓑ Ⓒ Ⓓ |
| 16 Ⓐ Ⓑ Ⓒ Ⓓ | 36 Ⓐ Ⓑ Ⓒ Ⓓ | 56 Ⓐ Ⓑ Ⓒ Ⓓ |
| 17 Ⓐ Ⓑ Ⓒ Ⓓ | 37 Ⓐ Ⓑ Ⓒ Ⓓ | 57 Ⓐ Ⓑ Ⓒ Ⓓ |
| 18 Ⓐ Ⓑ Ⓒ Ⓓ | 38 Ⓐ Ⓑ Ⓒ Ⓓ | 58 Ⓐ Ⓑ Ⓒ Ⓓ |
| 19 Ⓐ Ⓑ Ⓒ Ⓓ | 39 Ⓐ Ⓑ Ⓒ Ⓓ | 59 Ⓐ Ⓑ Ⓒ Ⓓ |
| 20 Ⓐ Ⓑ Ⓒ Ⓓ | 40 Ⓐ Ⓑ Ⓒ Ⓓ | 60 Ⓐ Ⓑ Ⓒ Ⓓ |

# Practice Exam 2 Answer Sheet for Section I

| | | |
|---|---|---|
| 1 Ⓐ Ⓑ Ⓒ Ⓓ | 21 Ⓐ Ⓑ Ⓒ Ⓓ | 41 Ⓐ Ⓑ Ⓒ Ⓓ |
| 2 Ⓐ Ⓑ Ⓒ Ⓓ | 22 Ⓐ Ⓑ Ⓒ Ⓓ | 42 Ⓐ Ⓑ Ⓒ Ⓓ |
| 3 Ⓐ Ⓑ Ⓒ Ⓓ | 23 Ⓐ Ⓑ Ⓒ Ⓓ | 43 Ⓐ Ⓑ Ⓒ Ⓓ |
| 4 Ⓐ Ⓑ Ⓒ Ⓓ | 24 Ⓐ Ⓑ Ⓒ Ⓓ | 44 Ⓐ Ⓑ Ⓒ Ⓓ |
| 5 Ⓐ Ⓑ Ⓒ Ⓓ | 25 Ⓐ Ⓑ Ⓒ Ⓓ | 45 Ⓐ Ⓑ Ⓒ Ⓓ |
| 6 Ⓐ Ⓑ Ⓒ Ⓓ | 26 Ⓐ Ⓑ Ⓒ Ⓓ | 46 Ⓐ Ⓑ Ⓒ Ⓓ |
| 7 Ⓐ Ⓑ Ⓒ Ⓓ | 27 Ⓐ Ⓑ Ⓒ Ⓓ | 47 Ⓐ Ⓑ Ⓒ Ⓓ |
| 8 Ⓐ Ⓑ Ⓒ Ⓓ | 28 Ⓐ Ⓑ Ⓒ Ⓓ | 48 Ⓐ Ⓑ Ⓒ Ⓓ |
| 9 Ⓐ Ⓑ Ⓒ Ⓓ | 29 Ⓐ Ⓑ Ⓒ Ⓓ | 49 Ⓐ Ⓑ Ⓒ Ⓓ |
| 10 Ⓐ Ⓑ Ⓒ Ⓓ | 30 Ⓐ Ⓑ Ⓒ Ⓓ | 50 Ⓐ Ⓑ Ⓒ Ⓓ |
| 11 Ⓐ Ⓑ Ⓒ Ⓓ | 31 Ⓐ Ⓑ Ⓒ Ⓓ | 51 Ⓐ Ⓑ Ⓒ Ⓓ |
| 12 Ⓐ Ⓑ Ⓒ Ⓓ | 32 Ⓐ Ⓑ Ⓒ Ⓓ | 52 Ⓐ Ⓑ Ⓒ Ⓓ |
| 13 Ⓐ Ⓑ Ⓒ Ⓓ | 33 Ⓐ Ⓑ Ⓒ Ⓓ | 53 Ⓐ Ⓑ Ⓒ Ⓓ |
| 14 Ⓐ Ⓑ Ⓒ Ⓓ | 34 Ⓐ Ⓑ Ⓒ Ⓓ | 54 Ⓐ Ⓑ Ⓒ Ⓓ |
| 15 Ⓐ Ⓑ Ⓒ Ⓓ | 35 Ⓐ Ⓑ Ⓒ Ⓓ | 55 Ⓐ Ⓑ Ⓒ Ⓓ |
| 16 Ⓐ Ⓑ Ⓒ Ⓓ | 36 Ⓐ Ⓑ Ⓒ Ⓓ | 56 Ⓐ Ⓑ Ⓒ Ⓓ |
| 17 Ⓐ Ⓑ Ⓒ Ⓓ | 37 Ⓐ Ⓑ Ⓒ Ⓓ | 57 Ⓐ Ⓑ Ⓒ Ⓓ |
| 18 Ⓐ Ⓑ Ⓒ Ⓓ | 38 Ⓐ Ⓑ Ⓒ Ⓓ | 58 Ⓐ Ⓑ Ⓒ Ⓓ |
| 19 Ⓐ Ⓑ Ⓒ Ⓓ | 39 Ⓐ Ⓑ Ⓒ Ⓓ | 59 Ⓐ Ⓑ Ⓒ Ⓓ |
| 20 Ⓐ Ⓑ Ⓒ Ⓓ | 40 Ⓐ Ⓑ Ⓒ Ⓓ | 60 Ⓐ Ⓑ Ⓒ Ⓓ |

# Chapter 11

# Practice Exam 1

## Section I: Multiple Choice

**60 questions**

**90 minutes**

Calculators are not allowed for Section I.

**Note:** For all questions, assume the temperature is 298 K, the pressure is 1.0 atm, and solutions are aqueous unless otherwise specified.

**Directions:** Each of the questions or incomplete statements below is followed by four suggested answers or completions. Select the one that is best in each case.

**1.** Which choice best describes the reaction represented by the particulate diagram?

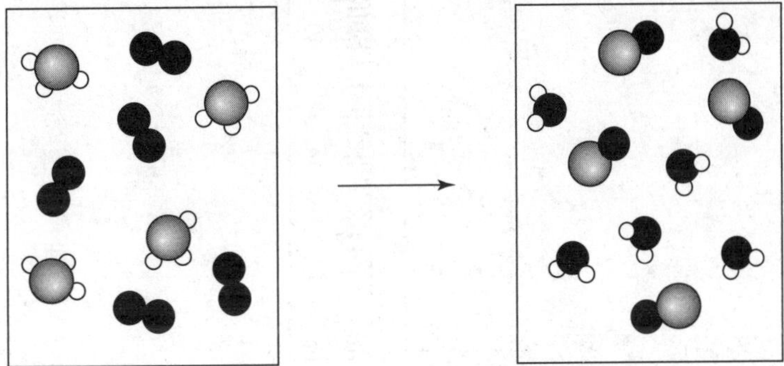

**A.** Phosphorus trichloride and oxygen react to form phosphite and chlorine monoxide.
**B.** Ammonia is oxidized to form nitrogen monoxide and water.
**C.** Methane, the simplest hydrocarbon, undergoes combustion to form carbon monoxide and water.
**D.** Chlorine reacts with hydronium to form chlorine dioxide and hydroxide.

**2.** Which properties are expected to be exhibited by solid potassium chloride?

**A.** It melts at a temperature that is lower than the melting point of pure potassium.
**B.** It is an electrical conductor.
**C.** It dissociates into molecules in an aqueous solution.
**D.** It has a higher melting point than solid potassium bromide.

*Questions 3–5 refer to the experiment described below.*

A student researcher performs an oxidation of borneol, $C_{10}H_{17}OH$, by treating it with $CrO_3$ and $H_2SO_4$ to produce camphor, $C_{10}H_{16}O$.

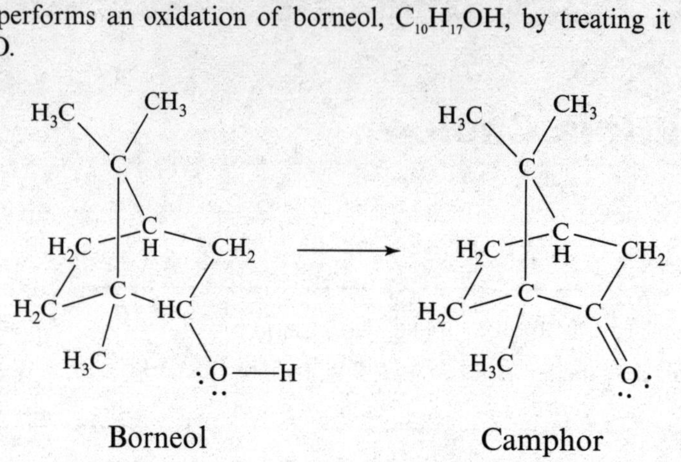

Borneol　　　　Camphor

The paper chromatograph below shows pure borneol in lane 1 and the reaction mixture in lane 2. Chloroform, $CHCl_3$, was used as the mobile phase.

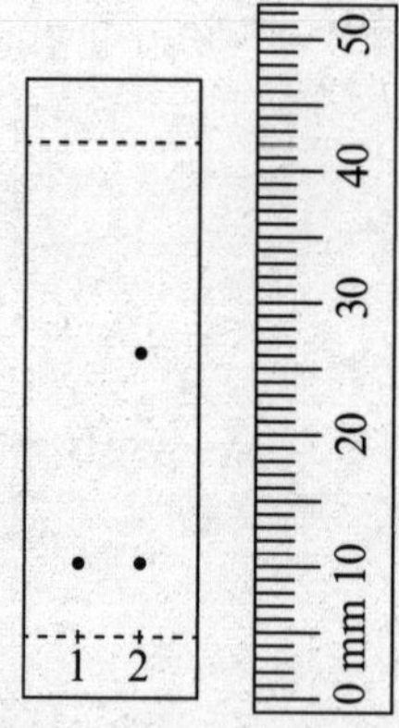


**3.** What can be concluded from this analysis?

**A.** Borneol experiences stronger intermolecular attractions with the mobile phase than camphor.
**B.** Borneol experiences weaker intermolecular attractions with the mobile phase than camphor.
**C.** Borneol bonds covalently with the stationary phase and does not move with the mobile phase.
**D.** Camphor bonds covalently with the stationary phase and does not move with the mobile phase.

**4.** Which choice correctly matches each compound with its $R_f$ in the chromatograph?

| | Borneol | Camphor |
|---|---|---|
| **A.** | 0.24 | 0.62 |
| **B.** | 0.62 | 0.24 |
| **C.** | 0.14 | 0.57 |
| **D.** | 0.57 | 0.14 |

**5.** Which statement about this experiment is correct?

**A.** The camphor was produced in a 50% yield.
**B.** The camphor was produced in a 100% yield.
**C.** The yield of the camphor was less than 100%.
**D.** No conclusion may be drawn about the reaction yield.

**6.** Which of the following correctly ranks the gaseous compounds in order of decreasing standard molar entropy?

A. $HCl > SiCl_4 > CF_4$
B. $CF_4 > SiCl_4 > HCl$
C. $HCl > CF_4 > SiCl_4$
D. $SiCl_4 > CF_4 > HCl$

**7.** Which of the following statements concerning interstitial alloys is correct?

A. The atomic radii of the elements in the mixture are similar, and the electron sea model of bonding serves to explain the conductivity and malleability of the alloy.
B. The atomic radii of the elements in the mixture are different, and the electron sea model of bonding serves to explain the conductivity and malleability of the alloy.
C. The atomic radii of the elements in the mixture are similar, and ionic bonding serves to explain the conductivity and malleability of the alloy.
D. The atomic radii of the elements in the mixture are similar, and ionic bonding serves to explain the conductivity and malleability of the alloy.

**8.** Use the data below to determine the enthalpy of reaction for the reaction of aluminum metal with chlorine gas.

$$2\ Al(s) + 3\ Cl_2(g) \rightarrow 2\ AlCl_3(s) \qquad \Delta H^\circ = ?$$

| Reaction | $\Delta H^\circ$ |
|---|---|
| $2\ Al(s) + 6\ HCl(aq) \rightarrow 2\ AlCl_3(aq) + 3\ H_2(g)$ | $w$ |
| $HCl(g) \rightarrow HCl(aq)$ | $x$ |
| $H_2(g) + Cl_2(g) \rightarrow 2\ HCl(g)$ | $y$ |
| $AlCl_3(s) \rightarrow AlCl_3(aq)$ | $z$ |

A. $w + 6x + 3y - 2z$
B. $-w - 6x - 3y + 2z$
C. $w + 3x + 2y - 3z$
D. $-w - 3x - 2y + 3z$

*Questions 9–10 refer to the complete, simulated photoelectron spectrum of a 1:1 mixture of two pure neutral elements, X and Y, shown below.*

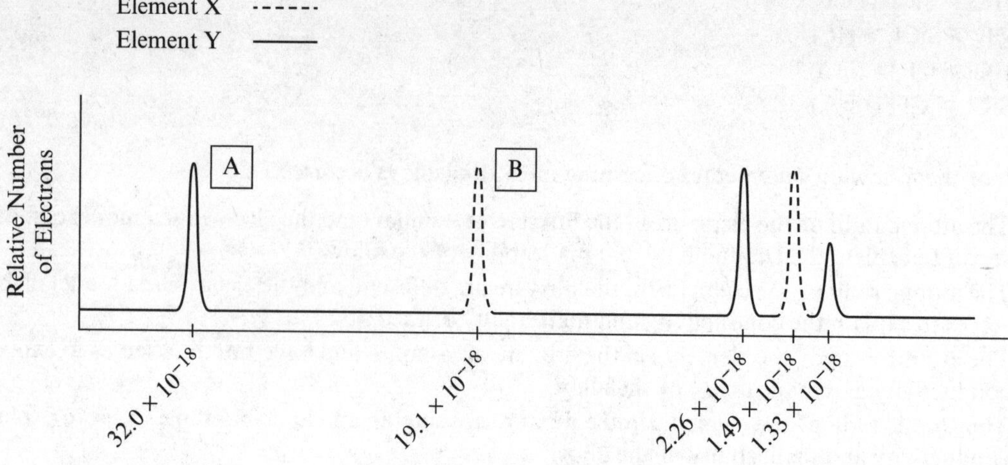


**9.** Why does peak A in the spectrum of element Y have a higher binding energy than peak B in the spectrum of element X?

**A.** The removal of an electron from a *2s* subshell requires more energy than the removal of the electron from a *2p* subshell.
**B.** The effective nuclear charge of element Y is greater than that of element X.
**C.** The electron-electron repulsions in the *s* sublevel of element X are greater than those of element Y.
**D.** A *2p* sublevel is higher in energy than a *1s* sublevel.

**10.** What wavelength of electromagnetic radiation is required to remove a valence electron from element X?

**A.** $6.21 \times 10^{-9}$ m
**B.** $1.04 \times 10^{-8}$ m
**C.** $1.33 \times 10^{-7}$ m
**D.** $1.49 \times 10^{-7}$ m

*Use the information below to answer questions 11–12.*

A 260-gram sample of chromium, Cr, is allowed to react completely with a 256-gram sample of sulfur, $S_8$.

$$16\ Cr + 3\ S_8 \rightarrow 8\ Cr_2S_3$$

| Compound | Molar Mass (g/mol) |
|---|---|
| $S_8$ | 256.48 |
| $Cr_2S_3$ | 200.18 |

**11.** Which species is the limiting reagent?

**A.** Sulfur is the limiting reagent because there are fewer grams of sulfur than there are grams of chromium.
**B.** Sulfur is the limiting reagent because chromium remains when the sulfur is completely consumed.
**C.** Chromium is the limiting reagent because fewer moles of product can be produced from the complete reaction of the chromium than can be produced from the complete consumption of the sulfur.
**D.** Neither reagent is limiting. When the reaction goes to completion, both reactants are entirely consumed.

**12.** If 400. grams of $Cr_2S_3$ are obtained, what is the percent yield of the reaction?

A. 60%
B. 75%
C. 80%
D. 100%

*Questions 13–15 refer to the following reaction.*

$$H_2AsO_4^-(aq) + H_2O(l) \rightleftharpoons HAsO_4^{2-}(aq) + H_3O^+(aq) \qquad K_a = 9 \times 10^{-8}$$

**13.** What is the approximate pH of a 1.00 $M$ solution of $H_2AsO_4^-$?

A. 1.5
B. 3.5
C. 5.5
D. 7.5

**14.** Which species are acting as Brønsted-Lowry acids in the reaction?

A. $H_2AsO_4^-$ and $H_2O$
B. $H_2O$ and $HAsO_4^{2-}$
C. $H_2O$ and $H_3O^+$
D. $H_2AsO_4^-$ and $H_3O^+$

**15.** Which species in the reaction has the largest $K_b$?

A. $H_2AsO_4^-$
B. $H_2O$
C. $HAsO_4^{2-}$
D. $H_3O^+$

**16.** Carbon dioxide from the atmosphere dissolves at the surface of a body of water and forms carbonic acid.

$$CO_2(g) + H_2O(l) \rightleftharpoons H_2CO_3(aq)$$

The carbonic acid is a weak, diprotic acid that dissociates in water according to the following equilibrium equations.

$$H_2CO_3(aq) + H_2O(l) \rightleftharpoons HCO_3^-(aq) + H_3O^+(aq)$$
$$HCO_3^-(aq) + H_2O(l) \rightleftharpoons CO_3^{2-}(aq) + H_3O^+(aq)$$

Which of the following processes are expected to decrease the pH of the body of water?

A. A decrease in the partial pressure of $CO_2$ in the atmosphere
B. A decrease in the partial pressure of $N_2$ in the atmosphere
C. A decrease in the temperature of the water
D. An increase in the dissolved carbonate concentration

**17.** The specific heat capacities for three metals are shown. A 1.00-gram sample of each metal is heated to 50 °C and added to 20.0 grams of water at 25 °C.

Rank the metals in order of lowest to highest temperature after thermal equilibrium is reached.

| Metal | Heat Capacity (J/(g·K)) |
|---|---|
| Al | 0.897 |
| Pb | 0.123 |
| Pt | 0.133 |

**A.** Al < Pt < Pb
**B.** Pt < Pb < Al
**C.** Al < Pb < Pt
**D.** Pb < Pt < Al

*Questions 18–19 refer to the following reaction.*

$$CH_3CH_2NH_2(g) \xrightarrow{\Delta} C_2H_4(g) + NH_3(g) \qquad \Delta H^\circ_{rxn} = 55\ \text{kJ·mol}^{-1}$$

**18.** Which of the following plots best represents the relationship between $\Delta G°$ and temperature for this reaction?

**A.**

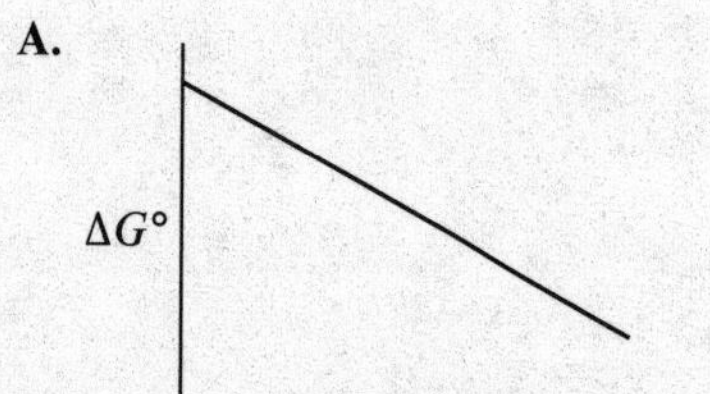


**B.**

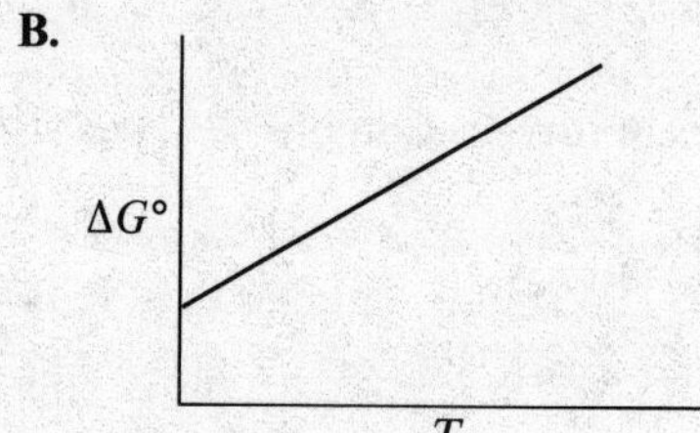


**C.**

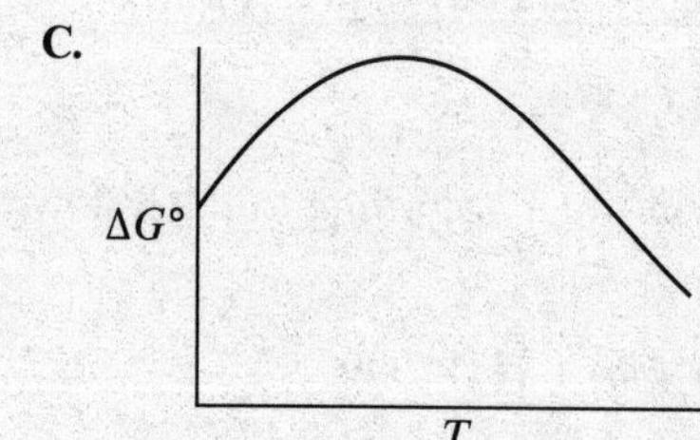


**D.**

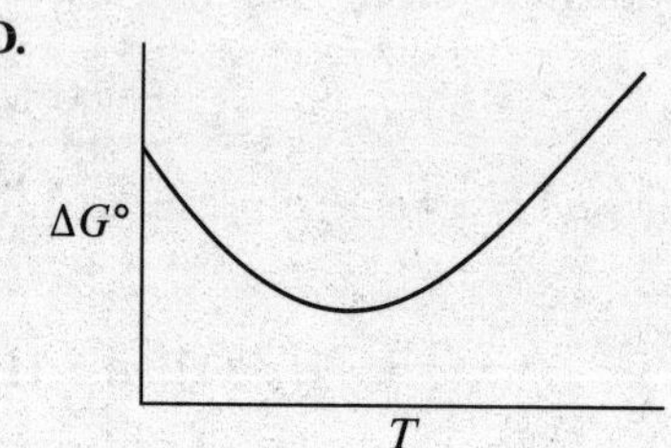


**19.** What is the approximate energy of a mole of C–N single bonds?

| Bond | Bond Energy (kJ/mol) |
|---|---|
| C–H | 412 |
| C–C | 347 |
| C=C | 611 |
| N–H | 389 |

**A.** 296 kJ
**B.** 379 kJ
**C.** 708 kJ
**D.** 1120 kJ

*Use the following equation for the combustion of propane to answer questions 20–21.*

$$1\ C_3H_8 + \_\ O_2 \rightarrow \_\ CO_2 + \_\ H_2O$$
$$\Delta H_{rxn} = -2.0 \times 10^3\ \text{kJ/mol}_{rxn}$$

**20.** Enough propane is burned in oxygen to release 12,000 kJ of heat. What volume does the $CO_2$ that is produced occupy at STP?

A. 0.13 L
B. 0.80 L
C. 400 L
D. 1.3 L

**21.** Which of the following sketches best represents the potential energy diagram for the reaction?

A.
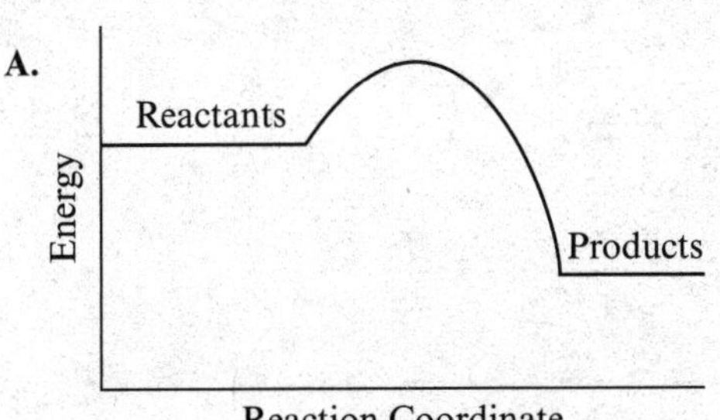


B.
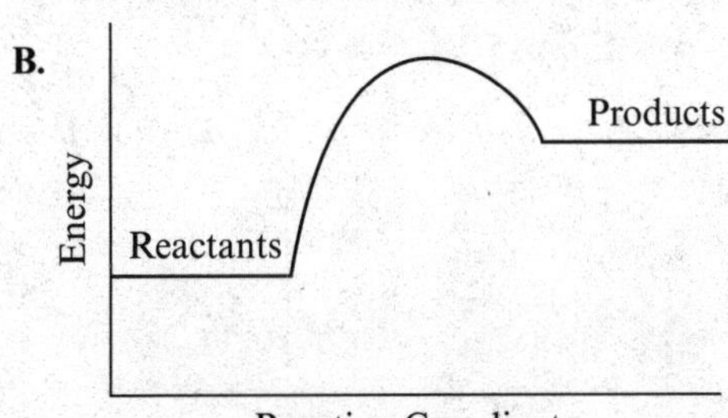


C.
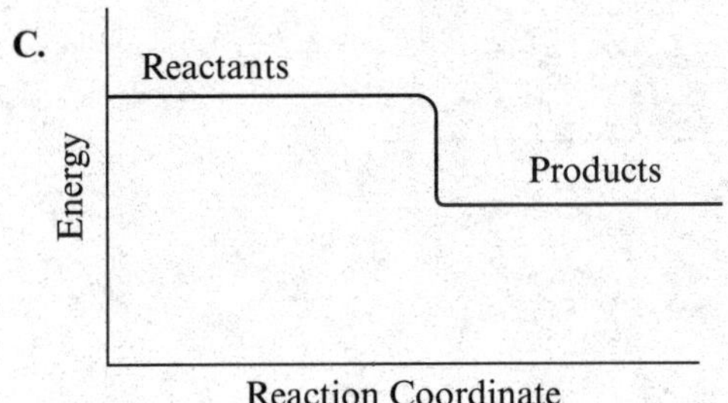


D.
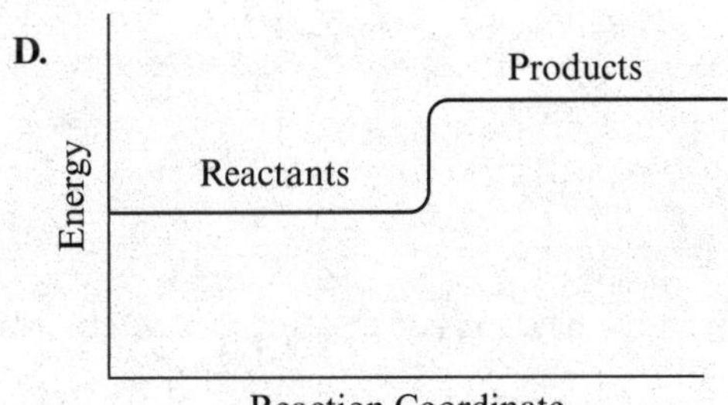


**22.** A sample of gas has a density of 2.00 g/L at 268 K and 1 atm. What is the identity of the gas?

A. $NH_3$
B. $N_2$
C. $CO_2$
D. $NO_2$

**23.** Which choice correctly shows the compounds in order of decreasing C–O bond length?

A. $H_3COH > (CH_3)_2CO > CH_3CO_2^-$
B. $(CH_3)_2CO > CH_3CO_2^- > H_3COH$
C. $CH_3CO_2^- > (CH_3)_2CO > H_3COH$
D. $H_3COH > CH_3CO_2^- > (CH_3)_2CO$

**24.** A cast iron skillet can be heated over an open flame without undergoing vigorous oxidation, but when powdered iron is exposed to a flame, it rapidly ignites to form $Fe_2O_3$. Which is the best explanation for this observation?

**A.** The activation energy for the reaction of the powdered iron is greater than that of the skillet.
**B.** The impurities in the powdered iron act as a catalyst for the combustion.
**C.** The greater surface area of the powdered iron compared to the skillet leads to an increased rate of oxidation.
**D.** The enthalpy change of the reaction of powdered iron is more negative than that of the skillet.

**25.** Which of the following correctly ranks the species in order of increasing radius?

**A.** $S^{2-} < Cl^- < Na^+ < Mg^{2+}$
**B.** $Cl^- < S^{2-} < Mg^{2+} < Na^+$
**C.** $Na^+ < Mg^{2+} < S^{2-} < Cl^-$
**D.** $Mg^{2+} < Na^+ < Cl^- < S^{2-}$

*Use the mass spectrum of chlorine, shown below, to answer questions 26–27.*

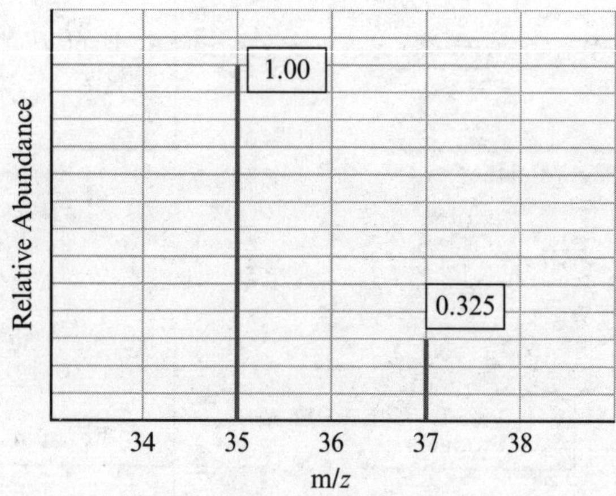


**26.** How many neutrons do the isotopes of chlorine contain?

**A.** 37 and 35
**B.** 20 and 18
**C.** Both isotopes contain 36 neutrons.
**D.** Both isotopes contain 17 neutrons.

**27.** What is the approximate mass of the least abundant naturally occurring chlorine molecule?

**A.** 37 amu
**B.** 70 amu
**C.** 72 amu
**D.** 74 amu

**28.** The plot below shows the first ionization energies for the elements with atomic numbers 5 to 9 (boron through fluorine). Which explanation best accounts for the deviation of the first ionization energy of oxygen from the overall ionization energy trend?

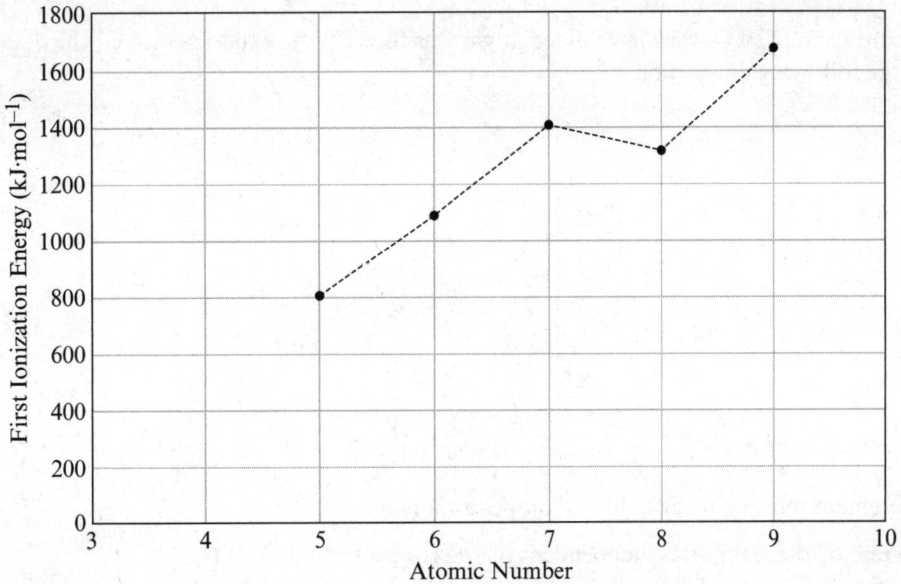


**A.** The valence electrons of oxygen experience greater shielding of the nucleus by core electrons than the valence electrons of nitrogen.
**B.** A repulsive interaction exists between the electrons in the doubly occupied oxygen 2*p* orbital.
**C.** The effective nuclear charge of oxygen is greater than that of nitrogen.
**D.** The atomic radius of oxygen is greater than that of fluorine, so the coulombic attraction between the valence electrons and the nucleus of oxygen is less.

**29.** What is the geometry of the $IF_2^-$ ion?

**A.** Linear, with a bond angle of 180°
**B.** Bent, with a bond angle greater than 120°
**C.** Bent, with a bond angle between 109.5° and 120°
**D.** Bent, with a bond angle less than 109.5°

**30.** The decomposition of dinitrogen pentoxide forms nitrogen dioxide and oxygen according to the following equation:

$$2\ N_2O_5(g) \rightarrow 4\ NO_2(g) + O_2(g)$$

The concentration of $N_2O_5$ was determined at various times in the experiment, and the data were used to generate the following linear plot.

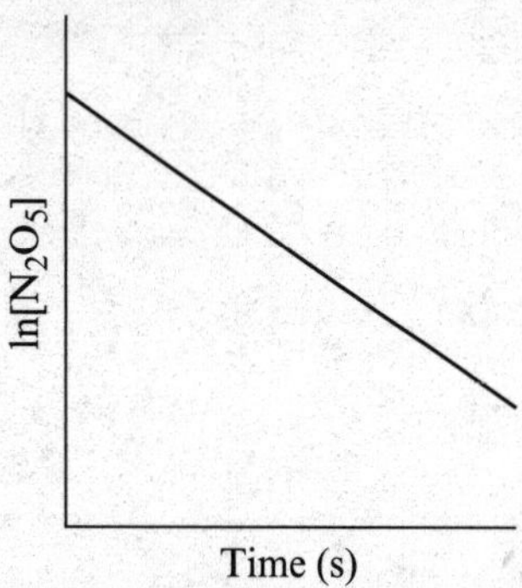


Which statement must be true of this decomposition reaction?

**A.** The rate of the reaction is independent of the concentration of $N_2O_5$.
**B.** The reaction is second order.
**C.** The rate has been accelerated by the addition of an appropriate catalyst.
**D.** The half-life of $N_2O_5$ is independent of its initial concentration.

*Questions 31–32 refer to the reaction below.*

$$SbCl_5(g) \rightleftharpoons SbCl_3(g) + Cl_2(g) \qquad K_c = 43 \text{ at } 373\ K$$

**31.** At a given point in an experiment, a 100-L rigid reaction vessel contains 20 moles of $SbCl_5$, 50 moles of $SbCl_3$, and 60 moles of $Cl_2$ at 373 K. Which of the following describes how the pressure in the vessel will change and why it will change as the reaction approaches equilibrium at constant temperature?

**A.** The pressure will increase because $K_c > Q_c$.
**B.** The pressure will increase because $K_c < Q_c$.
**C.** The pressure will decrease because $K_c > Q_c$.
**D.** The pressure will decrease because $K_c < Q_c$.

**32.** Consider two evacuated 100-L rigid reaction vessels, vessel 1 and vessel 2. Equal amounts of liquid $SbCl_5$ are added to the two vessels at room temperature. A small amount of solid $FeCl_3$ is added to vessel 2. Both vessels are sealed and heated to 373 K. The rates of the reactions are measured and tabulated below.

| Vessel | Time to Reach Equilibrium (seconds) |
|---|---|
| 1 | 1000 |
| 2 | 200 |

Which of the following describes the two systems?

**A.** The activation energy of the reaction in vessel 1 is greater than the activation energy of the reaction in vessel 2, and the total pressure at equilibrium is greater for vessel 1 than for vessel 2.
**B.** The activation energy of the reaction in vessel 2 is greater than the activation energy of the reaction in vessel 1, and the total pressure at equilibrium is greater for vessel 1 than for vessel 2.
**C.** The activation energy of the reaction in vessel 1 is greater than the activation energy of the reaction in vessel 2, and the total pressures at equilibrium are equal in the two vessels.
**D.** The activation energy of the reaction in vessel 2 is greater than the activation energy of the reaction in vessel 1, and the total pressures at equilibrium are equal in the two vessels.

**33.** Which structure contains an atom that is $sp^2$ hybridized?

**A.** $H_3C$–$CH_2$–$CH_2$–$N(=O)$–O

**B.** $H_3C$–$CH_2$–$CH_2$–C≡N

**C.** $H_3C$–$CH_2$–$CH_2$–O–H

**D.** $H_3C$–$CH_2$–$CH_2$–N(H)–H

**34.** The reaction between nitrogen dioxide and fluorine proceeds by the mechanism shown below.

$$2\ NO_2 + F_2 \rightarrow 2\ NO_2F$$

Mechanism:

$NO_2 + F_2 \rightarrow NO_2F + F$ (slow)
$NO_2 + F \rightarrow NO_2F$ (fast)

Which rate law is consistent with this mechanism?

**A.** Rate = $k[NO_2]^2[F_2]$
**B.** Rate = $k[NO_2][F_2]$
**C.** Rate = $k[NO_2]^2[F_2][F]$
**D.** Rate = $k[NO_2][F_2]^2$

**35.** Cyclohexane, $C_6H_{12}$, does not absorb electromagnetic radiation in the ultraviolet region of the spectrum, but benzene, $C_6H_6$, does. Which choice best describes the reason for this observation?

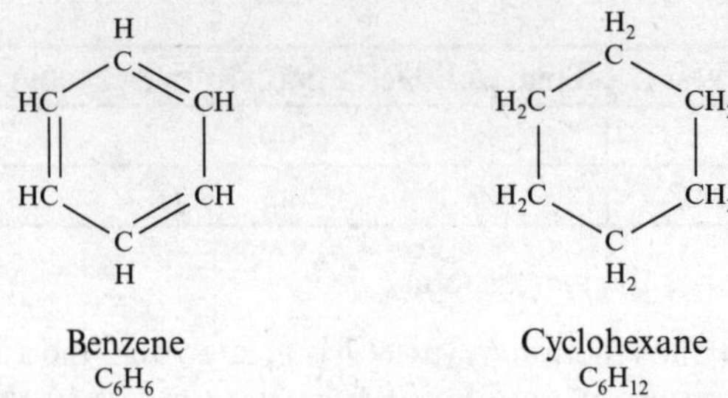


**A.** The $\pi$ electrons in benzene are capable of transitioning to higher-energy states by absorption of UV photons.
**B.** The carbon-hydrogen $\sigma$-bonds in benzene are capable of vibrating at frequencies corresponding to ultraviolet energies.
**C.** The nuclei of benzene are capable of transitioning to higher-energy states when ultraviolet photons are absorbed.
**D.** The $\pi$ C–C bonds in benzene vibrate with a frequency that corresponds to an ultraviolet energy.

*Questions 36–37 refer to the following data.*

Four 1-liter containers at 273 K contain equal numbers of moles of pure gases.

| Container | Gas | Pressure (atm) |
|---|---|---|
| 1 | $H_2$ | 1.0 |
| 2 | $CH_4$ | 1.0 |
| 3 | $CF_4$ | ? |
| 4 | $CCl_4$ | 0.68 |

**36.** The average speed of the gas particles is

**A.** greatest in container 1
**B.** greatest in container 2
**C.** greatest in container 4
**D.** equal in the four containers

**37.** The pressure in container 3 is expected to be

**A.** less than the pressure in container 4 because the C–F bond is more polar than the C–Cl bond
**B.** less than the pressure in container 2 because the C–F bond is more polar than the C–H bond
**C.** less than the pressure in container 4 because $CF_4$ experiences weaker London dispersion forces than $CCl_4$
**D.** less than the pressure in container 2 because $CF_4$ experiences stronger London dispersion forces than $CH_4$

**38.** Which of the following reactions does not involve an oxidation-reduction?

**A.** $NaBr + AgNO_3 \rightarrow AgBr + NaNO_3$
**B.** $2\,MoS_2 + 5\,O_2 \rightarrow 2\,MoO_3 + 4\,SO_2$
**C.** $2\,C_6H_6 + 15\,O_2 \rightarrow 12\,CO_2 + 6\,H_2O$
**D.** $2\,NH_4Cl + 2\,MnO_2 + Zn \rightarrow 2\,NH_3 + Mn_2O_3 + ZnCl_2 + H_2O$

**39.** Which of the following reactions is thermodynamically favored at low temperatures and unfavored at high temperatures?

| | Reaction | $\Delta H°$ (kJ/mol) | $\Delta S°$ (J/(mol·K)) |
|---|---|---|---|
| A. | $2\ N_2(g) + O_2(g) \rightarrow 2\ N_2O(aq)$ | 163 | –148 |
| B. | $2\ SO_2(g) + O_2(g) \rightarrow 2\ SO_3(g)$ | –198 | –187 |
| C. | $4\ H_3PO_4(s) \rightarrow P_4O_{10}(s) + 6\ H_2O(l)$ | 416 | 209 |
| D. | $C_6H_{12}O_6(s) + 6\ O_2(g) \rightarrow 6\ CO_2(g) + 6\ H_2O(g)$ | –2880 | 262 |

**40.** When ammonium chloride is dissolved in water, the temperature of the solution decreases. Which enthalpy change below is the largest?

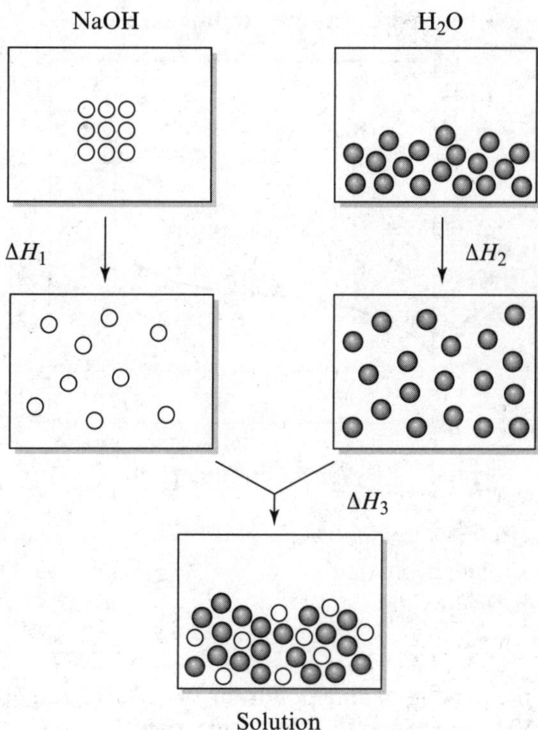


A. $\Delta H_3$
B. $\Delta H_1 + \Delta H_3$
C. $\Delta H_1 + \Delta H_2$
D. $\Delta H_2 + \Delta H_3$

**41.** The table below shows the $K_a$ values for two weak acids. Which of the following pairs of solutions results in a buffer with a pH closest to 7.4?

| Acid | $K_a$ |
|---|---|
| $HClO_2$ | $1.2 \times 10^{-2}$ |
| HOCl | $3.5 \times 10^{-8}$ |

A. 50 mL of 1.00 *M* $HClO_2$; 50 mL of 0.50 *M* NaOH
B. 50 mL of 1.00 *M* NaOCl; 50 mL of 0.50 *M* HCl
C. 50 mL of 1.00 *M* $HClO_2$; 50 mL of 1.00 *M* $NaClO_2$
D. 50 mL of 1.00 *M* HOCl; 50 mL of 1.00 *M* NaOH

**42.** A student sets up the same chemical reaction in two different test tubes and labels them A and B.

Test tube A is allowed to react at room temperature, 27 °C, and test tube B is placed in a 60 °C water bath.

The student measures a larger rate constant for the reaction in test tube B than for test tube A.

Which of the following is the best explanation for this observation?

**A.** The molecules in test tube B collide more frequently with sufficient average kinetic energy to overcome the activation barrier for the reaction.
**B.** More of the colliding particles in test tube B have the proper orientation for the reaction to occur.
**C.** The activation barrier for the reaction in test tube B is lower because of the relatively greater potential energy of the reactants.
**D.** The reactant molecules in test tube B have a lower average velocity so they collide more frequently and have a statistically greater chance of reacting.

**43.** The titration curve shown below best represents the titration of

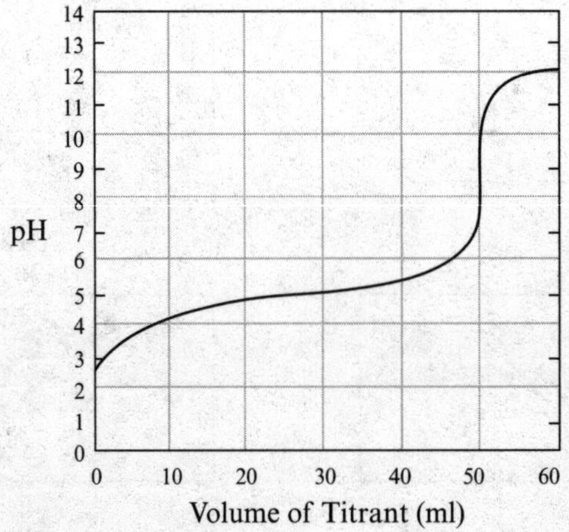


**A.** phosphoric acid with sodium hydroxide
**B.** hydrofluoric acid with sodium hydroxide
**C.** sulfuric acid with sodium hydroxide
**D.** hydrochloric acid with ammonia

**44.** The following graph shows the trend in boiling points for $PH_3$, $AsH_3$, and $SbH_3$. Does point X or point Y reflect the boiling point of $NH_3$, and why is this the correct point?

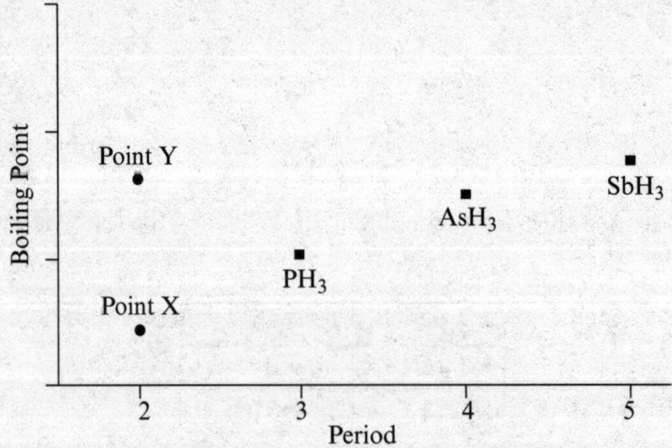


**A.** $NH_3$ boils at point X because it has the lowest polarizability of the group V hydrides.
**B.** $NH_3$ boils at point Y because it has the lowest polarizability of the group V hydrides.
**C.** $NH_3$ boils at point X because it is capable of hydrogen bonding.
**D.** $NH_3$ boils at point Y because it is capable of hydrogen bonding.

*Questions 45–47 refer to the following information.*

Isomers are compounds that have the same molecular formula, but different atomic connectivity. The compounds *n*-hexane and 2,3-dimethylbutane are isomers.

| Name | *n*-hexane | 2,3-dimethylbutane |
|---|---|---|
| Molecular Formula | $C_6H_{14}$ | $C_6H_{14}$ |
| Structural Formula | 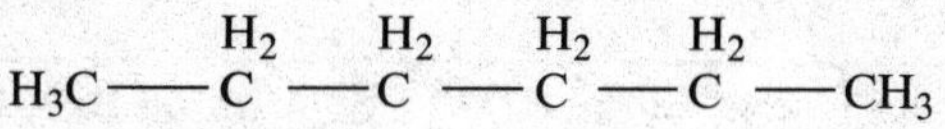 $H_3C-CH_2-CH_2-CH_2-CH_2-CH_3$ | 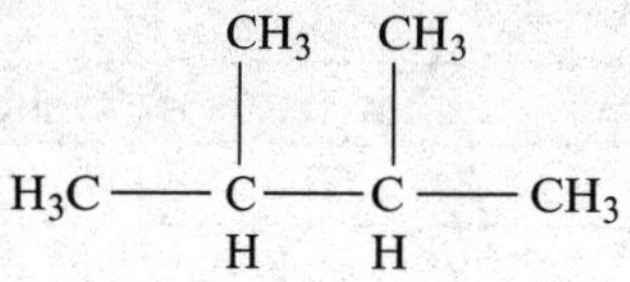 $H_3C-CH(CH_3)-CH(CH_3)-CH_3$ |

**45.** Which of the following compounds can exist as more than one isomer?

A. $CH_4$
B. $C_2H_2$
C. $C_2H_4$
D. $C_3H_6$

**46.** Geometric isomers are compounds that may exhibit isomerism due to restricted bond rotation. Which of the compounds below may exhibit geometric isomerism?

A. $H_2C=CH_2$

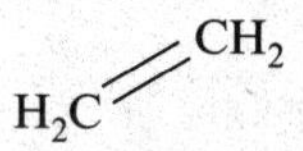

B. $H_2C=CH-CH_3$

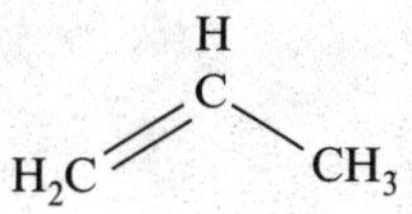

C. $H_3C-CH=CH-CH_3$

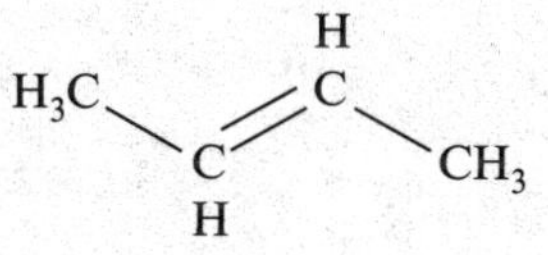

D.

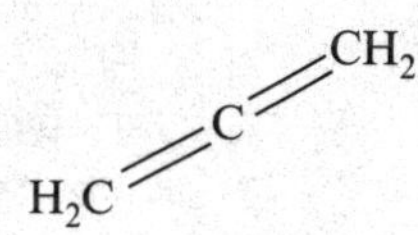


**47.** Which statement concerning isomers is true?

A. Only geometric isomers usually have identical physical and chemical properties.
B. Isomers usually have different chemical and physical properties.
C. Only changes in the number and type of atoms in a compound change the properties of the compound.
D. Neither of the isomers of hexane shown experience London dispersion forces.

**48.** A 5.0-L vessel initially contains 5.0 moles of CO(*g*) and 5.0 moles of $H_2O$(*g*). The following reaction occurs.

$$CO(g) + H_2O(g) \rightleftharpoons CO_2(g) + H_2(g)$$

When equilibrium is reached, there are 4.0 moles of $CO_2$ present. What is the equilibrium constant for the reaction?

A. 0.08
B. 0.8
C. 1.6
D. 16

*Questions 49–50 refer to the following reaction and initial rate data.*

Iodide is oxidized to hypoiodite in a basic solution of hypochlorite.

$$I^-(aq) + ClO^-(aq) \xrightarrow{OH^-} IO^-(aq) + Cl^-(aq)$$

The initial rates of the reaction were measured using various initial concentrations of the reactants.

| Trial | Initial $[I^-]$ ($M$) | Initial $[ClO^-]$ ($M$) | Initial $[OH^-]$ ($M$) | Initial Rate ($M{\cdot}s^{-1}$) |
|---|---|---|---|---|
| 1 | $5.00 \times 10^{-3}$ | $1.00 \times 10^{-2}$ | $5.00 \times 10^{-3}$ | $6.0 \times 10^{-2}$ |
| 2 | $1.00 \times 10^{-2}$ | $5.00 \times 10^{-3}$ | $5.00 \times 10^{-3}$ | $6.0 \times 10^{-2}$ |
| 3 | $5.00 \times 10^{-3}$ | $5.00 \times 10^{-3}$ | $5.00 \times 10^{-3}$ | $3.0 \times 10^{-2}$ |
| 4 | $5.00 \times 10^{-3}$ | $5.00 \times 10^{-3}$ | $1.00 \times 10^{-2}$ | $1.5 \times 10^{-2}$ |

**49.** What is the rate law for the reaction?

**A.** rate = $k[ClO^-][OH^-]$
**B.** rate = $k[I^-][ClO^-]$
**C.** rate = $k[ClO^-][I^-][OH^-]^{-1}$
**D.** rate = $k[ClO^-][I^-][OH^-]$

**50.** What are the units of the rate constant?

**A.** $M^{-1}{\cdot}s^{-1}$
**B.** $M{\cdot}s^{-1}$
**C.** $M^{-2}{\cdot}s^{-1}$
**D.** $s^{-1}$

*Questions 51–52 refer to the heating curve for benzene.*

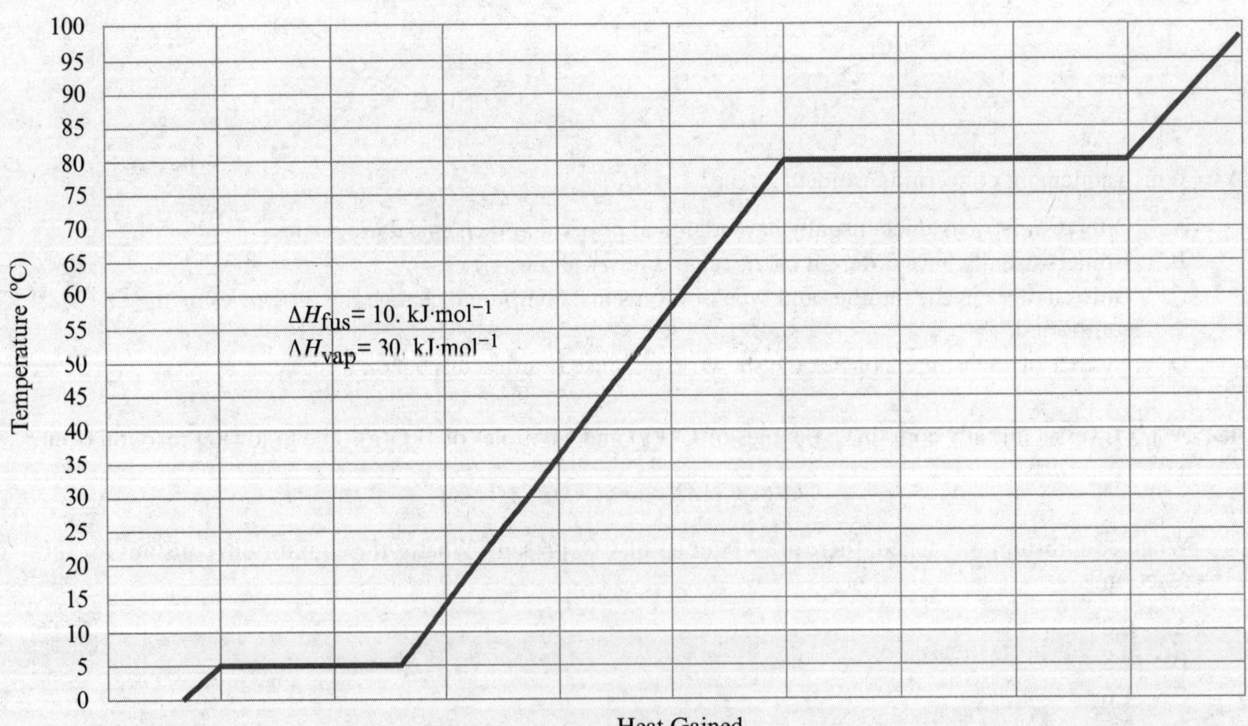

**51.** How much heat is required to convert 156 grams of liquid benzene, $C_6H_6$, at 80 °C to the gaseous state?

A. 10 kJ
B. 20 kJ
C. 30 kJ
D. 60 kJ

**52.** Why is the amount of energy required for the process occurring at 80 °C greater than the amount of energy required for the process occurring at 5 °C?

A. More energy is required to break the chemical bonds in benzene than to overcome intermolecular forces.
B. More energy is required to completely disrupt intermolecular attractions and separate the molecules in space than to partially overcome intermolecular attractions.
C. The heat capacity of solid benzene is greater than the heat capacity of liquid benzene.
D. More energy is required to overcome dipole-dipole interactions than to overcome London dispersion forces.

**53.** Determine whether the electrochemical reaction between tin and iron is thermodynamically favored as written under standard conditions.

$$Sn^{4+} + 2\ Fe^{2+} \rightarrow Sn^{2+} + 2\ Fe^{3+}$$

| Reaction | $E°$ (V) |
|---|---|
| $Sn^{4+} + 2\ e^- \rightarrow Sn^{2+}$ | 0.15 |
| $Fe^{3+} + e^- \rightarrow Fe^{2+}$ | 0.77 |

A. $E°_{cell} = -1.39$ V; the reaction is thermodynamically disfavored as written.
B. $E°_{cell} = -0.62$ V; the reaction is thermodynamically disfavored as written.
C. $E°_{cell} = -1.39$ V; the reaction is thermodynamically favored as written.
D. $E°_{cell} = -0.62$ V; the reaction is thermodynamically favored as written.

**54.** A current of 25.0 A is continuously applied through an aqueous solution of potassium chloride. The reaction that occurs at the cathode is

A. $2\ H_2O + 2\ e^- \rightarrow H_2 + 2\ OH^-$
B. $2\ Cl^- \rightarrow Cl_2 + 2\ e^-$
C. $2\ H_2O \rightarrow O_2 + 4\ H^+ + 4\ e^-$
D. $K^+ + e^- \rightarrow K$

**55.** Which pair of molecules is expected to exhibit the greatest dipole-induced dipole interactions?

A. $Br_2$ and $Cl_2$
B. HCl and HBr
C. HBr and $Br_2$
D. HCl and $Br_2$

**56.** What is the approximate pH of a 0.0100 *M* solution of caffeine? The $K_b$ for caffeine is $4 \times 10^{-4}$.

A. 1.7
B. 2.7
C. 11.3
D. 12.3

*Questions 57–58 refer to the following reaction of barium carbonate.*

Barium carbonate, $BaCO_3$, decomposes according to the reaction represented below.

$$BaCO_3(s) \rightleftharpoons BaO(s) + CO_2(g)$$

At 1500 K, the equilibrium constant, $K_p$, is $6.4 \times 10^{-2}$.

**57.** Some solid $BaCO_3$ is placed in a previously evacuated container and heated to 1500 K. When equilibrium is reached, some of the solid $BaCO_3$ remains. What is the pressure in the container at equilibrium at 1500 K?

- **A.** $2.5 \times 10^{-1}$ atm
- **B.** $1.3 \times 10^{-1}$ atm
- **C.** $6.4 \times 10^{-2}$ atm
- **D.** $4.1 \times 10^{-3}$ atm

**58.** The value of $\Delta G°$ for the decomposition of $BaCO_3$ at 1500 K is

- **A.** positive
- **B.** negative
- **C.** zero
- **D.** unpredictable from the information given

**59.** A plot of potential energy as a function of internuclear distance between two molecules, X and Y, is shown below. Which of the following choices shows the correct curve correspondence, along with the correct reasoning?

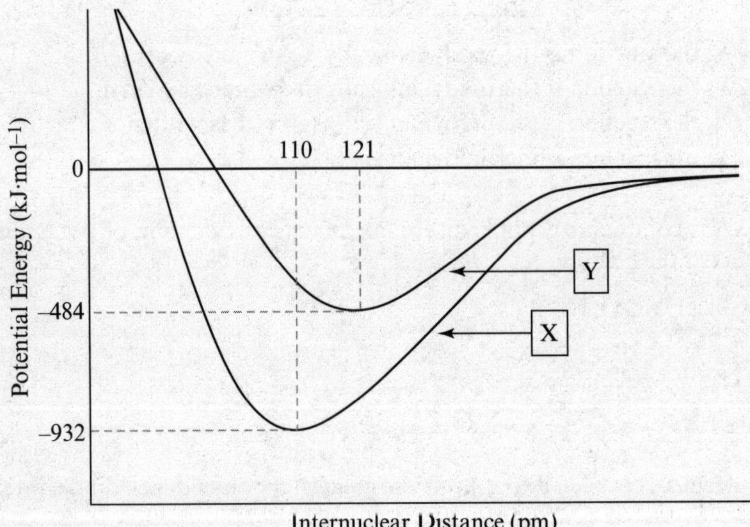


| | X | Y | Reason |
|---|---|---|---|
| **A.** | $N_2$ | $O_2$ | A triple bond is shorter and stronger than a double bond. |
| **B.** | $N_2$ | $O_2$ | A triple bond is longer and weaker than a double bond. |
| **C.** | $O_2$ | $N_2$ | A triple bond is shorter and stronger than a double bond. |
| **D.** | $O_2$ | $N_2$ | A triple bond is longer and weaker than a double bond. |

**60.** Some carboxylic acids form dimers, in which two molecules associate with one another. Which of the following sketches is the most likely structure of an acetic acid dimer?

**A.**

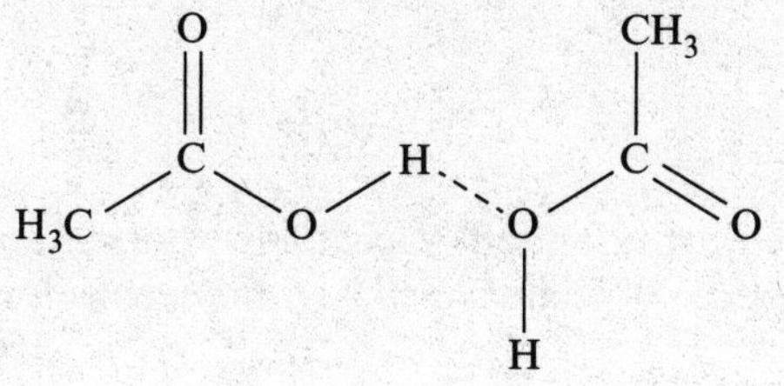

**B.**

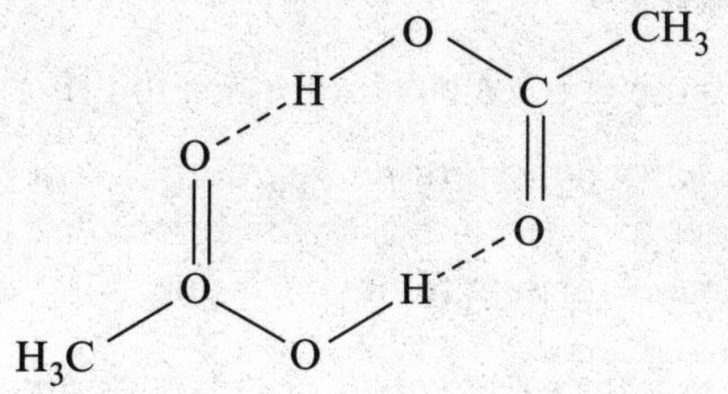

**C.**

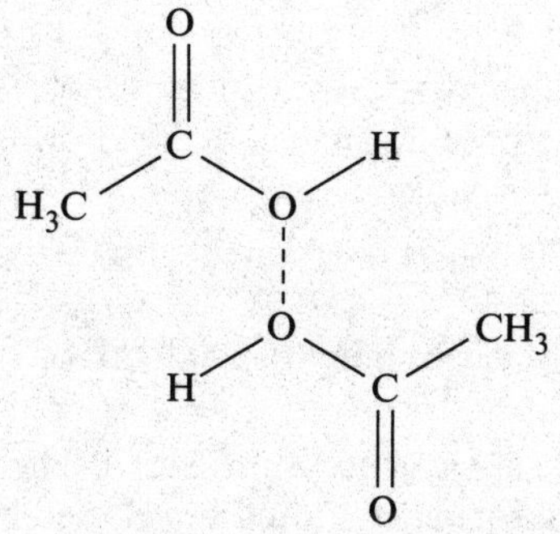

**D.**

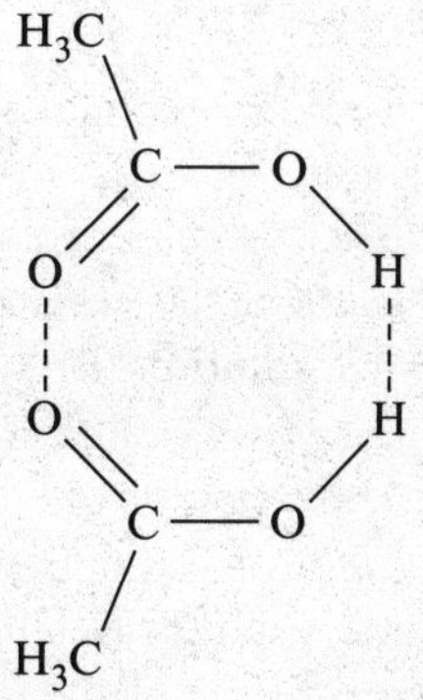

IF YOU FINISH BEFORE TIME IS CALLED, CHECK YOUR WORK ON THIS SECTION ONLY. DO NOT WORK ON ANY OTHER SECTION IN THE TEST.

# Section II: Free Response

**7 questions**

**105 minutes**

You may use your calculator for Section II.

**Directions:** Long free-response questions (questions 1–3) are worth 10 points each; allow yourself approximately 23 minutes for each one. Short free-response questions (questions 4–7) are worth 4 points each; allow yourself approximately 9 minutes for each one.

Include examples and equations in your responses where appropriate. Clearly show the method used and the steps involved in arriving at your answers. You must show your work to receive credit. Be mindful of significant figures.

**1.** **a.** A researcher is studying the effect of bleach on blue food dye, $C_{37}H_{34}N_2O_9S_3Na_2$ (MW = 792.86 g/mol). The researcher weighs 0.200 gram of the solid dye onto a piece of weigh paper. How many moles of dye molecules are present?

**b.** The researcher would like to make a stock solution of the dye by adding distilled water. Should the researcher use a volumetric flask or a graduated cylinder to prepare an accurately known concentration? Briefly explain your answer.

**c.** Water is added to the 0.200-gram sample of dye so that the final volume is 500. mL. What is the molarity of the dye solution? Be sure to show your calculation.

**d.** A 0.50-mL aliquot of the solution is diluted with distilled water to prepare 100.0 mL of stock solution. What is the concentration of the stock solution?

**e.** The stock solution is placed in a spectrophotometer and its absorbance is measured. The following plot of absorbance versus wavelength is generated. What absorbance should be selected in order to analyze the concentration of the food dye? Justify your response by referring to the data.

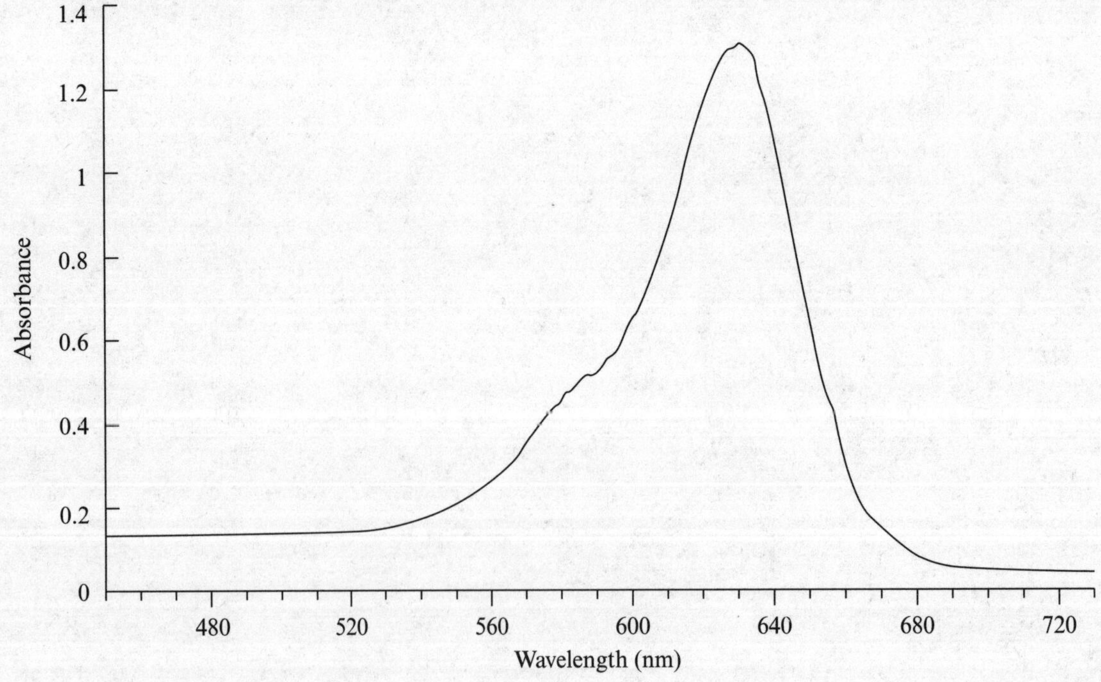

f. A known excess of bleach is added to a $5.0 \times 10^{-6}$ *M* sample of the dye solution. The solution is mixed well and placed in a dry cuvette. The absorbance of the solution is measured as a function of time, and the following data are obtained.

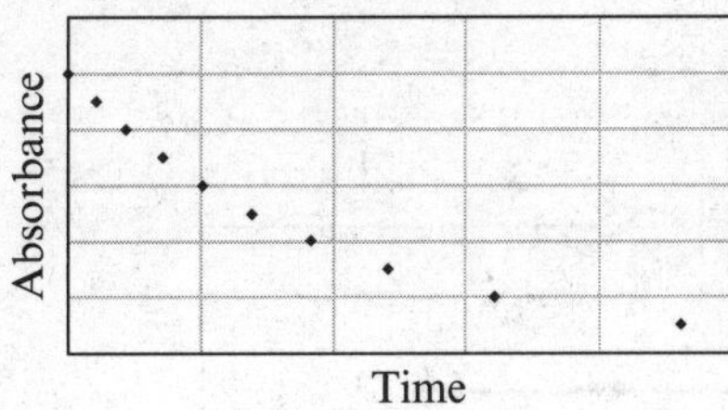


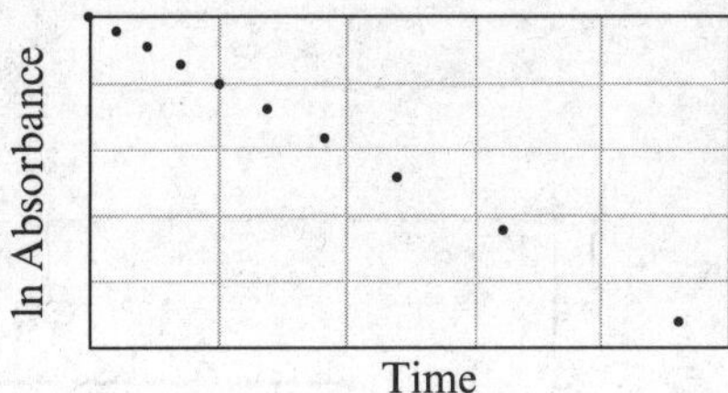


i. What is the order of the reaction with respect to the dye?

ii. When the concentration of dye is held constant and the concentration of bleach is doubled, the rate doubles. What is the rate expression for the reaction of bleach with dye?

**2.** Morphine, $C_{17}H_{19}O_3N$, is a weak base that is used to treat severe pain. The hydroxide ion concentration in a 0.0621 *M* aqueous solution of morphine is $3.17 \times 10^{-4}$ *M* at 25 °C. For the following questions, assume that the temperature remains constant and that the volumes are additive.

a. Write the equation for the dissociation of $C_{17}H_{19}O_3N(aq)$ in water.

b. Write the base dissociation constant expression for the reaction of $C_{17}H_{19}O_3N(aq)$ with water.

c. Determine the pH of the 0.0621 *M* morphine solution.

d. Determine the base dissociation constant, $K_b$, for morphine.

e. Determine the percent ionization of morphine in a 0.0621 *M* solution.

f. A chemist titrates a 25.0-mL sample of the 0.0621 *M* morphine solution with 0.0500 *M* HCl(*aq*).

i. Determine the volume of 0.0500 *M* HCl(*aq*) required to reach the equivalence point.

ii. Determine the pH of the solution at the equivalence point.

g. On the axes below, sketch the titration curve for the titration of morphine with HCl(*aq*). Label the pH values at the beginning of the titration, halfway to the equivalence point, and at the equivalence point. Label the volume of HCl added at the half-equivalence point and at the equivalence point.

**3.** An electrochemical cell is constructed with two cadmium electrodes. The cathode contains a 1.00 *M* solution of $Cd(NO_3)_2$, and the anode contains 1.00 *M* $Na_2S$.

$$CdS(s) + 2\,e^- \rightarrow Cd(s) + S^{2-}(aq) \qquad E° = -1.210\text{ V}$$

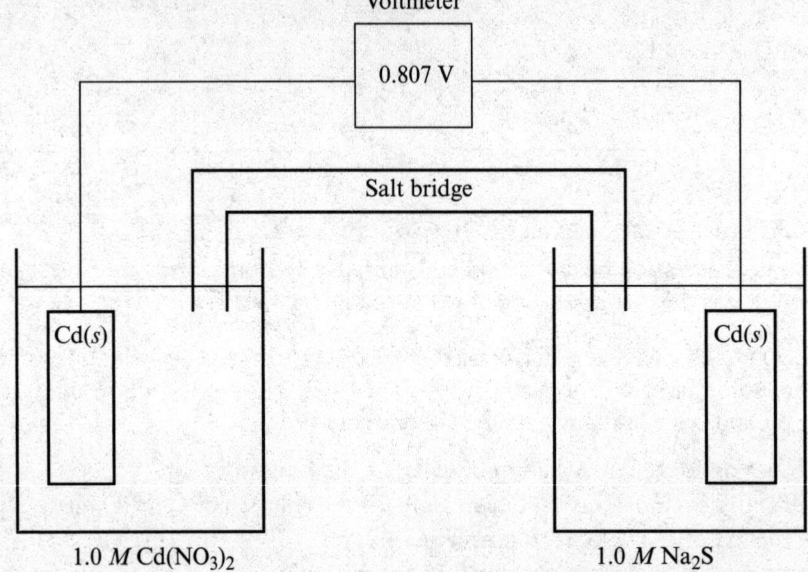


**a.** When the half-cells are connected, an overall potential of 0.807 V is measured. What is the standard reduction potential, *E*°, for the reduction of $Cd^{2+}$?

$$Cd^{2+}(aq) + 2\,e^- \rightarrow Cd(s) \qquad E° = ?\text{ V}$$

**b.** Clearly describe the function of the salt bridge in this cell.

**c.** Write the overall net ionic equation for the electrochemical cell.

**d.** If the cell is assembled with 0.500 *M* $Cd(NO_3)_2$ instead of 1.00 *M* $Cd(NO_3)_2$, what effect would this have on the cell potential? Provide a qualitative explanation of your response.

**e.** What is the standard free energy change, $\Delta G°$, at 25 °C for the overall reaction? Report your answer in units of kJ/mol.

**f.** Calculate the solubility product, $K_{sp}$, at 25 °C for CdS(*s*).

**g.** Solid CdS is added to a 0.100 *M* solution of $Na_2S$ at 25 °C. Calculate the solubility of CdS in 0.100 *M* $Na_2S$. Report your answer in units of mol/L.

**4.** The reaction between solid sodium hydride, NaH, and water produces hydrogen gas and sodium hydroxide. A sample of NaH was added to excess water and the reaction was allowed to proceed until gas production stopped. The gas collected in the cylinder is in thermal equilibrium with the water.

The depth of the cylinder is adjusted so that the water levels inside and outside of the cylinder are the same.

The barometric pressure is 757 torr and room temperature is 24 °C. The vapor pressure of water at 24 °C is 22.5 torr.

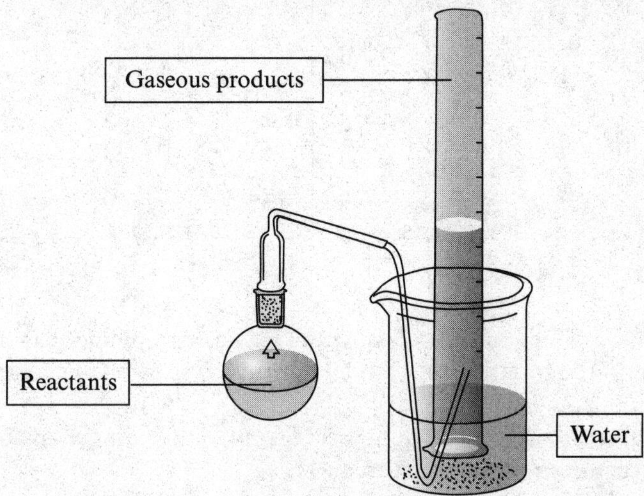


**a.** Write the balanced equation for the gas-forming reaction.

**b.** The collected gas occupies 19.2 mL. How many moles of gas have been collected?

**c.** The collected gas is dried over a desiccant, or drying agent. What volume is occupied by the dry gas at 757 torr? Express your answer in units of mL.

**d.** An equal number of moles of dry HCl gas is collected in a separate experiment. Is the volume occupied by the HCl equal to, less than, or greater than the volume occupied by the hydrogen? Explain your answer.

**5.** Propionic acid and methyl ethanoate are constitutional isomers. They have the same molecular formula, but the atoms have different connectivity.

**a.** Complete the Lewis electron-dot diagrams of propionic acid and methyl ethanoate below. Include any lone (nonbonding) pairs of electrons.

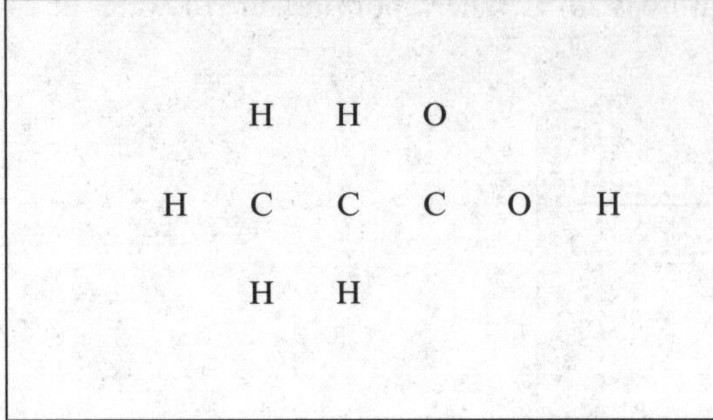


Propionic acid

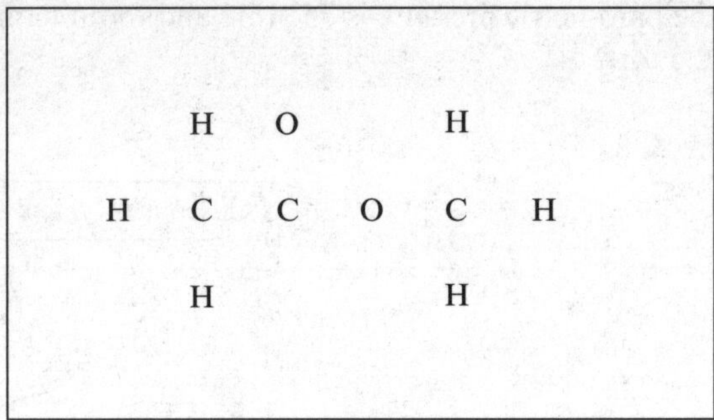


Methyl ethanoate

**b.** Which analytical technique could best distinguish between propionic acid and methyl ethanoate: mass spectrometry or infrared spectroscopy? Explain your answer by referencing the structures of the molecules.

**c.** Which of the compounds is expected to have a higher boiling point? Explain your answer by referencing the structures of the molecules.

**d.** How many sigma bonds and how many pi bonds are present in a molecule of methyl ethanoate? Label each carbon atom in methyl ethanoate as *sp*, $sp^2$, or $sp^3$ hybridized.

**6.** The balanced equation for the complete combustion of acetylene, $C_2H_2$, in oxygen is shown below.

$$C_2H_2(g)+\frac{5}{2}O_2(g)\rightarrow 2\,CO_2(g)+H_2O(g)$$

**a.** Use the data in the table below to determine $\Delta H°$ for the reaction. Report your answer in units of kJ/mol.

| Substance | $\Delta H°_f$ (kJ/mol) |
|---|---|
| $C_2H_2(g)$ | 227.4 |
| $CO_2(g)$ | –393.5 |
| $H_2O(g)$ | –241.8 |

**b.** Calculate the quantity of heat released when 500.0 g of acetylene react completely with oxygen. Report your answer in units of kJ.

**c.** The standard free energy change, $\Delta G°$, for the reaction is –1227 kJ/mol. Estimate the standard entropy change, $\Delta S°$, for the reaction at 298 K. Report your answer in units of J/(mol·K).

**d.** Is the reaction thermodynamically favorable at all temperatures, at high temperatures, at low temperatures, or at no temperature? Justify your response.

**7.** The results of a study of the rate of decomposition of $H_2O_2$ in the presence of a small amount of sodium iodide are shown in the table below.

$$2\ H_2O_2(aq) \xrightarrow{I^-} 2\ H_2O(l) + O_2(g)$$

| Experiment | Initial $[H_2O_2]$ $(mol{\cdot}L^{-1})$ | Initial $[I^-]$ $(mol{\cdot}L^{-1})$ | Initial Rate of Formation of $O_2$ $(mol{\cdot}L^{-1}{\cdot}s^{-1})$ |
|---|---|---|---|
| 1 | $3.7 \times 10^{-2}$ | $6.4 \times 10^{-3}$ | $2.72 \times 10^{8}$ |
| 2 | $1.1 \times 10^{-1}$ | $6.4 \times 10^{-3}$ | $8.16 \times 10^{8}$ |
| 3 | $3.7 \times 10^{-2}$ | $3.2 \times 10^{-3}$ | $1.36 \times 10^{8}$ |

**a.** What is the initial rate of disappearance of $H_2O_2$ in experiment 3?
**b.** Determine the rate law for the iodide-catalyzed decomposition of $H_2O_2$.
**c.** Determine the specific rate constant, *k*, and specify units.
**d.** Is the following mechanism consistent with the experimental results? Explain your response.

$$H_2O_2 + I^- \rightleftharpoons IO^- + H_2O \quad \text{(slow)}$$
$$H_2O_2 + IO^- \rightarrow I^- + H_2O + O_2 \quad \text{(fast)}$$

IF YOU FINISH BEFORE TIME IS CALLED, CHECK YOUR WORK ON THIS SECTION ONLY. DO NOT WORK ON ANY OTHER SECTION IN THE TEST.

STOP

# Answer Key

## Section I: Multiple Choice

| | | | | |
|---|---|---|---|---|
| 1. B | 13. B | 25. D | 37. D | 49. C |
| 2. D | 14. D | 26. B | 38. A | 50. D |
| 3. B | 15. C | 27. D | 39. B | 51. D |
| 4. C | 16. C | 28. B | 40. C | 52. B |
| 5. C | 17. D | 29. A | 41. B | 53. B |
| 6. D | 18. A | 30. D | 42. A | 54. A |
| 7. B | 19. A | 31. A | 43. B | 55. D |
| 8. A | 20. C | 32. C | 44. D | 56. C |
| 9. B | 21. A | 33. A | 45. D | 57. C |
| 10. C | 22. C | 34. B | 46. C | 58. A |
| 11. C | 23. D | 35. A | 47. B | 59. A |
| 12. C | 24. C | 36. A | 48. D | 60. B |

# Answers and Explanations

## Section I: Multiple Choice

**1.** **B.** Ammonia molecules, $NH_3$, are represented by the gray and white spheres on the reactant side of the figure. Oxygen molecules, $O_2$, are represented by the pairs of black spheres. Ammonia is oxidized to form nitrogen monoxide, NO, and water, $H_2O$, choice B. The overall equation can be elucidated by counting the spheres and studying their arrangement.

$$4\,NH_3 + 5\,O_2 \rightarrow 4\,NO + 6\,H_2O$$

The other choices describe reactions that do not occur or that cannot be explained by the figure.

$$PCl_3 + O_2 \neq HPO_3^{2-} + ClO$$
$$CH_4 + 2\,O_2 \rightarrow CO_2 + 2\,H_2O$$
$$Cl_2 + H_3O^+ \neq ClO_2 + OH^-$$

**2.** **D.** A bromide ion has a larger ionic radius than a chloride ion, so the distance between the cation and anion in potassium bromide is greater than the distance in potassium chloride. This leads to a lower coulombic attractive force between potassium and bromide compared to that between potassium and chloride. The lower attractive force results in a lower melting point for potassium bromide, making choice D correct.

**3.** **B.** In paper chromatography, substances that have stronger intermolecular forces with the mobile phase will move up the paper more quickly than substances that experience weaker intermolecular forces with the stationary phase. Since borneol (shown in lane 1) does not move as far up the paper as camphor (shown as the higher spot in lane 2), it experiences weaker intermolecular attractions with the mobile phase than camphor does, choice B.

**4.** **C.** The $R_f$ value for a compound in a particular solvent is calculated by taking the ratio of the distance traveled by the compound to the distance traveled by the solvent.

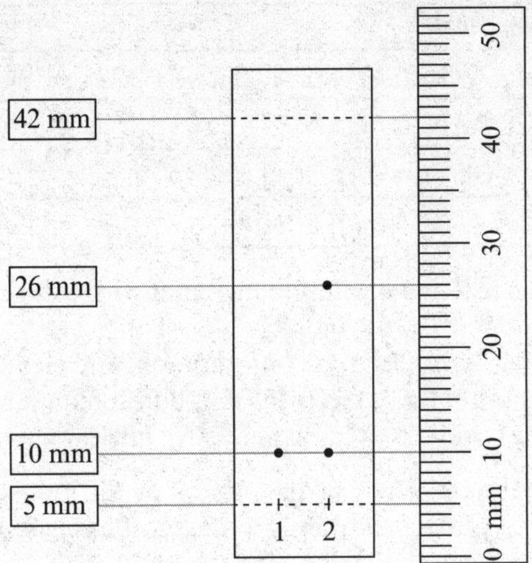


Borneol:

$$\frac{10-5}{42-5}=\frac{5}{37}\approx \text{around } \frac{1}{7} \text{ or } 0.14$$

Camphor:

$$\frac{26-5}{42-5}=\frac{21}{37}\approx\frac{3}{5} \text{ or } 0.6$$

The actual value of the $R_f$ of camphor is 0.57, so this is a good approximation.

$$\frac{21}{37}=0.57$$

Choice C is the correct match.

**5.** **C.** Since borneol was still present in the reaction mixture in lane 2, the reaction was not complete, and the yield was less than 100%, choice C.

**6.** **D.** Generally, the larger the number of electrons in a gaseous compound, the more complex the molecule is, and the higher its standard molar entropy. The $SiCl_4$ has the most electrons and the highest standard molar entropy, followed by $CF_4$. HCl is the smallest molecule, with the lowest standard molar entropy. Choice D correctly ranks the gaseous compounds in order of decreasing standard molar entropy: $SiCl_4 > CF_4 > HCl$.

**7.** **B.** Interstitial alloys are composed of mixtures of primarily metal atoms, such as iron, with smaller atoms such as carbon in the interstices of the crystal lattice. Even though the atomic radii of the elements in the mixture are different, the conductivity and malleability of the alloy can still be understood in terms of the electron sea model of metallic bonding, choice B.

**8.** **A.** The equations can be rearranged to add up to the overall transformation. The $\Delta H°$ values must be changed to reflect the changes made to the equations. The sum of the $\Delta H°$ values is the overall $\Delta H°$ for the reaction.

| Reaction | $\Delta H°$ |
|---|---|
| $2\ Al(s) + 6\ HCl(aq) \rightarrow 2\ AlCl_3(aq) + 3\ H_2(g)$ | $w$ |
| $[HCl(g) \rightarrow HCl(aq)] \times 6$ | $6x$ |
| $[H_2(g) + Cl_2(g) \rightarrow 2\ HCl(g)] \times 3$ | $3y$ |
| $[AlCl_3(s) \rightarrow AlCl_3(aq)] \times (-2)$ | $-2z$ |

The enthalpy of reaction for the reaction of aluminum metal with chlorine gas is $w + 6x + 3y - 2z$, choice A.

**9.** **B.** Element X is beryllium, with a $1s^2 2s^2$ electron configuration, and element Y is boron, with a $1s^2 2s^2 2p^1$ electron configuration. The removal of a $1s$ electron from boron requires more energy than the removal of a $1s$ electron from beryllium since boron has a higher effective nuclear charge, choice B.

**10.** **C.** The wavelength is related to the frequency of the photon of the lowest energy peak in the photoelectron spectrum of element X since that peak corresponds to the most easily removed valence electron of the element.

$$c = \lambda \nu$$

$$\lambda = \frac{c}{\nu}$$

The frequency is related to the energy of the photon.

$$E = h\nu$$

$$\nu = \frac{E}{h}$$

Combining the two equations gives

$$\lambda = \frac{c}{\left(\frac{E}{h}\right)} = \frac{hc}{E}$$

The precise calculation of the wavelength is as follows:

$$\lambda = \frac{(6.626 \times 10^{-34}\ \cancel{J} \cdot \cancel{s})(2.998 \times 10^{8}\ m/\cancel{s})}{1.49 \times 10^{-18}\ \cancel{J}} = 1.33 \times 10^{-7}\ m$$

Without a calculator, the wavelength can be estimated.

$$\lambda = \frac{(6.6 \times 10^{-34})(\cancel{3}^{2} \times 10^{8})}{\cancel{1.5} \times 10^{-18}} = 6.6 \times 2 \times 10^{(-34+8-(-18))} \approx 13 \times 10^{-8} = 1.3 \times 10^{-7}\text{, choice C}$$

**11.** **C.** The limiting reagent is determined by converting each mass of reagent to number of moles of product. The reagent that can produce fewer moles of product is limiting.

$$256\ \cancel{g\ S_8} \times \frac{1\ \cancel{mol\ S_8}}{256\ \cancel{g\ S_8}} \times \frac{8\ mol\ Cr_2S_3}{3\ \cancel{mol\ S_8}} = 2.66\ mol\ Cr_2S_3$$

$$260\ \cancel{g\ Cr} \times \frac{1\ \cancel{mol\ Cr}}{52\ \cancel{g\ Cr}} \times \frac{8\ mol\ Cr_2S_3}{16\ \cancel{mol\ Cr}} = 2.50\ mol\ Cr_2S_3$$

Therefore, chromium is the limiting reagent, choice C.

**12.** C. The actual yield is 400. grams. The theoretical yield is determined from the number of moles of product that can be produced.

$$2.50\ \text{mol } Cr_2S_3 \times \frac{200.18\ \text{g } Cr_2S_3}{1\ \text{mol } Cr_2S_3} = 500.\ \text{g } Cr_2S_3$$

$$\%\ \text{yield} = \frac{\text{actual yield}}{\text{theoretical yield}} \times 100\% = \frac{400}{500} \times 100\% = 80\%,\ \text{choice C}$$

**13.** B. It is clear from the small $K_a$ that $H_2AsO_4^-$ is a very weak acid. The pH of a weak acid solution is formally determined using a RICE table, although students with sufficient practice can perform this calculation without writing out the table.

| **Reaction** | $H_2AsO_4^-$ | $H_3O^+$ | $H_2AsO_4^{2-}$ |
|---|---|---|---|
| **Initial Concentrations ($M$)** | 1.00 | 0 | 0 |
| **Change** | $-x$ | $+x$ | $+x$ |
| **Equilibrium Concentrations ($M$)** | $1.00 - x$ | $x$ | $x$ |

The value of $x$ may be neglected in the denominator. Because the equilibrium constant is so small, subtraction of $x$ from 1.00 does not significantly change the value of the denominator.

$$K_a = \frac{[H_3O^+][HAsO_4^{2-}]}{[H_2AsO_4^-]}$$

$$9 \times 10^{-8} = \frac{x^2}{1.00 - x}$$

The pH of the solution is estimated to be between 3 and 4.

$$\sqrt{(9 \times 10^{-8})} = \sqrt{x^2}$$

$$x = 3 \times 10^{-4}\ M = [H_3O^+]$$

$$\text{pH} = -\log(3 \times 10^{-4}) \approx 3\text{–}4,\ \text{choice B}$$

**14.** D. Brønsted-Lowry acids are proton donors in solution. The species that behave as Brønsted-Lowry acids are $H_2AsO_4^-$ in the reactants and $H_3O^+$ in the products, choice D.

**15.** C. Brønsted-Lowry bases are proton acceptors in solution. The species that behave as bases in the reaction are $HAsO_4^{2-}$ in the products and $H_2O$ in the reactants. Because the acid dissociation constant is small, the equilibrium lies to the left. Because the equilibrium favors the side of the reaction with the weaker acid and base, $H_2O$ is a weaker base than $HAsO_4^{2-}$. Therefore, $HAsO_4^{2-}$ has the largest $K_b$, choice C.

**16.** C. Since the dissolution of a gas in water is an exothermic process, a decrease in the temperature of the water, choice C, increases the solubility of gaseous $CO_2$. This results in an increase in the concentration of $H_2CO_3$, which drives the acid dissociation reaction toward products. The resulting increase in the $H_3O^+$ concentration leads to a decrease in pH.

**17.** D. The substance with the highest heat capacity requires the greatest energy transfer in order to achieve a temperature change. When aluminum is placed in cool water, it transfers more energy to water with a smaller temperature change than the substances with lower heat capacities. Therefore, the metals ranked in order of lowest to highest temperature after thermal equilibrium is reached are shown in choice D: Pb < Pt < Al.

**18.** A. In this reaction, one mole of gas is converted to two moles of gas, so entropy increases. Because $\Delta H°$ and $\Delta S°$ are both positive, as temperature increases, the larger the $T\Delta S°$ term becomes in the equation $\Delta G° = \Delta H° - T\Delta S°$. As $T$ increases, $\Delta G°$ decreases, as plotted in choice A.

**19.** **A.** The difference between the sums of the bond energies of the reactants and the bond energies of the products gives an estimate of $\Delta H^\circ_{rxn}$.

$$\Delta H^\circ_{rxn} = (\Sigma \text{ bonds broken}) - (\Sigma \text{ bonds formed})$$

$$= [\cancel{5}^1(BE_{\text{C-H single}}) + 1(BE_{\text{C-C single}}) + \cancel{2(BE_{\text{N-H}})} + 1(BE_{\text{C-N single}})] - [1(BE_{\text{C-C double}}) + \cancel{4(BE_{\text{C-H single}})} + \cancel{3}^1(BE_{\text{N-H}})]$$

$$55 \text{ kJ·mol}^{-1} = 412 \text{ kJ·mol}^{-1} + 347 \text{ kJ·mol}^{-1} + BE_{\text{C-N single}} - [611 \text{ kJ·mol}^{-1} + 389 \text{ kJ·mol}^{-1}]$$

$$BE_{\text{C-N single}} = 296 \text{ kJ·mol}^{-1}, \text{ choice A}$$

**20.** **C.** The first step in determining the amount of $CO_2$ formed is to balance the chemical equation.

$$1\ C_3H_8 + 5\ O_2 \rightarrow 3\ CO_2 + 4\ H_2O$$

Next, determine the number of moles of $CO_2$ that must be produced to result in the amount of energy that is released.

$$12{,}000\ \cancel{\text{kJ}} \times \frac{3 \text{ mol } CO_2}{2{,}000\ \cancel{\text{kJ}}} = 18 \text{ mol } CO_2$$

Finally, the molar volume of a gas at STP may be used to determine the volume. This value is provided on the *AP Chemistry Equations and Constants* sheet.

$$18\ \cancel{\text{mol } CO_2} \times \frac{22.4 \text{ L}}{1\ \cancel{\text{mole of gas}}} = 400 \text{ L, choice C}$$

**21.** **A.** The negative value of $\Delta H^\circ$ indicates an exothermic process. The potential energy of the reactants is greater than the potential energy of the products. Choices A and C both fulfill this requirement, but only choice A shows the correct activation energy for the reaction.

**22.** **C.** The density, temperature, and pressure of a gas can be used to determine the molar mass of the gas. Three equations, all available on the *AP Chemistry Equations and Constants* sheet, can be combined to solve for the molar mass.

$$D = \frac{m}{V}$$

$$n = \frac{m}{\boldsymbol{M}}$$

$$PV = nRT$$

*D* is density, *m* is mass, *V* is volume, *n* is number of moles, ***M*** is molar mass, *P* is pressure, *R* is the gas constant, and *T* is temperature.

Substitution and rearrangement to solve for the molar mass gives

$$\boldsymbol{M} = \frac{dRT}{P} = \frac{(2.00 \text{ g}/\cancel{\text{L}})(0.08206\ \cancel{\text{L}} \cdot \cancel{\text{atm}} / (\text{mol} \cdot \cancel{\text{K}}))(268\ \cancel{\text{K}})}{1\ \cancel{\text{atm}}} = 44 \text{ g/mol}$$

This is the molar mass of $CO_2$, choice C.

**23.** **D.** Double bonds are shorter than single bonds, so the C–O bond length in $(CH_3)_2CO$, the right-most structure in the figure below, is less than that in $H_3COH$, the left-most structure in the figure. The bond in $CH_3CO_2^-$, a resonance hybrid as shown in brackets in the center of the figure, is of intermediate length. It is longer than a double bond, but shorter than a single bond. The compounds in order of decreasing C–O bond length are shown in choice D: $H_3COH > CH_3CO_2^- > (CH_3)_2CO$.

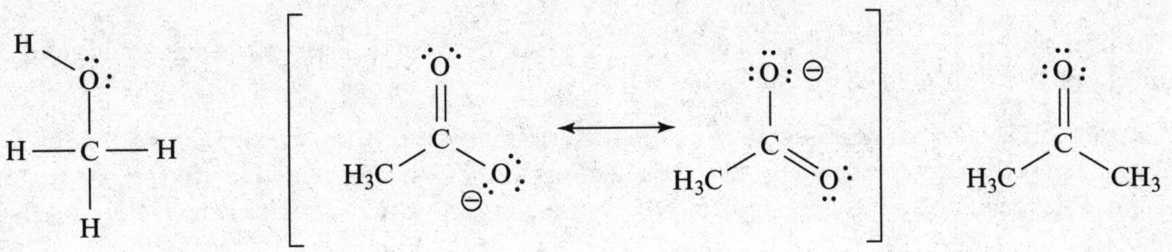

**24.** **C.** Increased surface area of a reactant, choice C, increases its exposure to the other reactant(s), allowing the reaction to proceed at a faster rate.

**25.** **D.** $Na^+$ and $Mg^{2+}$ are isoelectronic with neon. $Mg^{2+}$ has a higher effective nuclear charge than $Na^+$, so it is the smallest of the ions listed. $Cl^-$ and $S^{2-}$ are isoelectronic with argon. $S^{2-}$ has a greater degree of electron-electron repulsion, so it is the largest of the four ions. Choice D correctly ranks the species in order of increasing radius: $Mg^{2+} < Na^+ < Cl^- < S^{2-}$.

**26.** **B.** All isotopes of chlorine contain 17 protons. The isotope with the mass number of 37 must have 20 neutrons and the isotope with the mass number of 35 must have 18 neutrons, choice B.

**27.** **D.** Chlorine is a diatomic molecule, $Cl_2$. The least abundant chlorine molecule is the one in which both chlorine atoms are the least abundant isotope, $^{37}Cl$. The mass of the least abundant chlorine molecule is

$$2 \times 37 \text{ amu} = 74 \text{ amu, choice D}$$

**28.** **B.** The first ionization energy is slightly lower for oxygen than it is for nitrogen due to repulsion between the two electrons in the first doubly occupied *p*-orbital of oxygen, choice B.

**29.** **A.** The $IF_2^-$ ion has a trigonal bipyramidal electronic geometry, with three lone pairs and two bonds, as shown below. This leads to a linear molecular geometry with a 180° bond angle, choice A.

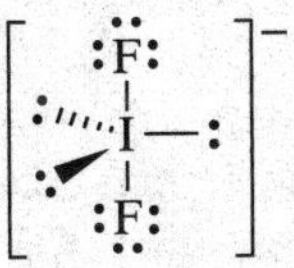

**30.** **D.** The linearity of the plot of time versus $\ln[N_2O_5]$ indicates that the reaction is first order with respect to $N_2O_5$. For any first order process, the half-life depends only on the rate constant, not on the initial concentration of the reactant. Choice D is the true statement.

**31.** **A.** The concentrations of the three species in the 100-L container are

$$[SbCl_5] = \frac{20 \text{ mol}}{100 \text{ L}} = 0.2\ M$$

$$[SbCl_3] = \frac{50 \text{ mol}}{100 \text{ L}} = 0.5\ M$$

$$[Cl_2] = \frac{60 \text{ mol}}{100 \text{ L}} = 0.6\ M$$

The equilibrium quotient can be compared with the equilibrium constant.

$$Q_c = \frac{[SbCl_3][Cl_2]}{[SbCl_5]} = \frac{(0.5)(0.6)}{0.2} = 1.5$$

Since $Q_c < K_c$, the reaction will shift toward products. Because one mole of gaseous reactant becomes two moles of gaseous products, the pressure will increase, choice A.

**32. C.** The solid catalyst increases the rate of the reaction in vessel 2 by decreasing the activation energy of the reaction but the equilibrium constant is unchanged, so the total equilibrium pressures are equal in both vessels, choice C.

**33. A.** The nitrogen structure in choice A has three regions of electron density (two single bonds and one double bond). It is $sp^2$ hybridized.

**34. B.** The rate law is based on the slow step, or rate-determining step, of the mechanism. When the first step of a mechanism is the rate-determining step, the reactants in the first step contribute to the rate law of the reaction. The reactants in the slow step are $NO_2$ and $F_2$. Therefore, choice B is correct: Rate = $k[NO_2][F_2]$.

**35. A.** Absorption of ultraviolet photons leads to electronic transitions in atoms and molecules. The $\pi$ electrons in benzene are capable of transitioning to higher-energy states by absorption of UV photons, choice A.

**36. A.** Because the four gases have equal temperatures, they have equal kinetic energies. For gases of similar kinetic energy, smaller molecular weight gases have greater average speeds. Because $H_2$ has the smallest molecular weight of the choices listed, the average speed of the gas particles is greatest in container 1, choice A.

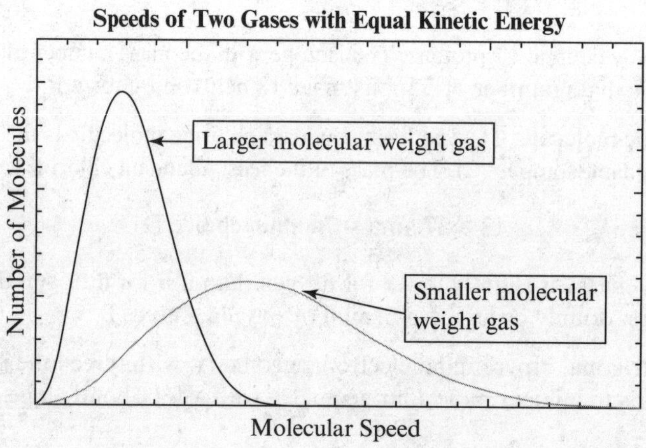


**37. D.** Fluorine atoms have more electrons and are more polarizable than hydrogen atoms, so the London dispersion forces experienced by $CF_4$ are greater than those experienced by $CH_4$. Greater intermolecular attractions decrease the pressure inside the container. Therefore, the pressure in container 3 is expected to be less than the pressure in container 2 because $CF_4$ experiences stronger London dispersion forces than $CH_4$, choice D.

**38. A.** The reaction shown in choice A, $NaBr + AgNO_3 \rightarrow AgBr + NaNO_3$, does not involve an oxidation-reduction. None of the elements in the reaction of NaBr with $AgNO_3$ undergoes a change in oxidation number. This is a double replacement reaction.

**39. B.** A thermodynamically unfavored process has a $\Delta G° > 0$, and a thermodynamically favored process has a $\Delta G° < 0$. As temperature increases, a more negative $T\Delta S°$ value is subtracted from $\Delta H°$ in the equation $\Delta G° = \Delta H° - T\Delta S°$, leading to an overall increase in $\Delta G°$. Because both $\Delta H°$ and $\Delta S°$ are negative in choice B, the resulting $\Delta G°$ becomes more positive as temperature increases. Therefore, choice B is correct.

**40. C.** The decrease in temperature indicates that the dissolution is endothermic. In order for $\Delta H = \Delta H_1 + \Delta H_2 + \Delta H_3$ to be positive overall, the magnitude of the sum of the two endothermic steps, $\Delta H_1 + \Delta H_2$, must be greater than the magnitude of the exothermic step, $\Delta H_3$. Therefore, choice C, $\Delta H_1 + \Delta H_2$, is correct.

**41. B.** A buffer is formed from approximately equal numbers of moles of a weak acid and its conjugate base. When one-half of an equivalent of HCl, a strong acid, is added to $OCl^-$, equal numbers of moles of HOCl and $OCl^-$ result. This produces a buffer solution. The pH of the buffer solution is equal to the $pK_a$ of the acid. For HOCl, the $pK_a$ is found as follows.

$$pK_a = -\log K_a = -\log(3.5 \times 10^{-8}) \approx 7.4$$

The pH of a buffer solution made with 50 mL of 1.00 *M* NaOCl and 50 mL of 0.50 *M* HCl, choice B, is approximately 7.4.

**42.** **A.** Temperature is a measure of average kinetic energy. The higher the average kinetic energy or temperature of a system, the more frequently the molecules collide with sufficient energy to overcome the activation barrier, choice A.

**43.** **B.** The pH at the equivalence point of the titration, when about 50 mL of titrant have been added, is about 9. The high pH at the equivalence point, in addition to the shape of the curve, shows that the titration is of a weak monoprotic acid with a strong base, such as hydrofluoric acid with sodium hydroxide, choice B.

**44.** **D.** The boiling point trend of the group 14 hydrides is generally increasing because as the size and polarizability of the electron clouds increase, the stronger the London dispersion forces are among the molecules. However, ammonia deviates from this general trend and has a much higher boiling point (point Y in the graph) than expected because it is capable of hydrogen bonding, choice D.

**45.** **D.** Of the choices given, only the compound with three carbons can exist as more than one isomer, choice D, $C_3H_6$. It can either contain a double bond or exist as a ring (see below).

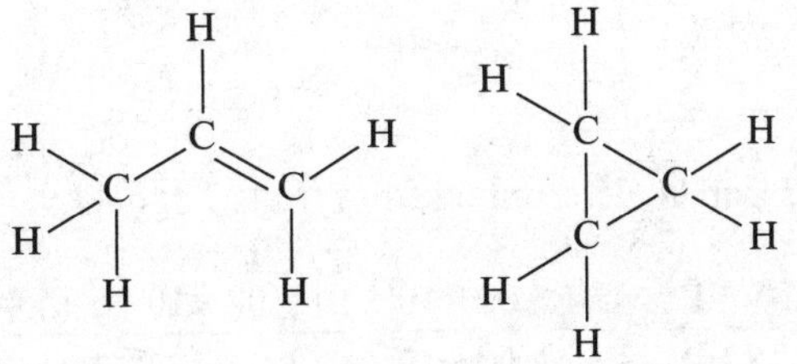

**46.** **C.** Only 2-butene, choice C, can exist as two geometric isomers due to restricted rotation about the π-bond. Rotation about the $\pi$-bond would lead to a different isomer (see below).

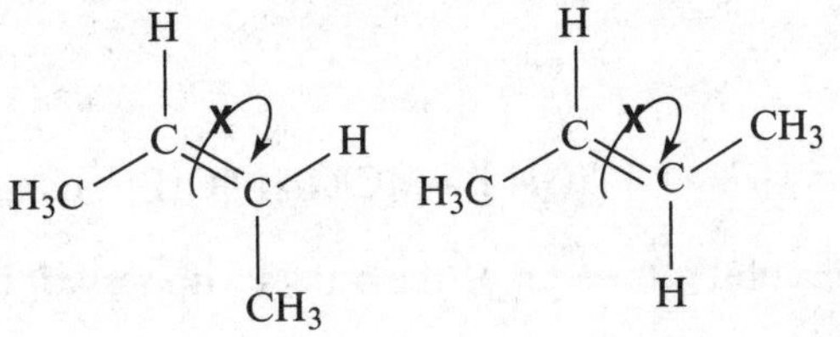

**47.** **B.** Isomers usually have different chemical and physical properties, choice B. For instance, *n*-hexane and 2,3-dimethylbutane have different boiling points because they have different surface areas, which leads to different degrees of intermolecular attractions.

**48.** **D.** The equilibrium concentrations can be determined from the initial concentration of CO and $H_2O$, 1.0 *M*, and from the final concentration of $CO_2$, 0.80 *M*.

| Reaction | CO | $H_2O$ | $CO_2$ | $H_2$ |
|---|---|---|---|---|
| Initial Concentrations (*M*) | 1.0 | 1.0 | 0 | 0 |
| Change | $-x$ | $-x$ | $+x$ | $+x$ |
| Equilibrium Concentrations (*M*) | 0.20 | 0.20 | 0.80 | 0.80 |

The equilibrium constant can then be determined from the equilibrium concentrations.

$$K_c = \frac{[CO_2][H_2]}{[CO][H_2O]} = \frac{(0.8)(0.8)}{(0.2)(0.2)} = 16\text{, choice D}$$

**49.** **C.** The rate law can be determined by comparing the trials.

The order with respect to $I^-$ can be determined from trials 2 and 3.

$$\frac{\text{rate}_2}{\text{rate}_3}=\frac{k[I^-]_2^x[ClO^-]_2^y[OH^-]_2^z}{k[I^-]_3^x[ClO^-]_3^y[OH^-]_3^z}$$

$$\frac{6.0\times10^{-2}}{3.0\times10^{-2}}=\frac{\cancel{k}(1.00\times10^{-2})^x(\cancel{5.00\times10^{-3}})^y(\cancel{5.00\times10^{-3}})^z}{\cancel{k}(5.00\times10^{-3})^x(\cancel{5.00\times10^{-3}})^y(\cancel{5.00\times10^{-3}})^z}$$

$$2=2^x$$

$$x=1$$

The order with respect to $ClO^-$ can be determined from trials 1 and 3.

$$\frac{\text{rate}_1}{\text{rate}_3}=\frac{6.0\times10^{-2}}{3.0\times10^{-2}}=\frac{\cancel{k}(\cancel{5.00\times10^{-3}})^x(1.00\times10^{-2})^y(\cancel{5.00\times10^{-3}})^z}{\cancel{k}(\cancel{5.00\times10^{-3}})^x(5.00\times10^{-3})^y(\cancel{5.00\times10^{-3}})^z}$$

$$2=2^y$$

$$y=1$$

The order with respect to $OH^-$ can be determined from trials 3 and 4.

$$\frac{\text{rate}_3}{\text{rate}_4}=\frac{3.0\times10^{-2}}{1.5\times10^{-2}}=\frac{\cancel{k}(\cancel{5.00\times10^{-3}})^x(\cancel{5.00\times10^{-3}})^y(5.00\times10^{-3})^z}{\cancel{k}(\cancel{5.00\times10^{-3}})^x(\cancel{5.00\times10^{-3}})^y(1.00\times10^{-2})^z}$$

$$2=\left(\frac{1}{2}\right)^z$$

$$z=-1$$

The rate law is written as

$$\text{rate}=k[ClO^-]^x[I^-]^y[OH^-]^z=k[ClO^-][I^-][OH^-]^{-1}\text{, choice C}$$

**50.** **D.** The overall order of the reaction is the sum of the orders with respect to the reactants.

$$x+y+z=1+1-1=1$$

The first order rate constant has units of $\text{time}^{-1}$, as shown in choice D, $s^{-1}$.

**51.** **D.** The molar mass of benzene, $C_6H_6$, is approximately 78 $g{\cdot}mol^{-1}$, so there are two moles of benzene.

$$n\Delta H_{vap}=2\text{ mol }C_6H_6\times\frac{30\text{ kJ}}{1\text{ mol}}=60\text{ kJ, choice D}$$

**52.** **B.** More energy is required to completely disrupt intermolecular attractions and separate the molecules in space than to partially overcome intermolecular attractions, choice B. During the transition from solid to liquid, the intermolecular forces between the molecules are disrupted; during the transition from liquid to gas, the intermolecular forces must be overcome completely as the molecules are separated completely.

**53.** **B.** The oxidation number of Sn changes from +4 to +2. It becomes less positive, so Sn is being reduced. The oxidation number of Fe changes from +2 to +3. It becomes more positive, so Fe is being oxidized.

$$E^\circ_{cell}=E^\circ_{reduction}-E^\circ_{oxidation}=0.15\text{ V}-0.77\text{ V}=-0.62\text{ V}$$

The negative cell potential of –0.62 V is indicative of a thermodynamically disfavored process, choice B.

**54.** **A.** The electrolysis of KCl will result in one of two reductions at the cathode.

$$K^+ + e^- \rightarrow K$$

or

$$2\,H_2O + 2\,e^- \rightarrow H_2 + 2\,OH^-$$

Because the alkali metal cations are very difficult to reduce (a neutral alkali metal atom achieves noble gas configuration easily by *losing* an electron), the reduction of water takes place at the cathode, as shown in choice A: $2\,H_2O + 2\,e^- \rightarrow H_2 + 2\,OH^-$.

**55.** **D.** Dipole-induced dipole attractions occur between a polar molecule and a nonpolar molecule. Proximity to the polar molecule causes the electrons of the nonpolar molecule to temporarily polarize, resulting in an intermolecular attraction. HCl is the most polar molecule included in the choices, and $Br_2$ is a nonpolar molecule, so the most important intermolecular force between the two molecules will be dipole-induced dipole attractions, choice D.

**56.** **C.** The base dissociation equation for caffeine is shown.

$$B + H_2O \rightleftharpoons BH^+ + OH^-$$

| Reaction | B | $H_2O$ | $BH^+$ | $OH^-$ | |
|---|---|---|---|---|---|
| Initial Concentrations ($M$) | 0.0100 | | 0 | 0 | |
| Change | $-x$ | | $+x$ | $+x$ | |
| Equilibrium Concentrations ($M$) | $0.0100 - x$ | | $x$ | $x$ | |

$$K_b = \frac{[BH^+][OH^-]}{[B]} = \frac{x^2}{0.0100 - \cancel{x}} = 4\times10^{-4}$$

$$x^2 = 4\times10^{-6}$$

$$x = 2\times10^{-3} = [OH^-]$$

$$pOH = -\log(2\times10^{-3}) \approx 2\text{–}3$$

The pH is found by subtraction of the pOH from 14, so the pH of the caffeine solution is between 11 and 12, choice C.

**57.** **C.** Carbon dioxide is the only gaseous product of the reaction, so the total pressure inside the container is equal to the partial pressure of $CO_2$, as shown in choice C.

$$K_p = P_{CO_2}$$

**58.** **A.** The change in Gibbs free energy is related to the equilibrium constant.

$$\Delta G° = -RT \ln K$$

Because $K$ is less than 1, $\ln K$ is negative and $-RT \ln K$ is positive, choice A.

**59.** **A.** Triple bonds, in general, are stronger and shorter than double bonds. An N–N bond in $N_2$ is a triple bond and an O–O double bond in $O_2$ is a double bond, so X corresponds to $N_2$ and Y corresponds to $O_2$, making choice A correct. (This is a simple explanation that does not take the diradical nature of $O_2$ into account; it is sufficient for AP-level justification.)

**60.** **B.** The structure in choice B shows the two molecules interacting at both available hydrogen bonding sites, making it the most likely structure of an acetic acid dimer.

# Section II: Free Response

**1. a.** The number of moles of dye is calculated using the molecular weight.

$$0.200\ \cancel{\text{g dye}} \times \frac{1\ \text{mol dye}}{792.86\ \cancel{\text{g dye}}} = 2.52 \times 10^{-4}\ \text{mol dye}$$

$$= (2.522513... \times 10^{-4})$$

**b.** A volumetric flask is a precisely calibrated piece of glassware that is designed to contain an accurately known volume of solvent at a particular temperature. A volumetric flask is more accurate than a graduated cylinder, so it should be chosen.

**c.** The molarity of the dye solution is found by dividing the number of moles by the volume. Remember to use the unrounded number of moles from part (a) in the calculation.

$$[\text{dye}] = \frac{2.52 \times 10^{-4}\ \text{mol}}{0.500\ \text{L}} = 5.05 \times 10^{-4}\ M$$

**d.** The dilution equation provides the new concentration.

$$M_1 \times V_1 = M_2 \times V_2$$

$$M_2 = \frac{M_1 \times V_1}{V_2}$$

$$M_2 = \frac{(5.05 \times 10^{-4}\ M) \times (0.50\ \cancel{\text{mL}})}{100.0\ \cancel{\text{mL}}} = 2.53 \times 10^{-6}\ M$$

**e.** The dye shows a maximum absorbance at about 630 nm, so this wavelength is a good choice for the absorbance measurements.

**f.** **i.** The reaction is first order with respect to the dye because the plot of ln absorbance versus time is linear.

**ii.** Because the reaction rate doubles when the concentration of bleach doubles, the reaction is first order in bleach. The rate expression is

$$\text{rate} = k[\text{dye}][\text{bleach}]$$

**2. a.** State symbols—(*aq*), (*l*), etc.—are not required for full credit unless otherwise stated on the exam.

$$C_{17}H_{19}O_3N(aq) + H_2O(l) \rightleftharpoons C_{17}H_{19}O_3NH^+(aq) + OH^-(aq)$$

**b.** The correct $K_b$ expression is

$$K_b = \frac{[C_{17}H_{19}O_3NH^+][OH^-]}{[C_{17}H_{19}O_3N]}$$

**c.** The pOH can be determined from the $OH^-$ concentration, and this can be subtracted from 14 to give the pH.

$$\text{pOH} = -\log\,[OH^-] = -\log\,(3.17 \times 10^{-4}) = 3.499$$

$$\text{pH} = 14.000 - 3.499 = 10.501$$

**d.** A RICE table can be used to determine the concentrations of the species needed to calculate the $K_b$.

| Reaction | $C_{17}H_{19}O_3N$ | $H_2O$ | $C_{17}H_{19}O_3NH^+$ | $OH^-$ |
|---|---|---|---|---|
| Initial Concentrations ($M$) | 0.0621 | | 0 | 0 |
| Change | $-3.17 \times 10^{-4}$ | | $+3.17 \times 10^{-4}$ | $+3.17 \times 10^{-4}$ |
| Equilibrium Concentrations ($M$) | 0.061783 | | $3.17 \times 10^{-4}$ | $3.17 \times 10^{-4}$ |

$$K_b = \frac{[C_{17}H_{19}O_3NH^+][OH^-]}{[C_{17}H_{19}O_3N]} = \frac{(3.17\times10^{-4})(3.17\times10^{-4})}{0.061783} = 1.63\times10^{-6}$$

Notice that the unrounded calculated value is $1.6264830 \times 10^{-6}$. This unrounded value should be used in question 2.f.ii. below.

**e.** The percent ionization is calculated as follows.

$$\%\text{ ionization} = \frac{3.17\times10^{-4}}{0.0621}\times100\% = 0.510\%$$

**f.** **i.** The volume of HCl required to reach the equivalence point is determined stoichiometrically.

$$HCl(aq) + C_{17}H_{19}O_3N(aq) \rightleftharpoons C_{17}H_{19}O_3NH^+Cl^-(aq)$$

$$25.0\ \text{mL morphine} \times \frac{1\ \text{L morphine}}{1000\ \text{mL morphine}} \times \frac{0.0621\ \text{mol morphine}}{1\ \text{L morphine}} \times \frac{1\ \text{mol HCl}}{1\ \text{mol morphine}} \times \frac{1\ \text{L HCl}}{0.0500\ \text{mol HCl}}$$
$$= 0.0311\ \text{L}$$

**ii.** Determination of the pH of the solution at the equivalence point requires that we first calculate the $K_a$ of the conjugate acid of morphine. The $K_a$ can then be used to determine the $H_3O^+$ concentration followed by the pH.

$$C_{17}H_{19}O_3NH^+(aq) + H_2O(l) \rightleftharpoons C_{17}H_{19}O_3N(aq) + H_3O^+(aq)$$

$$M_1 \times V_1 = M_2 \times V_2$$

$$M_2 = \frac{M_1 \times V_1}{V_2}$$

$$M_2 = \frac{(0.0621\ M)\times(25.0\ \text{mL})}{(25.0 + 31.1)\ \text{mL}} = 0.0277\ M$$

$$K_a = \frac{K_w}{K_b} = \frac{1.00\times10^{-14}}{1.6264830\times10^{-6}} = 6.15\times10^{-9}$$

| Reaction | $C_{17}H_{19}O_3NH^+$ | $H_2O$ | $C_{17}H_{19}O_3N$ | $H_3O^+$ |
|---|---|---|---|---|
| Initial Concentrations ($M$) | 0.0277 | | 0 | 0 |
| Change | $-x$ | | $+x$ | $+x$ |
| Equilibrium Concentrations ($M$) | $0.0277 - x$ | | $x$ | $x$ |

$$K_a = \frac{[C_{17}H_{19}O_3N][H_3O^+]}{[C_{17}H_{19}O_3NH^+]} = \frac{x^2}{0.0277 - x} = 6.15\times10^{-9}$$

Because $K_a$ is very small, $x$ can be neglected in the denominator. (Rounding takes place only after the final calculation.)

$$x = \sqrt{(0.0277)(6.15\times10^{-9})} = 1.31\times10^{-5} = [H_3O^+]$$
$$pH = -\log(1.31\times10^{-5}) = 4.884$$

**g.** The titration curve of morphine begins at a pH of 10.501, as we determined in question 2.c. The equivalence point, determined in question 2.f.ii., occurs at 4.884. Halfway to the equivalence point, when about 15.5 mL of HCl have been added, the pOH of the solution is equal to the $pK_b$.

$$pK_b = -\log(1.63\times10^{-6}) = 5.788 = pOH$$
$$pH = 14.000 - 5.788 = 8.212$$

The sketch should resemble the following.

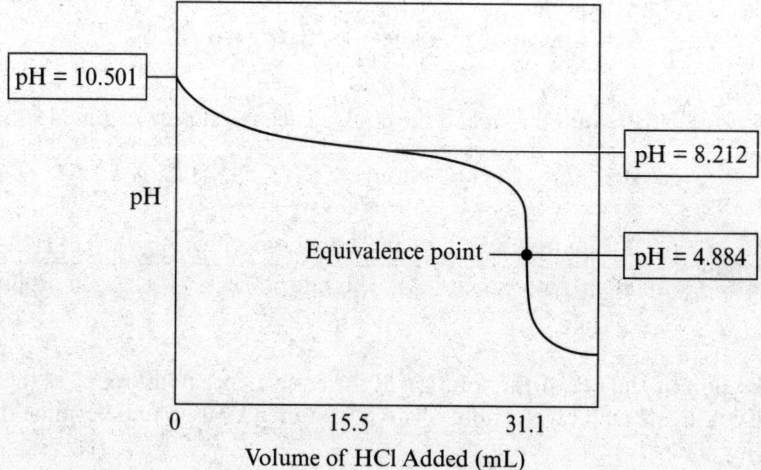


**3. a.** The reduction of $Cd^{2+}$ takes place at the cathode. At the anode, the solid cadmium reacts with sulfide, $S^{2-}$, to form CdS($s$), the reverse of the equation given in the question.

$$E^\circ_{cell} = E^\circ_{cathode} - E^\circ_{anode}$$
$$0.807\text{ V} = E^\circ_{cathode} - (1.210\text{ V})$$
$$E^\circ_{cathode} = 2.017\text{ V}$$

**b.** The salt bridge maintains electrical neutrality in the half-cells. It allows for the passage of cations into the cathode to replace the ones that are removed from solution during reduction, and it allows anions to migrate into the anode where cations are being released into the solution. The salt bridge also connects the two half-cells while preventing their contents from mixing.

**c.** The net ionic equation for the overall reaction is the sum of the reactions taking place at the cathode and at the anode.

$$Cd^{2+}(aq) + \cancel{2e^-} \rightarrow \cancel{Cd(s)}$$
$$\cancel{Cd(s)} + S^{2-}(aq) \rightarrow CdS(s) + \cancel{2e^-}$$
$$\overline{Cd^{2+}(aq) + S^{2-}(aq) \rightarrow CdS(s)}$$

State symbols are not required for credit on the AP Chemistry exam. If they are excluded, the equation becomes

$$Cd^{2+} + S^{2-} \rightarrow CdS$$

**d.** A decrease in the concentration of $Cd^{2+}$ would have the effect of decreasing the overall cell potential. The equilibrium quotient would be greater than the equilibrium constant, favoring the reactants. Although the Nernst equation is not required for this answer, it is helpful for checking your logic.

$$E = E° - \frac{0.06}{n} \log Q$$

$$Q = \frac{1}{[Cd^{2+}][S^{2-}]}$$

Because $[Cd^{2+}]$ is decreased, $Q$ is increased, and a larger value of log $Q$ is subtracted from $E°$.

**e.** The change in free energy is calculated as follows.

$$\Delta G° = -nFE° = -(2)(96{,}485\ \cancel{C}/\text{mol})\left(\frac{1\ \text{J}}{1\ \cancel{V}\cdot\cancel{C}}\right)(0.807\ \cancel{V})\left(\frac{1\ \text{kJ}}{1000\ \cancel{J}}\right) = -156\ \text{kJ/mol}$$

**f.** First, the reaction is reversed to show the dissociation of CdS; this changes the sign of $E°$.

$$CdS(s) \rightarrow Cd^{2+}(aq) + S^{2-}(aq)$$

$$E° = -0.807\ \text{V}$$

$$\Delta G° = -RT \ln K = -nF E°$$

$$-(8.314\ \cancel{J}/(\cancel{mol}\cdot\cancel{K}))(298\ \cancel{K})(\ln K_{sp}) = -(2)(96{,}485\ \cancel{J}/(\cancel{V}\cdot\cancel{mol}))(-0.807\ \cancel{V})$$

$$K_{sp} = e^{-62.9} = 5.042 \times 10^{-28}$$

**g.** Because $Na_2S$ dissociates completely, we are calculating the solubility of CdS in a 0.100 $M$ solution of the common ion $S^{2-}$. The solubility of CdS in $S^{2-}$ is calculated using a RICE table.

| **Reaction** | **CdS(*s*)** | **$Cd^{2+}(aq)$** | **$S^{2-}(aq)$** |
|---|---|---|---|
| **Initial Concentrations (*M*)** | | 0 | 0.100 |
| **Change** | | $+x$ | $+x$ |
| **Equilibrium Concentrations (*M*)** | | $x$ | $0.100 + x$ |

$$K_{sp} = [Cd^{2+}][S^{2-}]$$

$$K_{sp} = (x)(0.100 + x) = 5.042 \times 10^{-28}$$

Because the $K_{sp}$ is very small, the simplifying assumption can be made that $x$, the number of moles of CdS that dissolve in a 0.100 $M$ solution of $S^{2-}$, is negligibly small. In that case, it can be omitted from the $(0.100 + x)$ term.

$$5.042 \times 10^{-28} = x(0.100 + \cancel{x}) = 0.100x$$

$$x = 5.04 \times 10^{-27}\ \text{mol/L}$$

The solubility of CdS in a 0.100 $M$ $S^{2-}$ solution is $5.04 \times 10^{-27}$ mol/L.

**4.** **a.** The balanced equation for the reaction that produces $H_2(g)$ is

$$NaH(s) + H_2O(l) \rightarrow NaOH(aq) + H_2(g)$$

**b.** The number of moles is determined using the ideal gas law.

$$T = 24\ °C + 273 = 297\ K$$

$$P = 757\ \cancel{torr} \times \frac{1\ atm}{760\ \cancel{torr}} = 0.996\ atm$$

$$n = \frac{PV}{RT}$$

$$n = \frac{(0.996\ \cancel{atm})(0.0192\ \cancel{L})}{(0.08206\ (\cancel{L} \cdot \cancel{atm}/(mol \cdot \cancel{K}))(297\ \cancel{K})} = 7.85 \times 10^{-4}\ mol$$

**c.** In order to determine the number of moles of dry $H_2$, we need to take the vapor pressure of water into account. The total pressure, 757 torr, is composed of water vapor with a partial pressure of 22.5 torr. The rest is made up of $H_2$. We can first solve for the mole fraction of $H_2$.

$$P_{H_2} = P_{total} \times X_{H_2}$$

$$P_{H_2} = P_{total} - P_{H_2O}$$

$$757\ torr - 22.5\ torr = (757\ torr) \times X_{H_2}$$

$$X_{H_2} = 0.970$$

The number of moles of $H_2$ can be used with the ideal gas law to determine the volume.

$$n_{H_2} = (0.970)(7.85 \times 10^{-4}\ mol) = 7.61 \times 10^{-4}\ mol$$

$$V = \frac{nRT}{P} = \frac{(7.61 \times 10^{-4}\ \cancel{mol})(0.08206\ L \cdot \cancel{atm}/(\cancel{mol} \cdot \cancel{K}))(297\ \cancel{K})}{0.996\ \cancel{atm}} = 0.0186\ L = 18.6\ mL$$

**d.** The volume of HCl is less than the volume occupied by the $H_2$. HCl is a polar molecule in contrast to $H_2$, which is nonpolar. While $H_2$ experiences only London dispersion forces, HCl also experiences dipole-dipole attractions. This causes the volume of HCl to be less than the volume predicted by the ideal gas law.

**5.** **a.** Line structures and dot structures are both acceptable.

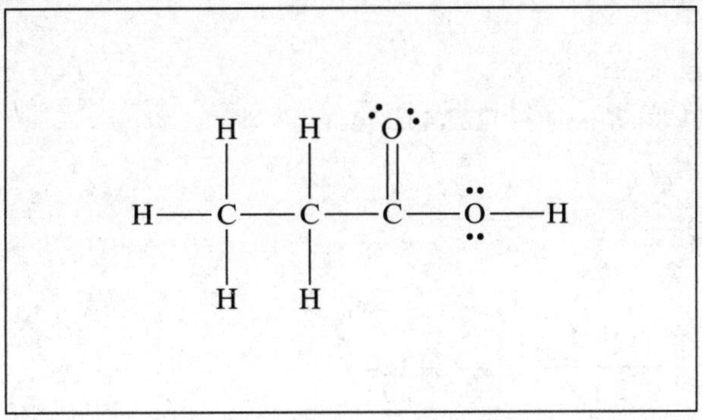

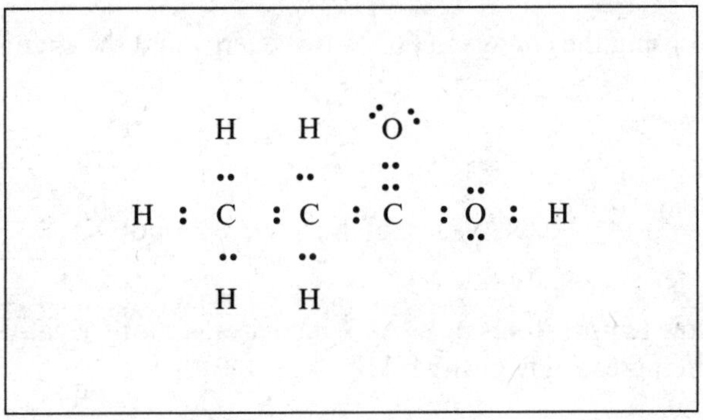

Propionic acid

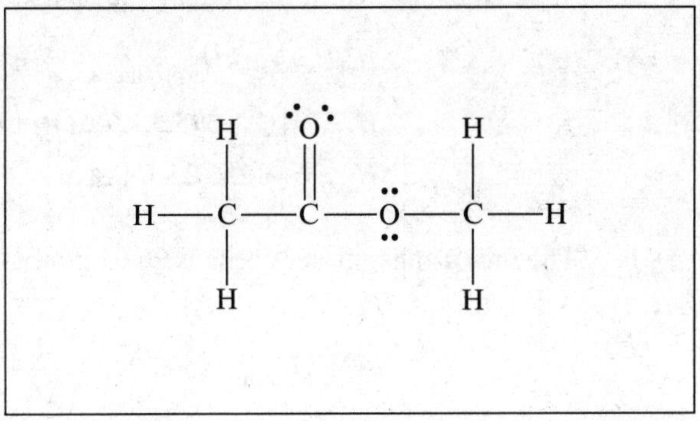

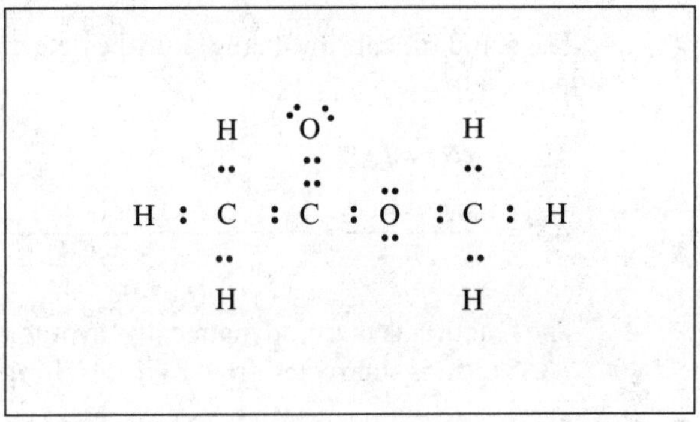

Methyl ethanoate

**b.** Infrared spectroscopy is a better choice to distinguish between propionic acid and methyl ethanoate because there are different bonds present in the two molecules. For example, propionic acid contains an O–H bond, while methyl ethanoate does not. Mass spectrometry would not be an appropriate choice because both compounds have three carbon atoms, six hydrogen atoms, and two oxygen atoms. In other words, both molecules have identical masses.

**c.** Propionic acid has a higher boiling point because it contains an O–H group capable of hydrogen bonding, while methyl ethanoate contains only polar and nonpolar bonds. Hydrogen bonds are the strongest intermolecular forces, and the stronger the intermolecular forces between molecules, the more energy is required to separate the molecules from one another.

**d.** There are ten sigma bonds and one pi bond in a molecule of methyl ethanoate. The hybridization of the carbon atoms is shown in the figure below.

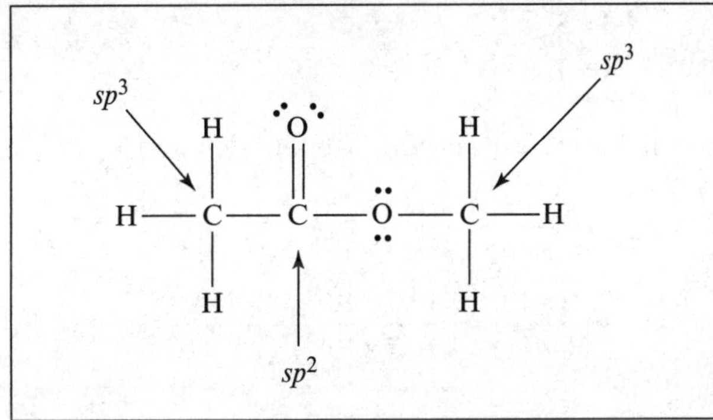

**6. a.** The enthalpy change for the reaction is determined by subtracting the sum of the heats of formation of the reactants from the sum of the heats of formation of the products.

$$\Delta H^\circ_{\text{rxn}} = \sum n\Delta H^\circ_{\text{f(products)}} - \sum n\Delta H^\circ_{\text{f(reactants)}}$$

$$\Delta H^\circ_{\text{rxn}} = [(2(-393.5 \text{ kJ/mol})) + (-241.8 \text{ kJ/mol})] - [227.4 \text{ kJ/mol}]$$

$$\Delta H^\circ_{\text{rxn}} = -1256.2 \text{ kJ/mol}$$

**b.** The molar mass of acetylene is 26.04 $\text{g}\cdot\text{mol}^{-1}$.

$$q = n\Delta H^\circ_{\text{rxn}} = (500.0 \cancel{\text{g}})\left(\frac{1 \cancel{\text{mol}}}{26.04 \cancel{\text{g}}}\right)\left(\frac{-1256.2 \text{ kJ}}{1 \cancel{\text{mol}}}\right) = -24{,}120 \text{ kJ}$$

**c.** The standard entropy change may be determined using the change in Gibbs free energy and the change in enthalpy.

$$\Delta G^\circ = \Delta H^\circ - T\Delta S^\circ$$

$$\Delta S^\circ = \frac{\Delta H^\circ - \Delta G^\circ}{T} = \frac{[-1256.2 \text{ kJ/mol} - (-1227 \text{ kJ/mol})]}{298 \text{ K}} = -0.0980 \text{ kJ/(mol}\cdot\text{K)} = -98.0 \text{ J/(mol}\cdot\text{K)}$$

**d.** The reaction is thermodynamically favorable at low temperatures only. As $T$ increases, a more negative $T\Delta S^\circ$ term is subtracted from $\Delta H^\circ$, resulting in an increasingly positive $\Delta G^\circ$.

**7. a.** The initial rate of disappearance of $H_2O_2$ is two times the initial rate of appearance of $O_2$.

$$-\frac{1}{2}\frac{\Delta[H_2O_2]}{\Delta t} = \frac{\Delta[O_2]}{\Delta t}$$

$$\frac{\Delta[H_2O_2]}{\Delta t} = -2(1.36\times10^8\ M\cdot\text{s}^{-1}) = -2.72\times10^8\ M\cdot\text{s}^{-1}$$

**b.** The method of initial rates can be used to determine the rate law. The order with respect to $H_2O_2$ can be determined from trials 1 and 2.

$$\frac{\text{rate}_2}{\text{rate}_1} = \frac{k[H_2O_2]_2^x[I^-]_2^y}{k[H_2O_2]_1^x[I^-]_1^y}$$

$$\frac{8.16\times10^8}{2.72\times10^8} = \frac{k(1.1\times10^{-1})^x(6.4\times10^{-3})^y}{k(3.7\times10^{-2})^x(6.4\times10^{-3})^y}$$

$$3 = 3^x$$

$$x = 1$$

The order with respect to $I^-$ can be determined from trials 1 and 3.

$$\frac{\text{rate}_1}{\text{rate}_3} = \frac{k[H_2O_2]_1^x[I^-]_1^y}{k[H_2O_2]_3^x[I^-]_3^y}$$

$$\frac{2.72\times10^8}{1.36\times10^8} = \frac{k(3.7\times10^{-2})^x(6.4\times10^{-3})^y}{k(3.7\times10^{-2})^x(3.2\times10^{-3})^y}$$

$$2 = 2^y$$

$$y = 1$$

The rate law is

$$\text{rate} = k[H_2O_2][I^-]$$

**c.** Any of the three trials may be chosen in order to calculate the rate constant. Using experiment 1, the rate law becomes

$$2.72 \times 10^{8}\ \text{mol}\cdot\text{L}^{-1}\cdot\text{s}^{-1} = k(3.7 \times 10^{-2}\ \text{mol}\cdot\text{L}^{-1})(6.4 \times 10^{-3}\ \text{mol}\cdot\text{L}^{-1})$$
$$k = 1.1 \times 10^{12}\ \text{mol}^{-1}\cdot\text{L}\cdot\text{s}^{-1}$$

**d.** The rate law is written based on the slow step of the mechanism. The first step of the mechanism is the slow step and it is the reaction of $H_2O_2$ with $I^-$, so this proposed mechanism is consistent with the experimental rate law.

## Chapter 12

# Practice Exam 2

## Section I: Multiple Choice

**60 questions**

**90 minutes**

Calculators are not allowed for Section I.

**Note:** For all questions, assume the temperature is 298 K, the pressure is 1.0 atm, and solutions are aqueous unless otherwise specified.

**Directions:** Each of the questions or incomplete statements below is followed by four suggested answers or completions. Select the one that is best in each case.

**1.** Combustion of which of the following in excess oxygen results in a 2:1 mole ratio of $CO_2$ to $H_2O$?

A. $CH_4$
B. $C_2H_2$
C. $C_2H_4$
D. $C_3H_6$

**2.** A conductivity experiment is conducted as shown in the figure below.

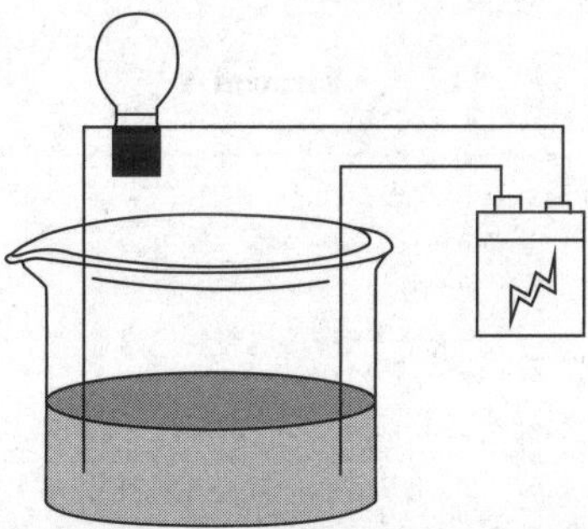

When a 1.00 *M* solution of ammonia, $NH_3$, and a 1.00 *M* solution of acetic acid, $CH_3COOH$, are mixed, the bulb glows brightly. Which particulate diagram best explains this result? Water is not shown.

- Carbon
- Nitrogen
- Oxygen
- Hydrogen

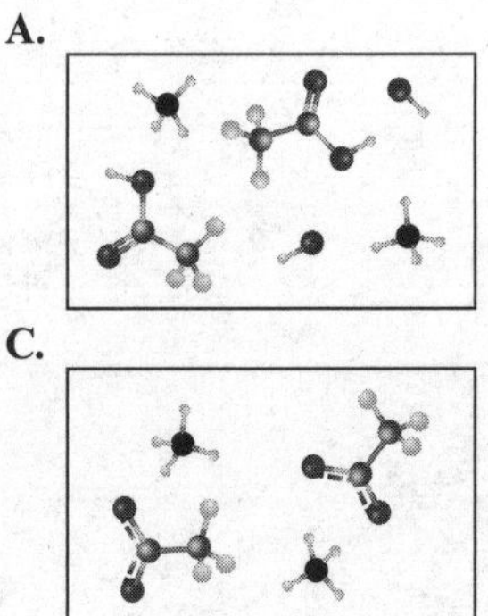


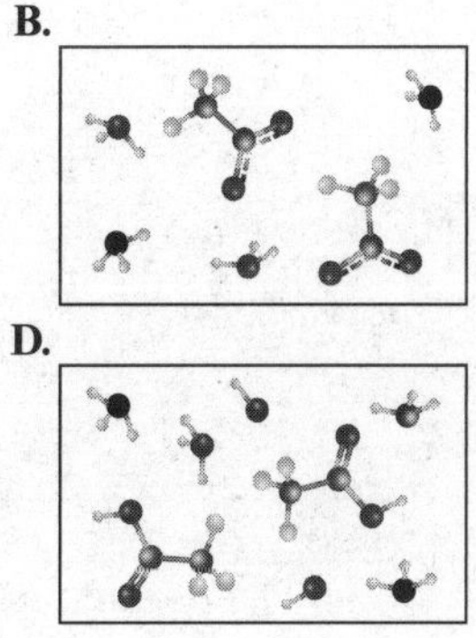

**3.** The $K_a$ values for $HPO_4^{2-}$ and $HSO_3^-$ are $4.8 \times 10^{-13}$ and $6.3 \times 10^{-8}$, respectively; $HPO_4^{2-}$ is a ________ acid than $HSO_3^-$ and $PO_4^{3-}$ is a ________ base than $SO_3^{2-}$.

**A.** stronger; weaker
**B.** weaker; stronger
**C.** stronger; stronger
**D.** weaker; weaker

*Use the photoelectron spectra for pure elements X, Y, and Z to answer questions 4–7.*

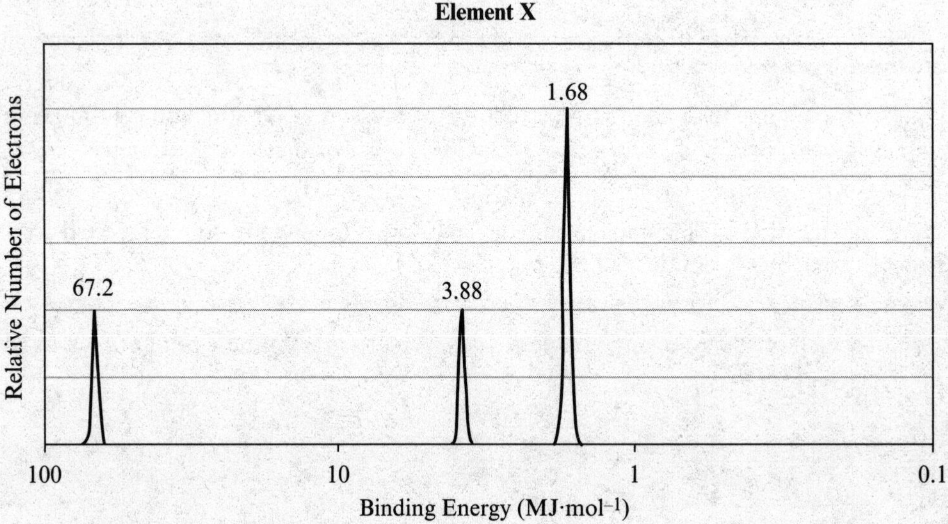


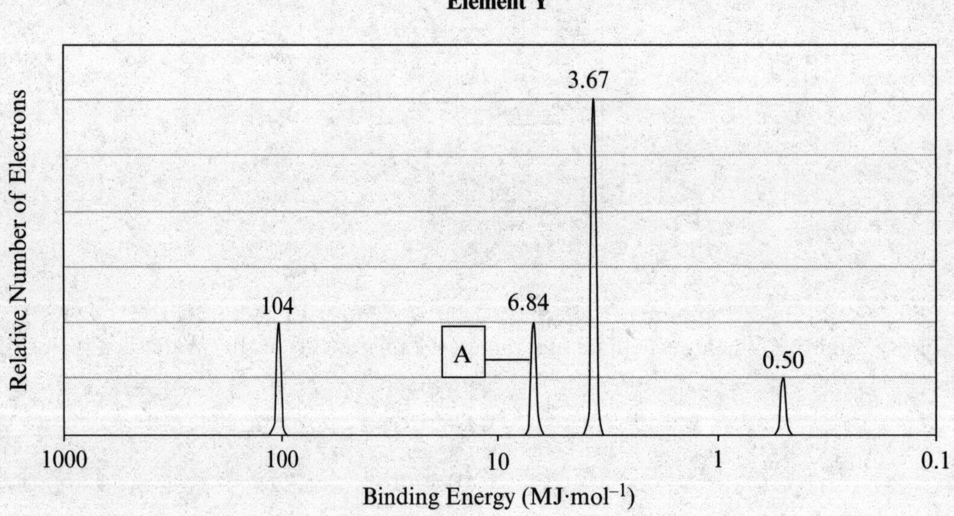

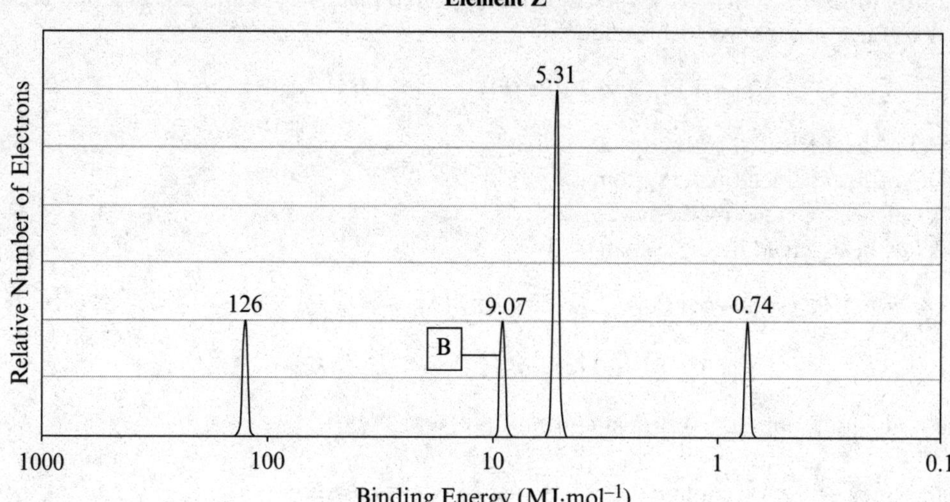


**4.** How would these three photoelectron spectra differ if the electronic energy levels in polyelectronic atoms were not divided into sublevels?

**A.** All the peaks would have lower binding energies due to equal shielding of the orbitals in a shell.
**B.** All but the highest-energy peaks would have lower binding energies due to equal orbital shielding.
**C.** The spectra would have fewer peaks because each orbital in a shell would have the same energy.
**D.** The spectra would have more peaks because each orbital in each subshell would have a slightly different energy.

**5.** Which element(s), X, Y, and/or Z, is/are expected to ionize by gaining electrons?

**A.** X only
**B.** Y only
**C.** Z only
**D.** Both Y and Z

**6.** Which neutral element is expected to have the largest atomic radius?

**A.** Element X, because it has the smallest nuclear charge of the three elements and its valence electrons experience the least shielding
**B.** Element Z, because it experiences the most shielding and the greatest electron-electron repulsions
**C.** Elements X and Y will have similar radii; both will be larger than element Z because they have smaller nuclear charges.
**D.** Element Y, because it has a smaller nuclear charge than element Z and its electrons experience more shielding than the valence electrons of element X

**7.** Compare the peak labeled "A" in the spectrum for element Y with the peak labeled "B" in the spectrum for element Z. Why does peak A have a lower binding energy than peak B?

**A.** Element Y has a lower effective nuclear charge than element Z.
**B.** The electrons contributing to the peak experience greater electron-electron repulsions in element Z than in element Y.
**C.** The coulombic attraction between the electrons and the nucleus is less in element Y than in element Z due to a smaller distance between the electrons and the nucleus.
**D.** Peak A arises from electrons in a $2s$ sublevel, while peak B arises from electrons in a $2p$ sublevel.

**8.** What quantity of heat is transferred when 2.00 grams of red phosphorus and excess liquid bromine react completely to form phosphorus tribromide?

$$2\ P(s) + 3\ Br_2(l) \rightarrow 2\ PBr_3(g) \qquad \Delta H° = -243\ kJ/mol$$

A. 7.85 kJ of heat released by the system
B. 7.85 kJ of heat gained by the system
C. 243 kJ of heat released by the system
D. 243 kJ of heat gained by the system

*Questions 9–10 refer to the experiment below.*

$$C(s) + CO_2(g) \rightleftharpoons 2\ CO(g) \qquad \Delta H > 0$$

The reaction is at equilibrium in a piston at constant pressure at 25 °C.

**9.** What would be the effect of cooling the reaction at constant pressure?

A. The reaction would shift toward reactants, and the volume in the piston would remain unchanged.
B. The reaction would shift toward products, and the volume in the piston would remain unchanged.
C. The reaction would shift toward reactants, and the volume in the piston would decrease.
D. The reaction would shift toward products, and the volume in the piston would increase.

**10.** What would be the effect of adding argon to the piston at constant total pressure and temperature?

A. The reaction would shift toward products, and the volume in the piston would increase.
B. The reaction would shift toward reactants, and the volume in the piston would decrease.
C. The reaction would shift toward reactants, and the volume in the piston would remain unchanged.
D. The reaction would shift toward products, and the volume in the piston would remain unchanged.

**11.** A sample of water of unknown mass has a temperature of 40.0 °C.

A 10.0-gram block of ice at –5.0 °C is added.

When thermal equilibrium is reached, the final temperature of the combined sample is 10.0 °C.

What is the approximate mass of the water that was originally at 40 °C?

| $\Delta H_{fus}$ (kJ/g) | $\Delta H_{vap}$ (kJ/g) | $c_{ice}$ (J/(g·°C)) | $c_{water}$ (J/(g·°C)) |
|---|---|---|---|
| 0.3 | 2 | 2 | 4 |

A. 2.8 g
B. 4.4 g
C. 25 g
D. 29 g

**12.** Which structure below is expected to make the greatest contribution to the resonance hybrid structure of $HClO_4$?

A.
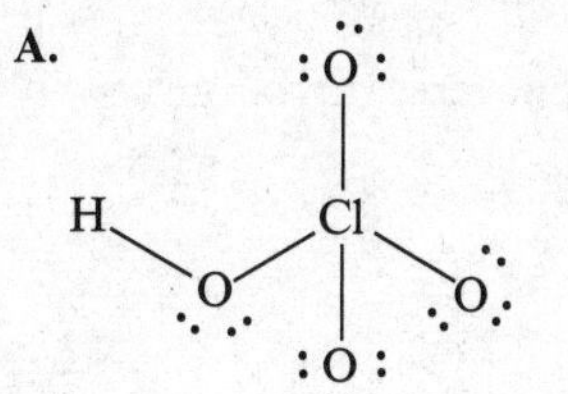

B.
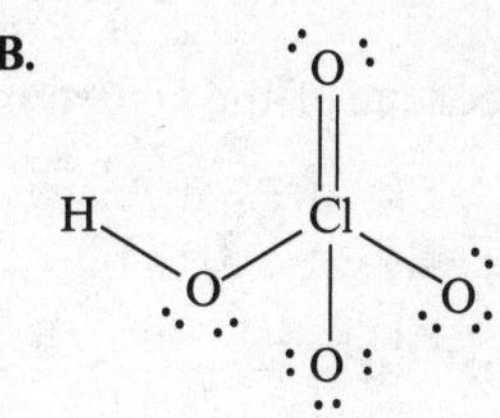

C.
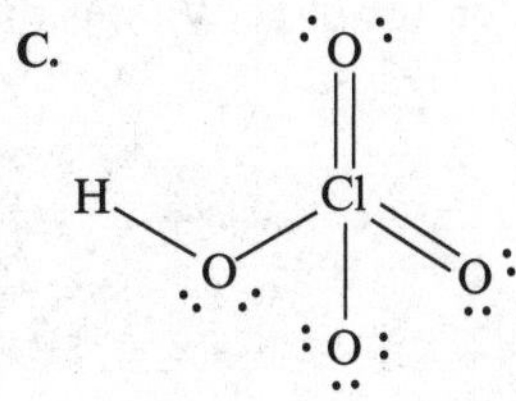

D.
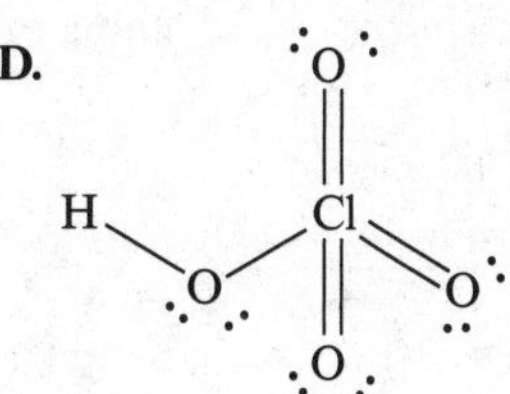

**13.** Which of the following correctly describes the shape of the $TeCl_5^-$ ion?

A. The ion has an electronic geometry similar to $SF_6$ and a molecular geometry similar to $BrF_5$.
B. The ion has an electronic geometry similar to $PF_5$ and a molecular geometry similar to $PF_5$.
C. The ion has an electronic geometry similar to $SF_6$ and a molecular geometry similar to $SbCl_5$.
D. The ion has an electronic geometry similar to $PF_5$ and a molecular geometry similar to $SbF_4^-$.

**14.** Which of the following correctly ranks the bonds according to increasing bond strength?

A. C–C double bond < Si–Si single bond < C–C single bond
B. C–C single bond < Si–Si single bond < C–C double bond
C. Si–Si single bond < C–C double bond < C–C single bond
D. Si–Si single bond < C–C single bond < C–C double bond

**15.** Which of the following processes is expected to be endothermic?

A. Ionization of a neutral atom due to absorption of a photon
B. Emission of a photon from a neutral atom due to relaxation of an electron from an excited state to the ground state
C. Condensation of gaseous helium atoms at 4 K and 1 atm
D. Dissolution of sodium hydroxide in distilled water

*Questions 16–17 refer to the hypothetical reaction below.*

At a given temperature, the equilibrium constant, $K_p$, for the following reaction is $1.0 \times 10^{10}$.

$$2\ AB_3(g) \rightleftharpoons A_2(g) + 3B_2(g)$$

**16.** At this temperature, what is the value of $K_p$ for the reaction below?

$$\frac{1}{2}\ A_2(g) + \frac{3}{2}\ B_2(g) \rightleftharpoons AB_3(g)$$

A. $1.0 \times 10^{5}$
B. $1.0 \times 10^{-5}$
C. $5.0 \times 10^{-9}$
D. $5.0 \times 10^{9}$

**17.** Consider the decomposition of gaseous AB to form gaseous A and B atoms.

$$AB(g) \rightarrow A(g) + B(g)$$

Which choice is likely to be true of the system in the reverse direction?

A. $\Delta S > 0$; $\Delta H > 0$
B. $\Delta S < 0$; $\Delta H < 0$
C. $\Delta S > 0$; $\Delta H < 0$
D. $\Delta S < 0$; $\Delta H > 0$

**18.** A saturated solution of $CaF_2$ has a $Ca^{2+}$ concentration of $x$ mol/L. What is the $K_{sp}$ of $CaF_2$?

A. $x^2$
B. $x^3$
C. $2x^3$
D. $4x^3$

**19.** The oxidation of glucose during the fermentation process is exothermic, and the equilibrium constant, $K_p$, is $9 \times 10^{39}$ at 298 K.

$$C_6H_{12}O_6(s) \rightleftharpoons 2\ C_2H_6O(l) + 2\ CO_2(g)$$

Which statement is true of the fermentation reaction?

A. The change in entropy for the reaction is negative.
B. The reaction is thermodynamically disfavored at high temperatures.
C. The equilibrium can be shifted toward products by increasing the temperature of the reaction.
D. The equilibrium constant can be increased by decreasing the temperature of the reaction.

**20.** Which of the following statements regarding enzyme catalysis is false?

A. The enzyme has a binding pocket in which the substrate may be positioned in a reactive orientation.
B. The enzyme raises the energy of the transition state of the reaction compared with the transition state of the uncatalyzed reaction.
C. The enzyme may provide an alternative reaction pathway with a transition state that is different from the uncatalyzed reaction.
D. The enzyme's three-dimensional structure is a result of intermolecular interactions between its amino acid residues and the aqueous surroundings.

**21.** The equation for the decomposition of nitryl chloride and the rate law for the reaction are shown below.

$$2\ NO_2Cl \rightarrow 2\ NO_2 + Cl_2$$
$$\text{rate} = k[NO_2Cl]$$

A researcher proposes two different mechanisms for the reaction.

| Mechanism 1 | | Mechanism 2 | |
|---|---|---|---|
| $NO_2Cl \rightarrow NO_2 + Cl$ | (slow) | $2\ NO_2Cl \rightleftharpoons N_2O_4 + Cl_2$ | (fast) |
| $Cl + NO_2Cl \rightarrow NO_2 + Cl_2$ | (fast) | $N_2O_4 \rightarrow 2\ NO_2$ | (slow) |

Based on this information, which of the following is true?

A. Only mechanism 1 is consistent with the rate law.
B. Only mechanism 2 is consistent with the rate law.
C. Both mechanism 1 and mechanism 2 are consistent with the rate law.
D. Neither mechanism 1 nor mechanism 2 is consistent with the rate law.

**22.** Which statement is false regarding crystalline solids, such as quartz, and amorphous solids, such as obsidian?

**A.** Crystalline solids have a sharp melting point, while amorphous solids melt over a temperature range.
**B.** Crystalline solids shatter into regularly shaped fragments, while amorphous solids do not.
**C.** A crystalline solid is composed of regularly spaced atoms, while the atoms of an amorphous solid are distributed randomly.
**D.** A crystalline solid, which has a regular long-range structure, is referred to as a glass.

**23.** Reactant A undergoes decomposition according to the reaction below. What conclusion regarding reactant A may be concluded from the data below?

$$A \rightarrow \text{products}$$

| Time (min) | [A] | ln [A] | $[A]^{-1}$ |
|---|---|---|---|
| 0.00 | 1.00 | 0 | 1.00 |
| 10.0 | 0.800 | –0.223 | 1.25 |
| 20.0 | 0.667 | –0.405 | 1.50 |
| 40.0 | 0.500 | –0.693 | 2.00 |

**A.** The decomposition is zero order.
**B.** The decomposition is first order.
**C.** The decomposition is second order.
**D.** No conclusion regarding the order of the reaction may be determined from the information given.

**24.** Consider the reaction below, for which $K = 4$ at a particular temperature.

$$A_2 + B_2 \rightleftharpoons AB$$

$A_2$

$B_2$

AB

In which of the containers represented below will the reaction shift toward reactants?

A.
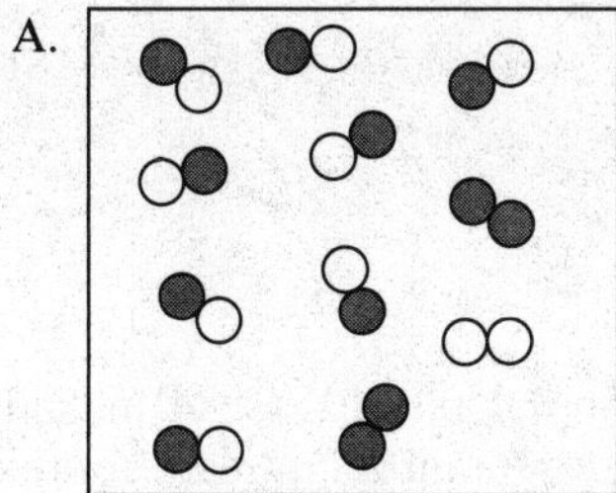

B.
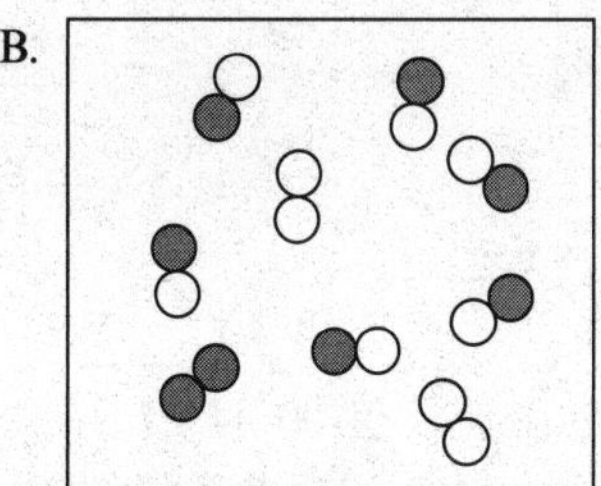

C.
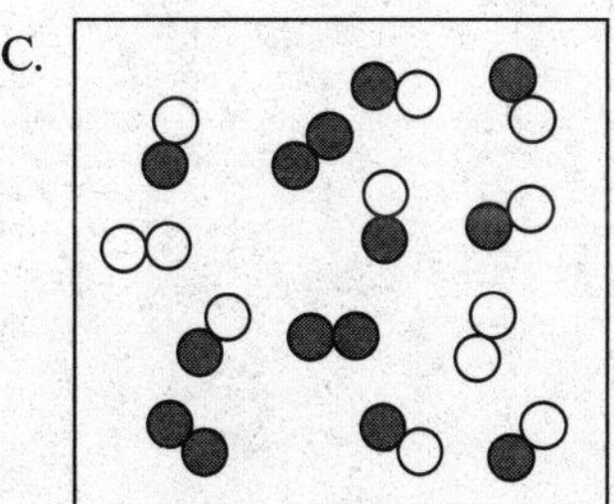

D.
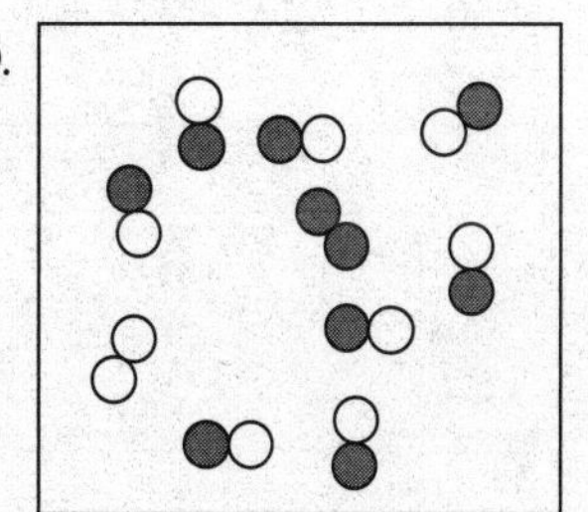

**25.** Use the $E°$ values in the table to determine which reaction would have the largest equilibrium constant as written.

| Reaction | $E°$ (V) |
|---|---|
| $Al^{3+} + 3e^- \rightarrow Al$ | –1.66 |
| $Fe^{2+} + 2e^- \rightarrow Fe$ | –0.44 |
| $Pb^{2+} + 2e^- \rightarrow Pb$ | –0.13 |

**A.** $3\ Pb^{2+} + 2\ Al \rightarrow 3\ Pb + 2\ Al^{3+}$
**B.** $3\ Fe^{2+} + 2\ Al \rightarrow 3\ Fe + 2\ Al^{3+}$
**C.** $3\ Pb + 2\ Al^{3+} \rightarrow 3\ Pb^{2+} + 2\ Al$
**D.** $3\ Fe + 2\ Al^{3+} \rightarrow 3\ Fe^{2+} + 2\ Al$

**26.** Rank the following liquids in order of decreasing molar enthalpy of fusion.

**A.** $SiCl_4 > CF_4 > H_2O$
**B.** $CF_4 > SiCl_4 > H_2O$
**C.** $H_2O > SiCl_4 > CF_4$
**D.** $H_2O > CF_4 > SiCl_4$

**27.** Two flasks are joined by a stopcock, as shown below. The 5.0-L flask on the left contains $H_2$ at 8.0 atm, and the 3.0-L flask on the right contains He at 4.0 atm. What is the total final pressure after the stopcock has been opened?

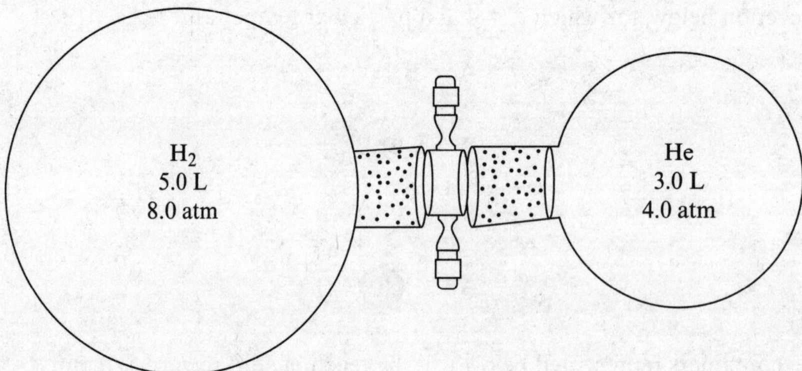


**A.** 1.5 atm
**B.** 6.0 atm
**C.** 6.5 atm
**D.** 12 atm

**28.** When a 50.0-mL solution of 0.100 *M* KI is titrated with 0.0500 *M* $K_2Cr_2O_7$ in acidic solution, the products of the reaction are $I_2$ and $Cr^{3+}$. What volume of $K_2Cr_2O_7$ is required to react completely?

$$6\ I^- = 14\ H^+ + Cr_2O7^{2-} \rightarrow 3\ I_2 + 2\ Cr^{3+} + 7\ H_2O$$

**A.** 6.67 mL
**B.** 16.7 mL
**C.** 30.0 mL
**D.** 60.0 mL

**29.** How many grams of aluminum would be produced if a sample of $Al_2O_3$ was electrolyzed for 96,500 seconds using a 3.00-amp current?

A. 9 g
B. 18 g
C. 27 g
D. 54 g

*Questions 30–31 refer to the unbalanced redox reaction below.*

$$MnO_4^- + SO_3^{2-} \rightarrow MnO_4^{2-} + SO_4^{2-}$$

**30.** Which ion in the equation above contains the element that is undergoing reduction as the reaction proceeds in the indicated direction?

A. $MnO_4^-$
B. $SO_3^{2-}$
C. $MnO_4^{2-}$
D. $SO_4^{2-}$

**31.** What is the oxidation number of sulfur in $SO_3^{2-}$?

A. 5
B. 4
C. 3
D. 2

*Use the following data to answer questions 32–33.*

| | Structure | Boiling Point (°C) | Molar Mass (g/mol) | Density (g/mL) |
|---|---|---|---|---|
| **Propanol** | 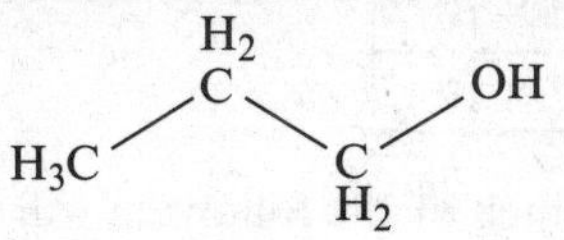 $H_3C-CH_2-CH_2-OH$ | 98 | 60.11 | 0.802 |
| **Propionaldehyde** | 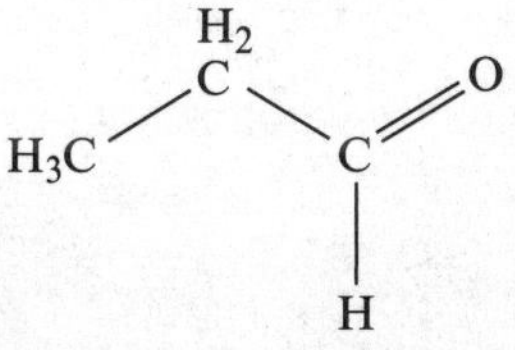 $H_3C-CH_2-C(=O)H$ | 50 | 58.09 | 0.807 |

**32.** A chemist has 100.0 mL of a clear, colorless liquid composed of 5.0 mL of propanol dissolved in 95.0 mL of propionaldehyde. Which method should the chemist use in order to separate the mixture?

A. Paper chromatography
B. Filtration
C. Distillation
D. Centrifugation

**33.** What is the approximate molarity of propanol in the solution?

A. 0.67
B. 0.83
C. 13
D. 16

**34.** Which choice provides the best explanation for the thermodynamic favorability of the freezing of benzene at 6 °C and 1 atm of pressure?

A. The decrease in the entropy of the system is offset by a sufficient increase in the entropy of the surroundings.
B. The increase in the entropy of the system is offset by a sufficient increase in the entropy of the universe.
C. The change in Gibbs free energy is negative because the process is highly endothermic.
D. The entropy of the system increases because of attractive London dispersion forces between the benzene molecules.

**35.** When a reactant's concentration is tripled, the rate of the reaction increases by a factor of 27. What is the order of the reaction with respect to the reactant?

A. 2
B. 3
C. 9
D. 27

**36.** A solution is produced by the addition of 250. mL of 0.100 *M* $CuNO_3$ and 250. mL of 0.100 *M* $AgNO_3$.

| Substance | $K_{sp}$ |
|---|---|
| CuBr | $4.2 \times 10^{-8}$ |
| AgBr | $7.7 \times 10^{-13}$ |

Solid sodium bromide is slowly added to the solution. Which of the following will be the first to precipitate?

A. CuBr
B. $NaNO_3$
C. AgBr
D. NaBr

**37.** Which of the following statements regarding real gas behavior is false?

A. A real gas behaves more ideally at high temperature and low pressure than it does at low temperature and high pressure.
B. Molecules of a real gas occupy significant volume.
C. Intermolecular attractions between gas molecules are most significant when the molecules are in the process of colliding.
D. Neon gas is expected to deviate very little from ideal behavior since neon atoms occupy volume but do not experience intermolecular attractions.

**38.** The autoionization of water is represented by the equation below. Which of the following statements is false?

$$2\ H_2O(l) \rightleftharpoons H_3O^+(aq) + OH^-(aq)$$

**A.** Since $K_w$ increases with increasing temperature, the autoionization of water is endothermic.
**B.** Pure water does not have a pH of 7.00 at 10 °C.
**C.** Since $pK_w$ decreases with increasing temperature, the autoionization of water does not occur at high temperature.
**D.** Pure water has a pH of 7.00 at 25 °C.

*Refer to the structures of toluene, ethylbenzene, chlorobenzene, and aniline to answer questions 39–40.*

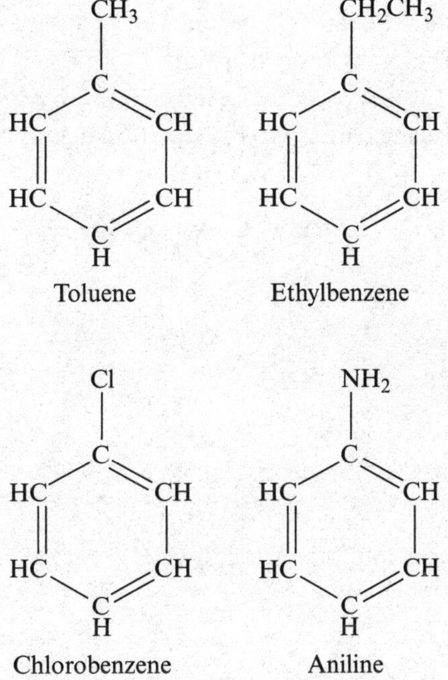


**39.** Which of the liquids has the greatest surface tension and why?

**A.** The surface tension of aniline is greatest because it is capable of hydrogen bonding.
**B.** The surface tension of ethylbenzene is greatest because it experiences the strongest London dispersion forces.
**C.** The surface tension of chlorobenzene is greatest because it has the highest molar mass and is thus most polarizable.
**D.** The surface tension of toluene is greatest because its small size results in the greatest coulombic attraction between molecules.

**40.** How is the surface tension of toluene expected to change with increasing temperature?

**A.** The surface tension will increase because the intermolecular forces increase at higher temperatures.
**B.** The surface tension will decrease because the intermolecular forces increase at higher temperatures.
**C.** The surface tension will increase because the kinetic energy of the molecules is greater at higher temperatures.
**D.** The surface tension will decrease because the kinetic energy of the molecules is greater at higher temperatures.

**41.** Hydrazoic acid, $HN_3$, has a $pK_a$ of 4.72. Which of the following solutions will have a pH greater than 4.72?

**A.** A solution that is 0.10 *M* $HN_3$ and 0.10 *M* $NaN_3$
**B.** A solution that is 0.10 *M* $HN_3$ and 0.15 *M* $NaN_3$
**C.** A solution that is 0.10 *M* $HN_3$ and 0.050 *M* NaOH
**D.** A solution that is 0.10 *M* $HN_3$ and 0.050 *M* HCl

**42.** Which choice shows the species listed in order of decreasing bond angle?

A. $NH_3 > NH_4^+ > NH_2^-$
B. $NH_4^+ > NH_3 > NH_2^-$
C. $NH_2^- > NH_3 > NH_4^+$
D. $NH_2^- > NH_4^+ > NH_3$

**43.** A sample of liquid ethanol and a sample of solid ethanol are at equilibrium at –114 °C and 1.0 atm. Which of the following statements is false?

A. The average kinetic energies of the liquid and the solid are equal.
B. The entropy of the solid is less than the entropy of the liquid.
C. The rate of melting is equal to the rate of freezing.
D. The enthalpy change for the melting process is zero.

**44.** When 4.00 moles of $Ca_3(PO_4)_2$ react according to the equation below, the number of moles of CO produced is

$$\_\_ Ca_3(PO_4)_2 + \_\_ SiO_2 + \_\_ C \rightarrow \_\_ CaSiO_3 + \_\_ P_4 + \_\_ CO$$

A. 4.00
B. 8.00
C. 10.0
D. 20.0

**45.** Mercury(II) ions form red complexes with the reagent $C_{12}H_7S_2N_3O_3$.

A 200.0-mL sample of dental wastewater was evaporated to dryness. It was re-dissolved in 1.0 mL of $H_2O$ and mixed with 1.0 mL of an acidic solution of excess $C_{12}H_7S_2N_3O_3$. The absorbance of the complex was compared with the Beer's law plot below.

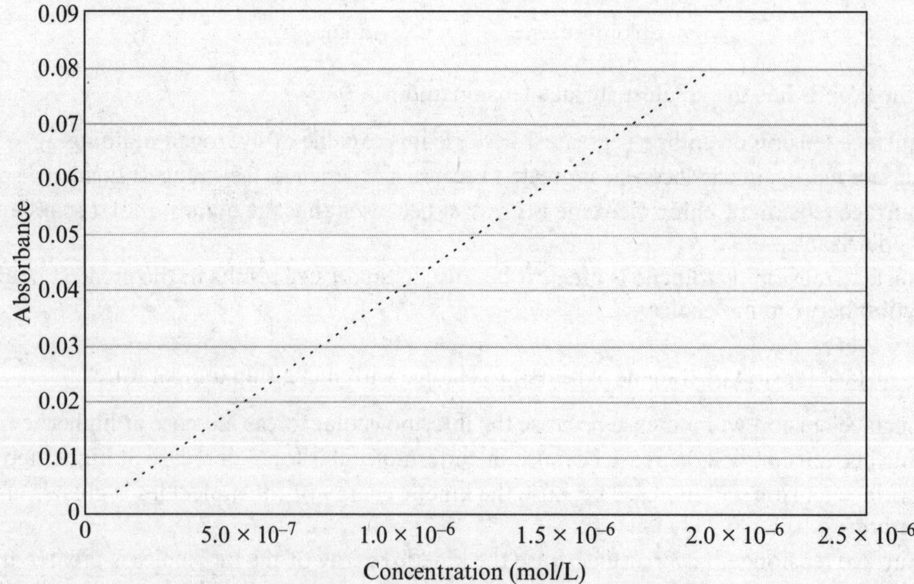


The absorbance of the final solution was 0.040. What is the concentration of $Hg^{2+}$ in the wastewater?

A. $5.0 \times 10^{-8}$ *M*
B. $1.0 \times 10^{-8}$ *M*
C. $5.0 \times 10^{-6}$ *M*
D. $1.0 \times 10^{-6}$ *M*

**46.** Which experimental result provides evidence for the assertion that not all boron atoms are identical?

A. There are three peaks in the photoelectron spectrum of a sample of pure boron.
B. The fourth ionization energy of boron is almost seven times as large as the third ionization energy.
C. There are two peaks in the mass spectrum of a sample of pure boron.
D. Boron can combine with oxygen to form boron trioxide, $B_2O_3$, or boron suboxide, $B_6O$.

**47.** Which of the following choices best describes the identities and temperatures of both Gas A and Gas B as represented in the Maxwell-Boltzmann distribution below?

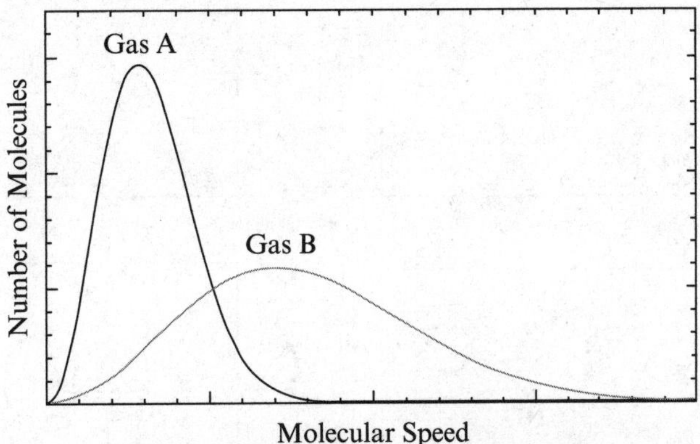


| | Gas A | Gas B |
|---|---|---|
| A. | He at 35 °C | $O_2$ at 35 °C |
| B. | $O_2$ at 35 °C | He at 35 °C |
| C. | He at 35 °C | He at 25 °C |
| D. | He at 35 °C | $O_2$ at 25 °C |

*Questions 48–49 refer to the following scenario.*

When KI(*aq*) and $Pb(NO_3)_2$(*aq*) solutions are combined, a yellow precipitate results. A 0.400 *M* solution of KI(*aq*) is added to a 20.0-mL solution of $Pb(NO_3)_2$(*aq*) until no additional precipitate forms. The resulting solid is filtered and thoroughly dried. The mass of the solid is determined to be 2.31 grams.

**48.** How many moles of $Pb(NO_3)_2$ were initially present?

A. $2.00 \times 10^2$
B. $2.00 \times 10^{-2}$
C. $5.01 \times 10^{-2}$
D. $5.01 \times 10^{-3}$

**49.** Which of the following diagrams best represents the filtrate solution?

**A.**

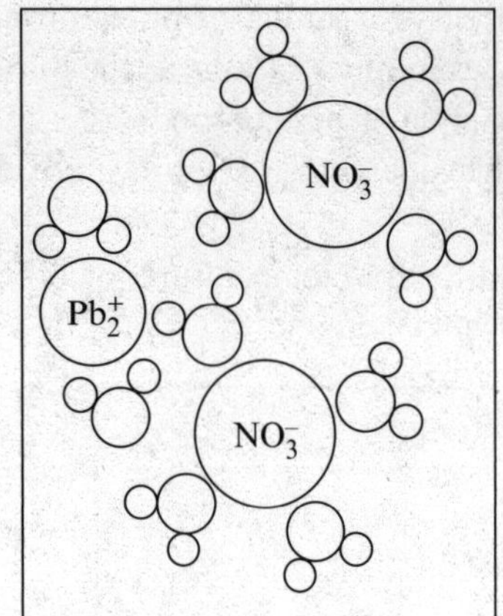


**B.**

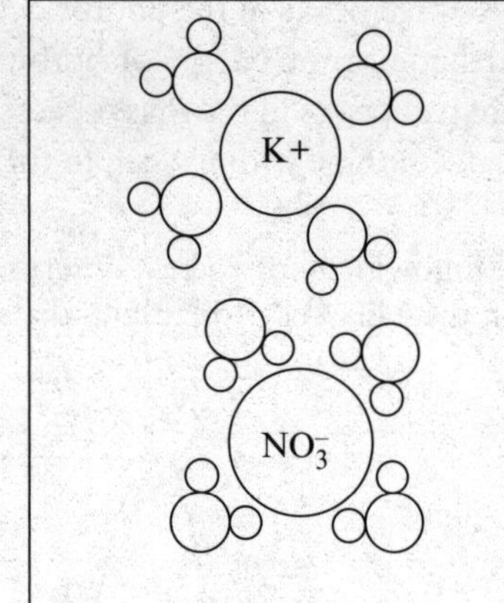


**C.**

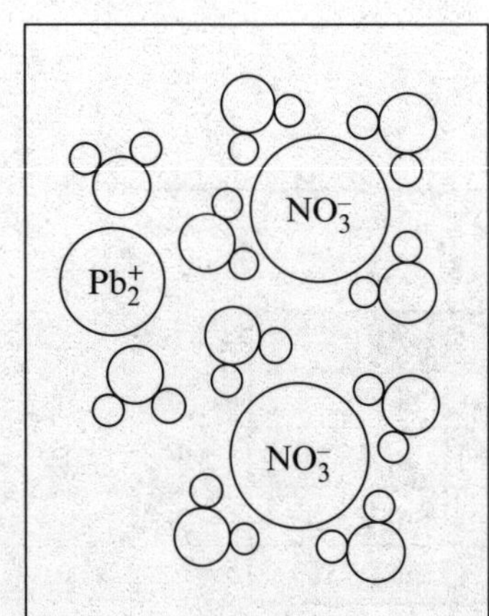


**D.**

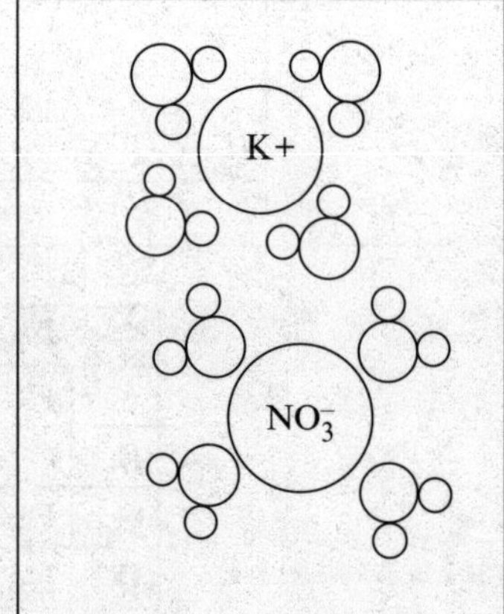


**50.** A general heating curve is depicted below. At what temperature does condensation of the substance occur?

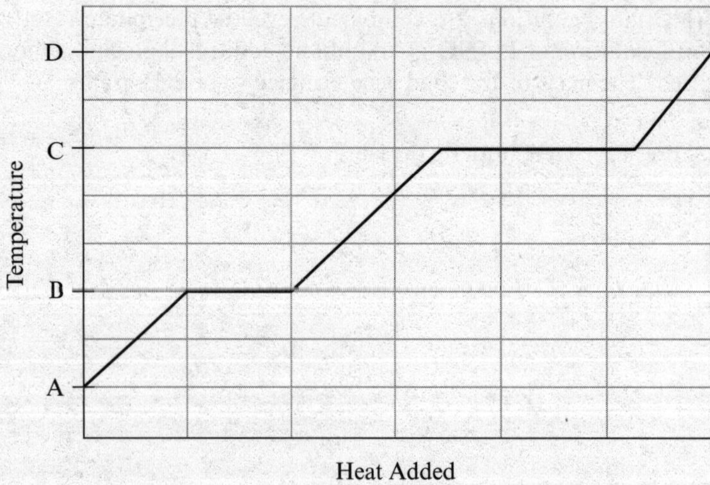


**A.** Temperature A
**B.** Temperature B
**C.** Temperature C
**D.** Temperature D

**51.** The hydrolysis of maltose at pH 7 and 298 K is exergonic.

$$\text{maltose} + H_2O \rightleftharpoons 2\ \text{glucose} \qquad \Delta G^\circ = -15.5\ \text{kJ/mol}$$

Which of the following statements is true?

**A.** $K < 1$ and the reaction is exothermic.
**B.** $K > 1$ and the reaction is endothermic.
**C.** $K < 1$ and no information about the change in enthalpy can be determined.
**D.** $K > 1$ and no information about the change in enthalpy can be determined.

*Questions 52–55 refer to the experiment described below.*

A 25.0-mL solution of phosphoric acid, $H_3PO_4$, with an unknown concentration is titrated with 0.100 *M* NaOH, and the following plot of pH versus volume of base added is obtained.

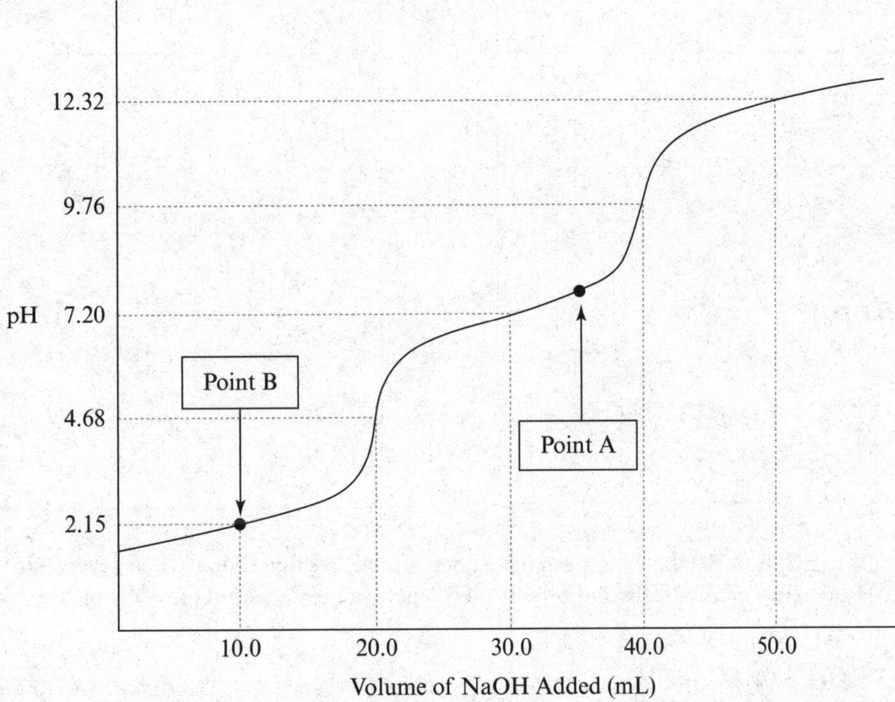


**52.** Using the plot to estimate, what is the $K_{a_2}$ of phosphoric acid?

**A.** $5.0 \times 10^{-12}$
**B.** $1.7 \times 10^{-10}$
**C.** $6.3 \times 10^{-8}$
**D.** $2.0 \times 10^{-5}$

**53.** Which laboratory scenario would cause the concentration of the acid at the first equivalence point to appear to be lower than the true value?

**A.** The chemist used a buret that was wet with distilled water before the sodium hydroxide solution was added.
**B.** The chemist neglected to drain sodium hydroxide solution into the tip of the buret before beginning the titration.
**C.** The chemist diluted the acid in the Erlenmeyer flask with an unknown quantity of water.
**D.** The chemist used an indicator that had a $pK_a$ of less than 3.5.

**54.** At point A in the titration, which particle diagram best represents the composition of the solution? Water is not pictured.

**A.**

$Na^+$ $Na^+$ $HPO_4^{2-}$ $Na^+$ $PO_4^{3-}$ $Na^+$ $Na^+$ $H_2PO_4^-$ $Na^+$

**B.**

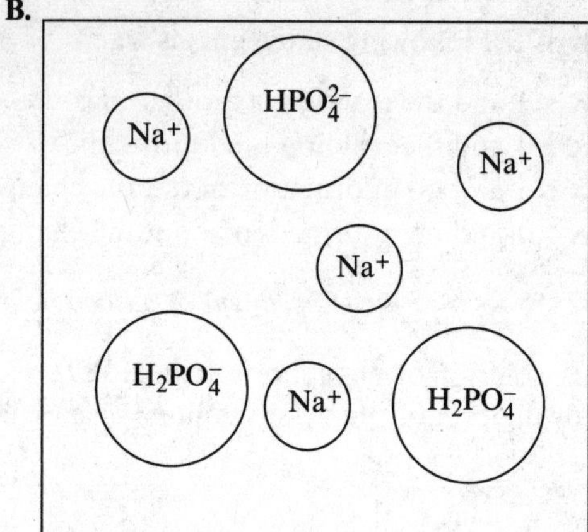


**C.**

$H_2PO_4^-$ $Na^+$ $Na^+$ $Na^+$ $OH^-$ $H_2PO_4^-$ $H_2PO_4^-$ $Na^+$ $Na^+$ $OH^-$

**D.**

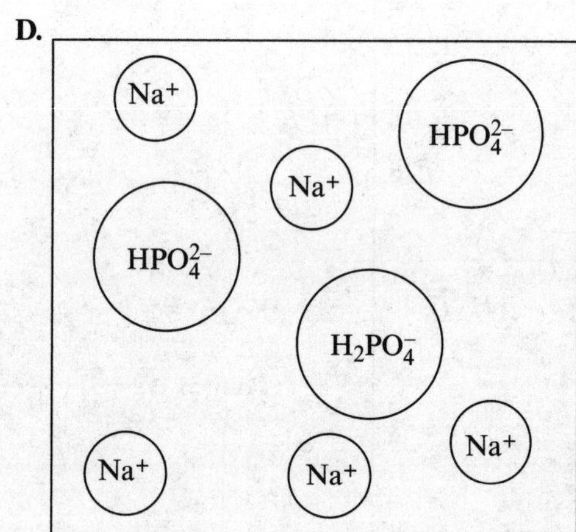


**55.** At point B in the titration, 10.0 mL of base have been added. At this point, when several drops of base are added, the pH does not change substantially. Which reaction best accounts for this observation?

**A.** $H_3PO_4 + OH^- \rightarrow H_2PO_4^- + H_2O$
**B.** $H_2PO_4^- + OH^- \rightarrow HPO_4^{2-} + H_2O$
**C.** $HPO_4^{2-} + OH^- \rightarrow PO_4^{2-} + H_2O$
**D.** $H_3O^+ + OH^- \rightarrow 2\ H_2O$

**56.** The reaction between $Hg_2Cl_4(g)$ and $Al_2Cl_6(g)$ to form $HgAlCl_5(g)$ has a heat of reaction that is close to zero.

$$Hg_2Cl_4 + Al_2Cl_6 \longrightarrow 2\ Cl\text{—}HgAlCl_5$$

Which choice provides the most likely structure of $Hg_2Cl_4$ and the reason for the near-zero enthalpy change?

| | Structure | Reason |
|---|---|---|
| **A.** | $Cl_2Hg\text{—}HgCl_2$ | One Hg–Hg bond is broken, two Al–Cl bonds are broken, two Al–Cl bonds are formed, and two Hg–Cl bonds are formed. |
| **B.** | Cl—Hg(μ-Cl)$_2$Hg—Cl | Two Al–Cl bonds are broken, two Hg–Cl bonds are broken, two Hg–Cl bonds are formed, and two Al–Cl bonds are formed. |
| **C.** | $Cl_2Hg\text{—}HgCl_2$ | One Hg–Hg bond is broken, four Al–Cl bonds are broken, two Al–Cl bonds are formed, and two Hg–Cl bonds are formed. |
| **D.** | Cl—Hg(μ-Cl)$_2$Hg—Cl | Four Al–Cl bonds are broken, two Hg–Cl bonds are broken, four Hg–Cl bonds are formed, and four Al–Cl bonds are formed. |

**57.** A reaction under kinetic control will form

A. the product with the lowest activation energy
B. the product that can be formed in the smallest number of steps
C. the product with the lowest potential energy
D. the product with the highest potential energy

*Questions 58–59 refer to the ionic radii for the four different ions tabulated below.*

| Ion | $A^+$ | $B^+$ | $X^-$ | $Y^-$ |
|---|---|---|---|---|
| Radius (pm) | 116 | 152 | 119 | 167 |

**58.** Based on this information, which choice correctly predicts the order of increasing melting point for the ionic compounds?

**A.** BX < BY < AX < AY
**B.** AY < AX < BY < BX
**C.** BY < AY < BX < AX
**D.** AX < BX < AY < BY

**59.** What are the identities of the four ions?

| | $A^+$ | $B^+$ | $X^-$ | $Y^-$ |
|---|---|---|---|---|
| **A.** | $Na^+$ | $K^+$ | $Cl^-$ | $F^-$ |
| **B.** | $K^+$ | $Na^+$ | $Cl^-$ | $F^-$ |
| **C.** | $Na^+$ | $K^+$ | $F^-$ | $Cl^-$ |
| **D.** | $K^+$ | $Na^+$ | $F^-$ | $Cl^-$ |

**60.** The solubility of elemental iodine in water at 20 °C is about 1.00 g per 3.5 L. Which intermolecular force is responsible for this solubility?

**A.** Ion-dipole
**B.** Dipole-induced dipole
**C.** Hydrogen bonding
**D.** Dipole-dipole

IF YOU FINISH BEFORE TIME IS CALLED, CHECK YOUR WORK ON THIS SECTION ONLY. DO NOT WORK ON ANY OTHER SECTION IN THE TEST.

# Section II: Free Response

**7 questions**

**105 minutes**

**Directions:** Long free-response questions (questions 1–3) are worth 10 points each; allow yourself approximately 23 minutes for each one. Short free-response questions (questions 4–7) are worth 4 points each; allow yourself approximately 9 minutes for each one.

Include examples and equations in your responses where appropriate. Clearly show the method used and the steps involved in arriving at your answers. You must show your work to receive credit. Be mindful of significant figures.

**1.** Iron(II) sulfate is a white solid. It forms a green hydrate when exposed to water. A 5.000-gram impure sample of the hydrated form of iron(II) sulfate is heated until the mass stops changing and only a white solid remains. The mass of the anhydrous sample is 2.755 grams.

**a.** How many moles of water were released from the hydrate?

**b.** The impure anhydrous iron(II) sulfate was diluted with approximately 100 mL of water in an Erlenmeyer flask and acidified with sulfuric acid. It was then titrated with a 0.0521 *M* standardized potassium permanganate solution according to the following equation.

$$5\ Fe^{2+} + 8\ H^{+} + MnO_4^{-} \rightarrow 5\ Fe^{3+} + Mn^{2+} + 4\ H_2O$$

Note that the impurity does not react with any other components of the titration.

A volume of 68.33 mL of $MnO_4^{-}$ is required to titrate the $Fe^{2+}$ solution to the equivalence point. How many moles of $Fe^{2+}$ are present?

**c.** How many water molecules are associated with each formula unit of iron(II) sulfate in the original crystal? Write the molecular formula of the hydrate.

**d.** How would the experimentally determined formula be affected if the mass of the sample had not stopped changing before the mass of water released was determined?

**e.** How many grams of the hydrated form of iron(II) sulfate were present in the original 5.000-gram sample?

**2.** The decomposition of $N_2O$ occurs according to the following equation.

$$2\ N_2O(g) \rightarrow 2\ N_2(g) + O_2(g)$$

**a.** Which structure below, 1 or 2, best represents a molecule of $N_2O$? Clearly explain your choice and support your reasoning by adding formal charges to the atoms in the structures below.

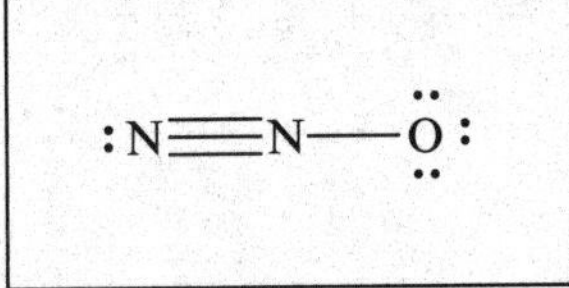


Structure 1

:N—N≡O:

Structure 2

**b.** A sample of $N_2O$ was heated to 1000 K, and the concentration of $N_2O$ was monitored as a function of time. The results are shown below. Determine the rate law for the decomposition of $N_2O$.

| Time (hours) | $[N_2O]$ | $\ln[N_2O]$ | $[N_2O]^{-1}$ |
|---|---|---|---|
| 0 | 1.00 | 0 | 1.00 |
| 1 | 0.223 | –1.50 | 4.81 |
| 2 | 0.0498 | –3.00 | 20.1 |
| 3 | 0.0111 | –4.50 | 90.0 |
| 4 | 0.00247 | –6.00 | 403 |
| 5 | 0.000553 | –7.50 | 1810 |
| 6 | 0.000123 | –9.00 | 8100 |

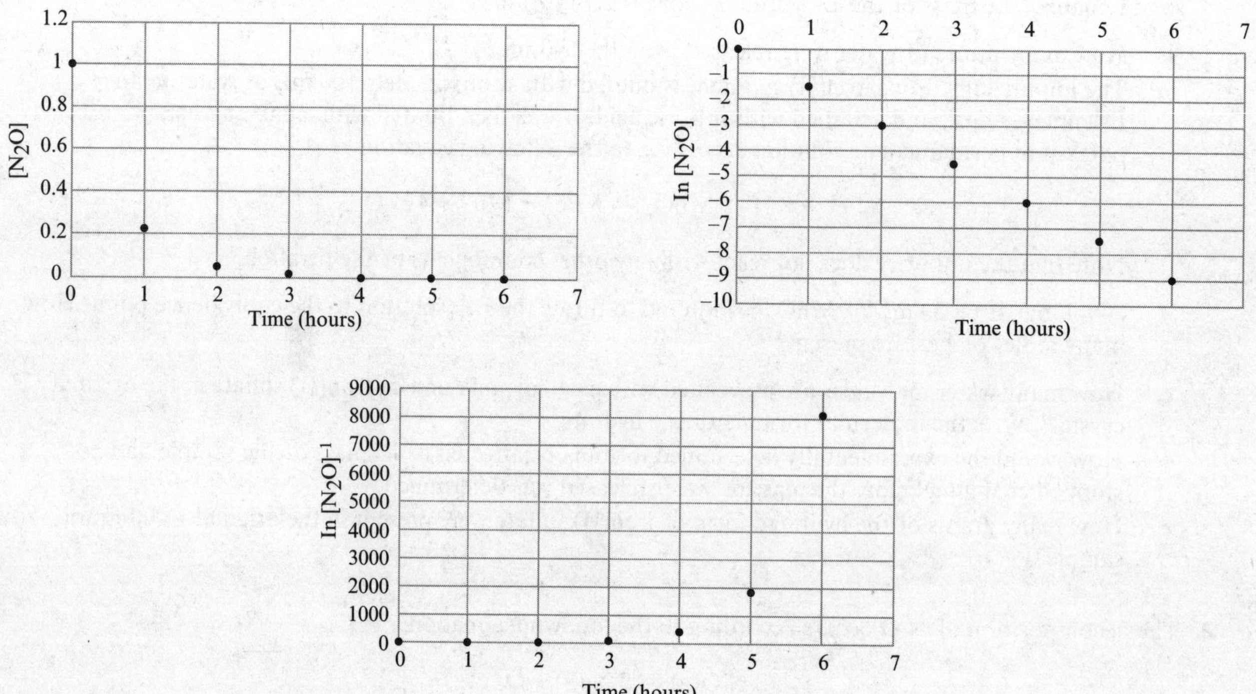

c. Using the standard entropies in the table below, calculate $\Delta S°$ for the reaction. Report your answer in units of J/(mol·K).

| Substance | $S°$ (J/(mol·K)) |
|---|---|
| $N_2O(g)$ | 220 |
| $N_2(g)$ | 192 |
| $O_2(g)$ | 205 |

d. The standard enthalpy of formation, $\Delta H°_f$, of $N_2O(g)$ is 81.6 kJ/mol. Determine the standard molar change in Gibbs free energy, $\Delta G°$, for the reaction at 298 K. Report your answer in units of kJ/mol.

e. On the axes below, sketch a potential energy diagram for the decomposition of $N_2O$. Label the energy of the reactants, energy of the products, activation energy, and enthalpy change for the process.

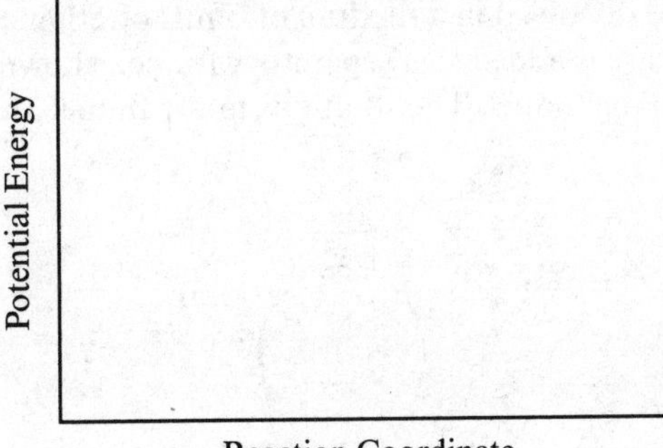

**3.** Azulene, $C_{10}H_8$, is a deep blue molecule that is present in botanicals such as wormwood.

H H
C C
HC C CH
HC C=CH
HC=CH

Azulene
$C_{10}H_8$
MW = 128.17 g·mol$^{-1}$

A 10.0-mg sample of azulene is dissolved in a mixture of 5 mL of ethanol and 25 mL of water to form a bright blue solution. This mixture is added to a separatory funnel, shown below. A 30-mL portion of cyclohexane, $C_6H_{12}$, is added to the funnel. The solutions in the funnel are mixed well and allowed to separate into two layers.

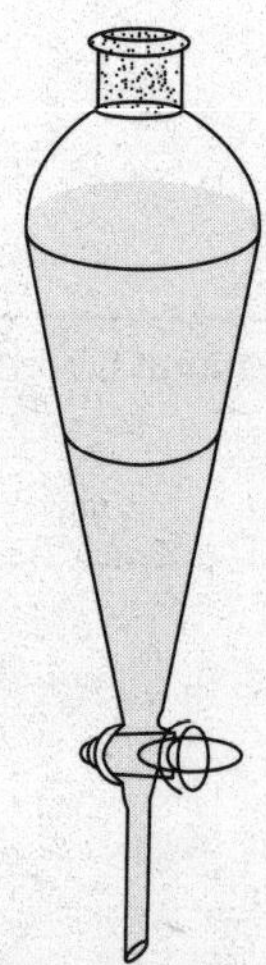

| Compound | Boiling Temperature (°C) | Melting Temperature (°C) | Density (g/mL) | Color |
|---|---|---|---|---|
| Azulene | 242 | 99 | 1.04 | Blue |
| Water | 100 | 0 | 1.00 | Colorless |
| Ethanol | 78 | –114 | 0.79 | Colorless |
| Cyclohexane | 81 | 6 | 0.78 | Colorless |

a. When the layers have separated, which solvent, water or cyclohexane, makes up the top layer and which makes up the bottom layer?

b. When the layers have separated, one layer is blue and the other layer is colorless. Which solvent layer is blue? Explain your response in terms of intermolecular forces.

c. How many $\pi$ bonds are present in a molecule of azulene?

d. What are the hybridization and approximate bond angles of the carbon atoms in azulene?

e. How might azulene be isolated from the solution in the separatory funnel? Briefly describe a procedure for recovering the 10.0 mg of pure azulene.

**4.** The reaction between $CuSO_4$ and NaOH was investigated using coffee cup calorimetry. The net ionic equation is shown below.

$$Cu^{2+}(aq) + 2\ OH^{-}(aq) \rightarrow Cu(OH)_2(s)$$

A volume of standardized 6.00 *M* NaOH is added to 25.0 mL of the $CuSO_4$ solution. Both solutions are initially at the same temperature. Assume that the density of the solution is 1.0 g/mL and the specific heat of the solution is 4.18 J/(g·°C). The results are shown in the table below.

| Volume of 6.00 *M* NaOH (mL) | Mass of $CuSO_4$ in 25.0 mL of Solution (g) | Initial Temperature of Solution (°C) | Final Temperature of Solution (°C) |
|---|---|---|---|
| 25.0 | 4.00 g | 25.2 | 30.2 |

**a.** What is the limiting reactant, $CuSO_4$ or NaOH? Justify your response with a calculation.

**b.** Calculate the magnitude of *q*, the thermal energy change, when the NaOH was added to the $CuSO_4$. Express your answer with appropriate units.

**c.** Determine the experimental value of $\Delta H°$ for the reaction between $CuSO_4$ and NaOH in units of $kJ/mol_{rxn}$.

**d.** Enthalpies of formation for the substances involved in the reaction are shown in the table below. What is the accepted value for the enthalpy of the reaction, and what is a likely reason for the discrepancy between the accepted value and the value determined experimentally?

| Substance | $\Delta H°_f$ (kJ/mol) |
|---|---|
| $Cu^{2+}(aq)$ | 64.8 |
| $OH^{-}(aq)$ | –230 |
| $Cu(OH)_2(s)$ | –450 |

**5.** When equimolar quantities of $SbCl_3$ and $GaCl_3$ are mixed under appropriate conditions, they react according to the equation below.

$$SbCl_3 + GaCl_3 \rightarrow SbGaCl_6$$

A research group would like to determine whether the structure of $SbGaCl_6$ has the form $[SbCl_2]^+[GaCl_4]^-$ or $[GaCl_2]^+[SbCl_4]^-$. They use a spectroscopic technique to determine that the cation has a bent molecular geometry.

**a.** Which of the two forms is the correct structure and why? Draw Lewis structures of both possible cations to support your response.

**b.** The spectroscopic technique gave the researchers information about the geometry of the molecule. Was the spectroscopic technique one that utilized infrared radiation or one that utilized ultraviolet/visible radiation? Explain your answer.

**c.** Draw the Lewis structure of the anion.

**d.** State the molecular geometry of the anion.

**6.** Sodium-24 is a radioactive isotope of sodium.

**a.** In terms of atomic structure, how do sodium-24 and the stable isotope of sodium, sodium-23, differ?

**b.** The rate of decay of $^{24}Na$ was determined experimentally and the results are plotted below.

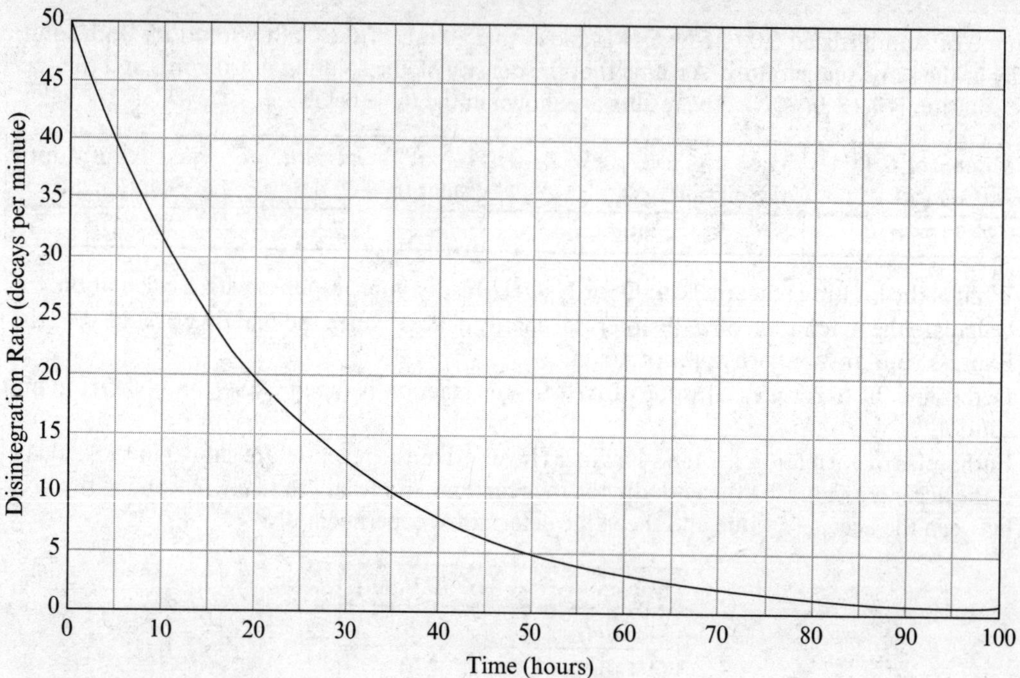


**i.** Determine the half-life of $^{24}Na$.

**ii.** Determine the rate constant for the decay of $^{24}Na$. Remember to include appropriate units.

**c.** The data can be used to show that the decay is first order, as indicated on the graph below. Label the vertical axis of the graph.

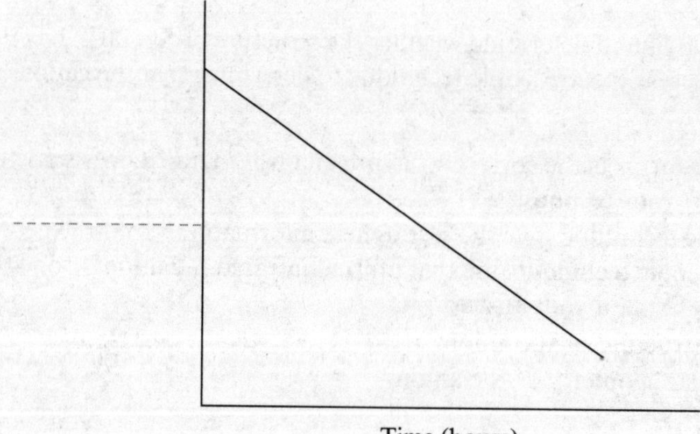

7. Consider two containers, each with a volume of 10.0 L at 298 K, as shown below. One container holds 1.0 mol He(*g*) and the other holds 1.0 mol $CH_4$(*g*). Assume that the He and $CH_4$ gases exhibit ideal behavior.

a. Is the pressure in the container of He greater than, equal to, or less than the pressure in the container of $CH_4$?

b. Is the average kinetic energy of the He molecules greater than, equal to, or less than the average kinetic energy of the $CH_4$ molecules? Explain your answer.

c. In which container do the molecules have a greater average speed? Justify your answer.

d. The sample of $CH_4$ is heated at constant pressure. On the graph below, sketch the expected plot of volume versus temperature.

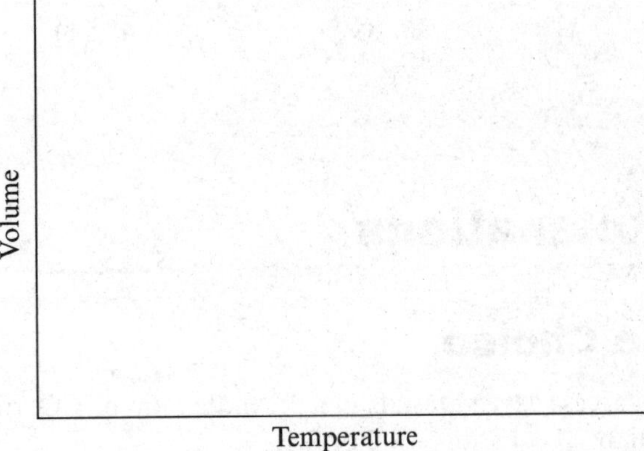

IF YOU FINISH BEFORE TIME IS CALLED, CHECK YOUR WORK ON THIS SECTION ONLY. DO NOT WORK ON ANY OTHER SECTION IN THE TEST.

# Answer Key

## Section I: Multiple Choice

| | | | | |
|---|---|---|---|---|
| 1. B | 13. A | 25. A | 37. D | 49. B |
| 2. C | 14. D | 26. C | 38. C | 50. C |
| 3. B | 15. A | 27. C | 39. A | 51. D |
| 4. C | 16. B | 28. B | 40. D | 52. C |
| 5. A | 17. B | 29. C | 41. B | 53. D |
| 6. D | 18. D | 30. A | 42. B | 54. D |
| 7. A | 19. D | 31. B | 43. D | 55. A |
| 8. A | 20. B | 32. C | 44. D | 56. B |
| 9. C | 21. A | 33. A | 45. B | 57. A |
| 10. A | 22. D | 34. A | 46. C | 58. C |
| 11. D | 23. C | 35. B | 47. B | 59. C |
| 12. D | 24. D | 36. C | 48. D | 60. B |

# Answers and Explanations

## Section I: Multiple Choice

1. **B.** Combustion of $C_2H_2$ in excess oxygen results in a 2:1 mole ratio of $CO_2$ to $H_2O$, choice B. The ratio of $CO_2$:$H_2O$ in the overall balanced equation is 4:2 or 2:1.

$$2\ C_2H_2 + 5\ O_2 \rightarrow 4\ CO_2 + 2\ H_2O$$

2. **C.** When ammonia, a weak base, is added to acetic acid, a weak acid, ammonium acetate, a strong electrolyte, is formed. The diagram in choice C shows $NH_4^+$ and $CH_3CO_2^-$ and is the correct choice.

$$NH_3(aq) + CH_3CO_2H(aq) \rightarrow NH_4^+(aq) + CH_3CO_2^-(aq)$$

3. **B.** The larger the $K_a$, the stronger the acid. $HPO_4^{2-}$ is a weaker acid than $HSO_3^-$ because it has a smaller $K_a$. The weaker the acid, the stronger the conjugate base, so $PO_4^{3-}$ is a stronger base than $SO_3^{2-}$. Therefore, choice B is correct.

4. **C.** If the shells were not divided into subshells, there would only be one peak per shell, resulting in fewer peaks in the spectra, choice C. There would also be different relative numbers of electrons per peak.

5. **A.** Element X has the electron configuration $1s^2 2s^2 2p^5$, which corresponds to an atom of fluorine. Elements Y and Z are sodium and magnesium, respectively. Of these three elements, only fluorine achieves an octet by gaining electrons, choice A.

6. **D.** Element Y is expected to have the largest atomic radius because it has a smaller nuclear charge than element Z and its electrons experience more shielding than the valence electrons of element X, choice D. Increasing nuclear charge causes the atomic radii to decrease from left to right across the periodic table.

7. **A.** Element Y has a lower effective nuclear charge than element Z, choice A. The lower effective nuclear charge of sodium results in a lower binding energy for its 2*s* electrons than that of the 2*s* electrons in magnesium.

**8.** **A.** The negative value of $\Delta H°$ indicates that heat is being released in an exothermic reaction. As shown in the following equation, 7.85 kJ of heat is released in this reaction, choice A.

$$2.00\ \cancel{g\,P} \times \frac{1\ \cancel{mol\,P}}{30.97\ \cancel{g\,P}} \times \frac{1\ \cancel{mol\,rxn}}{2\ \cancel{mol\,P}} \times \frac{-243\ kJ}{1\ \cancel{mol\,rxn}} = -7.85\ kJ$$

$$\frac{-240\ kJ}{30} \approx -8\ kJ$$

**9.** **C.** Cooling the reaction would cause the endothermic reaction to shift toward reactants, decreasing the equilibrium constant. A decreased equilibrium constant would shift the reaction from two moles of gaseous products to one mole of gaseous reactant. This decrease in the number of moles at constant pressure would cause the volume in the piston to decrease, choice C.

**10.** **A.** Since the addition of argon occurs under constant total pressure, it would cause the volume in the piston to increase, decreasing the partial pressures of the reactant and products in the piston. The increase in volume would cause the reaction to shift toward products, the side with more moles of gases, choice A.

**11.** **D.** The amount of heat gained by the ice as it warms to 0 °C (melts) and then warms to 10.0 °C (as liquid water) is equal to the amount of heat lost by the water that cools from 40.0 °C to 10.0 °C.

$$q = mc\Delta T$$

$$q_{lost} = -q_{gained}$$

$$mc_{water}\Delta T = -[mc_{ice}\Delta T + m\Delta H_{fus} + mc_{water}\Delta T]$$

$$m(4\ J/(\cancel{g}\cdot\cancel{°C}))(10.0\ \cancel{°C} - 40.0\ \cancel{°C}) = -(10.0\ \cancel{g})(2\ J/(\cancel{g}\cdot\cancel{°C}))(0\ \cancel{°C} - (-5.0\ \cancel{°C})) +$$

$$(10.0\ \cancel{g})(0.3\ \cancel{kJ}/\cancel{g})\left(\frac{1000\ J}{1\ \cancel{kJ}}\right) + (10.0\ \cancel{g})(4\ J/(\cancel{g}\cdot\cancel{°C}))(40.0\ \cancel{°C} - 0\ \cancel{°C})$$

$$m(-120\ J/g) = -(100 + 3000 + 400)\ J$$

$$m = \frac{3500\ \cancel{J}}{120\ \cancel{J}/g} = \frac{350}{12}\ g \approx 30\ g$$

Choice D, 29 g, is the closest answer choice.

**12.** **D.** When formal charges are assigned to the atoms in the choices, as shown in the figure below, only the structure in choice D has zero formal charges. All of the other structures have separation of charge, so they are predicted to be less important contributors to the structure.

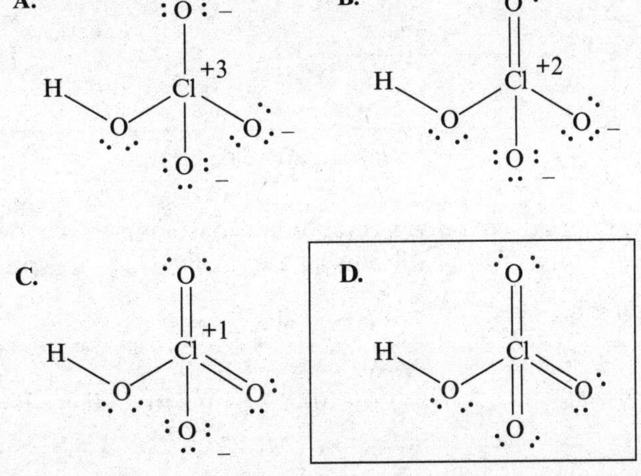

**13.** **A.** $TeCl_5^-$, $SF_6$, and $BrF_5$ all have octahedral electronic geometry. $TeCl_5^-$ and $BrF_5$ have square pyramidal molecular geometry. Therefore, choice A is correct.

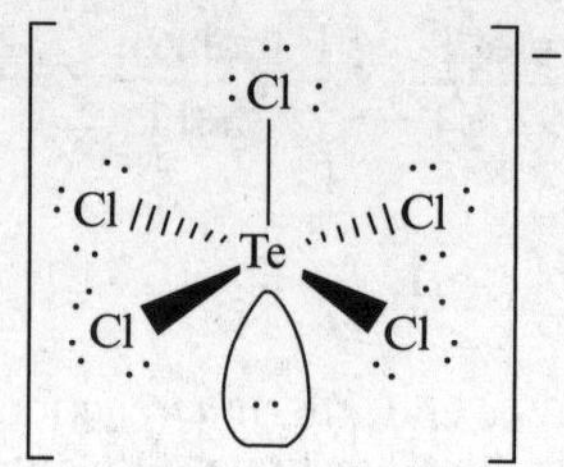


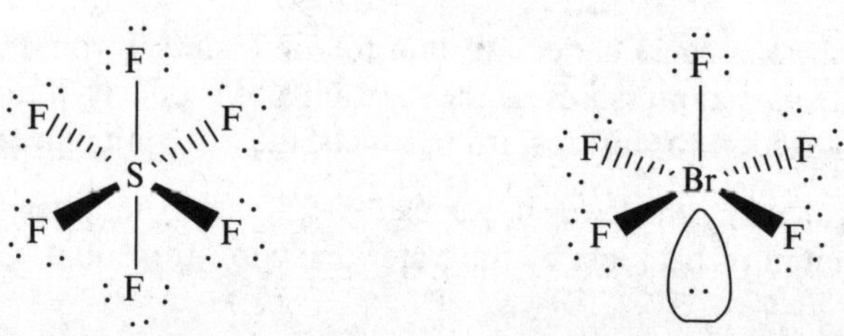


**14.** **D.** Silicon atoms are larger than carbon atoms; the bond between two silicon atoms is longer and, therefore, weaker than the bond between two carbon atoms. A double bond, in general, which is composed of one $\sigma$ and one $\pi$ bond, is shorter and stronger than a single bond. Therefore, choice D correctly ranks the bonds according to increasing bond strength: Si–Si single bond < C–C single bond < C–C double bond.

**15.** **A.** Ionization is endothermic, choice A. Absorption of a photon causes the potential energy of the products to be greater than the potential energy of the reactants.

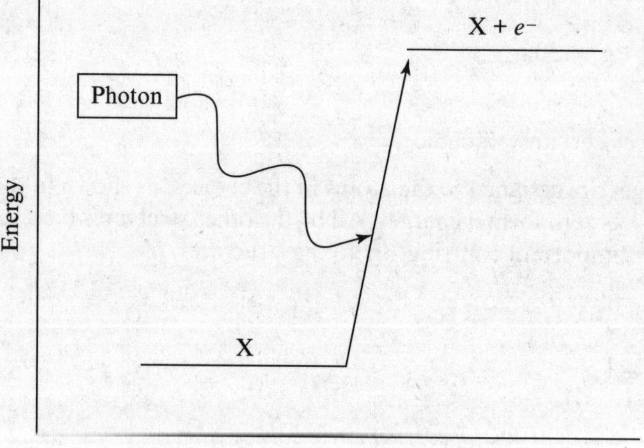


**16.** **B.** The reaction has been reversed, so the new equilibrium constant is the reciprocal of the equilibrium constant in the forward direction. The coefficients have been divided by 2, so the new equilibrium constant must be raised to the power of $\frac{1}{2}$.

$$K = (K_p^{-1})^{\frac{1}{2}} = K^{-\frac{1}{2}} = (1.0\times10^{10})^{-\frac{1}{2}} = 1.0\times10^{-5}, \text{ choice B}$$

Multiplication of the exponent by $-\frac{1}{2}$ gives the new equilibrium constant.

**17.** **B.** In the reverse reaction, two gaseous atoms come together to form one gaseous atom. This involves a decrease in entropy ($\Delta S < 0$). Bond formation releases energy, so the reverse process is exothermic ($\Delta H < 0$). Therefore, choice B is correct.

**18.** **D.** Calcium fluoride dissociates according to the equation

$$CaF_2(s) \rightleftharpoons Ca^{2+}(aq) + 2\,F^-(aq)$$

The solubility product expression is

$$K_{sp} = [Ca^{2+}][F^-]^2$$

Since $[Ca^{2+}] = x$, then $[F^-] = 2x$

$$K_{sp} = x(2x)^2 = 4x^3\text{, choice D}$$

**19.** **D.** In an exothermic process, heat can be thought of as a "product" of the reaction. As temperature is decreased, the equilibrium constant increases and more products must be formed to achieve equilibrium, choice D.

**20.** **B.** Enzymes lower the energy of the transition state compared with the transition state of the uncatalyzed reaction. Thus, choice B is the false statement.

**21.** **A.** The rate law for mechanism 1, based on the slow step, is

$$\text{rate} = k\,[NO_2Cl]$$

The rate law for mechanism 2 is based on the slow step.

$$\text{rate} = k\,[N_2O_4]$$

Because an intermediate is not included in the rate law, the equal rates of the forward and reverse reactions in the fast step are taken into account.

$$k_1[NO_2Cl]^2 = k_{-1}[N_2O_4][Cl_2]$$

The overall rate for mechanism 2 is

$$\text{rate} = k[NO_2Cl]^2[Cl_2]^{-1}$$

Only mechanism 1 is consistent with the experimental rate law, choice A.

**22.** **D.** A glass is an amorphous solid, not a crystalline solid. Therefore, choice D is the false statement.

**23.** **C.** Notice that for every 10.0-minute interval, $[A]^{-1}$ increases by 0.25. This linear relationship indicates a second order process, choice C.

**24.** **D.** For this reaction, the equilibrium expression is

$$K = \frac{[AB]}{[A_2][B_2]} = 4$$

In order to determine in which direction the reaction will shift, $Q$ must be compared with $K$. For choice D, $Q > K$, so the reaction will shift toward reactants.

$$Q = \frac{[AB]}{[A_2][B_2]} = \frac{8}{(1)(1)} = 8$$

**25.** **A.** The reaction with the most positive standard cell potential has the largest equilibrium constant. The value of $E°$ is determined by

$$E°_{cell} = E°_{reduction} - E°_{oxidation}$$

The combination of reactions that leads to the most positive overall cell potential is that of the reduction of $Pb^{2+}$ and the oxidation of Al, as shown in choice A.

**26.** **C.** The stronger the intermolecular attractions between molecules of a substance, the more energy is required to break the attractive forces during the process of melting. Of the choices given, molecules of $H_2O$ experience the strongest intermolecular forces, hydrogen bonding. Molecules of $CF_4$ and molecules of $SiCl_4$ both experience London dispersion forces. $CF_4$ is smaller, has fewer electrons, and is less polarizable than $SiCl_4$, so it requires a smaller amount of energy to melt. The correct order of decreasing molar enthalpy of fusion is shown in choice C: $H_2O > SiCl_4 > CF_4$.

**27.** **C.** The change in partial pressure of the gases upon expansion into a larger volume can be calculated as follows.

$$P_1V_1 = P_2V_2$$

$$P_2 = \frac{P_1V_1}{V_2} = \frac{(8.0\text{ atm})(5.0\text{ }\cancel{L})}{8.0\text{ }\cancel{L}} = 5.0\text{ atm } H_2$$

$$P_1V_1 = P_2V_2$$

$$P_2 = \frac{P_1V_1}{V_2} = \frac{(4.0\text{ atm})(3.0\text{ }\cancel{L})}{8.0\text{ }\cancel{L}} = 1.5\text{ atm He}$$

The total pressure is the sum of the partial pressures.

$$P_{total} = P_{He} + P_{H_2} = 6.5\text{ atm, choice C}$$

**28.** **B.** The volume may be determined from the balanced net ionic equation.

$$6\,I^- + 14\,H^+ + Cr_2O_7^{2-} \rightarrow 3\,I_2 + 2\,Cr^{3+} + 7\,H_2O$$

$$\cancel{0.050}\text{ }\cancel{L\,I^-} \times \frac{0.100\text{ }\cancel{mol\,I^-}}{1\text{ }\cancel{L\,I^-}} \times \frac{1\text{ }\cancel{mol\,Cr_2O_7^{2-}}}{6\text{ }\cancel{mol\,I^-}} \times \frac{1\text{ L }Cr_2O_7^{2-}}{\cancel{0.05}\text{ }\cancel{mol\,Cr_2O_7^{2-}}} = 0.0167\text{ L }Cr_2O_7^{2-} = 16.7\text{ mL, choice B}$$

**29.** **C.** The electrolysis of $Al^{3+}$ is a 3-electron transfer.

$$Al^{3+} + 3\,e^- \rightarrow Al$$

$$\cancel{96{,}500}\text{ }\cancel{s} \times \frac{\cancel{3.00}\text{ }\cancel{C}}{1\text{ }\cancel{s}} \times \frac{1\text{ }\cancel{mol\,e^-}}{\cancel{96{,}500}\text{ }\cancel{C}} \times \frac{1\text{ }\cancel{mol\,Al}}{\cancel{3}\text{ }\cancel{mol\,e^-}} \times \frac{27\text{ g Al}}{1\text{ }\cancel{mol\,Al}} = 27\text{ g Al, choice C}$$

**30.** **A.** The oxidation number of Mn is going from +7 to +6 as the reaction proceeds to the right; therefore, choice A is correct.

**31.** **B.** The oxidation number of oxygen in $SO_3^{2-}$ is –2, so in order for the ion to have an overall –2 charge, the oxidation number of sulfur must be +4, choice B.

**32.** **C.** Propionaldehyde and propanol are miscible liquids with similar densities. They have sufficiently different boiling points to be separated by distillation, choice C.

**33.** **A.** The molarity is the number of moles of solute per liter of solution.

$$5\text{ mL} \times \frac{0.802\text{ g}}{1\text{ mL}} \times \frac{1\text{ mol}}{60\text{ g}} \approx \frac{4}{60} = \frac{1}{15}\text{ mol}$$

$$\frac{\left(\frac{1}{15}\text{ mol}\right)}{\left(\frac{1}{10}\text{ L}\right)} = \frac{2}{3} = 0.67\ M\text{, choice A}$$

**34.** **A.** Freezing decreases the entropy of a liquid as it is converted to a solid, but this is offset by the increase in entropy of the surroundings, choice A.

**35.** **B.** Let [X] be the concentration of the reactant. The ratio of (rate 2):(rate 1) is 27.

$$\frac{\text{rate 2}}{\text{rate 1}} = \left(\frac{3[\text{X}]}{[\text{X}]}\right)^x$$

$$27 = 3^x$$

$$x = 3\text{, choice B}$$

**36.** **C.** AgBr, choice C, has a smaller $K_{sp}$ than CuBr. It has a lower solubility and precipitates first as the sodium bromide is added. Sodium salts are soluble in an aqueous solution.

**37.** **D.** Neon gas is expected to deviate very little from ideal behavior. Neon atoms occupy volume and they also experience intermolecular attractions, so choice D is the false statement. London dispersion forces are present between all molecules. The intermolecular forces in neon are weak, but they exist.

**38.** **C.** The p$K_w$ decreases with increasing temperature, but autoionization always occurs in liquid water, so choice C is the false statement. Instead, increasing temperature shifts the reaction toward products. This increases $K_w$ and decreases p$K_w$, which is equal to $-\log K_w$.

**39.** **A.** Hydrogen bonding is the strongest intermolecular force and has the greatest influence on surface tension. The $-NH_2$ group in aniline is capable of hydrogen bonding and, therefore, has the greatest surface tension, choice A.

**40.** **D.** As temperature, or average kinetic energy, of the molecules increases, they become increasingly capable of overcoming intermolecular forces. This increase in average kinetic energy results in a decrease in surface tension, choice D.

**41.** **B.** The Henderson-Hasselbalch equation may be used to determine the pH of a buffer solution. In order for the pH to be greater than the p$K_a$ of the weak acid, the $[A^-]/[HA]$ concentration must be greater than zero. For the solution described in choice B, $[A^-] = [N_3^-] = 0.15\ M$, and $[HA] = [HN_3] = 0.10\ M$.

$$\text{pH} = \text{p}K_a + \log\frac{[N_3^-]}{[HN_3]} = 4.76 + \log\frac{0.15}{0.10} > 4.76$$

**42.** **B.** All three molecules have tetrahedral electronic geometry, but $NH_4^+$, $NH_3$, and $NH_2^-$ have zero, one, and two lone pairs of electrons, respectively. Bond angle decreases with an increase in the number of lone pairs surrounding the central atom. Choice B shows the correct order: $NH_4^+ > NH_3 > NH_2^-$.

**43.** **D.** The process of melting is endothermic; energy is being absorbed by the system, so the enthalpy change is not zero. Choice D is the false statement.

**44.** **D.** The balanced chemical equation is

$$2\ Ca_3(PO_4)_2 + 6\ SiO_2 + 10\ C \rightarrow 6\ CaSiO_3 + P_4 + 10\ CO$$

$$4.00\ \cancel{\text{mol } Ca_3(PO_4)_2} \times \frac{10\text{ mol CO}}{2\ \cancel{\text{mol } Ca_3(PO_4)_2}} = 20.0\text{ mol CO, choice D}$$

**45.** **B.** The absorbance can be used to find the $Hg^{2+}$ concentration in the final solution. An absorbance of 0.04 corresponds to a concentration of $1.0 \times 10^{-6}$ *M*. The original concentration of mercury in the dental wastewater can be determined using the dilution equation.

$$M_1V_1 = M_2V_2$$
$$(1.0 \times 10^{-6}\ M)(2.0\ \text{mL}) = M_2(200.0\ \text{mL})$$
$$M_2 = 1.0 \times 10^{-8}\ M\text{, choice B}$$

**46.** **C.** The two peaks in the mass spectrum provide the proof that boron exists as two isotopes that differ by the number of neutrons in the nucleus, choice C.

**47.** **B.** The temperature of both gases in choice B is the same, so the curve corresponding to Gas A represents a gas with a larger molecular weight, because the molecules have a lower average velocity. The curve corresponding to Gas B represents a gas with a smaller molecular weight because the molecules have a higher average velocity.

**48.** **D.** The solid formed is $PbI_2$. The number of moles of $Pb(NO_3)_2$ can be determined using the stoichiometry of the chemical equation.

$$Pb(NO_3)_2(aq) + 2KI(aq) \rightarrow PbI_2(s) + 2KNO_3(aq)$$

The molar mass of $PbI_2$ is 461 g·mol$^{-1}$.

$$2.31\ \cancel{\text{g PbI}_2} \times \frac{1\ \cancel{\text{mol PbI}_2}}{461\ \cancel{\text{g PbI}_2}} \times \frac{1\ \text{mol Pb(NO}_3)_2}{1\ \cancel{\text{mol PbI}_2}} = 5.01 \times 10^{-3}\ \text{mol Pb(NO}_3)_2\text{, choice D}$$

**49.** **B.** When the precipitation reaction is complete, only the spectator ions remain. Notice that in diagram B, the large oxygen atoms of the water molecules are oriented toward the $K^+$ cations and the small hydrogen atoms are oriented toward the $NO_3^-$ anions.

**50.** **C.** Condensation occurs at the higher temperature plateau, temperature C, at which the phase change from liquid to gas or gas to liquid occurs.

**51.** **D.** An exergonic reaction is one in which $\Delta G° < 0$. Since $\Delta G° = -RT \ln K$, the equilibrium constant is greater than 1. No information is given about whether the reaction is endothermic or exothermic, choice D.

**52.** **C.** The $pK_{a_2}$ is equal to the pH halfway between the first and second equivalence points. The $pK_{a_2}$ is 7.20.

$$K_{a_2} = 10^{-7.20}$$

This is estimated to be close to $6.3 \times 10^{-8}$, choice C.

**53.** **D.** In order to accurately determine the equivalence point of a titration, the indicator selected must have a $pK_a$ within 1 pH unit of the pH of the equivalence point of interest. Therefore, if the chemist used an indicator that had a $pK_a$ of less than 3.5 , choice D, it would change color well below the equivalence point, at which the pH is 4.68, and the acid concentration determined in the titration would appear to be lower than the true value.

**54.** **D.** At point A in the titration, addition of $OH^-$ deprotonates $H_2PO_4^-$. The point is more than halfway to the second equivalence point, so $HPO_4^{2-}$ anions outnumber $H_2PO_4^-$ anions, as shown in the particle diagram in choice D.

**55.** **A.** At point B in the titration, within the first buffering region of the curve, addition of $OH^-$ deprotonates $H_3PO_4$, as shown in the reaction in choice A.

**56.** **B.** The correct structure is the one in which the bonds broken and the bonds formed are approximately equivalent, as in choice B.

**57.** **A.** A reaction under kinetic control will form the product with the lowest activation energy, choice A.

**58.** C. The attractions between positive and negative ions in an ionic crystal decrease with increasing ionic radius. The compound formed by the ions with the largest radii has the highest melting point, and the compound formed by the ions with the smallest radii has the lowest melting point. Choice C correctly predicts the order of increasing melting point for the ionic compounds: BY < AY < BX < AX.

**59.** C. Ionic radii increase down a group on the periodic table. Choice C correctly lists the identities of the four ions.

**60.** B. Iodine is a nonpolar solute and water is a polar solvent. Iodine dissolves in water due to dipole-induced dipole attractions, choice B.

## Section II: Free Response

**1.** **a.** The amount of water released can be calculated from the initial mass and the mass after heating.

$$(5.000 \text{ g impure sample}) - (2.755 \text{ g impure anhydrous sample}) = 2.245 \text{ g } H_2O$$

$$2.245 \cancel{\text{ g } H_2O} \times \frac{1 \text{ mol } H_2O}{18.02 \cancel{\text{ g } H_2O}} = 0.1246 \text{ mol } H_2O$$

**b.** The number of moles of $Fe^{2+}$ present may be determined using the stoichiometry of the balanced equation.

$$68.33 \cancel{\text{ mL } MnO_4^-} \times \frac{1 \cancel{\text{ L } MnO_4^-}}{1000 \cancel{\text{ mL } MnO_4^-}} \times \frac{0.0521 \cancel{\text{ mol } MnO_4^-}}{1 \cancel{\text{ L } MnO_4^-}} \times \frac{5 \text{ mol } Fe^{2+}}{1 \cancel{\text{ mol } MnO_4^-}} = 0.0178 \text{ mol } Fe^{2+}$$

**c.** The formula of the hydrate is determined by using the mole ration of $H_2O$:$Fe^{2+}$.

$$\frac{0.1246 \text{ mol } H_2O}{0.0178 \text{ mol } Fe^{2+}} = 7$$

$$FeSO_4 \cdot 7\, H_2O$$

**d.** When the mass stops changing, all of the water has been removed and the mass of the anhydrous sample can be accurately determined. Before the mass stops changing, water is still present, the sample is not anhydrous, and fewer $H_2O$ molecules per $FeSO_4$ formula would be obtained.

**e.** The formula weight of $FeSO_4 \cdot 7\, H_2O = 278.05 \text{ g·mol}^{-1}$. The original number of grams of hydrate present may be found by multiplying the moles of $FeSO_4 \cdot 7H_2O$ by the formula weight.

$$0.0178 \cancel{\text{ mol } FeSO_4 \cdot 7\, H_2O} \times \frac{278.05 \text{ g } FeSO_4 \cdot 7\, H_2O}{1 \cancel{\text{ mol } FeSO_4 \cdot 7\, H_2O}} = 4.95 \text{ g } FeSO_4 \cdot 7\, H_2O$$

**2.** **a.** Structure 1 is a better representation of the molecule. Structure 2 has adjacent formal positive charges on the central nitrogen and the oxygen. This is too repulsive an arrangement to make a significant contribution to the resonance hybrid.

| Structure 1 | Structure 2 |
|---|---|
| +1   −1 | −2   +1   +1 |
| :N≡N—Ö: (O with three lone pairs) | :N̈—N≡O: (terminal N with three lone pairs) |

Structure 1

Structure 2

**b.** The graph of ln [$N_2O$] versus time is linear, so the reaction is first order in $N_2O$.

$$\text{rate} = k[N_2O]$$

**c.** The entropy change is calculated as follows.

$$\Delta S^\circ = \sum S^\circ_{\text{products}} - \sum S^\circ_{\text{reactants}}$$
$$\Delta S^\circ = [2(192\ \text{J/(mol}\cdot\text{K)}) + (1(205\ \text{J/(mol}\cdot\text{K)})] - 2(220\ \text{J/(mol}\cdot\text{K)})$$
$$\Delta S^\circ = 149\ \text{J/(mol}\cdot\text{K)}$$

**d.** The standard change in Gibbs free energy is calculated using the enthalpy and entropy changes given.

$$\Delta G^\circ = \Delta H^\circ - T\Delta S^\circ = 81.6\ \text{kJ/mol} - \left[(298\ \cancel{\text{K}})(149\ \cancel{\text{J}}/(\text{mol}\cdot\cancel{\text{K}}))\left(\frac{1\ \text{kJ}}{1000\ \cancel{\text{J}}}\right)\right]$$
$$\Delta G^\circ = 37.2\ \text{kJ/mol}$$

**e.** Your sketch should resemble the following:

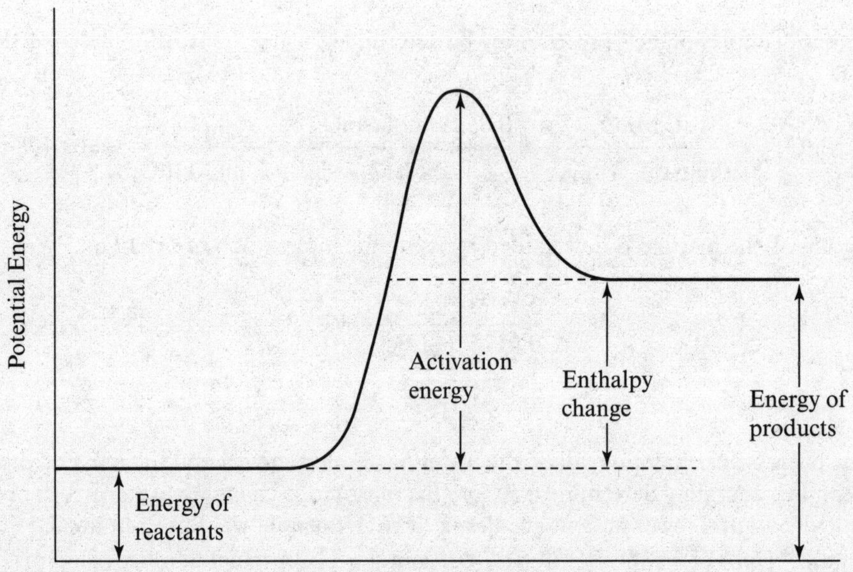


**3. a.** Because cyclohexane and water are the major components of the mixture, they are the solvents, and ethanol and azulene are the solutes. Cyclohexane is less dense than water, so cyclohexane makes up the top layer and water makes up the bottom layer.

**b.** The cyclohexane layer is blue because azulene is soluble in cyclohexane rather than water. Azulene is a nonpolar molecule. London dispersion forces are responsible for its solubility in cyclohexane, a nonpolar hydrocarbon. Azulene is much less soluble in water, which is polar and capable of hydrogen bonding.

**c.** Five π bonds are present in azulene.

**d.** The carbon atoms are $sp^2$ hybridized with approximate 120° bond angles.

**e.** The water layer could be drained from the bottom of the separatory funnel, leaving the cyclohexane layer. The cyclohexane layer could be poured into a distillation flask, and the cyclohexane could be removed from the solid azulene by distillation. (Alternatively, all the solvents could be evaporated.)

**4. a.** The limiting reactant may be determined from the number of moles of product that could be produced by each of the reactants.

$$4.00\ \text{g CuSO}_4 \times \frac{1\ \text{mol CuSO}_4}{159.61\ \text{g CuSO}_4} \times \frac{1\ \text{mol Cu}^{2+}}{1\ \text{mol CuSO}_4} \times \frac{1\ \text{mol Cu(OH)}_2}{1\ \text{mol Cu}^{2+}} = 0.0251\ \text{mol Cu(OH)}_2$$

$$0.0250\ \text{L NaOH} \times \frac{6.00\ \text{mol NaOH}}{1\ \text{L NaOH}} \times \frac{1\ \text{mol Cu(OH)}_2}{2\ \text{mol NaOH}} = 0.0750\ \text{mol Cu(OH)}_2$$

The $CuSO_4$ is the limiting reactant since fewer moles of product result from its complete consumption.

**b.** The heat released can be determined from the mass, the specific heat, and the temperature change of the water in the calorimeter.

$$m = 50.0\ \text{mL} \times 1\ \text{g/mL} = 50.0\ \text{g}$$

$$q = mc\Delta T = (50.0\ \text{g})\left(4.18\ \text{J/(g}\cdot{}^\circ\text{C)}\right)(30.2\,^\circ\text{C} - 25.2\,^\circ\text{C}) = 1050\ \text{J}$$

Heat is released (as evidenced by the increase in temperature of the solution), so $q = -1050$ J.

**c.** The enthalpy change per mole of reaction is determined by the amount of heat released as 0.0251 moles of $Cu(OH)_2$ are formed.

$$\Delta H^\circ = \frac{-1050\ \text{J}}{0.0251\ \text{mol}} \times \frac{1\ \text{kJ}}{1000\ \text{J}} = -41.8\ \text{kJ/mol}$$

**d.** The enthalpy of reaction may be determined using the sum of the enthalpies of formation of reactants and products.

$$\Delta H^\circ_{rxn} = \Sigma\Delta H^\circ_{f(products)} - \Sigma\Delta H^\circ_{f(reactants)}$$

$$\Delta H^\circ_{rxn} = 1(-450\ \text{kJ/mol}) - [(1(64.8\ \text{kJ/mol})) + (2(-230\ \text{kJ/mol}))]$$

$$\Delta H^\circ_{rxn} = -54.8\ \text{kJ/mol}$$

The accepted value is greater than the value that was determined experimentally. Heat was lost to the surroundings because the coffee cup calorimeter was not perfectly insulated.

**5. a.** $SbCl_2^+$ has a trigonal planar electronic geometry and bent molecular geometry; $GaCl_2^+$ has linear electronic and molecular geometry. The product must be $[SbCl_2]^+[GaCl_4]^-$.

$[\ddot{\text{Sb}}(\text{Cl})_2]^+$ (bent: Cl—Sb—Cl with a lone pair on Sb)    $[:\ddot{\text{Cl}}\text{—Ga—}\ddot{\text{Cl}}:]^+$

**b.** Infrared spectroscopy gives information about the types of bonds that are present in a molecule. The bond angle in the cation was determined by infrared spectroscopy rather than ultraviolet/visible spectroscopy, which gives information about electronic transitions.

**c.** The Lewis structure of the anion is as follows.

**d.** The molecular geometry of the anion, $GaCl_4^-$ is tetrahedral.

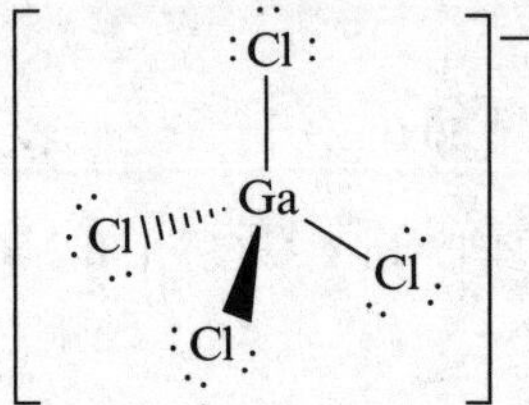


**6. a.** Sodium-24 has one more neutron in its nucleus than sodium-23.

**b.** **i.** According to the graph of disintegration rate versus time, half of the $^{24}$Na has decayed at 15 hours.

$$t_{½} = 15 \text{ h}$$

**ii.** Nuclear decay is always a first order process.

$$t_{½} = \frac{0.693}{k}$$

$$k = \frac{0.693}{t_{½}} = \frac{0.693}{15 \text{ h}} = 0.046 \text{ h}^{-}$$

**c.** The vertical axis of the graph should be labeled "ln A" or "ln (disintegrations)."

**7. a.** Equal numbers of moles of ideal gases occupying equal volumes at the same temperature exert the same pressure.

**b.** Both gases have the same average kinetic energy because both are at the same temperature.

**c.** The molar mass of He is less than the molar mass of $CH_4$, so the helium molecules have greater average speed.

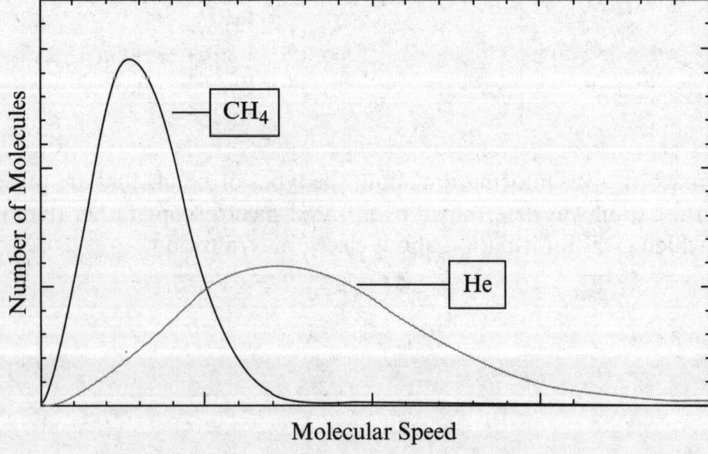

**d.** The volume of an ideal gas is directly proportional to temperature. The sketch should show a straight line with a positive slope.

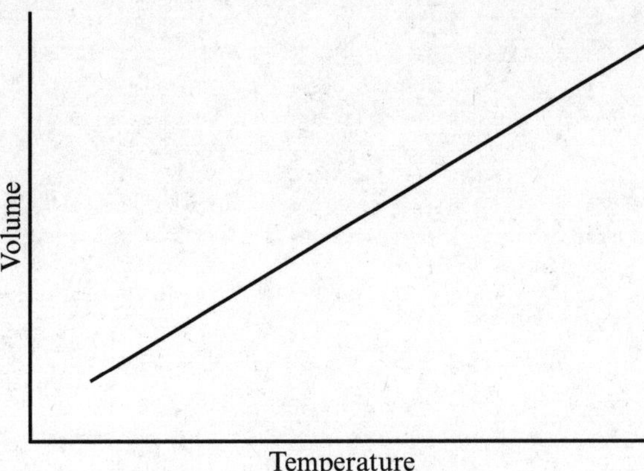